3rd Australasia and South East Asia Structural Engineering and Construction Conference (ASEA-SEC-3)

Integrated Solutions for Infrastructure Development

AA005354

Kuching, Malaysia
31 October - 4 November 2016

Editors:

H.H. Lau **C.K. Ng**
F.E. Tang **A. Singh**

ISBN: 978-1-5108-9999-5

Printed from e-media with permission by:

Curran Associates, Inc.
57 Morehouse Lane
Red Hook, NY 12571

Some format issues inherent in the e-media version may also appear in this print version.

Copyright© (2016) by The International Structural Engineering and Construction Society (ISEC)
All rights reserved.

Printed with permission by Curran Associates, Inc. (2020)

For permission requests, please contact The International Structural Engineering and Construction Society (ISEC) at the address below.

ISEC PRESS
PO Box 1334
Fargo, North Dakota 58107
USA

isec.sec@gmail.com

Additional copies of this publication are available from:

Curran Associates, Inc.
57 Morehouse Lane
Red Hook, NY 12571 USA
Phone: 845-758-0400
Fax: 845-758-2633
Email: curran@proceedings.com
Web: www.proceedings.com

TABLE OF CONTENTS

KEYNOTE PAPERS

IS PROJECT BASED LEARNING THE WAY TO GO? 1
Indubhushan Patnaikuni, Himabindu Mydaraboina

BEHAVIOR OF CONCRETE MATERIALS AND STRUCTURES

STANDARD SECTION DESIGN OF ICH RC TOWER SUPPORTING 5.0MW WIND TURBINE 7
Sungwon Kim, Taek Hee Han, Hyemin Hong

STANDARD SECTION DESIGN OF FRP DSCT TOWER SUPPORTING 3MW WIND TURBINE 13
Hyemin Hong, Sungwon Kim, Taek Hee Han

MATERIAL AND OUT-OF-PLANE FLEXURAL PROPERTIES OF TRADITIONAL MASONRY WALL 19
Noriko Takiyama, Akari Yamaguchi, Kohei Hara, Verdejo Jimenez Juan Ramon

A STUDY ON THE FACTORS AFFECTING ELECTRO-MIGRATION CHARACTERISTIC OF CHLORIDE IONS IN CONCRETE 25
Isao Ujike, Keiyu Kawaai

OPTIMIZING THE MIX DESIGN OF CLAY BASED GEOPOLYMER CONCRETE 31
Muhammad Mukhlesur Rahman, David W. Law, Indubushan Patnaikuni

INVESTIGATION OF THE COMPRESSIVE STRENGTH OF PORTLAND CEMENT MIXTURES BY APPLICATION OF BACTERIA - "SPOROSARCINA PASTEURII" 37
Samson Aregawi, Abiy Goshu, Bisrat Alemu, Dagmay Woldaregay, Nathnael Abdulkadir, Yonatan Mesfin

SELF-COMPACTING CONCRETE USING AGRO-INDUSTRIAL WASTES; RELATIVE HUMIDITY EFECT AND OPTIMUM MIXING METHOD 43
Brabha Nagaratnam, Ahmed Faheem, Muhammad Ekhlasur Rahman, Abdul Karim Mirasa, Mohammad Abdul Mannan, Md Saiful Islam

SUBMERGED CONCRETE CAISSON BREAKWATER CONSTRUCTION 49
Alvin Magallanes, Ben Sullivan, Andrew Moores, Amarjit Singh

RACKING PERFORMANCE OF LIGHT WEIGHT CONCRETE FILLED COLD-FORMED STEEL PANELS 55
Shahrzad Kasaeian, Hamid R. Ronagh, Reza Nikoonejad

PROJECT EXECUTION AND PRODUCTIVITY

INFRASTRUCTURE NEEDS IN SUB SAHARAN AFRICA WITH PARTICULAR REFERENCE TO SOUTH SUDAN 61
Ayman Y Nassif, Ann Stewart, Millie T Mutepfe, Petros Christou

ASSESSING THE STRENGTH AND WEAKNESS OF THE GHANAIAN CONSTRUCTION INDUSTRY 67
Bernard Martin Arthur-Aidoo, Nazeem Ansary, Clinton Aigbavboa

MANAGING MULTICULTURAL VIRTUAL ENGINEERING TEAMS 73
Rania Busada, Samer Ezeldin

INVESTIGATION INTO WH&S ISSUES WITHIN THE RESIDENTIAL CONSTRUCTION INDUSTRIES MIGRANT WORKFORCE 79
Swapan Saha, Andrew Mantoufeh

BUILDING CONTROL AND REGULATORY BALANCE: PERSPECTIVES OF BUILDING PROFESSIONALS 85
Jeff Clement Samasoni, James Olabode Bamidele Rotimi

THE INTERACTION BETWEEN LEADERSHIP STYLES AND MANAGEMENT LEVEL, AND THEIR IMPACT ON PROJECT SUCCESS 91
Hector Martin, Kisa Edwards

RISK ASSESSMENT ON DESIGN AND BUILD PROJECTS - A COMPARISON OF CONTRACTORS, CONSULTANTS AND OWNERS PERCEPTIONS 97
Sabihah Saaidin, Intan Rohani Endut, Siti Akmar Abu Samah, Ahmad Ruslan Mohd Rizduan, Nur Nabihah Abd Razak

FOUNDATION ENGINEERING AND CONSTRUCTION

THE STABILITY OF SLOPE USING CELL-BASED SMOOTHED FINITE ELEMENT METHOD AND SECOND ORDER CONE PROGRAMMING 103
Hai Le Nguyen, Hai Than Nguyen, Thien Vo Minh

ANALYTICAL STUDY ON PILED-RAFT FOUNDATION CONSIDERING NONLINEAR BEHAVIOR OF GIBSON SOIL 109
He Huang, Haruyuki Yamamoto

APPLICABILITY OF A NEW SURFACE PREPARATION METHOD FOR STEEL STRUCTURE USING LASER 115
Hiroshi Ogami, Katashi Fujii, Yukio Manabe, Kohei Ota, Asuo Yonekura

ESTIMATING LOAD-DISPLACEMENT BEHAVIOR OF SOFT CLAYS 121
Murat Tonaroglu, Cem Akguner, Murat E. Selcuk

PULL-OUT PROPERTIES OF GYPSUM BOARD ANCHORS 127
Yoshinori Kitsutaka, Fumiya Ikedo

EFFECT OF SURFACE PREPARATION GRADE ON REPAIR BY MEANS OF COVER PLATES AND ADHESIVE 133
Ryuichi Yokota, Katashi Fujii, Hisakazu Horii, Yuya Hidekuma

STUDY ON RETROFIT EFFECTS FOR HORIZONTAL PLANE OF TRADITIONAL WOODEN HOUSES BASED ON FULL-SCALE TEST 139
Mitsuhiro Miyamoto, Haruka Okuhiro

EVALUATION OF THE OPTIMAL ENGINEERING SOLUTION FOR SWIMMING POOL CONSTRUCTION 145
Ruta Miniotaite

SHEAR STRENGTH OF PERFORATED PLATE GIRDERS WITH INCLINED STIFFENERS 151
M. R. Azmi, M. Y. M. Yatim, W. H. Wan Badaruzzman

ANALYSIS OF STRUCTURES

NONLINEAR ANALYSIS OF JAPANESE TRADITIONAL WOODEN FRAMES WITH FITTING-TYPE JOINT 157
Naoto Idate, Yukimasa Yamada, Noriko Takiyama

AN APPLICATION OF HYPERPLASTICITY TO DETERMINE THE BEHAVIOR OF A PLANE STRESS CANTILEVER BEAM 163
Hai Than Nguyen, Lam Nguyen Sy

NUMERICAL ANALYSIS OF AXIALLY LOADED COLD-FORMED STEEL BUILT-UP BOX SECTION 169
T.C.H. Ting, H.H. Lau, M.S. Joo

NUMERICAL STUDIES OF CONCRETE PLATES UNDER LOCALIZED BLAST LOADS 174
Sofia W. Alisjahbana, Wiratman Wangsadinata, Irene Alisjahbana

PERFORMANCE CONFIRMATION TEST FOR TIMBER COLUMN-GROUND SILL JOINTS REINFORCED WITH ARAMID FIBER SHEETS 180
Akari Yamaguchi, Xinyan Chen, Noriko Takiyama

SHAPE OF YIELD SURFACE ESTIMATED BY USING NATURAL STRAIN UNDER PRE-DEFORMATION OF SHEAR AFTER UNIAXIAL TENSION 186
Yasuyuki Kato

A STUDY ON THE PUSHOVER TEST AND NUMERICAL ANALYSIS OF GFRP FRAMES 192
Yeou-Fong Li, Bo-Shiang Wang, Jian-Yu Lai

FIRE SPREADING ANALYSIS OF A GROUP OF WOODEN HOUSES IN TOWNSCAPE FORMATION DISTRICT 198
Tomiya Takatani

QUALITY, COST CONTROL, AND SUSTAINABILITY

ANALYSIS OF ECONOMIC DETERMINANTS OF AFFORDABLE HOUSING PRICES 204
Abdullateef Olanrewaju, Lee Lim Tat, Seong Yeow Tan, Mine Naoto, Zafarullah Nizamani, Abdul Rashid Abdul Aziz

CAUSES OF VARIATION ORDERS FOR MEGA CONSTRUCTION PROJECTS 210
Samer Ezeldin, Amr Khaled El-Sadek

SAFETY AND HEALTH SITE INSPECTIONS FOR ON-FIELD RISK ANALYSIS AND TRAINING 216
Marco L. Trani, Luca Beretta, Giada Nicolini, Manuele Cassano
CHALLENGES IN THE IMPLEMENTATION OF QUALITY CONTROL TECHNIQUES DURING SITE INSPECTION 222
Nur Nabihah Abd Razak, Intan Rohani Endut, Siti Akmar Abu Samah, Ahmad Ruslan Mohd Ridzuan, Sabihah Saaidin
NEAR ZERO ENERGY HOUSE (NZEH) DESIGN VARIABLES FOR HOUSING DEVELOPMENT 228
Yusuf Latief, Mohammed Ali Berawi, Leni Supriadi, Jade Petroceany, Ayu Herzanita, Ario Bintang Koesalamwardi
AN ENTROPY-BASED STUDY ON THE EFFECTIVENESS OF PREFABRICATION IN ACHIEVING SUSTAINABLE CONSTRUCTION 234
Peter S. P. Wong, Luke Edmonson, Michael Phillip Kanellopoulos
THE NEW TOLL ROAD FINANCING MODEL DETERMINATION 240
Ayomi Rarasati, Lukas Sihombing, Andreas Wibowo, Yusuf Latief
MULTIPLE CRITERIA EVALUATION OF ASSEMBLING BUILDINGS FROM STEEL FRAME STRUCTURES 246
Ruta Miniotaite

BRIDGE AND STRUCTURAL HEALTH MONITORING

DEVELOPMENT OF EMBEDDED MINI-SENSOR FOR DIAGNOSIS OF CORROSION OF STEEL REINFORCEMENT IN CONCRETE NONDESTRUCTIVELY 252
Mohamed A. Ismail, Han-Seung Lee, Mohd Warid Hussin
AN INVESTIGATION OF CORROSION ENVIRONMENT OF A BRIDGE CONSTRUCTED IN MOUNTAINOUS AREA 258
Kazuto Maruyama, Toshihiko Aso
DIAGNOSIS OF BRIDGE BEARINGS BASED ON FIELD MEASUREMENTS 264
Naho Shibasaki, Mariko Ikeda, Masahiro Sakano
REMAINING LIFE AND CORROSION LIMIT OF CORRODED TRUSS BRIDGE FOCUSING ON THE ULTIMATE STRENGTH 270
Hiroaki Fukuda, Naoyuki Asao, Katashi Fujii
THE BENEFIT OF THE NEW SUPER U-BEAM USED FOR SECTION 4 OF A WEST COAST EXPRESSWAY 276
Ong Jiundar, Lokman Hassan
SCISSORING ORIGAMI INSPIRED DEPLOYABLE BRIDGE FOR A DISASTER 282
Kotaro Adachi, Ichiro Ario, Yuki Chikahiro, Shuichi Ono
THE INFLUENCE OF TEMPERATURE AND HUMIDITY ON RUST GROWTH ON WEATHERING STEEL 288
Ryoya Nagata, Shusaku Tachibana, Toshihiko Aso
REMAINING STRENGTH EVALUATION OF PLATE GIRDERS WITH CORROSION NEAR SUPPORTS 294
Naoyuki Asao, Katashi Fujii, Hiroaki Fukuda, Ryuichi Yokota
SEISMIC PERFORMANCE OF AN OLD JAPANESE-STYLE TWO-STORY WOODEN HOUSE UNDER A STRONG EARTHQUAKE GROUND MOTION 300
Tomiya Takatani, Hayato Nishikawa
SEISMIC DAMAGE PREDICTION OF OLD JAPANESE-STYLE WOODEN HOUSE USING PREDOMINANT PERIOD DISTRIBUTION OF GROUND SURFACE LAYER 306
Tomiya Takatani, Hayato Nishikawa
INFLUENCE OF BASEBOARD HEIGHT ON RESISTANCE TO WIND FORCE OF SCAFFOLDS AT WINDWARD SIDE OF BUILDINGS 312
Hiroki Takahashi, Mizuki Aoki, Katsutoshi Ohdo, Kazuo Ohgaki
VIBRATION PROPERTIES OF TRADITIONAL MASONRY WALLS OF CORTES CHURCH 318
Noriko Takiyama, Akari Yamaguchi, Kohei Hara, Verdejo Juan Ramon Jimenez

DYNAMICS OF STRUCTURES

SEISMIC DESIGN VERIFICATION OF STEEL PRE-ENGINEERED BUILDINGS 324
Satish Kumar S Rajaram, Nibedita Sahoo

BASE ISOLATION IN SEISMIC RETROFITTING OF R/C BUILDINGS 330
Alberto Dusi, Marco Mezzi, Tan Teng Or
EXTREME ENVIRONMENTAL LOAD MODELS FOR WAVE AND CURRENT RESPONSE SURFACE METHOD 336
Zafarullah Nizamani, Yap Eng Ching, Mohamed Mubarak Bin Abdul Wahab, Abdullateef Olanrewaju
DYNAMIC ANALYSIS OF OFFSHORE PILES EMBEDDED IN ELASTOPLASTIC SOFT CLAY 342
Anis Mohamad Ali, Mohamad J. K. Essa, Abdulameer Qasim Hassan

CONSTRUCTION LAW AND TECHNOLOGY

THE DEVELOPMENT OF ANATOMY OF CONSTRUCTION DISPUTE 348
Felix Hidayat, Biemo W. Soemardi
PRACTICE EXPERIENCE AND MULTIDISCIPLINARY COLLABORATION IN PROJECT MANAGEMENT: A CASE STUDY 354
Silvia Mazzetto
WILL TRADESMAN COLLECTIVE WAGE AGREEMENT BE EXEMPTED FROM THE HONG KONG COMPETITION ORDINANCE? 360
Sai On Cheung
A CONCENTRATION STUDY OF THE PRIVATE NEW RESIDENTIAL UNITS MARKET 366
Sai On Cheung, Yuhan Niu
BUILDING THE EDUCATION REVOLUTION (BER) PROGRAM: GOVERNING THE IMPLEMENTATION OF PROJECT POLICIES 372
Duro Kolar, Derek H. T. Walker, Peter S. P. Wong
FACTORS AFFECTING THE FULL IMPLEMENTATION OF BUILDING INFORMATION MODELING AS A CONSTRUCTION MANAGEMENT TOOL 378
Swapan Saha, Scott Jessup
A DECISION SUPPORT SYSTEM FOR METHODS OF MEASUREMENT IN CONSTRUCTION PROJECTS 384
Abdelrahman Magdy, Samer Ezeldin

COST AND SCHEDULE

MITIGATING CONSTRUCTION PROJECT SCHEDULE OVERRUNS 390
Mulenga Mukuka, Clinton Aigbavboa, Wellington Thwala
AN ASSESSMENT OF CONSTRUCTION PROJECT COST OVERRUNS IN THE PUBLIC SECTOR 396
Jerminah Khabisi, Clinton Aigbavboa, Wellington Thwala
ECONOMIC AND FINANCIAL DEVELOPMENT OF SOLAR PHOTOVOLTAIC TECHNOLOGY 402
Herawati Zetha Rahman, Perdana Miraj, Jade Sjafrecia Petroceany

ORGANIZATIONAL BEHAVIOR

EFFECTS OF MENTORING ON SOCIALIZATION OF GRADUATE EMPLOYEES IN THE BUILDING ENVIRONMENT THROUGH KNOWLEDGE TRANSFER 408
Morena Wiilliam Nkomo, Wellington Didibhuku Thwala
CROSS TABULATION ANALYSIS: DETERMINANTS OF MATHEMATIC SUCCESS AMONG CIVIL AND BUILT ENVIRONMENT STUDENTS 414
Justus N. Agumba
DEVELOPING LEADERSHIP SKILLS OF PROJECT AND CONSTRUCTION MANAGERS IN THE CONSTRUCTION INDUSTRY 420
Murendeni Liphadzi, Clinton Aigbavboa
JOB DISSATISFACTION INFLUENCE ON CONSTRUCTION EMPLOYEE ABSENTEEISM AND TURNOVER 425
Portia Leasilane, Clinton Aigbavboa

CONSTRUCTION CONTRACTING AND SAFETY

PERFORMANCE OF REQUEST FOR INFORMATION (RFI) MANAGEMENT 431
Muhamad Abduh, Farrel Adam

USE OF UNMANNED AERIAL VEHICLES AND COMPUTER VISION IN CONSTRUCTION SAFETY INSPECTIONS 437
Mohamad Abbas, Bahaa Eddine Mneymneh, Hiam Khoury

MODIFIED CONCEPTUALIZED HEALTH AND SAFETY COMPLIANCE MODEL FOR SMALL AND MEDIUM-SIZED CONTRACTORS 443
Zakari Mustapha, Clinton Aigbavboa, Wellington Didi Thwala

DEVELOPING ALTERNATIVE CONCEPT FOR RAILWAY OPERATION AND MAINTENANCE CONTRACT 449
Herawati Zetha Rahman, Perdana Miraj, Jade Sjafrecia Petroceany

A REVIEW OF THE CHALLENGES OF RURAL ROAD NETWORK DEVELOPMENT 455
Justus N. Agumba

Author Index

ASEA-SEC-3
The Third Australasia and South-East Asia Structural Engineering and Construction Conference
Kuching, Sarawak, Malaysia, Oct 31-Nov 4, 2016
Integrated Solutions for Infrastructure Development

Welcome to ASEA-SEC-3

Dear Friends and Colleagues,

It is an honor to be entrusted to organize this prestigious Third Australasia and South-East Asia Structural Engineering and Construction Conference (ASEA-SEC-3) by the International Structural Engineering and Construction Society (ISEC).

On behalf of my Co-Chair, Prof. Ng Chee Khoon and the members of Local Organizing Committee, we are delighted to invite you to participate in the ASEA-SEC-3 in 2016 with the theme of "Integrated Solutions for Infrastructure Development".

ASEA-SEC-3 in 2016 will be held in Kuching, Sarawak, Malaysia from 31st October to 4th November 2016. This conference is organized by Curtin University Sarawak Malaysia and co-organized by University Malaysia Sarawak (UNIMAS). The conference proceedings will be Scopus indexed.

ASEA-SEC-3 will be expecting more than 200 contributions from round the world, bringing forth the latest technology and research on Structural Engineering and Construction Engineering. Keynote speakers are eminent researchers in their respective fields. The topics will be covered in the 2 plenary sessions and 16 technical sessions includes Infrastructure, Structures, Coastal Engineering, Energy, Construction, Construction Safety, Sustainability, Architecture & Architectural Engineering, Facilities and asset management, Geotechnical and foundation engineering, Cost and project management, Housing, Education & professional ethics, Law and dispute resolution, Materials, Policies for national development, Procurement, Quality, Risk analysis and disaster management, air and water.

We hope to see you in the Land of the Hornbills in 2016.

Yours sincerely,

Lau Hieng Ho
Local Organizing Committee Chairman
lau.hieng.ho@curtin.edu.my

Committees

Advisory Committee

- Amarjit Singh, President, ISEC Society; University of Hawaii at Manoa, USA
- Siamak Yazdani, Secretary, ISEC Society; North Dakota State University, USA
- Swapan Saha, University of Western Sydney, Australia
- Vanissorn Vimonsatit, Curtin University, Australia
- Sai On Cheung, City University of Hong Kong, China
- Indubhushan Patnaikuni, RMIT University , Australia

Technical Committee

- Syed Faiz Ahmad, Saudi Oger Ltd., Saudi Arabia
- Sofia W. Alisjahbana, Bakrie University, Indonesia
- Farah Abdul Aziz, Universiti Putra Malaysia, Malaysia
- Wan Hamidon Wan Badaruzzaman, National University of Malaysia, Malaysia
- Tak Ming Chan, Hong Kong Polytechnic University, Hong Kong
- Maged Georgy, RMIT University, Australia
- Luong Van Hai, Ho Chi Minh City University of Technology, Vietnam
- Harianto Hardjasaputra, University of Pelita Harapan, Indonesia
- Amr A. Kandil, Purdue University, USA
- Ruveyda Komurlu, Kocaeli University, Turkey
- Charles Chin Moy Loo, Xi'an Jiaotong Liverpool University
- Sherif Mohamed, Grifith University, Australia
- Deepa Nair, Cochin University of Science & Technology, India
- S. Thomas Ng, University of Hong Kong, Hong Kong
- Chiew Sing Ping, Singapore Institute of Technology, Singapore
- Stephen Wei Meng Quach, University of Macau, China
- Weerakaset Suanpaga, Kasetsart University, Bangkok, Thailand
- Priyo Suprobo, Sepuluh Nopember Institute of Technology, Indonesia
- Vivian Tam, Western Sydney University, Australia
- Cher Siang Tan, University Technology Malaysia
- Brian Uy, University of New South Wales, Australia
- Andrew Whyte, Curtin University, Australia
- Kenneth Tak Wing Yiu, University of Auckland, New Zealand
- Ben Young, University of Hong Kong, China

Local Organizing Committee

- Lau Hieng Ho, Chair, Curtin University Sarawak Campus, Sarawak, Malaysia
- Tina Ting Chui Huon, Curtin University Sarawak Campus, Sarawak, Malaysia
- Carrie Ho Lee Ing, Curtin University Sarawak Campus, Sarawak, Malaysia
- Mohamed Abdel Kader Ismail, Curtin University Sarawak Campus, Sarawak, Malaysia

- Jayakumar Muthuramalingam, Curtin University Sarawak Campus, Sarawak, Malaysia
- Brabha Hari Nagaratnam, Curtin University Sarawak Campus, Sarawak, Malaysia
- Ting Sim Nee, Universiti Malaysia Sarawak, Sarawak, Malaysia
- Chee Khoon Ng, Co-Chair, Universiti Malaysia Sarawak, Sarawak, Malaysia
- Muhammad Ekhlasur Rahman, Curtin University Sarawak Campus, Sarawak, Malaysia
- Fu Ee Tang, Curtin University Sarawak Campus, Sarawak, Malaysia

Sponsors

Curtin University Sarawak

Universiti Malaysia Sarawak

Sarawak Convention Bureau

Integrated Solutions for Infrastructure Development
Edited by Lau, H. H., Tang, F. E., Ng, C. K., and Singh, A.
Copyright © 2016 ISEC Press

IS PROJECT BASED LEARNING THE WAY TO GO?

INDUBHUSHAN PATNAIKUNI and HIMABINDU MYADARABOINA

School of Civil, Chemical & Environmental Engineering, RMIT University, Melbourne, Australia

Many educational institutions have moved or are moving to project-based learning (PBL) in Engineering. Project-based learning and problem-based learning have both advantages and also disadvantages. They are suitable for certain courses and may not be that helpful for certain other courses. In case of courses which require mathematical and analytical skills PBL does not seem to be very useful. As a matter of fact, many students do not seem to learn the basics/fundamentals if PBL method is used. Courses like statics, Mechanics, and structural analysis are the type of courses which can produce graduates without a proper fundamental understanding of the subject matter if PBL is used for the entire course.

Keywords: PBL, Statics, Structural analysis, Problem based learning.

1 INTRODUCTION

Project Based Learning (PBL) had emerged 40 years ago in the medical faculty at McMaster University in Canada and spread rapidly, predominantly in medical education, and also in the fields of law and engineering (Barret 2005, Ahern 2008).

Many universities are using project-based learning in the Engineering education programs (Bennet and Kelly 2010). Researchers around the world have investigated and suggested PBL as a major learning tool in civil engineering with some challenges in implementation (Gavin 2011, de Urena *et al.* 2003). Research by Nepal and Jenking (2011), suggested implementing a blended system which is a mixture of traditional and PBL which may give better results. But, there was not much research information shown on the individual student's performance.

The authors tried to study the advantages or disadvantages of using project-based learning. The course was designed to challenge the students to learn through problem-based learning and project based learning. The assessments were designed such that the students try to learn from the method of assessments like the quizzes which require the student to prepare well and understand the subject matter and assignments which require understanding the subject matter for completing the assignments. The students were given written quizzes once in every two weeks and the corrected answer sheets were usually returned the following week. For the students to know where they have gone wrong the quiz questions were answered by the academic when the answer scripts were returned. Despite too much time consumption for the Academic, especially for large classes and utilizing many resources the PBL is found to be either neutral or slightly beneficial only to a certain percentage of students who already have a sound basic knowledge and who are keen to learn and see the practical aspect of the theory. But, the rest of the students are likely to just take advantage of the teamwork and graduate without proper fundamental knowledge of the subject. Again it probably depends on the type of course. From the research, the authors believe that PBL may be useful for certain courses like design courses but may not be very useful for other mathematical/ analytical type of courses which need the student to understand the fundamentals and be able to apply these fundamentals to problem solving. At the same time, PBL can be very useful for student learning if only a small percentage of credit is given to this component in the assessment of the course. It is also important to design the PBL part in such a manner that each individual contributes to the group and has understood the parts of the project contributed by the other students in the group. This requires considerable time to be spent by the academic in designing the course such as interviewing the student and also varying the assessment from time to time so that the students do not have the possibility of copying from other students or from the past year student projects. But the resources put in may not justify the marginal benefit which may result finally. The aim of this paper is to bring out what is the best way of assessing the student learning.

2 RESEARCH METHOD

The first author taught courses in structural engineering area, mainly in statics, mechanics, structural analysis, Analysis of complex structures, finite element analysis, the theory of elasticity, plates and shells, plastic analysis etc. for several years. Problem-based learning was used for several years without giving much credit (marks) for this component which seems to have worked very well. But during the past 8 or 9 years, the requirement in many universities is to use project-based learning with a significant percentage of credit (marks) given to this PBL component. It was found that this does not achieve what was the intention of this PBL. To examine the students' performance in the traditional tests (tests include written tests, written quizzes, and examination) as well as project-based assessments, a study is carried out on three different academic year (2008, 2009, and 2010) scores of a course in the structural engineering area of students in a reputed University in Melbourne. The research methodology included both written tests and PBL assessment. The course contains different components, some are project based and the others are test/examination/quiz type. Both types of components were assessed for the grading. PBL assessment was a significant part of the total assessment. Both project based assessment and written test based assessment had an equal credit of 50% each.

3 DATA ANALYSIS & RESULTS

Though entire class of around 150 students is analyzed in three consecutive years as mentioned, this study mainly focuses on the percentage of marks obtained by 50 students from a class of around 150, who had poorly performed in written tests (WT) compared to their performance in the project work (PW). This was analyzed for three different academic years for the same course. The results for the entire 150 students also are shown in figures to get an idea of how the entire class performed.

Figure 1 shows an example of a student X, who scored different percentage marks in the five components of his/her course. Marks obtained in written tests & project works for student X from Class A in the year 2008 are also shown in Figure 1 and that the student scored very good percentage marks in Project Works, PW1, 2 & 3 though he/she got very less percentages marks in written tests, WT1, 2.

The cumulative percentages are shown in Figure 2, which also clearly show that the project-based learning fetches the student X high percentage even though the student is very weak in written examination/test which actually tests the student's basic understanding of the subject.

Hence, the analysis is carried out for 50 students in Class A of the year 2008 batch, and the average percentage marks of 50 poorly performed students in WT are calculated and are shown in

Figure 2. Also, the average % marks of PW and WT for 50 poorly performed (in WT) students are shown for academic years 2009 & 2010 in the Figures 2-b & 3-a respectively.

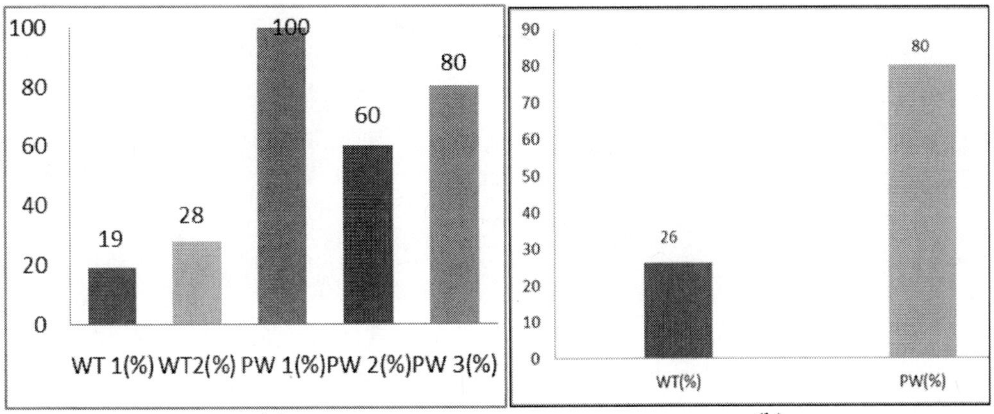

(a) (b)

Figure 1. a) Percentage of marks in all the components for student X in Class A -2008; b) WT% Vs PW% marks for student X in Class A-2008.

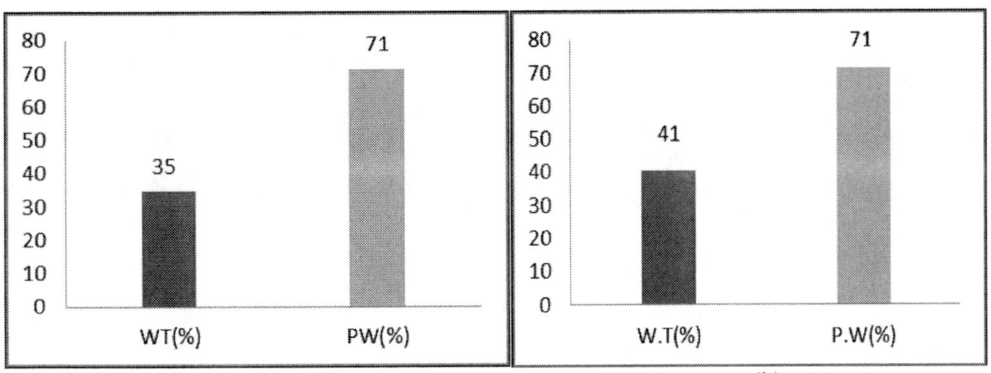

(a) (b)

Figure 2. a) Average % marks of 50 students in Class A-2008; b) in Class A-2009.

To show the entire picture of the majority of the students who have shown very good results in PW, irrespective of their performance in WT, an analysis was made for all the students in Class A for the year 2010 by showing their % marks in WT & PW and the graph is shown in Figure 3-b. The graph shows the PW% of the majority of the students always lies above the WT%. The WT% and PW% marks are calculated for all students for the same course in years 2008, 2009 and 2010 and the averages of each 25, 50, 100 and 150 students are shown in Table 1.

From Table 1, it can be seen that the % marks of WT was relatively higher in 2009 due to the fact that web-based quizzes instead of written quizzes were used as one of the components which could be attempted by the students in their own time and also the number of attempts allowed was more so that the students could learn the subject matter better if they fail to get the correct answer in the first or second attempt. It is possible that many students took the help of other students who had a better understanding of the subject matter.

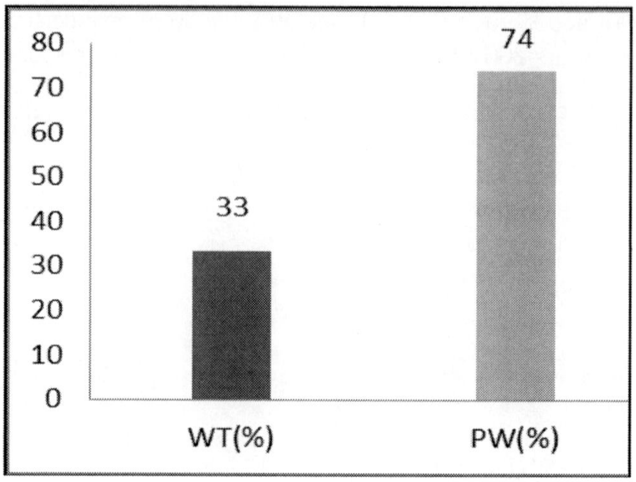

(a)

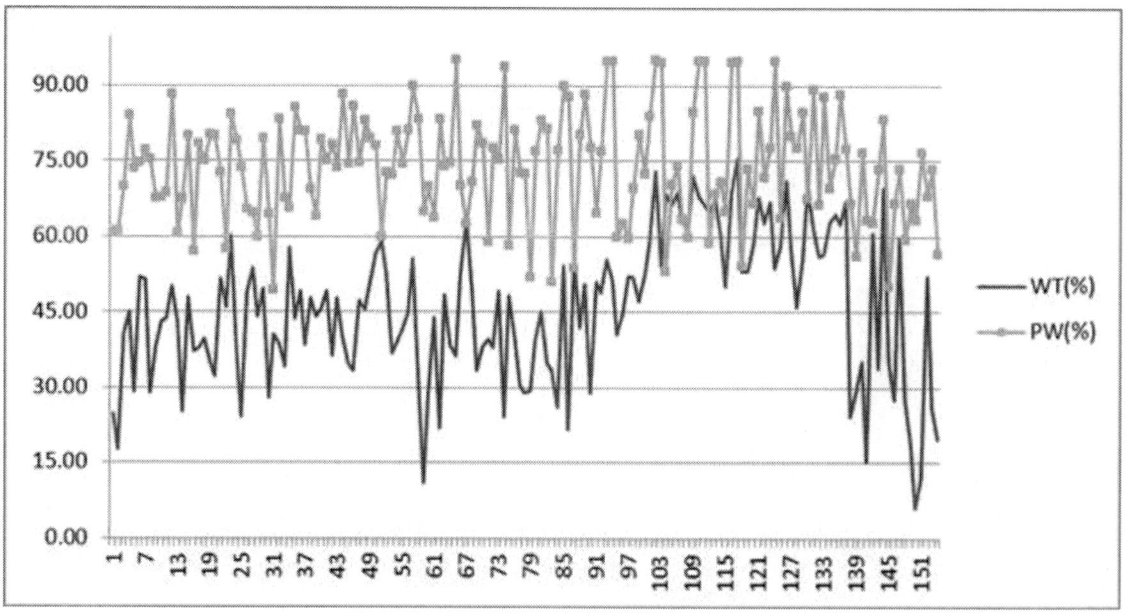

(b)

Figure 3. a) Average % marks of 50 students in Class A-2010;
b) PW% vs WT% marks for all the students in Class A-2010.

Table 1. Average % marks of students in Class A for years 2008, 2009, and 2010.

	25 students		50 students		100 students		150 students	
	WT%	PW%	WT%	PW%	WT%	PW%	WT%	PW%
2008	25.5	67.9	34.6	71.2	47.3	72.5	56.3	72.6
2009	30.5	69.3	40.5	71.2	51.3	72.3	60.0	74.2
2010	23.1	70.1	29.2	72.1	37.5	73.0	45.0	73.6

Table 2. The ratio of PW% and WT% marks of students in Class A for years 2008, 2009, and 2010.

	25 students	50 students	100 students	150 students
2008	2.66	2.06	1.53	1.29
2009	2.27	1.76	1.41	1.24
2010	3.03	2.47	1.95	1.64

According to Table 2, it can be observed that in each year and for every set of students, the PW% is more than double that of the WT% for the first 50 students except in 2009. The group of students in the year 2010 has performed relatively worse compared to other two years. The first 25 students received 166%, 127% and 203% more marks in PW compared to WT in the year 2008, 2009 and 2010 respectively. Similarly, the first 50 students received 106%, 76% and 147% more marks in PW compared to WT in the year 2008, 2009 and 2010 respectively.

4 CONCLUSIONS

As mentioned earlier, this study is mainly focused on the performance of the students in Project Work who got very less percentage in conventional tests i.e. Written Tests (which include written test, written quizzes, and examination).

From all the above analysis the authors derive the following conclusions:

- Despite very less percentage marks in WT, many students were getting overall pass mark (50%) due to the high percentage of marks in PW which is not desirable in terms of development of basic knowledge for the future courses the student might undergo and the knowledge they require as engineers
- To avoid this problem, there is a need to reduce the weight of the PBL part of the courses in the overall assessment. PBL should be considered as a useful learning tool but should not be given higher weightage as given in current educational system in engineering as it is not actually measuring the student's performance as each individual contributes differently in a team.
- If project based learning is used as a main component, then the projects should be designed for each individual student which may not be practical and cost effective. The requirement of resources will be prohibitive.
- The authors believe that project based learning is not suitable for all types of courses. Especially so in the case of the mathematical and analytical type of courses. PBL may be useful for design type of courses. PBL can be very useful for learning if it is designed in such manner as not to distort the results.

5 RECOMMENDATIONS

- Credit for project-based components needs to be minimal to achieve better learning by the students. Also, the courses need to be designed in such a manner that the students are required to solve problems and attempt projects with minimum credit to these components.
- There should be a hurdle mark for the written test components to avoid students passing without proper learning.

- If web-based tests are conducted they should not be conducted by allowing the students to attempt the tests in their own time but should be conducted like other tests at a particular time.
- Project based learning may be used for courses like design courses with some written test component also. Though the results are not included here, the results of a structural design course where only PBL I used was compared.

References

Ahern, A., What are the perceptions of lecturers towards using cooperative learning in civil engineering? *European Journal of Engineering Education*, 32 (5), 517–526, 2008.

Bennett, Gareth J. and Kevin Kelly., Implementation of Project Based Learning in a Large Engineering Programme. *3rd International Symposium for Engineering Education*, University College Cork, Irland, 2010.

Barrett T., "*Understanding the Problem Based Learning (PBL).* In: Barrett, T., Mac Labhrainn, I., Fallon, H. (Eds). Handbook of Enquiry & Problem Based Learning: Irish case studies and International Perspective, 2005.

de Urena, J. M., Menendez, J. M., and Coronado, J. M., Project/Problem Based Learning in Civil Engineering: the Ciudad Real (Spain) Experience. *International Conference on Engineering Education*, Valencia, Spain, 2003.

Gavin, K., Case Study of project-based learning course in civil engineering design, *European Journal of Engineering Design*, 36;6, 547-558, 2011.

Nepal, K. P. and Jenkins, G. A., Blending project-based learning and traditional lecture-tutorial-based teaching approaches in engineering design courses. *Proceedings of the 2011 AAEE Conference*, Western Australia, 2011.

Integrated Solutions for Infrastructure Development
Edited by Lau, H. H., Tang, F. E., Ng, C. K., and Singh, A.
Copyright © 2016 ISEC Press

STANDARD SECTION DESIGN OF ICH RC TOWER SUPPORTING 5.0MW WIND TURBINE

SUNGWON KIM, TAEK HEE HAN, and HYEMIN HONG

Korea Institute of Ocean Science & Technology, Ansan, Korea

A new-type wind power tower, internally confined hollow reinforced concrete (ICH RC) tower was developed and its standard sections were designed. Its targeting turbine capacity was 5.0 MW and the large displacement effect by turbine weight was considered. CoWiTA which is a non-linear analysis program was used for analyzing each section. Initial bottom diameter of the tower was 5.5 m with hollow ratios; 0.9, 0.85, 0.80, and 0.75 and 10% smaller diameter case was also considered with the same hollow ratios. In addition, the ICH RC tower could be separated into some modules for convenience of construction and installation. Each cross section at top and bottom of modules was designed and analysis of the sections was performed as well. As a result, all design cases except one case showed enough performance to support a 5.0 MW turbine. Also, it was confirmed that the modularized ICH RC wind tower satisfied with required load condition at each height by analyzing modules.

Keywords: Wind power, Confined, Composite, Concrete.

1 INTRODUCTION

Among the renewable energies, the wind energy has been growing fast over the world during the last decades (Watts *et al.* 2016). World wind energy association announced that 63 GW was added within the year 2015 and the total wind capacity of the world has reached 432.9 GW, representing cumulative market growth of more than 17% (GWEC 2016). 186 governments have agreed with the Paris agreement in late 2015 and it makes the wind energy market to grow continuously.

According to Klaassen *et al.*, the capacity of wind turbine is growing not only to increase energy efficiency but also to reduce the unit cost of construction by public research and development (Klaassen *et al.* 2005). By this trend, longer blades and bigger towers have been required to operate wind turbine system. This trend makes tall steel tower which has high slenderness ratio could not be safe enough from risk of buckling failure. About the buckling failure of the wind towers were conducted by Kimopoulos and Gantes (2012). For the concrete structures, they have different stress-strain relation between confined concrete and unconfined concrete structure. Popovics conducted study about experimental justification about confined concrete (Popovics 1973). Based on the confining effect of the concrete, Han *et al.* found that the inner tube and transverse reinforcements make the concrete in a tower to be triaxially confined state so that the strength of the tower could be enhanced (Han *et al.* 2010). In this study, the ICH RC wind tower which is composed of inner tube, reinforcements, and concrete is suggested, sectionally designed and analyzed. Figure 1 shows composition of ICH RC wind tower and it was applied confining effect by the components.

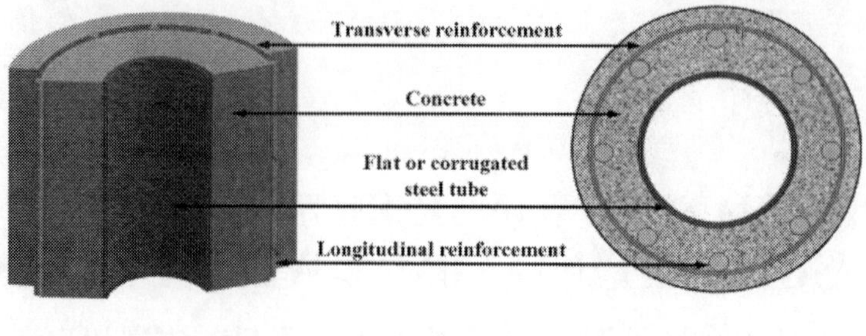

Figure 1. Composition of ICH RC tower.

2 SECTION DESIGN

Cross section of the ICH RC tower for supporting 5.0 MW wind turbine was designed. Load condition was based on literature review of existing steel tower which supports a 5.0 MW turbine. It indicates that the moment at the tower bottom, the moment at the tower top, and the axial force by turbine weight are 159,000 kN-m, 31,800 kN-m, and 4,185 kN respectively (Kim *et al.* 2016). The diameters at the tower bottom and at the tower top were selected as 5.5 m and 3.9 m, respectively. The diameter at the tower top was considered connecting area where the bottom of a nacelle assembly placed. The tower height was set as 95 m during section design of the ICH RC wind tower. Based on the case 01: 5.5 m (DB_0) bottom diameter, the case02: 90 percent of the DB_0 (5.5 x 0.9 = 4.95 m) was considered. In addition, various hollow ratios; 90, 85, 80, and 75 were applied to each case, and specific cases were represented as CaseAA_BB. AA should be 01 or 02, and they mean 5.5 m of the tower bottom diameter case and 4.95 m of the bottom diameter case, and BB indicates hollow ratio. For example, Case01_75 means the case of the tower has 5.5 m of bottom diameter with 75% of hollow ratio. Finally, total eight cases of tower section design were designed and analyzed to find optimum section design. The calculation process for tower section design is shown at figure 2. All required equations, design condition, and design factors of the ICH RC wind tower followed Korean concrete structure design standard (Korea Concrete Institute, 2012).

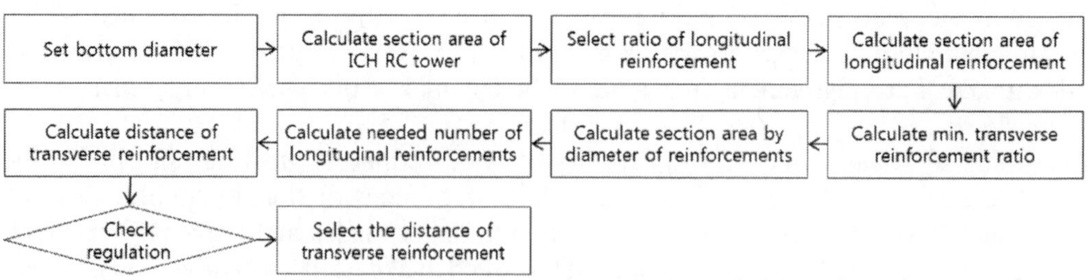

Figure 2. Design process of ICH RC tower.

The section design results of the ICH RC wind tower is shown at table 1. Safety factor 1.2 was applied and 60 mm of concrete cover was selected. The ICH RC wind tower was assumed operating in land and the 60 mm of concrete cover thickness is regulated in the Korean concrete structure design standard. It provides the thickness of concrete cover depending on environmental conditions and construction situations.

Table 1. Section design of the ICH RC wind tower.

	D'(mm)	H/R	D_i(mm)	A_g(mm^2)	A_s(mm^2)	A_c(mm^2)	$\rho_{s,min}$	R_t/No.
Case01 (1.0x DB_0)	5500	0.90	4,950.0	5,562,111	66,745	5,495,365	0.000524	D41/52
		0.85	4,675.0	7,640,962	91,691	7,549,270	0.000524	D41/68
		0.80	4,400.0	9,601,021	115,212	9,485,809	0.000524	D51/60
		0.75	4,125.0	11,442,289	137,307	11,304,981	0.000524	D51/68
Case02 (0.9x DB_0)	4950	0.90	4,455.0	4,600,764	55,209	4,545,555	0.000524	D41/44
		0.85	4,207.5	6,284,633	75,415	6,209,217	0.000524	D51/40
		0.80	3,960.0	7,872,281	94,467	7,777,814	0.000524	D51/48
		0.75	3,712.5	9,363,708	112,364	9,251,343	0.000524	D51/56

where, D': tower diameter without concrete cover, D_i: tower inner diameter without inner tube, A_g: area of ICH RC tower, A_s: area of longitudinal reinforcement, A_c: area of concrete, $\rho_{s,min}$: distance ratio of transverse reinforcement, R_t: diameter of longitudinal reinforcement, No.: number of longitudinal reinforcement

The longitudinal reinforcement ratio was applied as 0.012 that is a regulated minimum value 0.01 with safety factor 1.2. By choosing the minimum value of the ratio, the lowest construction cost was expected. The ICH RC wind tower could be divided into some modules for the convenience of construction and transportation. The number of longitudinal reinforcements was calculated by following process in figure 3, but it was selected multiples of four in range between 30 and 70. The number of longitudinal reinforcement can be reduced from lower module to upper module for cost efficiency and for keeping the ratio of longitudinal reinforcement. About the reducing longitudinal reinforcements is mentioned at module analysis part below. For the transverse reinforcements, 16 mm-diameter rebar was selected. The distance between transverse reinforcements was decided as 660.8 mm. For the convenience of construction and for enhancing safety of the tower, it was set as 600 mm. Thickness of inner tube can be obtained from CoWiTA program (Han 2015). It was 4 mm, but 5 mm of inner tube was selected to use factory-made products and to reduce manufacturing cost.

3 ANALYSIS

With the section design results, the ICH RC wind tower was analyzed by non-linear program CoWiTA. The needed input data of material properties were yield stress, ultimate stress, and elastic modulus of the reinforcements and the inner tube, and they were 313 MPa, 490 MPa, and 206,101 MPa respectively. 30 MPa of compressive strength was used as the concrete input data.

An additional moment by large displacement effect (LDE) was considered when the ICH RC tower was analyzed with CoWiTA. As figure 3 represents, tower top moves from the left to the right by wind thrust. This displacement and the weight of nacelle parts including turbine and blades generate additional moment.

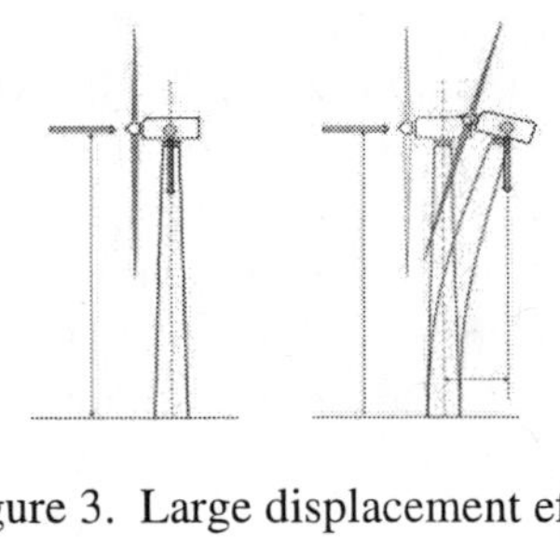

Figure 3. Large displacement effect.

4 RESULTS

4.1 Single Body Type of ICH RC Tower

Figure 4 and figure 5 represent the P-M interaction curves and the moment-curvature curves of the designed ICH RC towers, respectively. In case01, with any of hollow rations the ICH RC wind tower were satisfied with required moment load. When it comes to the case02, the tower has 4.95 m of bottom diameter also could be resisted safe to its demand load except the case of 90% hollow ratio. By comparing figure 4 and figure 5, the lines were located slightly above the required load line in figure 5 than in figure 4. The large displacement effect mentioned above could be a reason of it. The P-M interaction curves indicate its own property of the ICH RC wind tower so that they do not include the large displacement effect. The moment-curvature curves in figure 5, however, applied the large displacement effect so that the analysis results lines were located more close with the required moment load line.

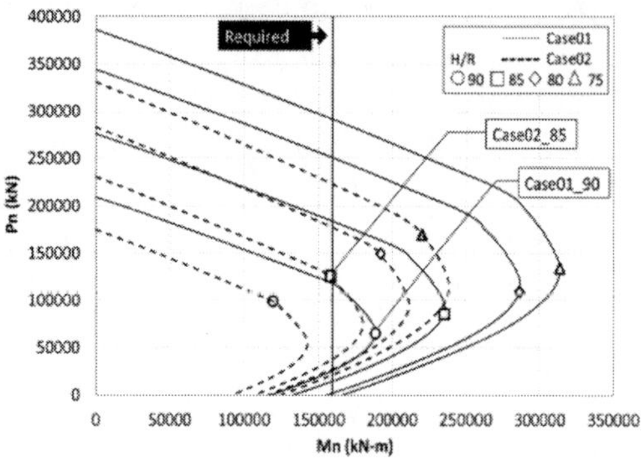

Figure 4. P-M interaction curve.

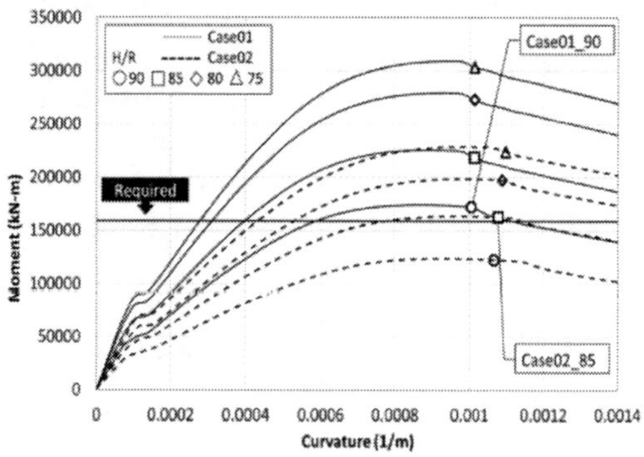

Figure 5. Moment-curvature curve.

4.2 Modular ICH RC Tower

The tall towers such as offshore wind tower could have difficulties of construction, installation, and transportation. The modularized ICH RC wind tower can be an answer of them. In this study, two cases were chosen; case01_90 and case02_85 to design modularized ICH RC wind tower. They not only satisfied with required moment but have the greatest hollow ratio among having different tower diameter cases. The height of each module was determined based on the road traffic law in Korea (Korean National Police Agency, 2016) which presents maximum length of the transporting structure. Each module height was set about 15 m but the height of the lowest module was selected different value as 5 m as shown in figure 6. The module 1 located at the lowest part of the ICH RC wind tower has plastic hinge area so that it was designed to have larger diameter and more transverse reinforcements than any other modules.

The cross-sectional area of the modularized ICH RC wind tower decreases from bottom to top. By reducing the amount of longitudinal reinforcements, the modular ICH RC wind tower can have uniform ratio of them and it increases economic feasibility. As module height goes up, four longitudinal reinforcements were eliminated double-symmetrically for convenience of construction and the balance of bending stiffness. However, if the ratio of longitudinal reinforcement was dropped under 0.012 chosen before, numbers of the transverse reinforcement was maintained.

Table 2. Section design of the modular ICH RC wind tower.

Case01_90		D' (mm)	D_i (mm)	No.	Dist. (mm)	Case02_85		D' (mm)	D_i (mm)	No.	Dist. (mm)
Module1	Bottom	5,500	4,950	52	300	Module1	Bottom	4,950	4,208	40	300
	Top	5,416	4,874				Top	4,895	4,161		
Module 2	Bottom	5,416	4,874	52	600	Module 2	Bottom	4,895	4,161	40	600
	Top	5,247	4,723				Top	4,784	4,067		
Module 3	Bottom	5,247	4,723	48	600	Module 3	Bottom	4,784	4,067	36	600
	Top	4,995	4,495				Top	4,618	3,926		
Module 4	Bottom	4,995	4,495	44	600	Module 4	Bottom	4,618	3,926	32	600
	Top	4,742	4,268				Top	4,453	3,785		
Module 5	Bottom	4,742	4,268	40	600	Module 5	Bottom	4,453	3,785	28	600
	Top	4,473	4,025				Top	4,276	3,634		
Module 6	Bottom	4,473	4,025	36	600	Module 6	Bottom	4,276	3,634	24	600
	Top	4,203	3,783				Top	4,099	3,484		
Module 7	Bottom	4,203	3,783	32	600	Module 7	Bottom	4,099	3,484	24	600
	Top	3,900	3,510				Top	3,900	3,315		

where, Dist.: distance between transvers reinforcements

Table 2 shows the result of section design of each module. CoWiTA was used for analysis of each section at its height and figure 7 shows the results.

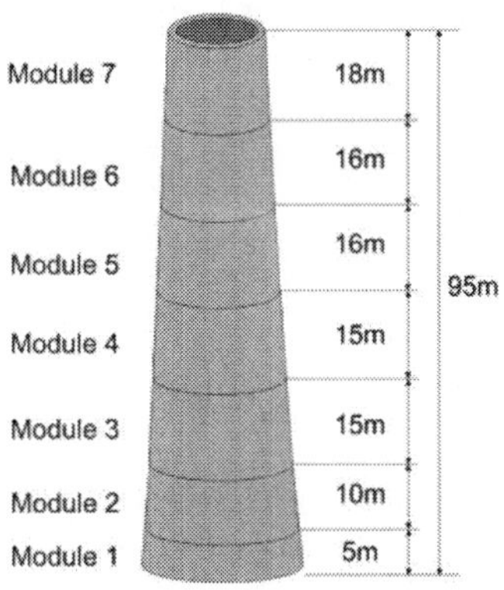

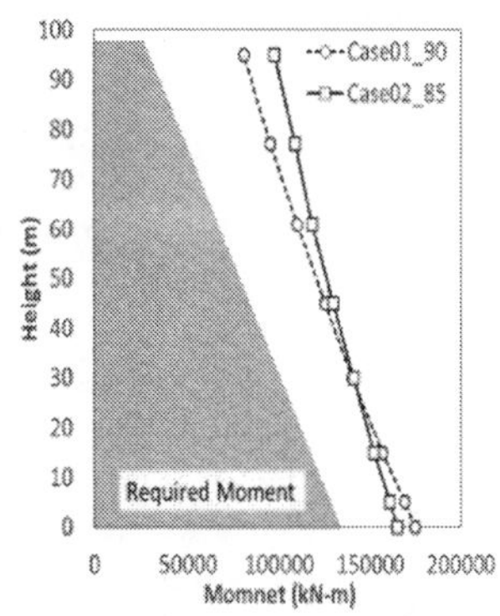

Figure 6. Section height. Figure 7. Results of modular ICH RC tower.

The maximum moment of the modular ICH RC tower at the tower bottom were 174,974 kN-m on case01_90 and 164,976 kN-m on case02_85. They were 110% and 104% of the required moment load respectively. The results of modular ICH RC wind tower indicated that moments at tower top could be safe until 255% and 307% of required moment on case01_90 and case02_85. Finally, all modularized tower sections were confirmed to satisfy with the required moment at each height.

5 CONCLUSION

An ICH RC tower was suggested for supporting 5.0 MW wind turbine. The different sections which have two different tower diameters with various hollow ratios were designed and analyzed by CoWiTA. The large displacement effect was applied in analysis and all design cases except case02_90 showed enough performance to support a 5.0 MW turbine. Two cases among the 8 cases of section design were chosen to modularize. They satisfied with required moment and they have the greatest hollow ratio among having different tower diameter cases. The modularized ICH RC wind tower which has advantages of construction and transportation were sectionally designed and analyzed. By analyzing each module, it was confirmed that the modularized ICH RC wind tower satisfied with required load condition at each height. As a result, the ICH RC tower could support 5.0 MW wind turbine and could be modularized. Finally, the modularized ICH RC wind tower could be an alternative wind power tower instead of the steel towers having risk of buckling failure, and could be applied as offshore wind tower which requires large capacity and tall height.

Acknowledgments

This research was financially supported by the Ministry of Land, Infrastructure and Transport (MOLIT) of the Korea government (code 12 Technology Innovation E09) and Korea Institute of Ocean Science & Technology (project no. PE99421).

References

GWEC, *Global wind report-annual market update,* April, 2016

Han, T. H., Stallings, S. K., and Kang, Y. J., Behaviour of a hollow RC column with an internal tube, *Magazine of Concrete Research*, 62(1), 25-38, January, 2010.

Han, T. H., *Korea Institute of Ocean Science & Technology*, CoWiTA manual, 2015.

Kim, S., Hong, H., Han, T. H., and Yi, J., Standard section design and analysis of 5.0MW ICH RC wind tower by using FAST, *Korea Institute for Structural Maintenance and Inspection*, 165-168, April, 2016.

Kimopoulos, C.A., and Gantes, C.J., Experimental investigation of buckling of wind turbine tower cylindrical shells with opening and stiffening under bending, *Thin-Walled Structures,* 54, 140-155, May, 2012.

Klaassen, G., Miketa, A., Larsen, K., and Sundqvist, T., The impact of R&D on innovation wind energy in Denmark, Germany and the United Kingdom, *Ecological Economics, 54, 227-240,* March, 2005.

Korea Concrete Institute, Concrete Structure Design Standard, Seoul, Korea, 2012.

Korean National Police Agency, The road traffic law, 2016.

Popovics, S., A numerical approach to the complete stress-strain curves of concrete, *Cement and concrete Research,* 3(5), 583-599, September, 1973.

Watts, D., Oses, N., and Perez, R., Assessment of wind energy potential in Chile: A project-based regional wind supply function approach, *Renewable Energy,* 96, 738-755, May, 2016.

Integrated Solutions for Infrastructure Development
Edited by Lau, H. H., Tang, F. E., Ng, C. K., and Singh, A.
Copyright © 2016 ISEC Press

STANDARD SECTION DESIGN OF FRP DSCT TOWER SUPPORTING 3MW WIND TURBINE

HYEMIN HONG, SUNGWON KIM and TAEK HEE HAN

Korea Institute of Ocean Science &Technology, Ansan, Republic of Korea

A modular FRP DSCT (Fiber Reinforced Polymer Double-Skinned Composite Tubular) wind power tower was suggested and its standard section supporting 3MW wind turbine was designed. An FRP DSCT tower is composed of two concentric FRP tubes and concrete filled between the inner and outer tubes. A DSCT structure has hollowed section which reduces its self-weight. Due to inner and outer tubes, concrete is confined to increase strength and ductility. The performance of the new-type wind power tower was assessed via structural analysis. Design and analysis were carried with AutoDSCT and CoWiTA programs with consideration of confining effect of concrete and large displacement by wind turbine weight itself. The bottom diameter of the tower was set as 4.5 m as a standard. The towers which have 100%, 95%, 90% and 85% of bottom diameter were considered for design. Top diameter of the tower was set as 3.6m. For the consideration of constructing and installing the tower, each tower was divided into modules. The module sections were designed on the basis of the idea that module sections should satisfy the required bending moment and axial load. As a result, bottom diameter of 4.5m with hollow ratios below 0.85 satisfied the required load capacity and the bottom diameter of 95%, 90% and 85% with the hollow ratios below 0.82, 0.76 and 0.7 satisfied the required capacity.

Keywords: Wind power tower, Concrete, Column, CoWiTA, AutoDS.

1 INTRODUCTION

Wind power energy is one of the most efficient renewable energies and is economically feasible. Nowadays, the wind turbine tower needs to be high enough to support the turbine size of which is directly proportional to the energy generated. As a result, the slenderness of ratio increases and it is the main cause of buckling in the event of heavy weather that can eventually make the tower collapse and the residents live in fear (Han 2015). Thus, the wind turbine tower must be strong enough to withstand the buckling phenomena. Another problem is the corrosion caused by saltwater reacting with steel made offshore wind turbine tower.

In this research, an FRP DSCT structure is applied to wind power tower to suggest a new type of tower section design supporting 3MW wind turbine. FRP as a material is known to have small weight-strength ratio and corrosion resistance. It is also durable and long-lasting. A DSCT column which was researched by Shakir-Khalil and Illouli in 1987 is composed of two concentric tubes and concrete (Shakir-Khalil and Illouli 1987). Such composition is actively being studied in order to maximize the function of the hollowed section to decrease the weight of the tower. Recently, Han *et al.* (2013) suggested the nonlinear material column model of DSCT structure (Han *et al.* 2013).

2 SECTION DESIGN

2.1 Automatic Design Program, AutoDSCT

AutoDSCT (Han 2014) is a program based on the strain compatibility of a DSCT tower developed from the section analysis and the interactions between axial load and bending moment considering material nonlinearity and confining effect of concrete (Han *et al.* 2013). In this research, AutoDSCT was used for generating bottom design sections of DSCT tower. The design process of AutoDSCT is shown in Figure 1. It automatically creates bottom sections of the tower with variety of hollow ratios, inner and outer tube thicknesses by processing input including outer diameter, the mechanical properties of materials, required bending moment and axial force. AutoDSCT not only creates most economical design sections, but also provides axial load-bending moment (P-M) interaction curves and lateral force-lateral displacement (P-Δ) curves which can help understanding in terms of strength of the tower. At every run, ten cases of the sections are produced with hollow ratios varying from 97% to 70% by decreasing steps of 3% (Han *et al.* 2010). Decision of outer diameter of inner tube depends on hollow ratio.

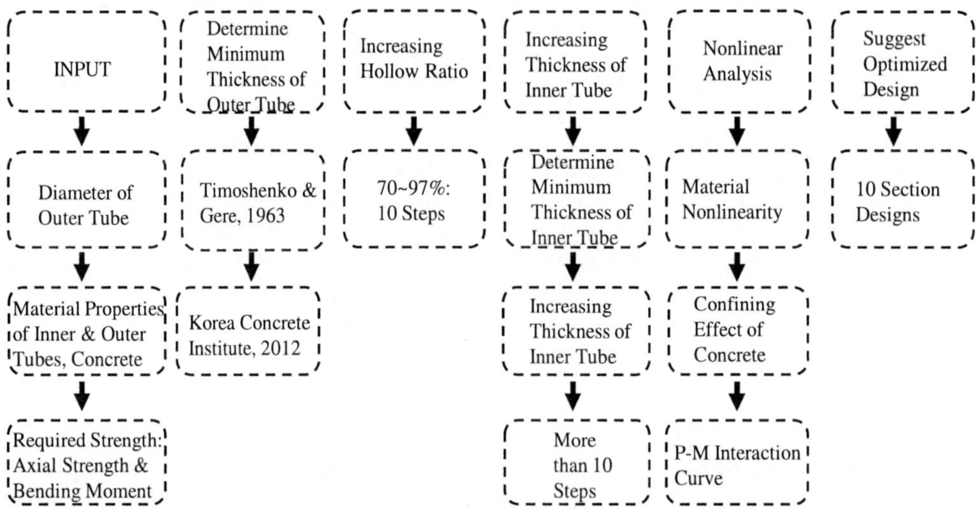

Figure 1. Design process in AutoDSCT program.

2.1.1 *Design sections generated by AutoDSCT*

In this research, basis of the steel tower suggested by Ljjj and Gravesen in 2008 was considered and its bottom diameter was used (Ljjj and Gravesen 2008). The bottom and top diameters of the tower were set as 4.5 m and 3.6 m respectively. The height of the tower was set as 90 m. For the thickness of the FRP outer and inner tubes, 35mm was used as a random value. The axial force of 2,835kN and bending moment of 88,000kN-m were calculated by considering 3 MW wind turbine applied on the top of the tower. The FRP mechanical property calculated based on FRP tension test is shown in Table 1 and the forty cases of bottom sections generated from AutoDSCT are shown in Table 2.

Table 1. Material mechanical properties.

Material Property Information	
Nominal Compressive Strength of Unconfined Concrete	30 MPa
Density of Concrete	2,300 kg/m^3
Ultimate Strain of Concrete	0.003
Yield Strength of Outer/Inner Tube	48.685 MPa
Ultimate Strength of Outer/Inner Tube	230.25 MPa
Modulus of Elasticity of Outer/Inner Tube	1,5436.7 MPa
Ultimate Strain of Outer/Inner Tube	0.0244 MPa
Density of FRP	1,500 kg/m^3

Table 2. Design sections generated by AutoDSCT.

Design Case (4.5m)	Case 1	Case 2	Case 3	Case 4	Case 5	Case 6	Case 7	Case 8	Case 9	Case 10
	3F100/97	3F100/94	3F100/91	3F100/88	3F100/85	3F100/82	3F100/79	3F100/76	3F100/73	3F100/70
D	4500	4500	4500	4500	4500	4500	4500	4500	4500	4500
D_i	4365	4230	4095	3960	3825	3690	3555	3420	3285	3150
t	31	19.5	13	12	12	12	12	12	12	12
t_i	36.386	27.966	22.105	20.538	19.838	19.138	18.437	17.737	17.037	16.337
Design Case (4.275m)	3F95/97	3F95/94	3F95/91	3F95/88	3F95/85	3F95/82	3F95/79	3F95/76	3F95/73	3F95/70
D	4275	4275	4275	4275	4275	4275	4275	4275	4275	4275
Di	4146.75	4018.5	3890.25	3762	3633.75	3505.5	3377.25	3249	3120.75	2992.5
t	36	24	17	13	11.5	11.5	11.5	11.5	11.5	12
t_i	38.412	30.24	24.638	20.835	18.928	18.26	17.592	16.924	16.256	15.923
Design Case (4.05m)	3F90/97	3F90/94	3F90/91	3F90/88	3F90/85	3F90/82	3F90/79	3F90/76	3F90/73	3F90/70
D	4050	4050	4050	4050	4050	4050	4050	4050	4050	4050
Di	3928.5	3807	3685.5	3564	3442.5	3321	3199.5	3078	2956.5	2835
t	40	30	22	17.5	15	13.5	13	13	13.5	14
t_i	42.68	32.907	27.281	23.529	21.041	19.257	18.205	17.514	17.143	16.74
Design Case (3.825m)	3F85/97	3F85/94	3F85/91	3F85/88	3F85/85	3F85/82	3F85/79	3F85/76	3F85/73	3F85/70
D	3825	3825	3825	3825	3825	3825	3825	3825	3825	3825
Di	3710.3	3595.5	3480.8	3366	3251.3	3136.5	3021.3	2907	2792.3	2677.5
t	44.5	35.5	28.5	23	19.5	17.5	17	16.5	16.5	17.5
ti	47.482	36.707	30.176	26.214	23.315	21.307	20.232	19.176	18.419	18.189

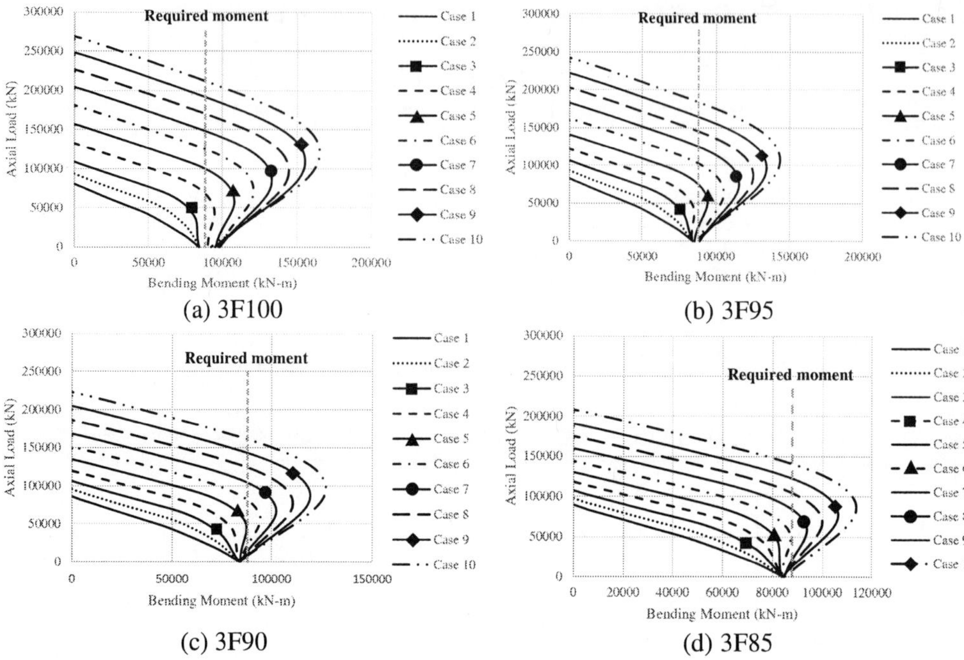

Figure 2. Axial load-bending moment curves (P-M interaction curves) for DSCT tower supporting 3MW wind turbine.

2.2 Nonlinear Analysis by CoWiTA

To prevent buckling failure, each bottom section designed from AutoDSCT was performed for nonlinear analysis. The performances of the designed sections were conducted via CoWiTA (Han 2015) which evaluates bending performance of a column considering material nonlinearity, confining effect of concrete, and large displacement effect. Cases with satisfying the required moment strength from P-M interaction curves shown in Figure 2 were selected for nonlinear analysis through CoWiTA. Figure 3 presents the lateral load-lateral displacement relation curves of the DSCT tower. As the lateral load becomes larger, the lateral displacement increases. In the case of the tower with bottom diameter, 4.5m, design sections with hollow ratios from 85% to 70% have higher moment strength than the required strength. This means that the tower with designed sections of those hollow ratios can support the 3MW wind turbine without in the event of buckling. For the cases of 3F95, 3F90 and 3F85, the hollow ratios of 82% to 70%, 79% to 70% and 76% to 70% exceeded the required bending moment.

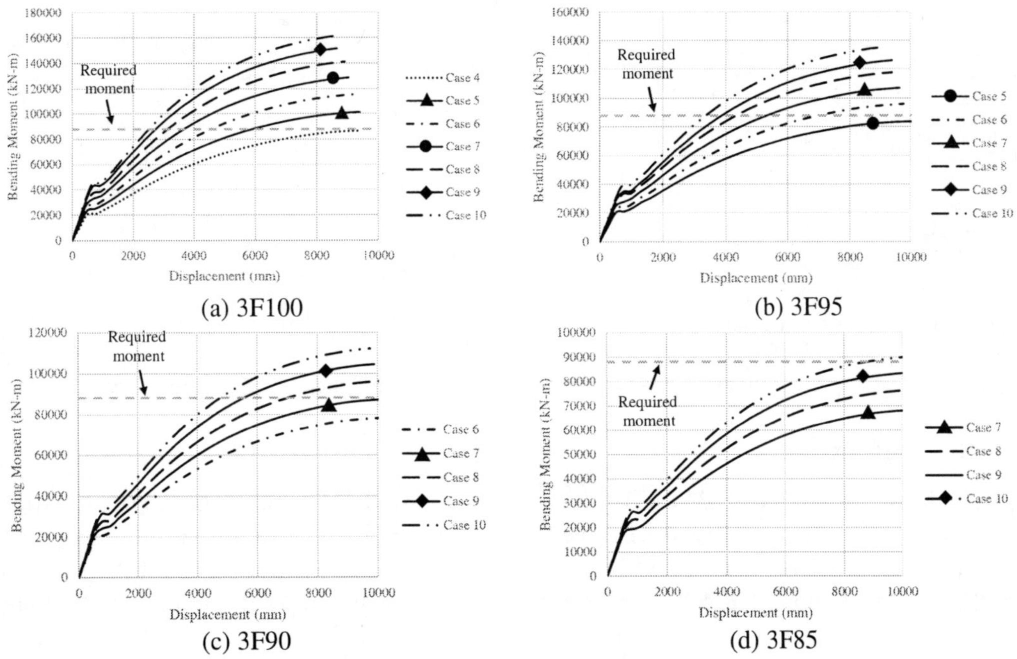

Figure 3. Lateral load-lateral displacement curves for DSCT tower supporting 3MW wind turbine.

2.2.1 *Modular FRP DSCT tower*

For the consideration of installing and transporting the tower, each tower was split into seven modules and Table 3 shows dimensions of each module for four towers. Figure 4 appears that the bending moment of the tower modularized according to module height from the bottom to the top of the whole tower by considering large displacement effect and without large displacement effect.

Table 3. Section design of modular FRP DSCT.

		Module height (m)	4.5m		4.275m		4.05m		3.825m	
	Hollow ratio		1	0.85	1	0.82	1	0.76	0.7	
Module 1	Bottom	5	4500	3825	4275	3505.5	4050	3078	3825	2677.5
	Top		4450	3782.5	4237.5	3474.75	4025	3059	3812.5	2668.75
Module 2	Bottom	5	4450	3782.5	4237.5	3474.75	4025	3059	3812.5	2668.75
	Top		4400	3740	4200	3444	4000	3040	3800	2660
Module 3	Bottom	15	4400	3740	4200	3444	4000	3040	3800	2660
	Top		4250	3612.5	4087.5	3351.75	3925	2983	3762.5	2633.75
Module 4	Bottom	15	4250	3612.5	4087.5	3351.75	3925	2983	3762.5	2633.75
	Top		4100	3485	3975	3259.5	3850	2926	3725	2607.5
Module 5	Bottom	16	4100	3485	3975	3259.5	3850	2926	3725	2607.5
	Top		3940	3349	3855	3161.1	3770	2865.2	3685	2579.5
Module 6	Bottom	16	3940	3349	3855	3161.1	3770	2865.2	3685	2579.5
	Top		3780	3213	3735	3062.7	3690	2804.4	3645	2551.5
Module 7	Bottom	18	3780	3213	3735	3062.7	3690	2804.4	3645	2551.5
	Top		3600	3060	3600	2952	3600	2736	3600	2520

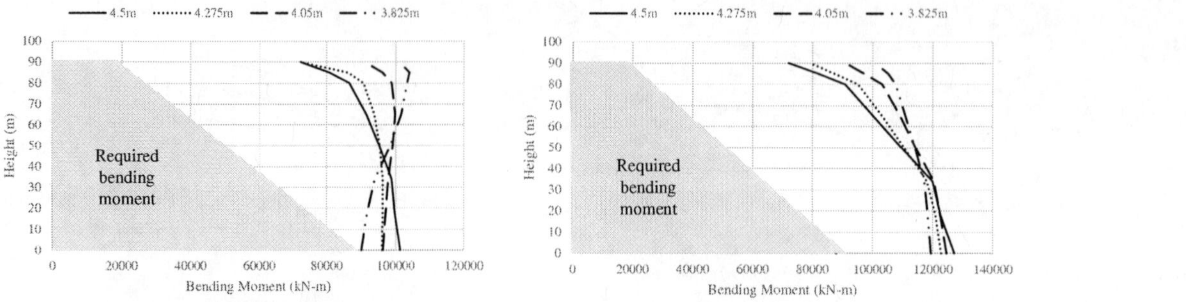

(a) Large displacement effect (b) No large displacement effect

Figure 4. Tower height-bending moment curves.

3 CONCLUSION

Design sections of modular FRP DSCT tower were suggested. The tower of the bottom diameter of 4.5m with hollow ratios below 0.85 satisfied the required load capacity. For the towers of the bottom diameter of 95%, 90% and 85%, the hollow ratios below 0.82, 0.76 and 0.7 satisfied the required capacity to support 3MW wind turbine. The suggested design sections of modular FRP DSCT tower can withstand the required moment strength which means it is proven to be safe in the event of buckling.

Acknowledgments

This research was financially supported by the Ministry of Land, Infrastructure and Transport (MOLIT) of the Korea government (code 12 Technology Innovation E09) and Korea Institute of Ocean Science & Technology (project no. PE99421).

References

Han, T. H., Korea Institute of Ocean Science & Technology, AutoDSCT manual Ver. 1.1, 2014.

Han, T. H., Korea Institute of Ocean Science & Technology, CoWiTA manual, 2015.

Han, T.H., Stallings, J.M. and Kang, Y.J., "Nonlinear Concrete Model for Double-Skinned Composite Tubular Columns", Construction and Building Materials, Elsevier, 4(12), 2542-2553, 2010.

Han, T. H., Won, D. H., Kim, S. and Kang, Y.J., Performance of A DSCT Column under Lateral Loading: Analysis, Magazine of Concrete Research, 65(2), 121-135, 2013.

Han, T. H., Park, Y. H., Won, D. H. and Lee, J. H., Design Feasibility of Double-Skinned Composite Tubular Wind Turbine Tower, Wind and Structures, 21(6):727-753, 2015

Ljjj L.B.G. and Gravensen, H., Kriegers Flak Offshore Wind Farm –Design Basis Foundations, Vattenfall Vindkraft AB, 2008.

Shakir-Khalil, H. and Illouli, S., Composite Columns of Concentric Steel Tubes, Proceeding of Conference on the Design and Construction of Non-Conventional Structures, 73-82, 1987.

Integrated Solutions for Infrastructure Development
Edited by Lau, H. H., Tang, F. E., Ng, C. K., and Singh, A.
Copyright © 2016 ISEC Press

MATERIAL AND OUT-OF-PLANE FLEXURAL PROPERTIES OF TRADITIONAL MASONRY WALL

NORIKO TAKIYAMA[1], AKARI YAMAGUCHI[1], KOHEI HARA[1], VERDEJO JIMENEZ[2], and JUAN RAMON[2]

[1]*Division of Architecture and Urban Studies, Tokyo Metropolitan University, Tokyo, Japan*
[2]*Graduate School of Environmental Planning, University of Shiga Prefecture, Shiga, Japan*

The October 2013 Earthquake hit the Bohol Island of the Philippines. Many traditional churches were damaged by the earthquake. In this study, material tests were conducted to assess the strength of the Philippines' traditional masonry wall, which is composed of various materials. Additionally, out-of-plane flexural experiments were conducted to elucidate the mechanical characteristics of masonry walls found in traditional churches. The wall specimens used in this study were made of a variety of materials with different compositions. The major findings of this study are summarized as follows: (a) compressive tests and splitting tensile tests were conducted for various test samples having distinct compositions. For both tests, the strength of the cylinders without additional materials was greater than that with contamination, expect for those containing egg whites; (b) out-of-plane flexural experiments were conducted on four specimens to identify their structural characteristics.

Keywords: The 2013 Bohol Earthquake, Masonry structure, Material composition, Material test, Out-of-plane flexural experiments.

1 INTRODUCTION

The October 2013 Earthquake that hit Bohol Island in the Philippines killed 222 people; more than 3 million people were affected in the Philippines (Ministry 2016). According to a previous disaster investigation, 23 churches with masonry structures in Bohol Island were completely destroyed or partially damaged (Japan 2014) due to their poor seismic tolerance.

The walls of traditional masonry churches in Bohol are fixed using a rammed-earth method; in this method, mud containing stones, shells, gravel, and lime paste is piled up into layers and tamped down (Hanazato 2000, Rammed 2016).

These churches are a place of assembly; hence, they must be able to withstand enough seismic activity to minimize earthquake damage. However, they were built without any information regarding their seismic performance. Moreover, some constructions were damaged by the earthquake, whereas some were not.

In this study, material tests were conducted to assess the strength of the walls composed of various materials. Furthermore, out-of-plane flexural experiments were conducted to elucidate the mechanical characteristics of the masonry walls present in traditional churches. The wall specimens were unique compositions of several materials.

2 MATERIAL STRENGTH TEST OF MUD MORTAR

Mud mortar is made up of stones, shells, gravel, and lime paste. In this study, we created test cylinders having various proportions of different materials and conduct strength tests to understand the difference of strength from material combination and proportion.

2.1 Test Cylinders

The material combinations and proportions of the test cylinders are indicated in Table 1. The main materials are slaked lime, sand, and water. Furthermore, we selected cement, egg yolk, egg white, palm fiber, and straw as additional materials. The volume ratio of materials was changed while a combination of 01 to 11 indicated only main materials, and a combination of 12 to 21 indicated main materials with one or two additional materials. The number of each cylinder combination was four or six. Combination 02 was named 'PALETADA' and represented the composition of the traditional wall. The curing period was nominally 28 days based on JIS A 1132, and when it was short, the period was a multiple of 28 days.

Table 1. Combination and proportion of materials (volume ratio).

Comb.	Number	Main materials			Additional materials				
		Slaked lime	Sand	Water	Cement	Egg yolk	Egg white	Palm fiber	Straw
01	6	0.8	3.2	1	0	0	0	0	0
02	6	1	3	1	0	0	0	0	0
03	6	1.17	2.1	1	0	0	0	0	0
04	6	1.47	1.97	1	0	0	0	0	0
05	6	1.5	2.5	1	0	0	0	0	0
06	6	2	2	1	0	0	0	0	0
07	6	2.2	1.64	1	0	0	0	0	0
08	4	2.5	2	1	0	0	0	0	0
09	6	2.94	1.31	1	0	0	0	0	0
10	4	1	3	0.6	0	0	0	0	0
11	4	1.47	1.97	0.8	0	0	0	0	0
12	4	1	3	0.8	1	0	0	0	0
13	4	1.47	1.97	1	0.32	0	0	0	0
14	4	1	3	0.6	0	0	0	0.02	0
15	4	1.47	1.97	1	0	0	0	0.2	0
16	4	1	3	0.6	0	0	0	0	0.02
17	4	1.47	1.97	1	0	0	0	0	0.2
18	4	1.47	1.97	1	0.32	0	0	0	0.2
19	4	1	3	0.6	0	0.5	0	0	0
20	4	1.47	1.97	1	0	0.5	0	0	0
21	4	1	3	0.6	0	0	0.5	0	0

2.2 Material Strength

Material tests were conducted based on JIS A 1108 and JIS A 1113. The compressive strength test and splitting tensile test and main damage are depicted in Figures 1 and 2. The strength of each cylinder is illustrated in Figure 3, excluding cases containing cement. Because the strength of the cylinders with cement were found to be twenty times that of cylinders without cement, the results with cement were excluded from this paper.

Almost all cylinders were found to be very brittle. The horizontal axis in Figure 3 is the ratio of slaked lime to sand. For compressive strength, the cylinders without additional materials were stronger than contaminated ones, expect for egg white. For the splitting tensile test, the strength increased as the ratio of slaked lime to sand increased up to a value of 1.0, but with high variation. Furthermore, like the compressive test, the cylinders without additional materials were stronger than ones with contamination, expect for those with egg whites. The strength of the cylinder containing palm fiber was uncertain.

(a) Compressive test (b) Compressive fracture (c) oblique fracture

Figure 1. Compressive test and main fracture.

(a) Splitting tensile test (b) Fracture surface (c) Peripheral fracture

Figure 2. Splitting tensile test and main fracture.

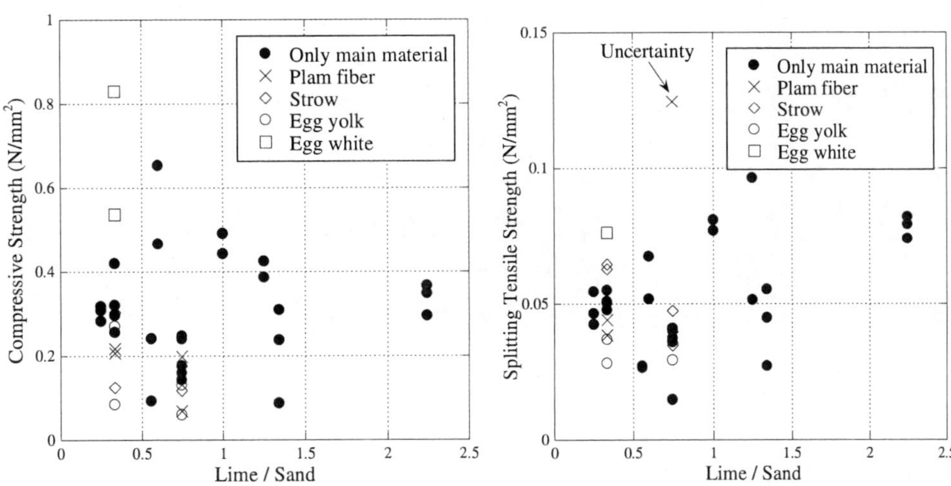

(a) Compressive strength (b) Splitting tensile strength

Figure 3. Relationship between material strength and the ratio of lime to sand.

3 STATICALLY DESTRUCTIVE LOADING TEST

Based on former cylinder tests, a statically destructive loading test was conducted on four full-scale wall specimens.

3.1 Wall Specimens

Four wall specimens were constructed at San Carlos University in Philippines on September 15th to 21st, 2015. The walls were 1500 mm high, 650 mm thick, and 1000 mm long. 300 mm were constructed each day, and the curing period was four months. The materials used included lime mortar, slaked lime, sand, water, and rock as indicated in Table 2, based on PALETADA. Lime mortar is commonly used in the Philippines, and is typically cured lime in water for three months at a very high concentration. However, this was very expensive, so we substitute slaked lime, in part, for lime mortar, assuming that slaked lime 2 was equal to lime mortar 1.

We constructed wall specimens A to D as shown in Figure 4. First, we build a raising mold out of timber. Next, coral stone plates were arranged on both sides of the wall. Then, we kneaded the lime mortar, slaked lime, and sand using a mixer. Furthermore, we install knead materials and rock among the stone plates as shown in Figure 4 (a). This same process was repeated five times over as shown in Figure 4 (b). The completed specimen is shown in Figure 4 (c). Additionally, five material test cylinders were made at the same time from the material used to build each wall specimen.

Table 2. Proportion of materials of the wall specimens (volume ratio) and test results.

Wall	Volume ratio of material					Material ratio	Cylinder strength [N/mm^2]		Loading test [kN]
	Lime Mortar	Slaked lime	Sand	Rock	Water	Lime / Sand	Compressive	Splitting tensile	Maximum Load
A	1	0	3	4	0	0.33	1.33	0.29	Uncertainty
B	0	2	2	3	0.75	0.50	1.14	0.39	17.4
C	1	2	1	3	0.75	2.0	0.73	0.11	10.1
D	0	3	1.5	3	1	1.0	1.07	0.14	9.4

(a) Containing stone plates and materials (b) Raising mold _ (c) Wall specimen

Figure 4. Construction of wall specimens.

3.2 Material Strength Test

Compressive strength tests and splitting tensile strength tests were conducted. However, some test cylinders broke prior to testing. For splitting tensile strength, when the ratio of slaked lime for sand was 0.5, Wall B, showed the highest strength as indicated in Table 2.

3.3 Loading of Wall

We designed a loading system as shown in Figure 5. Two wall specimens were laid face-to-face of out-of-plane on the steel member, the chain block was used to connect both tops of the walls, and they were pulled against each other. The pulling loads are measured by load cell.

The maximum loads are indicated in Table 2 and the destruction is shown in Figure 6. For Wall B to C, bending destruction was found, but for only Wall A, the foot of the wall was broken, so the maximum loading could not be measured.

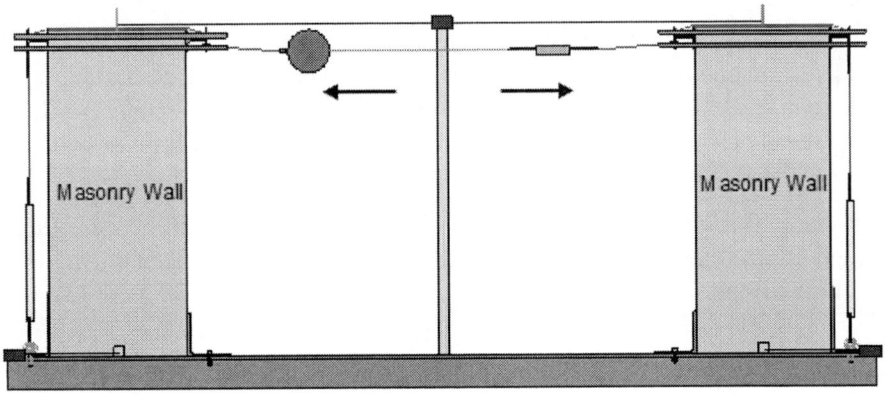

Figure 5. Loading system.

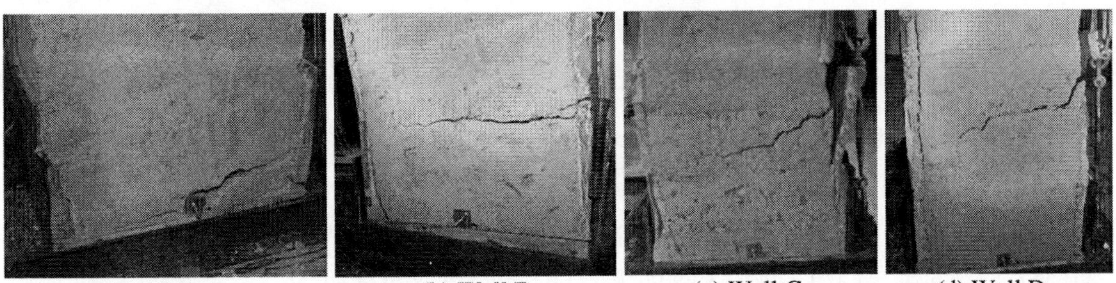

(a) Wall A (b) Wall B (c) Wall C (d) Wall D

Figure 6. Main damage of wall specimens.

4 DISCUSSIONS

The major findings from this study and discussions on them are summarized as follows:

(a) In material tests for various compositions of materials, for compressive tests, the strength of the cylinders without additional materials was greater than that with contamination, expect for the case with egg whites and mortar. And even if the ratio of slaked lime to sand is any value, it is almost constant in 0.3 kN/mm^2. On the other hands, for the splitting tensile test, the strength increased as the ratio of slaked lime to sand increased up to a value of 1.0, but demonstrated high variability.

(b) In case of out-of-plane flexural experiments on four wall specimens, when the ratio of slaked lime to sand was 0.5, the maximum load was the highest

5 CONCLUSIONS

In this study, material tests were conducted to assess the strength of Philippines' traditional masonry walls, which are composed of various materials including slaked lime, sand, and water as the main materials, and cement, egg yolk, egg white, palm fiber, and straw as additional materials. Out-of-plane flexural experiments were conducted to elucidate the mechanical characteristics of masonry walls in traditional churches. The wall specimens were composed of main materials with different compositions. The major findings of this study are summarized as follows:

(a) Compressive tests and splitting tensile tests were conducted for various compositions of materials. For both tests, the strength of the cylinders without additional materials was greater than that with contamination, expect for the case with egg whites. For the splitting tensile test, the strength increased as the ratio of slaked lime to sand increased up to a value of 1.0, but demonstrated high variability.

(b) Out-of-plane flexural experiments were conducted on four wall specimens, and their structural characteristics were investigated. In case of this experiment, when the ratio of slaked lime to sand was 0.5, the maximum load was the highest.

Acknowledgments

We are grateful to Special Research Grant of University of Shiga Prefecture, "Method of damage assessment, preservation and restoration of stone building that was affected by the Bohol earthquake in the Philippines", The Japan Foundation Asia Center, Asia Center Fellowship Program "JAPANESE-PHILIPPINE RESTORATION TALLER FOR BOHOL ISLAND'S CULTURAL HERITAGES", Jesuit House 1730, Parian Cebu and Escuela Taller de Filipinas Foundation inc. We thank Ar. Troy Dino S. ELIZAGA, Professor of San Carlos University, Ikumi OSAWA and Fumiaki OKAMURA, the students of Tokyo Metropolitan University, and the students of San Carlos University for their assistance with making wall specimens and the statically destructive loading tests. Naoto IDATE, the students of Tokyo Metropolitan University, for his assistance with designing loading system.

References

Hanazato, T., Monitoring to the Paoay Church, the Philippines, ICOMOS, 2000.

Japan Consortium for International Cooperation in Cultural Heritage, *Survey Report on the Protection of Cultural Heritage in Republic of the Philippines*, Japan, 2014.

Ministry of Land, Infrastructure, National Spatial Planning and Regional Policy Bureau, An Overview of Spatial Policy in Asian and European Countries. Retrieved from http://www.mlit.go.jp/kokudokeikaku/international/spw/general/philippines/ on June 1, 2016. (in Japanese)

Rammed earth construction. Retrieved from http://www.rammedearthconstructions.com.au/ on June 1, 2016.

Integrated Solutions for Infrastructure Development
Edited by Lau, H. H., Tang, F. E., Ng, C. K., and Singh, A.
Copyright © 2016 ISEC Press

A STUDY ON THE FACTORS AFFECTING ELECTRO-MIGRATION CHARACTERISTIC OF CHLORIDE IONS IN CONCRETE

ISAO UJIKE and KEIYU KAWAAI

Dept of civil and environmental engineering, Ehime University, Matsuyama, Japan

As one of the methods to evaluate the durability performance of reinforced concrete structure for the deterioration due to chloride attack, there is a migration test method. The effective diffusion coefficient by migration test method fluctuates with changes in the pore structure of concrete. In this study, focused on the changes of the pore structure with elapse of age of concrete, the objective of this study is to assess the influence exerted on electro-migration characteristic caused by the type of aggregates and with the condition of early age curing. It is confirmed that the resistibility to the movement by the migration test method of the chloride ion was increased by using non-ferrous slag fine aggregates. In the test specimen using sandstone of 40-65% of water-cement ratio, it is presumed that the increase in the effective diffusion coefficient of chloride ion is caused by the fact that the volume of continuous pore or pore size by the curing method and age of concrete become bigger.

Keywords: Reinforced concrete, Chloride attack, Effective diffusion coefficient, Pore structure, Type of aggregates, Non-ferrous slag, Curing condition, Age of concrete.

1 INTRODUCTION

Degradation of reinforced concrete structure owing to chloride attack is a phenomenon causing reinforcement corrosion by penetrating of chloride ions in concrete and leading to the deterioration in performance of structure such as safety. Verification of durability performance against chloride attack of reinforced concrete structure predicts chloride ions concentration at location of reinforcement in order to determine the presence or absence of corrosion occurrence of concrete. Based on Fick's diffusion law, chloride ions concentration distribution of cover concrete is predicted by using diffusion coefficient of chloride ions.

As one of the effective methods to obtain the diffusion coefficient of concrete, it is established that the method for estimating an effective diffusion coefficient of chloride ions in concrete by means of an electro-migration phenomenon (JSCE 2013). An effective diffusion coefficient obtained by the electro-migration indicates the easiness of electro-migration of an ion existing in the pore solution of concrete. It is different from an apparent diffusion coefficient that shows the diffusion rate of moving in the pore solution by a driving force from the concentration gradient. However, this method is very effective in evaluating the diffusion performance of concrete in a short time.

The purpose of this study is to assess the influence exerted on electro-migration characterization, focused on the type of the fine aggregates using in concrete and the condition of early age curing of concrete, from the viewpoint of pore structure in concrete. In addition, the effective diffusion coefficient obtained by migration test is compared with an apparent diffusion coefficient obtained by the immersion test.

2 EXPERIMENTAL OUTLINE

2.1 Materials and Specimens

Ordinary Portland cement was used. Table 1 shows the physical property of aggregates used in this experiment. Copper slag fine aggregates and manganese slag fine aggregates were used for non-iron slag fine aggregates. Table 2 shows the mix proportions of concrete. In the case (1), fine aggregates ratio was 48%. Non-ferrous slag fine aggregates replaced 30% natural aggregates in a volume ratio. In the case (2), sand-total aggregates ratio was 46%.

The concrete curing were comprised of two ways; one is that the concrete was released for 1 day and cured in water until 28 days and the other was that released for 3 days and cured in t air curing until 28 days after casting the concrete respectively. In the case (1), specimens are tested at age of 28 days, and in the case (2) specimens are tested at age of 28 days, 6 months and 1 year.

Table 1. Properties of aggregate.

Aggregate	Designation	Types	Density (g/cm^3)	Absorption capacity (%)
Fine	CSS	Crushed sand obtained from sandstone	2.61	1.06
	CSA	Crushed sand obtained from andesite	2.60	1.74
	CSL	Crushed sand obtained from limestone	2.66	0.58
	CUS	Copper slag fine aggregate	3.55	0.04
	MNS	Manganese slag fine aggregate	2.91	0.97
Coarse	CSG	Crushed stone obtained from sandstone	2.63	1.15
	CAG	Crushed stone obtained from andesite	2.61	1.97

Table 2. Mix Proportions of concrete.

Case	Name of mix	W/C (%)	Unit weight (kg/m^3)								
			Water	Cement	Aggregate						
					Fine					Coarse	
					CSS	CSA	CSL	CUS	MNS	CSG	CAG
(1)	SOPC	55	175	318	851					929	
	AOPC	55	175	318		847					922
	SCUS	55	175	318	595			347		929	
	SMNS	55	175	318	595				285	929	
	ACUS	55	175	318		593		347			922
	AMNS	55	175	318		593			285		922
(2)	SOPC40	40	175	438	382		394			912	
	SOPC50	50	175	350	395		412			957	
	SOPC65	65	175	269	428		428			994	

2.2 Outline of Migration Test and Immersion Test

The migration test method was conducted according to JSCE-G 571-2013 (JSCE 2013) using the device in Figure 1. A disc specimen used for the migration test was shaped into 50 mm thickness by cutting off a 50mm from the placed face by a diamond cutter from φ100×200mm cylindrical specimen. The sealed disc specimen was immersed in the vacuum desiccator injected distilled water, and depressurized with a vacuum pump for 24 hours and was saturated with water. The following parameters were measured; current value, a potential difference of the surface of the specimen, solution temperature and chloride ions concentration of cathode side and anode side every 24 hours. When the chloride ions concentration of the anode side solution was changed with time and the increase rate of chloride ions concentration became the same by measuring at least 5 times, it was judged to be steady state. Then, effective diffusion coefficient (D_e) of chloride ions of concrete (cm^2/yaer) was calculated as follows:

$$D_e = \frac{J_{cl} RTL}{|Z_{cl}| F C_{cl} (\Delta E - \Delta E_c)} \times 100 \tag{1}$$

where, J_{cl} : flux of chloride ion under the steady state (mol/(cm^2·year)), R: gas constant (8.31J/(mol·K)), T: absolute temperature (K), Z_{cl}: charge of chloride ion (= -1), F: Faraday constant (96500C/mol), C_{cl}: chloride ions concentration of cathode side (mol/l)、$\Delta E - \Delta E_c$: potential between the test specimen surfaces (V), L: specimen thickness (mm).

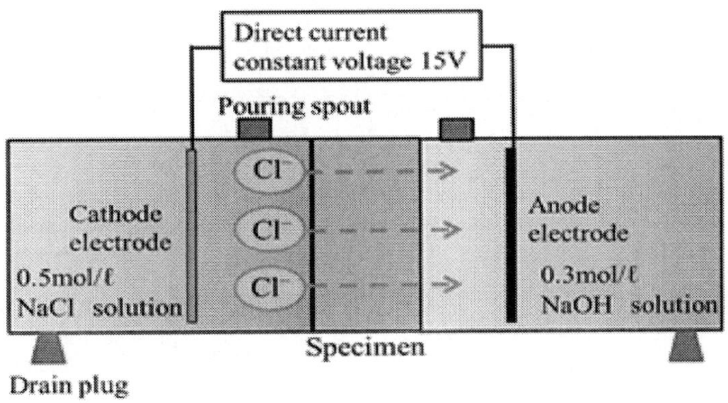

Figure 1. Apparatus for migration test.

The immersion test method calculates apparent diffusion coefficient of chloride ions under non-steady state by immersing the concrete in salt water for 90 days. Two prism specimens of 100×100×400mm are made for each mix proportion; one is cured in water and the other is cured in air 28 days. The surface concrete was left immersed and the others were sealed. Then, the surface was immersed into the salt water having 3% concentration. Subsequently, small specimens were cut 5mm to 25mm down from the surface, and chloride ions contents were measured. Apparent diffusion coefficient was calculated from the chloride ions content.

3 TEST RESULTS AND DISCUSSION

3.1 Effects of Type of Aggregate and Measuring Age on Electro-migration Characteristic

Figure 2 shows the effects between type of fine aggregates and curing condition on the effective diffusion coefficient. Based on the concrete used sandstone (CSS and CSG), the effective diffusion coefficient of the concrete remained virtually unchanged while the coarse aggregates was replaced with coarse aggregates from andesite or was under air curing. However, as 30% sandstone fine aggregates was replaced with manganese slag fine aggregates, the effective diffusion coefficient decreased by about 20%. On the other hand, when 30% sandstone fine aggregates was replaced with copper slag fine aggregate, it didn't show an influence on the effective diffusion coefficient. However, when aggregates from andesite was replaced with copper slag fine aggregates, the effective diffusion coefficient decreased around 16%.

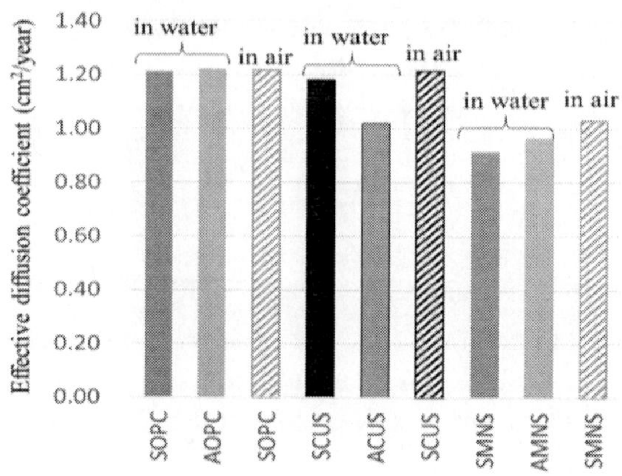

Figure 2. Effective diffusion coefficient of concrete using non-iron slag fine aggregate.

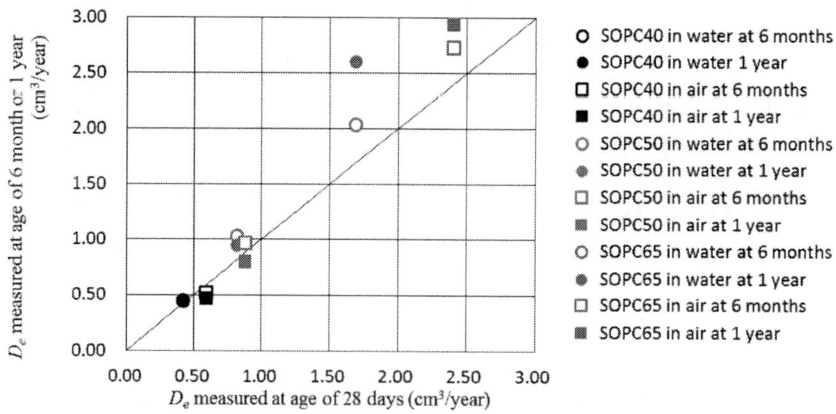

Figure 3. Effect of measured age on effective diffusion coefficient.

Figure 3 shows the result both age of 6 months and 1 year compared to the one of 28 days. For concrete with W/C40% and W/C50%, the effective diffusion coefficient of concrete after the lapse of 6 months or 1 year is almost the same as that of concrete measured at 28 days. However, for concrete with W/C65%, the effective diffusion coefficient of concrete cured in air become larger than that of concrete cued in water. Furthermore, the effective diffusion coefficient increases with the lapse of measuring gage.

3.2 Consideration from the Viewpoint of Pore Structure in Concrete

Figure 4 shows the relationship between pore volume measured at 28days and that measured at 1 year. The pore volume was measured by a mercury penetration method. Compared to pore volume measured by age of 28 days and 1 year specimens, the result showed no differences between the age of 28 and 1 year specimens of W/C 40%. Besides except for the result which was W/C50% under air curing, pore volume became larger after the elapse of 1 year for the age of concrete. Especially, it was noted that pore volume became larger with elapse of age of concrete in each curing for the specimens of W/C65%.

Figure 5 shows the relationship between the effective diffusion coefficient and pore volume. It was confirmed that large water-cement ratio results in large effective diffusion coefficient.

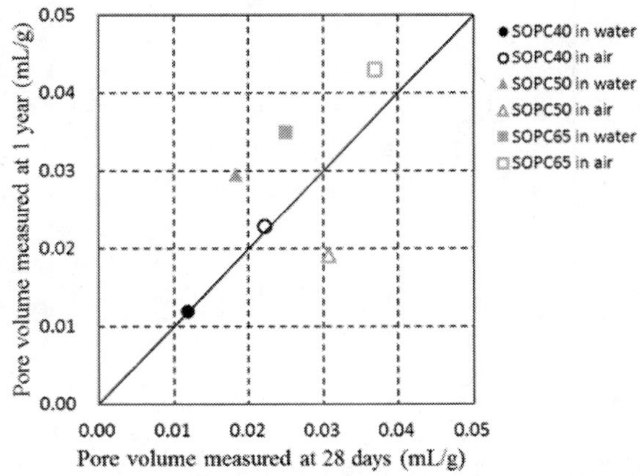

Figure 4. Relationship between pore volume measured at 28days and that measured at 1 year.

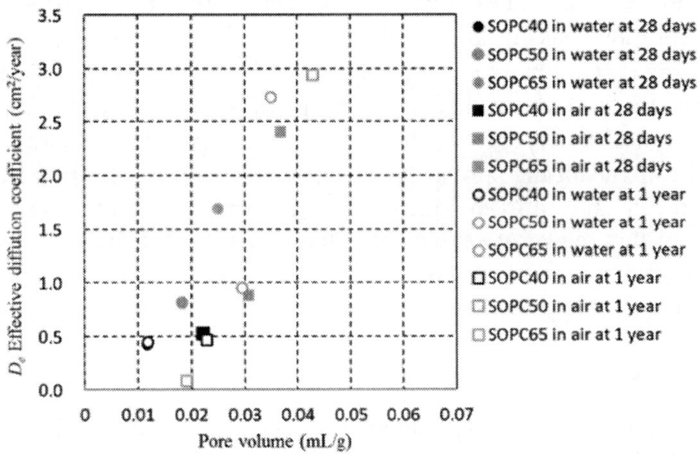

Figure 5. Relationship between pore volume and effective diffusion coefficient.

3.3 Relation between Apparent Diffusion Coefficient and Effective Diffusion Coefficient

Figure 6 shows the apparent diffusion coefficient by the immersion method and the effective diffusion coefficient of age of 28 days. As it is commonly known, apparent diffusion coefficient of concrete became small under the water-curing than air curing. Moreover focused on the differences from fine aggregates, in spite of the same water-cement ratio, apparent diffusion coefficient became smaller about 67.3% and 65.1% respectively by replacing 30% manganese slag fine aggregate with copper slag fine aggregates in water curing. Relatively, there was good correlation between the apparent diffusion coefficient and the effective diffusion coefficient including the result from the concrete using non-ferrous slag fine aggregates.

Previous study (JSCE 2013) has been reported that the apparent diffusion coefficient is calculated from the effective diffusion coefficient obtained from electro-migration test as follows:

$$D_a = k_1 \cdot k_2 D_e \tag{2}$$

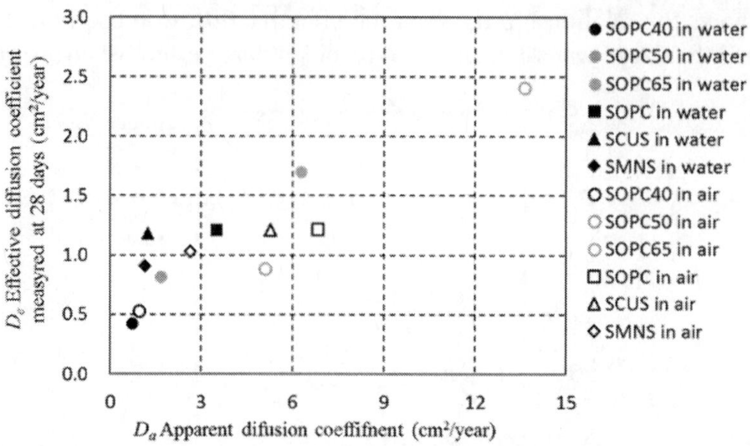

Figure 6. Relationship between apparent diffusion coefficient and effective diffusion coefficient.

Then, k_1: function with the balance of chloride ions concentration on the surface of the concrete side, cathode side and solution side respectively. k_2: coefficient with solidification phenomenon of chloride ions in the cement hydrates. The coefficient $k_1 \cdot k_2$ of concrete under water curing became larger when the water-cement ratio became larger. Besides coefficient $k_1 \cdot k_2$ became larger when it was under air curing except for 40%water-cement ratio. It is noted that coefficient k_1 theoretically coincides with the inverse of porosity of concrete. However, it came out the opposite of the result the pore volume shown in Figure 5. It is noted that coefficient k_2 is determined based on the ratio of fixed chloride ions concentration to all chloride ions concentration, from this reason it seemed that the concrete W/C50% and 65% under air curing has small amount of chloride ions in the cement hydrates. On the other hand, coefficient $k_1 \cdot k_2$ used non-ferrous slag fine aggregates was smaller than that of the natural aggregates concrete.

4 CONCLUSIONS

The conclusions derived from this experimental program are summarized as follows:

1. It was confirmed that using non-ferrous slag fine aggregates contributed to the improved resistance of the movement by the electro-migration of chloride ions.
2. For concrete with W/C40%, the diffusion coefficient of chloride ions remained virtually unchanged while the initial curing was changed from water curing to air curing, moreover even if the test launched 1 year later it was almost unchanged.
3. For concrete with W/C50% and 65%, the diffusion coefficient of chloride ions became larger when the initial curing was changed from water curing to air curing.
4. The effects of initial curing condition and age of concrete on the diffusion coefficient of chloride ions of concrete without ferrous slag fine aggregates can be explained from the viewpoint of changes of pore structure in concrete.

References

Japan Society of Civil Engineers (JSCE), *STANDARD SPECIFICATINS FOR CONCRETE STRUCTURES – 2013 Test Methods and Specifications*, Maruzen Publishing Co. Ltd, Tokyo, 2013.

Integrated Solutions for Infrastructure Development
Edited by Lau, H. H., Tang, F. E., Ng, C. K., and Singh, A.
Copyright © 2016 ISEC Press

OPTIMIZING THE MIX DESIGN OF CLAY BASED GEOPOLYMER CONCRETE

MUHAMMAD MUKHLESUR RAHMAN[1], DAVID W. LAW[2] and INDUBUSHAN PATNAIKUNI[2]

[1] *School of Engineering, RMIT University,Melbourne,Australia and Ministry of Road Transport and Bridges, Bangladesh*

[2] *School of Engineering, RMIT University, Melbourne, Australia*

Although considerable research has been undertaken on alkali-activated binders for geopolymer concrete production, few of these studies have focused on clay based geopolymer concrete. This paper reports the results of an experimental investigation to optimize the mix design of geopolymer concrete prepared solely from clay material. Four Activator Modulus (AM); 1.0, 1.25, 1.5 and 1.75 for each of two Sodium Oxide (Na_2O) dosages of 10% and 15% were selected for this investigation. A total of eight mix ratios comprising 72 specimens were tested. A comparison of physical behavior of clay based geopolymer mortar to that of fly ash based geopolymer mortar has been drawn throughout this paper. A Na_2O dosage of 10% provided consistent strength for all AM tested, however the specimens prepared with Na_2O dosage of 15% showed a reduction with time for all but the AM 1.0 specimens. Compressive strength of specimens prepared with AM of 1 for both of the Na_2O dosage of 10% and 15% exhibited the superiority over other AM. The investigation showed potential for clay material to be an alternative to ordinary Portland cement (OPC).

Key words: Geopolymer, Concrete, Clay, Compressive strength.

1 INTRODUCTION

Geopolymer concrete, an environmental friendly material, has come in to use as a substitute material for OPC concrete in the construction industry in recent times. The production of geopolymer not only limits the emissions of contaminants but also consumes less energy and thus plays a significant role in reducing the global warming problem (Heath *et al.* 2014, Zhang 2013). Different types of raw materials such as fly ash, kaolinite, palm oil fuel ash, rice husk ash, shale, natural zeolite, types of sludge, lignite bottom ash, clay and mud are being used for the production of geopolymer. While discussing the potential of geopolymeric materials Heath *et al.* (2014) pointed out that the use of geopolymer concrete offers a significant advantage in managing the problems associated with aluminosilicate by-products from industrial processes, as well as in reducing the environmental impact arising from the use of OPC concrete. Recent studies reveal that geopolymer concrete made from different industrial waste material display comparable or even better mechanical and durability properties than those associated with OPC concrete (Bernal *et al.* 2012, Ferone *et al.* 2015, Zhang *et al.* 2015).

Research conducted to date indicates that fly ash is the main source of binder material for the production of geopolymer concrete, although other source materials are being investigated. A concern is that fly ash production will decrease in the future as power generation will switch from coal to other renewable forms of electricity production. Clay, which is a naturally occurring material, and abundant throughout the world, could be a suitable alternate source material to fulfil this requirement. The potential for global warming could be reduced up to 40% if clay based geopolymer is used as an alternative material of OPC (Heath *et al.* 2014). Perna *et al.* (2014) have also identified the necessity for alternate substitute source material for the production of geopolymer concrete. At present researchers are studying a range of alternate raw materials for geopolymerization such as natural zeolites (Villa *et al.* 2010), calcinated paper sludge (Santa *et al.* 2013), rice husk ash (He *et al.* 2013), lignite bottom ash (Sathonsaowaphak *et al.* 2009). Recently, a new category of geopolymer i.e., blended geopolymer which is derived from the mixture of two or more industrial by-products has drawn attention from a number of researchers (Bhutta *et al.* 2014, He *et al.* 2013, Xu *et al.* 2014). However, research on the utilization of clay as an alternate binder material for geopolymerization system is still in its infancy.

Research carried out in developing clay based geopolymer indicates that this new type of clay based binder possess significant potential to be an alternative material to OPC. This paper reports the results of an experimental investigation to optimize the mix design of geopolymer concrete prepared solely from clay material instead of fly ash. Four Activator Modulus (AM), namely 1, 1.25, 1.5 and 1.75, for each of two Na_2O dosages of 10% and 15% were selected for this investigation. A total of eight mix ratios comprising 72 specimens were tested. AM and Na_2O Dosage is defined in Eq. (1) and Eq. (2) as follows:

$$\text{AM} = (\text{SiO}_2 \text{ in Alkaline Activator Solution}) / (\text{Na}_2\text{O in Alkaline Activator Solution}) \quad (1)$$

$$\text{Na}_2\text{O Dosage (\%)} = (\text{Na}_2\text{O in Alkaline Activator Solution}) / (\text{Mass of Binder}) \quad (2)$$

2 EXPERIMENTAL PROCEDURE

2.1 Material Used

In this study, the clay material used as a binder material was supplied by The Selkirk Company from the Enfield area about 30 km from Ballarat, Victoria, Australia. The chemical composition along with phase composition and particle size distribution supplied by Selkirk Company are presented in Tables 1 and 2. A mixture of 15 Molar NaOH solution and D-grade liquid Na_2SiO_3 was used as activator solution. River sand of specific gravity of 2.5 and fineness modulus of 3.0 was used as fine aggregate.

Table 1. Chemical composition of clay material.

Oxide	%
Al_2O_3	14.83
SiO_2	69.88
CaO	0.24
Fe_2O_3	3.88
K_2O	2.75
MgO	1.09
MnO	0.015
Na_2O	0.23
TiO_2	0.82
Cl	475 ppm

Table 2. Phase composition of clay material.

PHASE	% Weight
Illite	24.98
Kaolinite	27.17
Quartz	43.05
Rutile	1.67
Chlorite	1.42
Montmorillite	0.17
Albite	0.87
Goethite	0.6

2.2 Proportioning of the Materials

The mass proportioning of ingredients of different AM for both Na_2O dosages of 10% and 15% has been presented in the Tables 3 and 4. The proportioning of ingredients was calculated following an established mix design of Adam (2009). The mass ratio of sand to binder was fixed at 2.75 (ASTM C109/ C109M-07). The w/s (water/solid) ratio of 0.37 was used to ensure consistent workability of the geopolymer mortars. The quantity of water in the mix is the sum of the water contained in the sodium silicate, sodium hydroxide and the added water, while the quantity of solid is the sum of the mass of clay, and the solid contained in the alkaline activator solution.

Table 3. Mix design details based on AM for Na_2O dosage = 15%.

AM	Clay (kg)	Sand (kg)	Na_2SiO_3 (kg)	NaOH (kg)	Added Water (kg)	Na_2SiO_3/NaOH
0.5	1	2.75	0.25	0.36	0.123	0.69
0.75	1	2.75	0.38	0.30	0.096	1.25
1	1	2.75	0.51	0.24	0.07	2.12
1.25	1	2.75	0.64	0.17	0.044	3.61
1.5	1	2.75	0.76	0.11	0.018	6.65
1.75	1	2.75	0.89	0.06	0	14.24

Table 4. Mix design details based on AM for Na2O dosage = 10%.

AM	Clay (kg)	Sand (kg)	Na_2SiO_3 (kg)	NaOH (kg)	Added Water (kg)	Na_2SiO_3/NaOH
0.5	1	2.75	0.17	0.24	0.205	0.70
0.75	1	2.75	0.25	0.2	0.187	1.27
1	1	2.75	0.34	0.15	0.17	2.17
1.25	1	2.75	0.42	0.11	0.153	3.69
1.5	1	2.75	0.51	0.07	0.135	6.95
1.75	1	2.75	0.6	0.04	0.118	14.63

2.3 Specimen Preparation and Testing

The clay was dried in oven at 80^0C for 24 hours. After drying, the clay material was powdered using Ball Mill Grinder Machine. The mill comprised of a total of 15 balls and 5000 cycles were applied to make the clay powder. The powdered clay was sieved through a 106 μm size sieve. This sieved powder was mixed with sand for 4 minutes using 5-liter Hobert mixer. A 15 Molar

NaOH solution and D-grade liquid Na_2SiO_3 of 1.52 g/cc density with composition of 14.7% Na_2O, 29.4% SiO_2 and 55.90% water were premixed 15 minutes before the additional water was added to the activator solution. This activator solution was then added to the mixture of clay powder and sand. After manual mixing for 1 minute, the whole mix was blended by Hobert mixer machine with two rotating speeds: 150 and 300 rev/min for 4 and 2 minutes respectively. The mix was then placed in 50x50x50 mm Teflon moulds followed by 30 second vibration on vibrating table. After allowing 24 hours at room temperature, the moulds were kept in oven at 120^0C for another 24 hours. The moulds were then demoulded and cured at room temperature until testing. Compressive strength measurements of mortars were performed on a TCM Testing Machine in accordance with ASTM C109/C109M-13 with a loading rate of 0.34 $N/mm^2/S$. Three cubes were tested for each data point.

3 RESULTS AND DISCUSSION

The specimens were tested for compression at 7, 14, 28 and 40 days. As a significant fall in strength was observed between 14 and 28 days, the 40 days test was done to assess the rate of strength reduction. Testing at 3 days was not possible as the specimens had not achieved structural integrity. This indicates that clay based geopolymer mortar takes longer duration to harden than that of normal cement mortar or fly ash based geopolymer mortar. A normal Portland cement would be expected to set within 24 hours (Neville 2011), while research using elevated curing for geopolymers has shown that structural integrity is achieved following the elevated curing (Sagoe-Crentsill *et al.* 2010). He *et al.* (2013) also found that red mud (RM) and rice husk ash (RHA) based geopolymer paste took at least 35 days to achieve complete geopolymerization. They mentioned three possibilities: dominant crystalline solid phase acts as unreactive filler, larger particle size slows down the dissolution rate and the presence of impurities may have a detrimental effect on the rate of geopolymerization process. Hanjitsuwan *et al.* (2014) pointed out that at higher NaOH concentration (up to 18 M) the geopolymerization process occurs at a slower rate resulting in a longer setting time.

Table 5. Compressive strength of specimens for Na2O dosage = 15%.

AM	7 days strength (MPa)	14 days strength (MPa)	28 days strength (MPa)	40 days strength (MPa)
1	9.58	10.80	9.40	11.72
1.25	10	11.34	9.74	7.83
1.5	12.86	13.45	9.45	7.64
1.75	20.02	11.96	12.04	11.05

Table 6. Compressive strength of specimens for Na2O dosage = 10%

AM	7 days strength (MPa)	14 days strength (MPa)	28 days strength (MPa)	40 days strength (MPa)
1	9.68	14.5	10.42	13.88
1.25	10.87	14.21	10.06	13.25
1.5	11.36	10.97	9.75	12.5
1.75	14.92	10.36	9.96	9.69

The 7, 14, 28 and 40 days compressive strength of geopolymer mortars prepared with Na_2O dosages of 15% and 10% at different AM are presented in Tables 5 and 6. The results show that the strength generally increases from 7 to 14 days for both the 10% and 15% Na_2O dosages. The only exception being the 15%, 1.75 AM which achieves the maximum strength (20.02 MPa) at 7 days which reduces significantly to 14 days (11.96 MPa). The 15%, AM 1.25 and 1.5 both reduced in strength from 14 to 40 days while the AM 1.0 gained approximately a further 1 MPa from 14 to 40 days. Similar behaviour has been observed for FA-rice husk bark ash (RHBA) geopolymer where it was found that the compressive strength of specimens cured at higher temperature(i.e., > 90^0C) started to decrease after certain period of time (Nazari *et al.* 2011). They reported that higher temperature destroyed the granular structure of geopolymer which caused dehydration of geopolymer matrix and excessive shrinkage due to contraction of geopolymeric gel. Sukmak *et al.* (2013) observed the existance of micro-cracks at a higher curing temperature (85^0C) with a short curing duration (24 h) and even at a lower curing temperature (75^0C) with a longer curing duration (75 h) for clay-FA geopolymer bricks. Microcracking has also been observed in alkali activated slag geopolymers following elevated curing, which can negatively impact compressive strength (Collins and Sanjayan 2001).

For Na_2O dosage 15%, the 7 days compressive strength was highest (20.02 MPa) for AM of 1.75 whereas AM of 1.5 provided the highest strength (13.45MPa) at 14 days. The compressive strength at 28 days was similar for all the AMs except for AM of 1.75. The 15% Na_2O dosage AM 1.0 provided the highest strength (11.72 MPa) at 40 days. For Na_2O dosage 10%, AM 1.0 showed the optimum strength for 14, 28 and 40 days whereas AM of 1.75 showed the optimum strength at 7 days. A Na_2O dosage of 10% provided more consistent strength than specimens prepared from a Na_2O dosage of 15%. Compressive strength of specimens prepared with AM of 1.0, for both of the Na_2O dosage of 10% and 15%, exhibited superior strength to other specimens. Tables 5 and 6 also show that the compressive strengths at different AM of Na_2O dosage of 10% are greater than corresponding strengths at different AM of Na_2O dosage of 15% except for the AM of 1.75. The results indicate that Na_2O content is having a detrimental effect i.e., the more is the Na_2O content the lower the strength with time. A possible reason may be the formation of Na_2O-CaO-SiO_2-H_2O, as the solubility of Na+ ion is very low in alkali activated binder material as reported by Malolepszy (1993).

4 CONCLUSIONS

The following conclusions were drawn from the above discussion:

- The setting time of clay based geopolymer mortar is longer than that of FA based geopolymer mortars or normal OPC mortars.
- The clay based geopolymers generally showed an increase in strength up to 14 days. After this a slight increase in strength was observed for the specimens with a 10% Na2O dosage, while a decrease was observed for those with a 15% Na2O dosage, other than with an AM of 1.0.
- The results indicate that excess Na2O content has a detrimental effect on the strength gain of clay based geopolymer mortar i.e. the higher the Na2O content, the lower will be the strength with time.
- Na2O dosage of 10% provided consistent strength compared to the specimens prepared from Na2O dosage of 15%.

- Compressive strength of specimens prepared with AM 1.0 for both of the Na2O dosage of 10% and 15% exhibited superiority over other AM.

References

Adam, A. A., Strength and durability properties of alkali activated slag and fly ash-based geopolymer concrete, PhD Thesis, RMIT University, Melbourne, Australia, 2009.

ASTM C109/ C109M-(07-13): Standard Test Method for Compressive Strength of Hydraulic Cement Mortars [Using 50-mm] Cube Specimens.

Bernal, SA., de Gutiérrez, RM., and Provis, JL., Engineering and durability properties of concretes based on alkali-activated granulated blast furnace slag/metakaolin blends, *Construction and Building Materials*, vol. 33, pp. 99-108, 2012.

Bhutta, MAR., Hussin, WM., Azreen, M., and Tahir, MM., Sulphate resistance of geopolymer concrete prepared from blended waste fuel ash, *Journal of Materials in Civil Engineering*, vol. 26, no. 11, p. 04014080, 2014.

Collins, F., and Sanjayan G. J., Microcracking and strength development of alkali activated slag concrete, *Cement and Concrete Composites,* 23(4-5), 345-352, 2001.

Ferone, C., Liguori, B., Capasso, I., Colangelo, F., Cioffi, R., Cappelletto, E.,and Di Maggio, R., Thermally treated clay sediments as geopolymer source material, *Applied Clay Science*, vol. 107, pp. 195-204, 2015.

Hanjitsuwan, S., Hunpratub, S., Thongbai, P., Maensiri, S., Sata, V., and Chindaprasirt, P., Effects of NaOH concentrations on physical and electrical properties of high calcium fly ash geopolymer paste, *Cem Concr Compos,* 45:9–14, 2014.

He, J., Jie, Y., Zhang, J., Yu, Y., and Zhang, G., Synthesis and characterization of red mud and rice husk ash-based geopolymer composites', *Cement and Concrete Composites*, vol. 37, pp. 108-18, 2013.

Heath, A., Paine, K.., and McManus, M., Minimising the global warming potential of clay based geopolymers, *Journal of Cleaner Production*, vol. 78, pp. 75-83, 2014.

Malolepszy J., Some aspects of alkali activated cementitious materials setting and hardening. In *3rd Beijing international symposium on cement and concrete*, vol. 2, 1043–6, 1993.

Nazari, A., Bagheri, A., and Riahi S., Properties of geopolymer with seeded fly ash and rice husk bark ash. *Mater Sci Eng A*, 528(24),7395–401, 2011.

Neville, A. M., *Properties of concrete.* England, Longman Scientific & Technical, 2011.

Perná, I., Hanzlíček, T., and Šupová, M., The identification of geopolymer affinity in specific cases of clay materials, *Applied Clay Science*, vol. 102, pp. 213-9, 2014.

Sagoe-Crentsil, K., Brown T., and Yan S., Medium to Long Term Engineering Properties and Performance of High Strength Geopolymer for Structural Applications, *Advances in Science and Technology*, 9, 135-142, 2010.

Santa, RAAB., Bernardin, AM., Riella, HG., and Kuhnen, NC., Geopolymer synthetized from bottom coal ash and calcined paper sludge, *Journal of Cleaner Production*, vol. 57, pp. 302-7, 2013.

Sathonsaowaphak, A., Chindaprasirt, P., and Pimraksa, K., Workability and strength of lignite bottom ash geopolymer mortar, *Journal of Hazardous Materials*, vol. 168, no. 1, pp. 44-50, 2009.

Sukmak, P., Horpibulsuk, S.,and Shen S-L., Strength development in clay-fly ash geopolymer. *Construction and Building Materials*, 40, 566–74, 2013.

Villa, C., Pecina, E., Torres, R., and Gómez, L., Geopolymer synthesis using alkaline activation of natural zeolite, *Construction and Building Materials*, vol. 24, no. 11, pp. 2084-90, 2010.

Xu, H., Gong, W., Syltebo, L., Izzo, K., Lutze, W., and Pegg, IL., Effect of blast furnace slag grades on fly ash based geopolymer waste forms', *Fuel*, vol. 133, pp. 332-40, 2014.

Zhang, L., Production of bricks from waste materials–a review, *Construction and Building Materials*, vol. 47, pp. 643-55, 2013.

Zhang, Z., Zhu, H., Zhou, C., and Wang, H., Geopolymer from kaolin in China: An overview, *Applied Clay Science,* 2015.

Integrated Solutions for Infrastructure Development
Edited by Lau, H. H., Tang, F. E., Ng, C. K., and Singh, A.
Copyright © 2016 ISEC Press

INVESTIGATION OF THE COMPRESSIVE STRENGTH OF PORTLAND CEMENT MIXTURES BY APPLICATION OF BACTERIA – '*SPOROSARCINA PASTEURII*'

SAMSON AREGAWI, ABIY GOSHU, BISRAT ALEMU, DAGMAY WOLDAREGAY, NATHNAEL ABDULKADIR, and YONATAN MESFIN

School of Civil and Environmental Engineering, Addis Ababa Institute of Technology (AAiT), Addis Ababa University (AAU), Addis Ababa, Ethiopia

The concept of microbial concrete is one of the recent advances in concrete technology. In the past two decades, concrete technologies are working towards developing high performance concrete. Researches over the globe are being carried out in the wake of promising results found on the improvement of cementitious mix performances due to the application of live microorganisms. In this research live microorganism named *Sporosarcina pasteurii*, soil bacterium, has been used. Different set of experiments were carried out to investigate the effect of the bacterial medium, bacterial nutrient and bacterial concentration. From the test results it was found out that the bacterial medium had little effect, while the bacterial nutrient, whose main constituent is yeast extract, significantly reduced the compressive strength and increased the flow table as well as the slump in both mortar and concrete. Those samples with aforementioned bacteria together with the bacterial nutrient showed an improved compressive strength. The micro behaviors observed in terms of compressive strength indicate that this gain of strength was due to the calcite precipitation induced by the bacteria. The paper concludes by stating the need of further investigation, especially with regards to finding a better substitute for yeast extract of the bacterial nutrient.

Keywords: Sporosarcina Pasteurii, Bacterial-concrete, Microbiologically induced calcite precipitation, Portland cement mixes, Compressive strength.

1 INTRODUCTION

In the past two decades researchers have been working towards improving the performance behavior of concrete. And recently a novel technique is adopted in re-mediating cracks and fissures in concrete by utilizing microbiologically induced calcite or carbonate (CaCO3) precipitation (MICP) (Kumar *et al.* 2013a). Microbiologically induced calcite precipitation (MICP) is a technique that comes under a broader category of science called *bio-mineralization*. It is a process by which living organisms form inorganic solids. Bio-mineralization is a promising technology for producing crack resistance/highly self-healing concrete (Kumar *et al.* 2013a, Verma *et al.* 2015).

Microbiologically induced calcite precipitation is one of the more promising techniques which have been used in geotechnical engineering for soil and sand cementation (Martinez *et al.* 2013). The idea of microbiologically induced calcite precipitation has also been investigated for its application in Portland cement mixtures. Considerable researches have been conducted on different bacteria, which are considered to have a potential for calcite precipitation (Kumar *et al.* 2013a, Verma *et al.* 2015, Martinez *et al.* 2013, Chahal *et al.* 2011, Gupta *et al.* 2013).

The main objective of this research is to identify and highlight the possibility of microbiologically induced calcite precipitation by indirectly observing its effect on compressive strength of Portland cement mixes.

Sporosarcina pasteurii (formerly known as *Bacillus pasteurii)* is a bacterial species that is well known for producing urease and hydrolyzing urea to form ammonia and bicarbonate ions, and was thus used in this study (Martinez *et al.* 2013, ATCC 2013). As a microbial sealant, CaCO3 exhibited its positive potential in selectively consolidating simulated fractures and surface fissures in granites and in the consolidation of sand.

1 EXPERIMENTAL PROGRAM

1.1 General

Bacteria species called *Sporosarcina pasteurii* (previously known as *Bacillus pasteurii*), obtained from Ethiopian Institute of Biodiversity – Microbiology Department, was used in this study. One potential reason for selecting *S. Pasteurii* for its application in concrete for calcite precipitation is that *S. pasteurii* is a common bacterium naturally occurring in the subsurface soil containing CaCO3. Finding these bacteria in CaCO3 reach soil leads us to the assumption that presence of CaCO3 in the soil might be due to microbiologically induced calcite precipitation by the aforementioned bacteria. *Sporosarcina pasteurii* is an aerobic microorganism that uses oxygen for propagation. 13 g/l nutrient broth, 5 g/l peptone, 8.5 g/l yeast extract, 4.5 g/l beef extract, 25 mM $CaCl_2$ and 20 g/l Urea solution medium were used to prepare the media mixture needed to cultivate *S. Pasteurii.* The pH of the medium was then adjusted before sterilization to 6.5 by 1 N HCl. Growth was measured using Optical Density (OD) measurements which were carried out at different intervals using a Spectrophotometer Equipment.

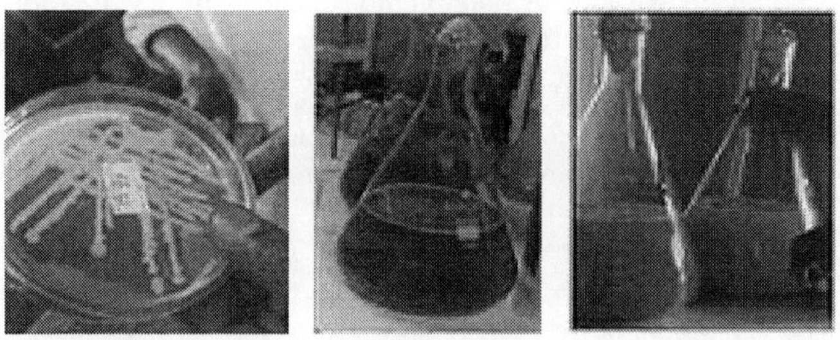

(a) (b) (c)

Figure 1. *Sporosarcina pasteurii* on (a) nutrient, (b) Clear vs. (c) turbid growth media.

Due to the fact that *Sporosarcina pasteurii* is an aerobic bacterium, it was hypothesized that the effectiveness of calcite precipitation can be affected by the available voids in the specimen. As a result, concrete mixture with low and high water cement ratio was prepared during the experimental investigation.

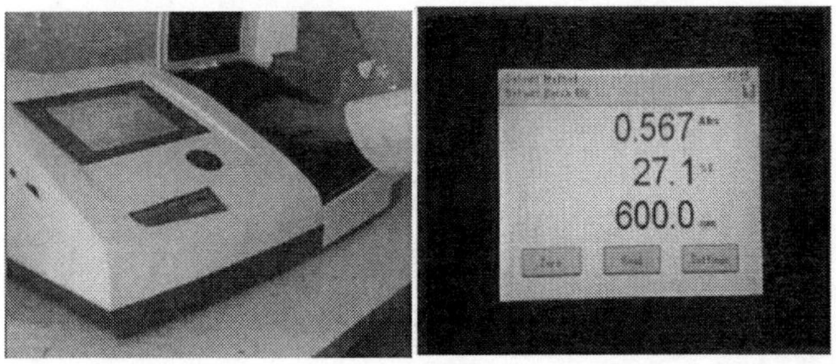

Figure 2. Optimum Density (Bacteria Concentration) measurement using Spectrophotometer.

1.2 Materials

While conducting this research, Ordinary Portland Cement of grade 42.5, which was available in the market, crushed basaltic coarse aggregate, Natural River sand and commercially available high range water reducer was used. Table 1 summarize physical properties of the fine and coarse aggregate.

Table 1. Physical properties of constituent materials.

Properties	Fine aggregate	Coarse aggregate
Specific Gravity (SSD)	2.5	2.76
Absorption	4.8%	0.82%
Unit Weight	1534	1553
Silt Content	3.45%	

1.3 Mix Proportion

The ACI mix design manual, ACI 211.1-91 for concrete with w/c of 0.575 and ACI 211.4R-08for w/c of 0.35 has been used. Table 2 and 3, summarize the quantity of each constituent material required for the mortar and concrete respectively. The description for the mix codes is as shown in result and discussion part.

Table 2. Mix Proportion for Mortar (kg/m^3).

Mix Code	Cement [kg]	W/C ratio	Fine aggregate [kg]	Bacterial Concentration	Other
MW	525	0.485	1505	0	None
MM	525	0.485	1505	0	**
MN	525	0.485	1505	0	***
MB5	525	0.485	1505	0.5 OD	None
MB15	525	0.485	1505	1.5 OD	None

Mixing and casting was done in accordance with ASTM C305-99 for the mortar and ASTM C 192-99 for the concrete. The control specimens were cured in tap water while the rest of the mixes were cured in a solution containing Urea (20gm/l) and $CaCl_2$ (2.75gm/l).

Finally, Compressive strength test on both mortar and concrete cubes was conducted using load - gauged type test machine with a capacity of 300kN and 3000kN respectively. The testing was in accordance with ASTM C39-99 for concrete and ASTM C109-99 for mortar cubes.

Table 3. Mix Proportion for Concrete [kg/m^3].

Mix Designation	Cement [kg]	W/C ratio	Coarse aggregate	Fine aggregate	Bacteria Concentration	Other	Super plasticizer
YW	330	0.575	938	894	0	None	0
YB0	330	0.575	938	894	0	**	0
YB5	330	0.575	938	894	0.5 OD	None	0
XW	545	0.35	938	720	0	None	1lit/100kg of cement
XB0	545	0.35	938	720	0	**	
XB10	545	0.35	938	720	1.0 OD	None	

**20gm/l Urea and 2.75gm/l $CaCl_2$; ***13 g/l nutrient broth, 5 g/l peptone, 8.5 g/l yeast extract, 4.5 g/l beef extract, 25 mM $CaCl_2$ and 20 g/l Urea

2 RESULTS AND DISCUSSIONS

This research commenced with identifying and highlighting the potential of bio-mineralizing bacteria for calcite precipitation by indirectly testing the compressive strength of concrete and mortar. The bacteria, *Sporosarcina pasteurii*, were preserved until all other mixing conditions were fulfilled. Once everything was ready, the bacteria were checked for contamination before inoculating in their growth medium. The fresh and ready bacteria underwent turbidity measurements using a Spectrophotometer before their application into the mix. Figure 3-7summarize results of both fresh and hardened properties of mortar and concrete.

The different mortar and concrete mixes are described as follows: (M: Mortar, Y & X for Concrete)

- MW, YW, XW – Control sample group for mortar, concrete with w/c of 0.575 and concrete with w/c of 0.35 respectively,
- MM, YB0, XB0 – The medium mix group: the water used had 20gm/l Urea and 2.75gm/l $CaCl_2$,
- MN – The nutrient mix group: the water containing all the nutrients; yeast, beef extract, nutrient bronze, peptone, urea, $CaCl_2$
- MB5, YB5 – The bacterial mix group: the mix proportion with bacteria of 0.5 OD for mortar and concrete respectively
- MB15 – The bacterial mix group: the mix proportion with bacteria of 1.5 OD
- XB10 – The bacterial mix group: the water with the solution having OD 1.0

As it can be seen from figure 3(a), the highest slump was recorded on mortar containing all nutrients; yeast is the main constituent of the nutrients. As the bacterial concentration increases, reduction in slump is also observed; the reason can be due the fact that the bacteria consumed the yeast and lessening its effect in creating voids. The other slump test conducted was on the concrete mixture. From figure 3(b), the first three charts, the Y series, are of concrete without admixture and the rest three, which is the X-series, are concrete mixtures with super plasticizer. The mix, YB5, which consists of yeast, beef extracts and bacteria, showed the highest slump from the Y-series mixes. During the slump test it is also observed that the mix is like foamed concrete (figure 4(a) &4(b), side and front view respectively). This might be the effect of the yeast.

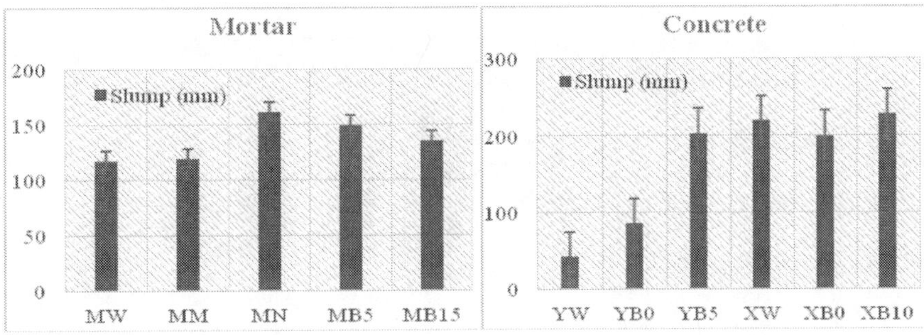

(a) (b)

Figure 3. Summary of Slump Results.

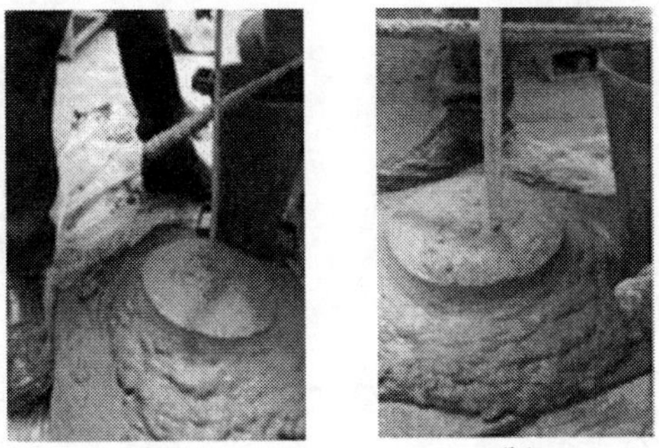

(a) (b)

Figure 4. "Foam-Like" Slump (Bacterial Concrete).

The other result observed, in addition to the slump, was compressive strength in both mortar and concrete. Figure 5 and 6, summarize these results.

From the test results, it was observed that cement-mortar cubes containing all the nutrients (MN mix group) showed a significant reduction in compressive strength up to 70% at 28th day, compared with the control group. On the other hand, mortar cubes with only medium showed a significant increase in compressive strength up to 20%. However, with the inclusion of bacteria (*Sporosarcina pasteurii*), the reduced compressive strength due to the nutrient effect, which was 70% reduction, has improved. The reduction in strength is only 18% for bacterial concentration of 0.5 OD and 11% for 1.5 OD when it is compared with the control group. This improved compressive strength can be due to microbiologically induced calcite precipitation. And when we see results of compressive strength test of concrete cubes (figure 6), the effect of medium is not significant as it was observed in mortar cubes. There is a slight increase in compressive strength up to 2% which was 20% in the case of mortar. However, a significant reduction in compressive strength was observed on concrete cubes with w/c of 0.575 and bacterial concentration of 0.5 OD. As it was observed during the mixing process and slump test result, the yeast extract created numerous void spaces and increases the volume during mixing.

However, as the bacterial concentration increased (1.0 OD), in the case of water-cement ratio of 0.35, the compressive strength was improved compared with the control group. The higher bacterial concentrations resulted in increased compressive strength compared with control specimen.

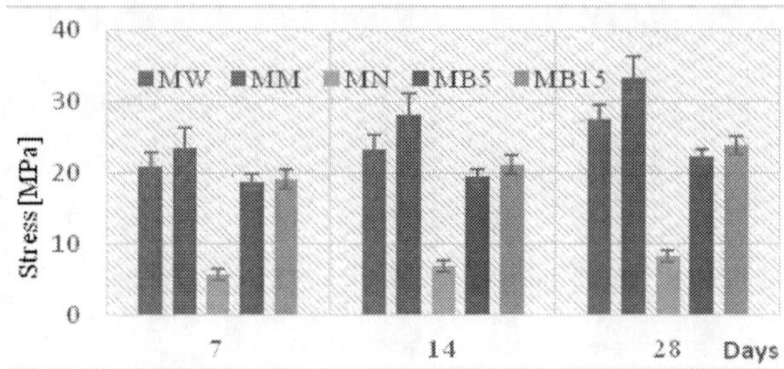

Figure 5. Summary of Compressive Strength of mortar cubes.

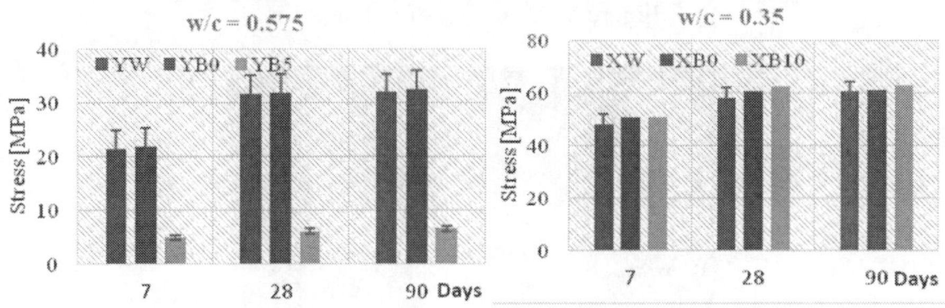

Figure 6. Summary of Compressive Strength of Concrete.

3 CONCLUSION

From the test results, it can be concluded that, bio-mineralization has potentially occurred by *Sporosarcina pasteurii*; but this statement is only backed by results obtained from compressive strength tests. This can be strongly proven with the employment of direct tests on the microstructure of crushed samples. For the bio- mineralization to be efficient the yeast in the nutrient should be optimized or an alternative nutrient has to be used.

References

American Type Culture Collection, ATCC, 2013. "*Sporosarcina pasteurii*" (ATCC® 11859™), Retrieved November 2014, from www.atcc.org

B. C. Martinez, J. T. DeJong A, T. R. Ginn, B. M. Montoya, T. H. Barkouki, C. Hunt, and B. Tanyu, D. Major. 2013. "Experimental Optimization of Microbial-Induced Carbonate Precipitation for Soil Improvement." ASCE Journal of Geotechnical and Geo-Environmental Engineering, 139(4): 587-598

Jagadesha Kumar B G, R. Prabhakara, and Pushpa H.2013a. "Effect of Bacterial Calcite Precipitation on Compressive Strength of Mortar Cubes."International Journal of Advances in Engineering and Technology, 2(3): 486-491

Navneet Chahal, Anita Rajor, and Rafat Siddique. 2011. "Calcium Carbonate Precipitation by Different Bacterial Strains." African Journal of Biotechnology, 10(42): 8359-8372. Doi: 10.5897/AJB11.345

Rajesh K. Verma, Leena Chaurasia, Vishkha Bisht, and Manisha Thakur.2015. "Bio-mineralization and Bacterial Carbonate Precipitation in Mortar and Concrete" American Institute of Science, Bioscience and Bioengineering, 1(1): 5-11

Shivani Gupta, Chhavi Rathi, and Susman Kapur. 2013, "Biologically Induced Self-Healing Concrete: A Futuristic Solution for Crack Repair" International Journal of Applied Sciences and Biotechnology, 1(3): 85-89. Doi: 10.3126/ijasbt.v1i3.8582

Integrated Solutions for Infrastructure Development
Edited by Lau, H. H., Tang, F. E., Ng, C. K., and Singh, A.
Copyright © 2016 ISEC Press

SELF-COMPACTING CONCRETE USING AGRO-INDUSTRIAL WASTES; RELATIVE HUMIDITY EFFECT AND OPTIMUM MIXING METHOD

BRABHA NAGARATNAM[1], AHMED FAHEEM[2], MUHAMMAD EKHLASUR RAHMAN[1], ABDUL KARIM MIRASA[3], MOHAMMAD ABDUL MANNAN[4], and MD SAIFUL ISLAM[5]

[1]*School of Engineering and Science, Curtin University Sarawak, Sarawak, Malaysia*
[2]*School of Civil Engineering, The University of Sydney, NSW, Australia*
[3]*School of Engineering & Information Technology, Universiti Malaysia Sabah, Kota Kinabalu Sabah Malaysia*
[4]*Dept of Civil Engineering, Universiti Malaysia Sarawak, Sarawak, Malaysia*
[5]*Faculty of Science, University Putra Malaysia, Selangor, Malaysia*

This research identifies the relative humidity effect and the optimum mixing method for Self-Compacting Concrete (SCC) containing OPC mixed with agro-industrial wastes of palm oil fuel ash (POFA) and low calcium based fly ash (FA). The effects of temperature and relative humidity on the fresh state properties were assessed under three conditions; moderate humidity, H1 (RH of 70 to 80% with temperature of 32°C) and low humidity, H2 (RH of less than 70% with temperature of 36°C) and extremely high humidity, monsoon season, H3 (RH of more than 80% with ambient temperature around 23 ±2°C). Six types of mixing method were evaluated for the ternary blended mix (TNY40) with two types of super-plasticiser (SP) additions: simultaneous and delayed, in order to discover the optimum mixing time and loading sequence. Furthermore, the slump flow retention was recorded every 15 minutes after batching. In conclusion, it was found that the increased temperature with lower relative humidity accelerates the reduction of workability (up to 15%) especially in POFA SCC as compared to FA and TNY (8%) samples. From the evaluation of various mixing method, the optimum mixing method observed was using the delayed addition of SP with a total mixing time of 12 minutes. This method resulted in excellent workability properties and with mixes retaining its slump within 15 minutes of batching.

Keywords: Palm oil fuel ash, Fly ash, Binary blend and Ternary blend.

1 INTRODUCTION

SCC is a complex mixture. The more complex the concrete, the more sensitive it is to the material variations and temperature fluctuations during production (Ghafoori and Diawara 2010). Temperature and humidity is an important parameter to be considered when developing SCC. Brameshuber *et al.* (2003) reported that a temperature variation of 1^0C can lead SCC to lose its' self-compacting ability. In a tropical climate like Malaysia the temperature fluctuations does not vary as much as other countries (four seasons). However, the humidity variance can be very high (up to 100%). With high humidity, there is more moisture in the air and it will be essential to understand if this has any effect on the workability of SCC.

The loading sequence and mixing duration is an essential step in order for SCC to achieve a uniform, cohesive and homogeneous mix without segregating or bleeding (Ferraris 2001, Rahman *et al.* 2011) . This mainly depends on the type of mixer, loading sequence, and the duration and power of mixing. Irrespective of the type of mixer used, a longer mixing time than for normal concrete is necessary to fully activate the SP and to reduce frictional forces of the SCC mixes (EGSCC 2005). The use of POFA as agricultural and FA as industrial waste is not common in SCC production technology locally as there is a lack of research that comprehensively presents the effects of using these materials. The purpose of this paper is to investigate the effect of mixing method and sequence as well as humidity effect in tropical climatic conditions on the fresh properties of SCCs containing these wastes.

2 DETAILS OF MATERIALS USED

2.1 Cement

Ordinary Portland Cement (OPC) grade 42.5 based on ASTM C150 was used in the concrete as cementitious material and Its chemical composition is shown in Table 1.

2.2 Fly Ash (FA)

Low calcium FA (FA), Class F as per ASTM C618, was obtained from Sejingkat Coal Fired Power Station in Sejingkat, Kuching. The chemical composition of the FA is given in Table 1 and the particle size distribution (Figure 2) demonstrates a D50 of 13μm.

2.3 Palm Oil Fuel Ash (POFA)

Palm oil fuel ash (POFA) was obtained from a palm oil mill at Lambir, Miri, Sarawak, Malaysia. Samples were grinded in a ball mill for 8 hours. The chemical composition of POFA is given in Table 1. In addition, Figure 2 shows a D50 of 10μm.

2.4 Coarse Aggregates

10mm nominal size coarse aggregates (crushed quartzite) was used here. It is composed of particles within the range of 5mm to 10mm. Sieve analysis of 2000 gram sample showed that the entire sample (100%) passed through 9.5mm sieve while only 5% passed the 4.75mm sieve.

2.5 Fine Aggregates

The fine aggregates consisted of locally available aggregates with a maximum size of 5 mm. Two categories of fine aggregate were chosen after various bulk density tests. Type 1 was crushed quartzite that had a size range of 600 μm to 5.0 mm as per Table 2. The other category (Type 2) had a nominal size of 600 μm (uncrushed river sand as per Table 3) and a small amount of micro-fines (Table 3).

2.6 Super-plasticiser (SP)

The SP used in this research was Glenium Ace 389, supplied by BASF (Malaysia) Sdn Bhd. This SP is categorized as type F in accordance to ASTM C494 and BS EN 934-2 European Standard.

Table 1. Chemical Composition of OPC, FA and POFA (%).

Compound	FA	POFA	OPC
SiO2	57.8	52.3	20.0
Al2O3	20.0	-	5.20
Fe2O3	11.7	6.87	3.30
SO3	0.08	1.78	2.40
MgO	1.95	5.43	0.80
CaO	3.28	10.8	63.2
K2O	3.88	11.4	-
TiO2	2.02	0.61	-
Na2O	0.30	-	-
LOI	0.32	-	2.50

Table 2. Crushed quartzite gradation.

Sieve Size	% Finer	AS 2758.1 Requirements
4.75 mm	99	90 -100
2.36 mm	50	60 -100
1.18 mm	20	30 -100
600 μm	2	15 - 80

Table 3. River sand gradation.

Sieve Size (μm)	% Finer	AS 2758.1 Requirements
600	100	15 - 100
300	58	5 - 50
150	10	0 - 20

3 MIX PROPORTIONS AND MIXING METHODS

The mix proportions of SCC are summarized in Table 4. In addition Table 5 outlines the various methods investigated. 6 batches of SCC with different mixing time and sequence were investigated in this research with SCC incorporating 40% ternary cementitious material. The mixing time was summarized as in Table 5. All the mixes were rested not less than 2.5 minutes before fresh state tests were carried out. In addition, the slump flow retention was measured every 15 minutes.

For the evaluation of relative humidity, three batches using Method 5 (M5) were evaluated as in Table 6 using a hygrometer. All mixing was done with a pan mixer in a laboratory and an average of three samples was conducted for each testing.

The three essential workability requirements for SCC are filling ability, passing ability and segregation resistance. This is measured as per specification provided by the European Guidelines for SCC (EGSCC 2005).

Table 4. Mix proportions.

	C	FA		POFA		Ternary (POFA & FA)	
Percentage Replacement (%)	0	30	40	30	40	30	40
W/P	0.38	0.38	0.38	0.44	0.44	0.38	0.38
SP Content (%)	0.86	0.66	0.66	1.2	1.4	1.1	1.1
FA(kg/m3)	0	162	216	0	0	81	108
POFA(kg/m3)	0	0	0	162	216	81	108
Cement(kg/m3)	540	378	324	378	324	378	324
0-600μm (Sand)	600	600	600	600	600	600	600
600μm-5mm (Quarry Dust)	550	550	550	550	550	550	550
Course Agg (5-10mm)	400	400	400	400	400	400	400

Table 5. Mixing methods.

Method	Procedure
M1	The aggregates were blended for 5 minutes followed by the addition of dry powders (cement, FA, and POFA) and run for another 5 minutes. 50% of water was added to about 50% of the SP in a bucket and then added into the mixer. Total running of 8 minutes. The remaining water and SP was added to the end and run for another 3 minutes before discharge.
M2	The aggregates were blended for 5 minutes followed by the addition of dry powders (cement, FA, and POFA) and run for another 5 minutes. All the required amount of water and SP was added directly into the mixer. The mixer is kept running for 8 minutes in total.
M3	All the required amount of water was added in the concrete mixer followed by addition of cement, FA and POFA. The mixer was kept running for 5 minutes after the addition of the powders. Pre-blend aggregates were added to the mixer and run for 3 minutes. SP was added at the end and run for another 2 minutes.
M4	The aggregates were blended for 5 minutes followed by the addition of dry powders and run for another 5 minutes. All the required amount of water was added to into the mixer. The mixer is kept running for 5 minutes in total. All the required amount of SP was added to the end and run for 3 minutes before discharge.
M5	The aggregates were blended for 5 minutes followed by the addition of dry powders and run for another 5 minutes. All the required amount of water was added to into the mixer. The mixer is kept running for 9 minutes in total. All the required amount of SP was added to the end for 3 minutes before discharge.
M6	The aggregates were blended for 5 minutes followed by the addition of dry powders (cement, FA, and POFA) and run for another 5 minutes. All the required amount of water was added. Total running of 15 minutes. All the required amount of SP was added to the end for another 3 minutes before discharge.

4 Results and Discussion

4.1 Mixing Method

Table 6 presents the results for mixing method. It is observed that M1 gave good workability properties, however the samples loses its' self-compacting ability within 15 minutes of batching. This was also the case for M2. With M3, when water was added to cementitious material in the beginning causes the samples to be too fluid and highly segregated. In M4, the time allocated for the addition of liquid and final mixing was not adequate which led to inefficient mix. Mixing the samples for 9 minutes in M5 gave the optimum mixing method, with highly workable sample and lower segregation ratio. In addition, Figure 5 shows that with M4 and M5, the mix could retain its slump flow within 15 minutes of mixing. However, longer mixing time after the introduction of water (M6) will result in 'balling effect' due to drying of concrete and eventually loses its workability.

4.2 Evaluation of Relative Humidity with Ambient Temperature

As seen in Table 7, increased ambient temperature, with low relative humidity, accelerated the reduction of the workability properties. A slump variance of up to 15% was seen in POFA SCC, whereas FA and TNY samples exhibited a slump variance of up to 8%, under varying temperature and relative humidity changes (Figure 1). When exposed to lower humidity in tropical weather, SCC incorporating POFA and FA, demonstrated an increase in T500 and V-funnel flow time, while the segregation ratios remained stable.

Table 6. Results for mixing methods.

Fresh State Properties			M1	M2	M3	M4	M5	M6
Filling Ability	Slump flow	T500 (s)	2.63	4.58	0.54	1.39	2.09	2.47
		Max. spread (mm)	790	800	850	820	820	820
	V-Funnel	T10 (s)	5.63	13.58	2.84	9.38	5.48	6.48
Passing Ability	J-Ring	T500J (mm)	770	710	820	800	750	780
		Step height, sh (mm)	13	10	5	7	7	7
Segregation Resistance	VSI15min	Index (0.5 - 3)	n/a-high viscosity	n/a-high viscosity	3	3	1	0.5
	Sieve Stability	Sieved portion, (%)	0.5	0.5	18	10	1.5	5

Remarks on the mixes: M1 – Mix loses consistency within 15 min, M2 – Mix loses consistency within 10 min, M3 – High amount of bleeding, M4 – Highly segregated mix, M5 – Nice & consistent mix, M6 – 'Balling Effect'.

Table 7. Fresh properties of SCC under various relative humidity.

Mix	Environmental	RH (%)	Slump Flow (mm)	T500s (sec)	J-ring Flow, (mm)	V- Funnel (sec)	Sieve Segregation (%)
Control	H3	82	800	2.54	750	9.14	12
	H2	65	740	3.71	680	11.48	9
	H1	75	790	2.02	740	9.24	11
POFA30	H3	81	770	2.00	730	6.87	7
	H2	60	670	3.39	590	16.58	4
	H1	70	750	2.48	720	7.19	6
POFA40	H3	80	770	2.64	710	8.41	10
	H2	51	650	4.09	550	17.04	3
	H1	79	760	2.28	710	7.19	10
FA30	H3	80	730	2.70	640	7.82	14
	H2	58	665	3.46	600	8.66	10
	H1	70	710	2.90	630	8.32	13

Table 7. Fresh properties of SCC under various relative humidity (Continued)

Mix	Environmental	RH (%)	Slump Flow (mm)	T500s (sec)	J-ring Flow, (mm)	V- Funnel (sec)	Sieve Segregation (%)
FA40	H3	83	780	2.01	650	5.68	18
	H2	66	700	3.00	610	7.5	15
	H1	77	760	1.49	640	5.52	18
TNY30	H3	87	780	2.27	800	8.68	12
	H2	55	700	2.66	640	10.13	6
	H1	70	755	2.19	700	8.56	10
TNY40	H3	83	755	1.81	680	9.47	5
	H2	59	700	3.15	660	12.48	2
	H1	77	745	2.39	690	10.12	6

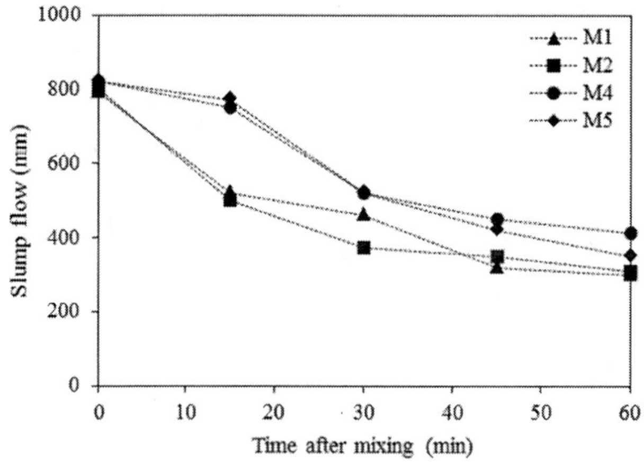

Figure 1. Slump flow retention of SCC under varying mixing methods.

5 CONCLUSION

Based on this study on SCC with agro-industrial wastes, a total mixing time of 9 min is recommended. In addition, when exposed to lower humidity in tropical weather accelerates the reduction of the workability properties of SCC especially in SCC incorporating POFA.

References

Brameshuber W., Uebachs S., Wallevik O. and Nielsen I., "The influence of the temperature on the Rheological Properties of SCC," in Proc. 3rd Int. RILEM Symposium, 2003.

EGSCC. (2004). The European Guide lines for Self-Compacting Concrete EGSCC. Concrete, T.E.G.S.C.

Ferraris C.F. Concrete Mixing, Methods and Concrete Mixers: State of the Art. *Journal of Research of the National Institute of Standards and Technology*, Vol. 106(2), pp. 391-399, 2001.

Ghafoori N., and Diawara H., Influence of temperature on fresh performance of self-consolidating concrete, *Construction and Building Materials*, Vol. 24(6), pp. 946-955, 2010.

Rahman M, Rashid M.H., Hossain M.A., Adrita F.S. and Hossain T. Mixing Time Effects on properties of Self-Compacting Concrete. *ARPN Journal of Engineering and Applied Sciences*, Vol. 6(8), 2011.

Integrated Solutions for Infrastructure Development
Edited by Lau, H. H., Tang, F. E., Ng, C. K., and Singh, A.
Copyright © 2016 ISEC Press

SUBMERGED CONCRETE CAISSON BREAKWATER CONSTRUCTION

ALVIN MAGALLANES[1], BEN SULLIVAN[2], ANDREW MOORES[2], and AMARJIT SINGH[1]

[1]*Dept of Civil and Environmental Engineering, University of Hawaii at Mānoa, Honolulu, USA*

[2]*HOV Environment, LLC, Las Vegas, USA*

After numerical and physical modeling, and alternate analysis, the proposed solution to mitigate wave energies entering Kahului Harbor, Maui was to construct a submerged concrete caisson breakwater, just outside the harbor that would economically improve the harbor tranquility and be environmentally friendly. One of the major challenges of this project is that construction is to be completed over a five-month period, during the summer months, when there are no swells. This requires as much prefabrication as possible, detailed planning, and precise scheduling. Construction of the individual caisson units is proposed to take place in a dry dock, cast in a single pour to avoid cold joints. On completion, it will be floated and tugged away from the dry dock and placed onto a heavy lift vessel and transported to its installation site at Kahului. Various alternate methods of transportation were studied and discarded. Also proposed is the transplantation and seeding of coral fragments on the caisson units to enhance the durability and increase the longevity of the submerged breakwater. This submerged breakwater caisson project will be the first of its kind for mitigating wave energies entering a harbor. This paper aims to present a viable construction methodology.

Keywords: Coastal engineering, Harbor protection, Wave energy mitigation, Heavy lift vessel, Scour protection units, Coral seeding.

1 BACKGROUND

Kahului Harbor is the primary commercial port for the island of Maui that experiences high amounts of traffic. This, along with the seasonal swells, makes for undesirable wave propagations inside the harbor, leading to possible safety violations and operational delays. The proposed solution: construct a submerged concrete caisson breakwater, just outside the harbor.

Being the first of its kind, this construction project will need a uniquely developed construction system. The following sections will provide insight to the strategies and construction methods.

2 SITE-SPECIFIC CONSTRUCTION CONSTRAINTS

After numerical modeling and physical verification, the proposed installation site for the breakwater is 280 m north of the existing eastern breakwater of Kahului Harbor (Figure 1). Construction and deployment of the caisson units near the installation site would be ideal. However, it was discovered that construction of this magnitude could not be undertaken on or from the island of Maui. Hence, a dry dock in Honolulu, on the neighboring island of Oahu was selected as the seat of the construction operations.

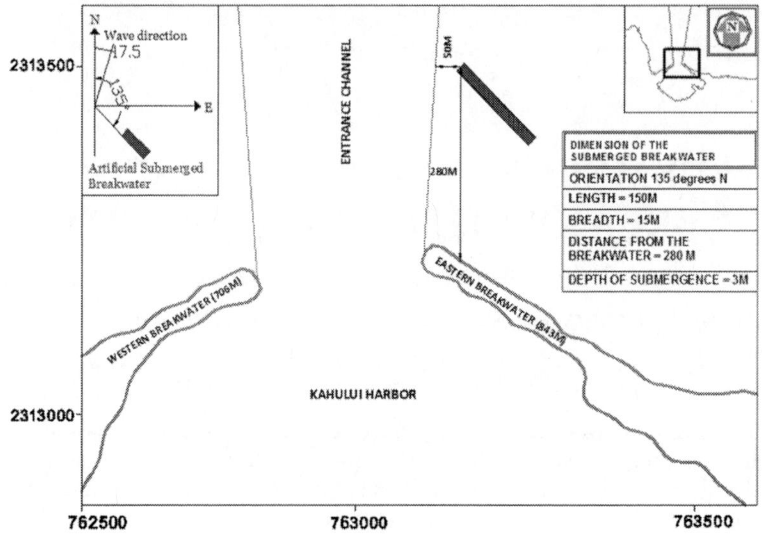

Figure 1. Final layout of the submerged caisson breakwater (Foley *et al.* 2015).

2.1 Heavy Lift Vessel Transport

Alternatives were considered in the transport of the caisson units to the construction site. However, tugging the individual caisson units is not the most secure mode of mobilization as there is a risk of capsizing and the floating dry docks cannot accommodate a caisson unit. After consulting with the marine contractor, it was recommended that a heavy lift vessel (HLV) be used in the transport of the caisson units.

2.2 Optimal Construction Window

Construction times will be limited due to the seasonal wave heights at the site. There are no specific guidelines for the maximum wave height at which caisson installation can take place. After simulations, Esteban *et al.* (2009) found wave height could be up to 1.5 m without causing risk or damage during installation to the caisson, or health and safety regulations. Thus, the only practical time period for construction of the caisson units would be in the summer season (May to September), when the waves are small.

3 PROPOSED CONSTRUCTION METHOD

The proposed method can be separated into 17 phases, as outlined in the following sections.

3.1 Phase 1: Pre-Construction Surveys

Prior to construction, specific surveys will be conducted that include a bathymetric survey, sea bed dive survey, and marking the site with high-visibility navigation buoys.

3.2 Phase 2: Mobilization of the Dry Dock

The mobilization of the dry dock will take a year of planning and permitting. The elements to be mobilized for casting in the dry dock are two long-reach cranes for loading/unloading all equipment and formwork, two concrete pumps, tool/maintenance workshop, rebar stockpile, site office/cabin, slip formwork, and base layer formwork.

3.3 Phase 3: Supply of Material

The materials required for this project will be delivered to the two main delivery sites: (1) the dry dock (materials: ready-mix concrete, reinforcing steel, slipform and formwork, and associated tools) and (2) Kahului Port (materials: rock or concrete mattresses for foundation layer, fill material to submerge caissons, pre-cast scour blocks, and crushed-rock leveling layer).

3.4 Phase 4: Excavation of Seabed for Foundation

The sea bed is required to be excavated 0.5 m across the footprint of the structure. The method for efficient and effective excavation would be with the use of a cutter suction dredger. On completion, the sea bed will be inspected and compacted samples will be collected. There might be a requirement to compact the excavated bed by an underwater dive team and a manually operated grader to achieve the required ground consistency (Mann 1999).

3.5 Phase 5: Mobilization of Plant and Manpower for Foundation

The required resources for the installation of the foundation are as follows: split hopper barge; tug boat, if barge is not self-propelled; commercial dive team; 30 metric ton crane barge; manual underwater grader; and remote operated grader.

3.6 Phase 6: Placement of Pre-Cast Foundation Units

Pre-cast concrete mattresses will be used as a foundation layer as it would dramatically decrease installation time and be less labor intensive. A fine layer of rubble is recommended to be deposited prior to the installation of a foundation unit. These operations will use a barge mounted crane and be overseen by a dive team.

The suggested length of each mattress is 10 m, derived from recommendations of the United States Army Corps of Engineers (USACE). The USACE guideline for mattress height/thickness is 60 cm, and in order to achieve the required 1.3 m height, two layers of mattresses will be required. The foundation layer must support an area of 126 m in length by 27.75 m wide; the total area that the caisson structure and scour protection blocks will occupy. Therefore, the total number of mattresses required would be 466.

3.7 Phase 7: Concrete Pouring of Caissons

Two main elements make up the caisson unit: (1) the bottom slab, and (2) the outer and inner walls. As the bottom slab is designed with an outer toe wider than the outer caisson walls, two techniques of concrete casting are required. A traditional formwork will be constructed for the bottom slab and the rebar will be placed per the design requirements. As soon as the pour for the bottom slab has been completed, a slipform for the outer and inner walls will be used, and concrete will be poured immediately to ensure no cold joints (Hibbs *et al.* 2009). The caisson is designed to be completely submerged and will have to be made watertight.

Another critical aspect is the installation of 12 anchor points (using pad eyes) into the structure during casting, as they will be used as point of connection for towing and positioning. The 12 points are located as follows: one for each corner at the top of the structure; one for each corner in the bottom of the structure; one on top of the caisson in the center of each length. Once the concrete is cured, the calculated water line (the buoyancy of the caisson) will need to be marked on the structure at each corner as a guide to assess the stability and keel of the caisson when floating in the dry dock and during ballasting.

3.8 Phase 8: Pre-Cast Concrete Caps

Pre-cast caps will be installed at the last phase of construction and will be used as a platform for the casting of top slab. The pre-cast concrete caps need to be constructed with a 4" to 6" flexible pipe inserted into it and should match the same point on the top slab. There will be two pipes installed for each cell: one as an inlet port and the second with a geotextile membrane for the output, allowing water to be expelled during the sand filling stage. *Bauer* connections, or similar, are recommended so they are easy to operate by a diver.

3.9 Phase 9: Transport of Caisson Units to Kahului Site

Once the first caisson unit has been completed, the dry dock will be prepared for flooding, during which the caisson will be monitored to assess there is an even keel. In the instance that the keel is uneven, sand ballasting in the correct cells will take place until it is at the acceptable level. The caissons will be connected to tugs via the anchor points and then towed to the HLV.

3.10 Phase 10: Unloading of Caisson Unit

Once the HLV is in place at the construction site, the vessel will ballast itself to approximately 40 ft; the caisson's working draft is 29 ft. This will ensure adequate clearance for the caisson to be towed away and moved to above the foundation layer. The heavy lift vessel will then navigate back to the dry dock to collect the next caisson unit.

3.11 Phase 11: Positioning and Ballasting of Caisson Unit onto Foundation

The positioning and ballasting of the caisson unit is vital to the success of the project. Once the caisson is removed from the HLV and is floating on its own buoyancy, the next stage of installation is to move the caisson into position over the prepared foundation. The installation of the first caisson unit is vital to the alignment of the other caisson units.

In order to move the caisson unit in the most controlled approach, a minimum number of four heavy tug boats will be used. Two tugs will be positioned at the front of the caisson and one at the back, pushing slowly when required. A fourth tug will be on standby for any adjustments on either side of the caisson. Outside the perimeter of the foundation, a four point anchoring system will be used, to guide the caisson into place with marine grade rope and pulleys to allow for finer control, until the position is satisfactory and checked with GPS (Hibbs *et al.* 2009).

During the ballasting stage, movement of the caisson may occur due to wave/wind forces or uneven ballasting to the caisson. In this situation, the tugs and mooring lines will have to readjust the caisson and check with the RTK (real-time kinetic) GPS, having an accuracy to 2 cm, enabling informed operational control. To prevent any heavy loading on the caisson cells as well as any roll of the unit during ballasting, the cells should be loaded uniformly (Tsinker 2014). The proposed approach is to use a total of eight water pumps with a discharge of 50 m^3 per hour. Each cell is 99 m^3 and 80% of all cells need to be filled for the caisson to be on the foundation layer.

3.12 Phase 12: Pumping of Fill Material in Caisson Unit

Filling should be carried out as soon as the caisson is correctly positioned and seaward compartments should always be completely filled for stability under wave loading (BS3649-7:1991). Directly after the caisson unit has been ballasted and on the foundation layer, the filling activities can take place. This will involve the use of four hydraulic pumps and the same stage-wise approach for filling. The dive team will connect the outlet hose to the inlet port on top of the caisson. The hydraulic pumps will be placed into a barge containing 1,000 m^3 of crushed rock/sand fill material. Water will be pumped into the barge and a slurry mix will be created. The hydraulic pumps will be mounted from an *A-Frame/Davit Arm* and lowered into the slurry and pump the slurry from the barge into the caisson cells via the 4" delivery hose. During the filling process, constant checks with RTK GPS will be made and the tugs may have to re-align the caisson. If it is noticed that the caisson is out of position, water can be pumped out of cells allowing it to float and for the tugs/guidance system to reposition the unit.

3.13 Phase 13: Caisson Unit Connection

The caisson design has been incorporated with an interlocking arrangement between each Caisson unit. This will assist with acting as a guide and allow accurate confirmation that the caisson unit is in the correct X- and Y-axis positions in relation to the previous unit.

Prior to installation of the remaining three caisson units, the four point anchor system will need to be moved with the crane mounted on the barge. Additionally, the connections on Caisson 1 will be used to pull and position the Caisson 2 into place. When the male connection of Caisson 1 smoothly fits with the female connection of Caisson 2, the ballasting will take place. The dive team and support boat will need to monitor to make sure that the connection whilst ballasting is flush between Caissons 1 and 2. If a gap is noticed one of the support tugs will have to push against Caisson 2 to remove any gap. With the possibility of horizontal movement, the tugs will again be used to readjust any difference and caissons secured with guidelines to prevent further movement.

3.14 Phase 14: Casting of Scour Protection Units

A traditional formwork mold will be constructed and pouring will be conducted on Oahu. As the unit's weight will be 19.2 tons, they can be lifted by crane and transported by barge to Kahului Port for loading. Once at the port, they will be loaded onto the crane barge used to assist with the caisson installation. The caissons will be lowered into place using the crane with guidance from the dive team and the use of the RTK GPS system for accuracy.

3.15 Phase 15: Installation of Scour Protection Units

Placement of the scour protection blocks will take place immediately after a caisson unit has been set. Scour protection (Figure 2) on the seaward side should be placed and completed as soon as possible after positioning of the caisson (BS 6349-7: 1991). During the time the heavy lift vessel returns, the scour protection units for the first section needs to be completed. There are a total of 162 scour units per caisson section and with the aim of installing 15 units per day; it would take approximately 10.8 days to complete each section of scour protection.

3.16 Phase 16: Post Construction Surveys

On completion of the construction project, final surveys will be conducted along with as-built drawings. The surveys will include: RTK GPS survey to assess final structures position and crest height; bathymetric survey; videographic survey recording any damage to the structure during construction (e.g., any gaps between the caisson units or other construction abnormalities).

3.17 Phase 17: Coral Seeding/Propagation

An important element of this project is to conduct coral seeding and coral propagation activities to enhance the local ecology, contribute to the local environment, rather than destroying it, and to increase the durability of the caisson units. The seeding is done on the caisson units at the dry dock, after concrete pouring has been completed and the formwork removed. Studies indicate that coral seeding is feasible and that coral are indifferent to basalt, limestone, or recycled concrete (Foley *et al.* 2015).

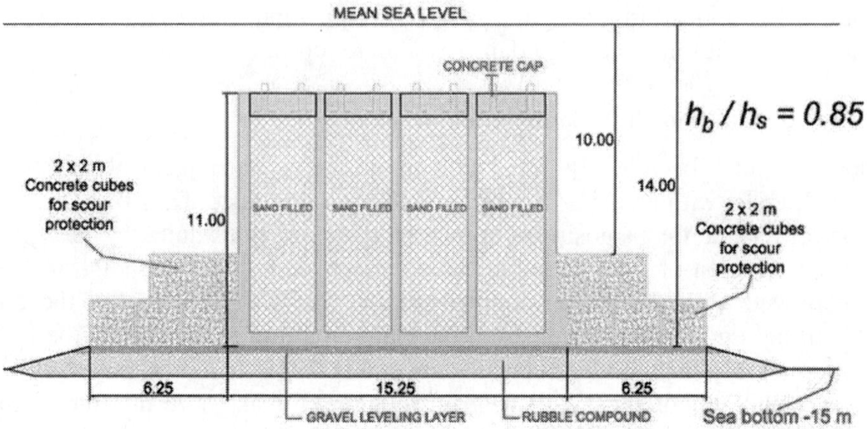

Figure 2. Elevation view of caisson breakwater showing scour protection units (Alvarez *et al.* 2015).

4 CONCLUSION

The construction of this submerged concrete caisson breakwater is feasible with the addressing of the construction constraints and implementation of the proposed construction method.

Acknowledgments

Prepared in cooperation with the State of Hawaii, Dept. of Transportation Highways Division and US Dept. of Transportation Federal Highway Administration. The contents of this report reflect the views of the authors, who are responsible for the facts and accuracy of the data presented herein. The contents do not necessary reflect the official views or policies of the State of Hawaii Dept. of Transportation or the Federal Highway Administration. This report does not constitute a standard, specification, or regulation.

References

Alvarez, E., Solis, A., Singh, A., Foley, M., and Magallanes, A., *Conceptual Design of Concrete Caisson Submerged Breakwater for Kahului Harbor: Structural Design and Construction Procedure Phase I*, Technical Report, HDOT Contract Award No. TA 2010-1R, Honolulu, October 2015.

British Standards Institution, Maritime Structures - Part 7: Guide to the design and construction of breakwaters, BS63497:1991, BSI Publishing, 1991.

Esteban, M., Takagi, H., and Shibayama, T., Methodology for the Simulation of the Construction of a Breakwater Taking into Account Climate and Construction Accident Risks, Coastal Engineering Journal, Vol. 51, No. 01 : pp. 49-68, March 2009.

Foley, M. J. H., Singh, A., Jokiel, P. L., and Stender, Y., *Submerged Breakwater Modeling and Coral Reef Ecological Analysis for Harbor Protection*, University of Hawaii, College of Engineering, Civil and Environmental Engineering, Research Report UHM/CEE/15-03, June 2015.

Hibbs, W., Bowers, P., Young, M., and Dingwell, J., Installation of Caisson Breakwater at Costa Azul, Mexico, Thomas Telford Limited, 2009.

Mann, S., A Floating Factory for the Maasvlakte 2 Caisson Breakwater, TUDelf University, 2009. Retrieved from repository.tudelft.nl/assets/uuid:841bb0d2-fa8f-4b9c-a5ec-4ad900c9ac0a/Mann.pdf.

Tsinker, G., Handbook of Port and Harbor Engineering: Geotechnical and Structural Aspects, p. 499, Springer Publishing, 2014.

Integrated Solutions for Infrastructure Development
Edited by Lau, H. H., Tang, F. E., Ng, C. K., and Singh, A.
Copyright © 2016 ISEC Press

RACKING PERFORMANCE OF LIGHT WEIGHT CONCRETE FILLED COLD-FORMED STEEL PANELS

SHAHRZAD KASAEIAN[1], HAMID R. RONAGH[2], and REZA NIKOONEJAD[3]

[1]*School of Civil Engineering, University of Queensland*
[2]*Institute for Infrastructure Engineering, University of Western Sidney*
[3]*Civil Engineering, Azad University of Yazd*

The implementation of Cold-Formed Steel (CFS) as a structural element is almost new. As a light weight material, CFS members in the earthquake impose low inertia force to the structure. So greatly increased worldwide demand for such structures. Due to the thin-walled nature of CFS they are susceptible to buckling. Great efforts had been made to promote the lateral load resistance of CFS panels. The common bracing methods are not capable of economically resisting the high demands imposed on the system in highly seismic regions. In some instances, all panels are to be covered with Oriented Strand Board or steel sheeting in order to adequately address the anticipated earthquake load, and this renders the system too expensive. A promising method to combat this deficiency is to fill the cavity in between the panel studs with concrete. Results showed that a panel made in such a manner was able to resist the lateral loads three times more effectively than a similar configuration panel but with strap bracing. The strength, ductility and earthquake response factor of such a system were the major concerns. The experimental tests were performed on a 1.2 m × 2.4 m wall with three different configurations of studs and tracks.

Keywords: Thin-walled, Lateral load, Strength, Ductility, Earthquake response factor.

1 LITERATURE REVIEW

The construction of structures framed with CFS panels are becoming more common throughout the world. These panels are obtained through the assemblage of internal and external profiles (studs) with lipped channel cross-sections. The studs interconnect at each end by members (tracks) with un-lipped channel cross sections, in such a way as to realize a Cold-Formed Steel Frame Panel (CFSP). Panels with one- or two-sided sheathing have been widely used to construct the external and internal walls (CFSW) of buildings. For external walls, the panels support the structural load according to building regulations. For internal walls, some are load bearing, whereas others are only used for partitioning purposes and thus, are only required to offer secondary structural support. The main problem with these walls is their low brace capacity which hinders their use in medium to high seismic regions despite their advantageous light weight. Bracing capacities, provided by normal bracing methods such as steel straps or gypsum board cladding, are in the range of 2-7.4 kN/m (Vieira *et al.* 2012). For example, according to AISI S213-07, for a sheathed CFSP with ½" gypsum board on one side and screws placed 200 mm on the edges and 300 mm within the board, its nominal shear capacity under seismic loads is 230 lb/ft (3.35 kN/m) and once factored with the capacity reduction factor of 0.6, an ultimate capacity of 2 kN/m would be the result. However, while making it double-sided gives about 4 kN/m, it is still small. Although decreasing the screw spacing to 100 mm can create approximately twice this capacity, it is impractical and too labor intensive. With steel straps a similar problem exists; strap capacity is small and cannot be relied upon as it is often limited to the capacities of the end connecting screws. An alternative to these would be using an expensive Oriented Strand Board (OSB) that allows higher capacities (12 kN) to be achieved. However, transferring these forces through the bottom plate of the panel to the foundation is very much questionable and would often hinder the use of the theoretically available capacity (Fülöp and Dubina 2004, Landolfo *et al.* 2004).

Comparing the above lateral strength to the demand, a sharp contrast becomes evident. For a typical 300 m^2 two storey cold-formed house in a seismic region with a base acceleration of around 0.3 g, one would expect base shears of around 80 to 120 kN. Base shear could rise to 160 kN for a three storey. It is then difficult to control these lateral seismic forces through sheathed panels with low lateral capacities of 4 kN/m. One would need 40 m long panels to resist 160 kN. In many instances, this creates an architectural nightmare as openings must be cut short. Therefore, study a new structural wall is needed. As studs to evaluate the lateral load carrying capacity of concrete filled CFSP. As studs naturally form a good support for the formwork and placement of concrete, this research focuses on the study of the behavior of concrete filled CFS wall. The main behavior that was studied is the maximum lateral load bearing capacity of the specimens, their deformation behavior, the earthquake response factor (R factor) and different failure modes of system.

2 MATERIAL AND METHODS

The nine concrete filled cold-formed steel wall (CFCFSW) specimens, which are represented in three different configurations (A, B and C) shown in Table 1, were tested in the structures laboratory at University of Taft, using a testing frame designed specifically for in-plane shear loading. The cold-formed steel properties are provided in Table 2. All walls were 2400 × 1200 mm in size with light weight concrete (shown in Figure 1). The light weight concrete properties and mix design are tabulated in Table 3. All the members were connected using wafer head self-drilling/self-tapping screws. The screws shear and tensile strength are 3.3 kN and 3.8 kN, respectively.

Table 1. CFCFSW configuration.

Specimen type	Corner studs (single)	Middle stud (single)	Noggin	Number
A	√	-	-	3
B	√	√	-	3
C	√	√	√	3

Table 2. Cold-formed steel properties.

Grade (MPa)	Thickness (mm)	E Modulus (GPa)	F_y (MPa)	F_u (MPa)	ε_y (%)	ε_u (%)	F_u/F_y
550	1	169	592	617	0.45	2.86	1.04

Table 3. Light weight concrete properties.

F'$_c$ (MPa)	Specific Weight (kN/m^3)	Aggregate	Cement	Slump (mm)
13	15	LECA	Portland	110

All the specimens were installed next to the side studs in the middle of the tracks on both sides of the wall between the upper (fixed) and lower loading rigid beams of the laboratory frame, using four M16 high strength bolts, which were located in frame before the pouring of the concrete. Measurements consisted of bottom wall displacement, acceleration of the loading beam assembly, the shear load at the top of the wall, as well as the uplift force in the hold down anchor rods. The LVDTs, strain gauges, load cells and accelerometer were connected to a computer and the Lab View Signal Express Software to obtain the load-displacement curve for each specimen.

Figure 1. Light weight (LECA) concrete-filled cold formed steel wall specimen in the test frame.

3 LOADING PROTOCOL

The cyclic loading regime that has been used in this study is based on Method B of ASTM-E2126 Standard (Table 4). The Method B test protocol is intended to produce data that sufficiently describes elastic and inelastic cyclic properties, and the typical failure mode that is expected in earthquake loading.

Table 4. Cyclic loading regime, Method B - ASTM E2126-07.

Pattern	Step	Minimum Number of Cycles	Amplitude, % Δ_m
1	1	1	1.25
	2	1	2.5
	3	1	5
	4	1	7.5
	5	1	10
2	6	3	20
	7	3	40
	8	3	60
	9	3	80
	10	3	100
	11	3	Additional increments of 20 (until wall failure)

*Where Δ_m is the ultimate displacement, monotonic which is the displacement corresponding to the failure limit state in a monotonic test.

4 RESULTS

4.1 Lateral Load Capacity

Light weight concrete is a brittle material, especially in tension, so naturally it cannot sustain large deformation like the CFS does. However, at small drifts a considerable amount of lateral resistance from the use of CFCFSW was found. As shown in Figure 2, the hysteretic response of the CFCFSW demonstrates pinching behavior with stiffness and strength degradation. The main failure mode in these specimens was the crack and rupture of concrete in connection with studs. The failure continued by stud-to-track connecting screw pull-out and rivet tilting and then caused some distortional buckling with wing rupture in the lower track. Thereafter, the lateral load was reduced and the frame lost its load bearing capacity. According to the experimental tests, it is crystal clear that middle stud and noggin not only do not play any important role in the lateral force capacity of CFCFSW but depending on the separation caused in the concrete which dropped the lateral force capacity and ductility of the wall. Overturning decreased the lateral load capacity in specimens Type A and Type B. The limit of the lateral strength associated with overturning depends on the presence of hold down anchor rods. In test specimens the rods are located in the bottom corners and in specimens Type B and C the lateral load transferred in two parts of the concrete separately, the tension in the middle of the track did not transfer properly and subsequently caused track distortion and concrete cracking and rupture.

To make the comparison between the proposed wall and conventional walls, in Table 5 the lateral load capacity, which AISI-S213 provides, is presented. The walls are a limited configuration of gypsum board sheathed and oriented strand board sheathed CFS panels. The mentioned shear walls can have openings to resist wind and seismic loads. The lateral load resistance of this wall can be reduced by multiplying in C_a (presented in Table 6).

Table 5. Nominal shear strength (Rn) for wind and seismic loads for shear walls faced with gypsum board or fiberboard. (kN/m) (United States and Mexico).

Assembly Description	Maximum Aspect ratio (h/w)	Fastener Spacing at panel Edges/Field (inches) 7/7	4/4	4/12	8/12	4/6	3/6	2/6
1/2" gypsum board on one side of wall; studs max. 24" o.c.	2:1	4.3	6.3	4.4	3.4	-	-	-
1/2" fiberboard on one side of wall; studs max. 24" o.c.	1:1	-	-	-	-	6.3	9.1	9.9

Table 6. Shear resistance adjustment factor (C_a).

Percent Full-Height Sheathing	Maximum Opening Height Ratio 1/3	½	2/3	5/6	1
	Shear Resistance Adjustment Factor				
10%	1	0.69	0.53	0.43	0.36
20%	1	0.71	0.56	0.45	0.38
30%	1	0.74	0.59	0.49	0.42
40%	1	0.77	0.63	0.53	0.45
50%	1	0.8	0.67	0.57	0.5
60%	1	0.83	0.71	0.63	0.56
70%	1	0.87	0.77	0.69	0.63
80%	1	0.91	0.83	0.77	0.71
90%	1	0.95	0.91	0.87	0.83
100%	1	1	1	1	1

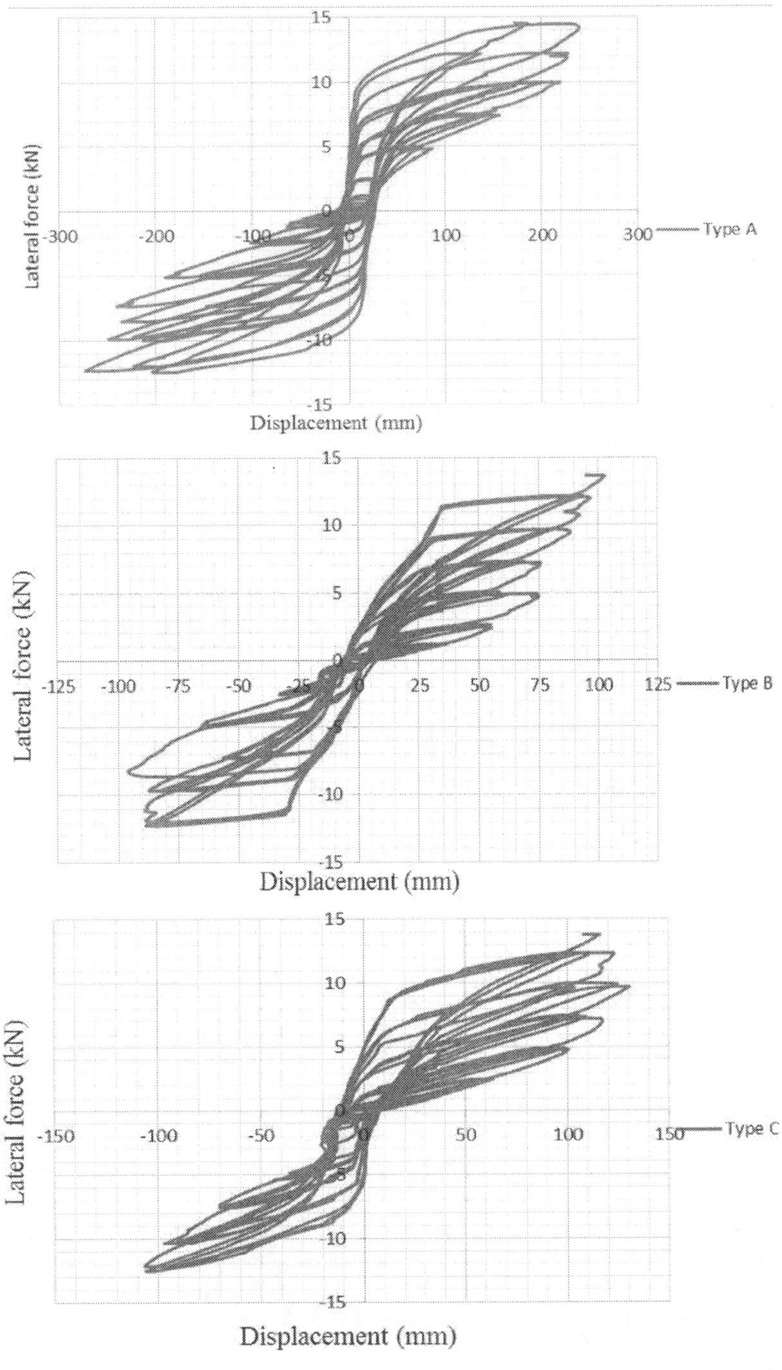

Figure 2. Hysteretic diagram of test specimens (Types A, B and C).

4.2 Earthquake Response Factor (R Factor)

To calculate the response factor, the hysteresis curve was utilized, using the FEMA 356 method. The results are shown in Table 7. Each provided result is the mean value of three specimens with the same configuration.

The achieved R factor illustrates the high ductility of CFCFSW in comparison with conventional CFS walls that are provided in AS/NZS4600 and the National Research Council of Canada.

Table 7. R factor.

Spec.	R
A	8.11
B	6.21
C	7.25

5 CONCLUSION

The two most important parameters that can be discovered from Figure 2 and Tables 4 and 5 are the higher lateral force capacity and ductility of CFCFSW in comparison with conventional CFS walls. The only configuration that has almost the same lateral capacity as CFCFSW is OSB sheathed CFSW. However, construction of a CFS wall filled with concrete is much easier than a CFS wall sheathed with OSB that has 2inch fastener spacing in field. On the other hand, some countries that are located in the desert do not have access to wood and the only way to achieve a desired lateral resistance is to apply other materials. Thus, for an effective construction of a light weight building that can resist earthquake and wind load properly, filling of cold formed steel panels with light weight concrete is suggested.

References

American Iron and Steel Institute, AISI-S213, North American Standard for Cold-Formed Steel Framing – Lateral Design 2007.

American Iron and Steel Institute, AISI S213-07, North American Standard for Cold-Formed Steel Framing – Lateral Design, Washington DC, USA, 2007.

AS/NZS4600, Cold-formed steel structures, Australian Building Codes Board, 2005.

ASTM-E2126-07, Standard test methods for cyclic (reversed) load test for shear resistance of walls for buildings, p. 13, USA, 2007.

FEMA-356, Prestandard and commentary for the seismic rehabilitation of buildings NEHRP, USA: Building Seismic Safety Council, 2000.

Fülöp, L. A. and Dubina, D., Performance of wall-stud cold-formed shear panels under monotonic and cyclic loading, *Thin-Walled Structures*, 42, 339-349, 2004.

Landolfo, R., Corte, G. D., and Fiorino, L., Testing of sheathed cold-formed steel stud shear walls for seismic performance evaluation, *13th World Conference on Earthquake Engineering Vancouver*, BC, Canada. 2004.

National Research Council of Canada, National Building Code of Canada, 12th Edition, Ottawa, On, Canada, 2005a.

Vieira Jr., L. C. M. and Schafer, B.W., Lateral stiffness and strength of sheathing braced cold-formed steel stud walls, *Engineering Structures*, 37, 205-213, 2012.

Integrated Solutions for Infrastructure Development
Edited by Lau, H. H., Tang, F. E., Ng, C. K., and Singh, A.
Copyright © 2016 ISEC Press

INFRASTRUCTURE NEEDS IN SUB SAHARAN AFRICA WITH PARTICULAR REFERENCE TO SOUTH SUDAN

AYMAN Y. NASSIF[1], ANN STEWART[2], MILLIE T. MUTEPFE[3], and PETROS CHRISTOU[4]

[1]*School of Civil Engineering and Surveying, University of Portsmouth, UK*
[2]*Business School, University of Portsmouth, UK*
[3]*Havant Borough Council, UK*
[4]*Dept of Civil Engineering, Frederick University, Cyprus*

This paper focuses on South Sudan's infrastructure delivery in an efficient and effective way. South Sudan, though endowed with considerable natural resources, still lacks the necessary infrastructure for improvement in the standard of living for the population. This paper presents aspects of qualitative and quantitative data obtained using questionnaires and interviews. It was found that most of the population had access to health facilities within a 10 km radius and 65% of the respondents stated that they are able to walk such distances. Regional connectivity infrastructure was rated very highly by the respondents. As far as soft infrastructure needed for education, there is a crucial need to invest in adult literacy especially in the field of science and engineering to ensure appropriate participation of the citizens in the developmental agenda for South Sudan. Developing the human capital is crucial in attracting local and foreign investors in infrastructure projects. Tertiary institutions of learning in South Sudan will need to focus on education and training to ensure that the country will have locally qualified people to design, construct and manage the necessary soft and hard infrastructure. The findings can be relevant to Asian countries of similar needs in developing their infrastructures.

Keywords: International development, Health, Education, Transportation.

1 INTRODUCTION

The paper reports some aspects of a wider project focusing on the provision and maintenance of public infrastructure in South Sudan. The investigation included transportation, water and sanitation facilities, irrigation systems, power supply and communication systems. Skills availability pertaining to engineering and science were also investigated.

After almost five decades of civil war, in July 2011 the people of South Sudan completed the process of independence from Sudan. South Sudan then became the United Nations' 193rd member and the 55th African nation (Salman 2011). Although there is limited data about South Sudan, the United Nations Development Program (UNDP) has ranked Sudan as 171 out of 186 on the Human Development Index (HDI) 2012 ranking. South Sudan has rich soils and good rainfall patterns which should provide a basis for a thriving economy but the area is chronically underdeveloped due to its' historical marginalization from the north (Jooma 2011). Poverty in the south is estimated at 90% (Jooma 2011). This was not helped by South Sudan's most recent history of conflict which broke out in Juba, December 2013 and lasted for nearly 2 years. The financial impact of the conflict in the capital Juba in 2013-15 led to the decline in gross oil revenue from $29.35 million in December 2013, falling to 10.8 million in January 2016 (The World Bank 2016). The Government of South Sudan (GoSS), with the support of donor countries, UN and Non-Governmental Organizations (NGOs) have worked diligently to ensure that security, roads, health, education and agriculture are prioritized (Government of South Sudan 2013).

The neighbors of South Sudan, shown in Figure 1, have varying degrees of interest in the political developments in their neighborhood. Ethiopia, Kenya and Uganda have been developing ties with Juba. They welcome the presence of a state with an African identity as opposed to Arab identity (Belloni 2011). Juba, the capital, offers a strategic location serving the local and foreign investors in the region. Juba is reported to be rapidly expanding with the fastest rate of urbanization in history (Grant and Thompson 2013).

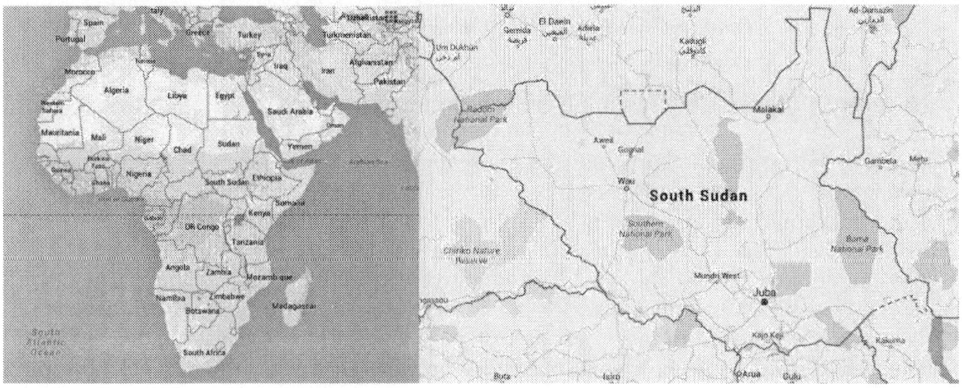

Figure 1. Map of South Sudan.

The strategic development plan of the GoSS until 2040 includes very ambitious social and economic aspirations (Government of South Sudan 2013). The GoSS has identified infrastructure as one of the pillars for its' economic development. South Sudan is one of the least-developed and the most oil-dependent countries of the world with oil accounting for over 95% of its' national budget.

The Gross Domestic Product (GDP) per capita in SS in 2010 was $1546 (Salman 2011). This was much higher compared to Kenya, Rwanda, Tanzania, Uganda and Ethiopia which had GDP per capita of $769, 548, 527, 503 and 350 respectively. This was attributed to the higher revenue from the oil in SS which was used to support the war efforts while the basic services were still not sufficiently available. With the oil infrastructure of the pipeline, the refineries, the export facilities, as well as the ports themselves all situated in northern Sudan, the two countries are poised for intricate and difficult negotiations on this matter. South Sudan is forming strategic cooperative agreements with Kenya and Ethiopia to gain access to skills and capability in exchange for natural resources (Thompson and Strickland 2004).

2 RESEARCH PROGRAM

During this research, questions related to mode of transportation infrastructure in particular relief logistics to provide humanitarian relief supplies as well as issues related to water supply, sanitation and public health, food security, and energy resources were explored. Qualitative and Quantitative methods were used for triangulation (Fellows and Liu 2009). The triangulation was achieved by combining the answers from respondents who agreed with suggestions made by the author and open ended questions which gave an insight to the author in certain aspects. A literature review of the data collected by aid organizations, the World Bank, the UN and the South Sudan infrastructure reports was carried out to identify knowledge gaps.

Primary data were collected through questionnaires from contacts in South Sudan. Raw data were also collected from the residents of South Sudan in the form of an open ended questionnaire (qualitative data) and interviews in order to gather their perceptions about infrastructure issues that impact on their lives. The questionnaires also asked the type of work they did so as to assess the respondents' bias. Through the questionnaire, the respondents were also asked to help in ranking or prioritizing infrastructure that is of importance to them as opposed to imposing what the researchers think is best. Secondary data were mainly collected from the GoSS development plan and case studies such as reported by Buys *et al.* (2010) and Yoshino *et al.* (2011). The importance of an integrated road network to stimulate the economy was highlighted.

3 FINDINGS

Figure 2-A shows that about 58% of the respondents walked if the distance was less than 10km. When the distance is greater than 10km, 67% travelled by private care while approximately 33% travelled by bus for longer distances. It seems that the river Nile was not exploited for transportation purposes. Use of public transportation is very limited due partially to security issues.

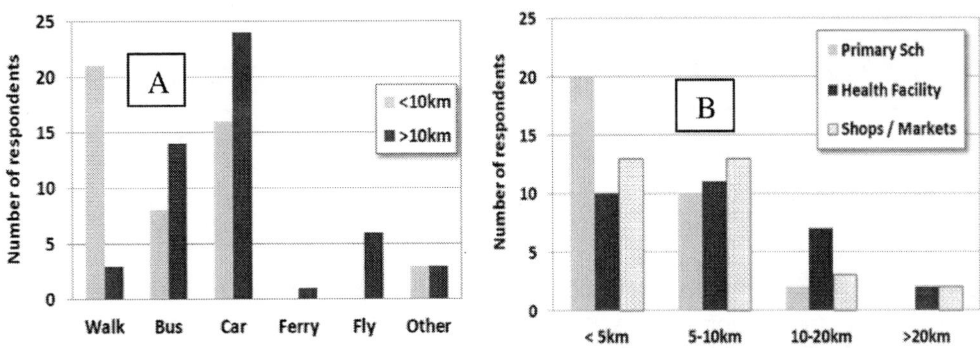

Figure 2. Transportation issues: A) Mode of travel and B) Distance to amenities.

Responding to the question related to the distance to the nearest primary school, 63% said they were within a 5km distance which is generally considered as walking distance for primary school children while 31% said the primary school was more than 5km but less than 10km away as shown in Figure 2-B. In terms of the proximity to a health facility, 33% responded that they were within 5km while 37% said the nearest health facility was more than 5km but less than 10km away. It is noteworthy that 7% said the nearest health facility was more than 20km away. Regarding distances to nearest shops and markets, 40% responded that they lived within 5km of the shops, for 43% of them the shops were 5 to 10km away. Over 16% said the amenities were over 10km away.

Regarding fresh water and sanitation issues, Figure 3-A shows that about 50% said they obtained their water from a well or borehole within their village and 31% said their fresh water supply was from a river. Regarding sanitation, 72% said they used pit latrines while only 22% said they had toilets with a flushing system. Interviews highlighted obvious absence of flush toilets throughout the country. Regarding sources of energy and fuel for domestic use, 75% used firewood and 44% used electricity while at least 28% used petroleum or paraffin while 13% used gas as shown in Figure 4-A.

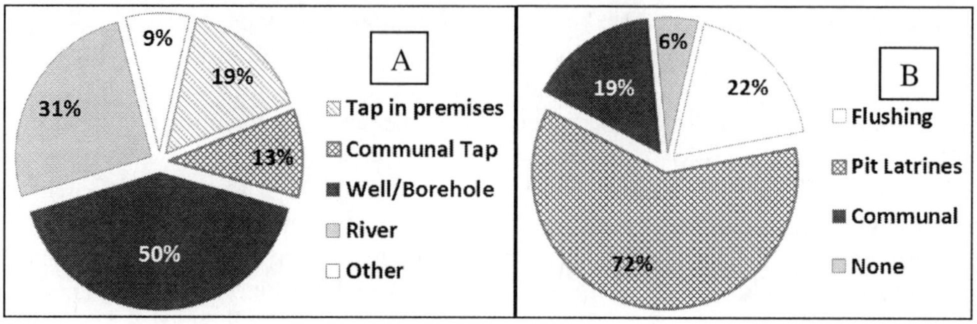

Figure 3. Water and sanitation issues: A) Sources of water and B) Types of toilets.

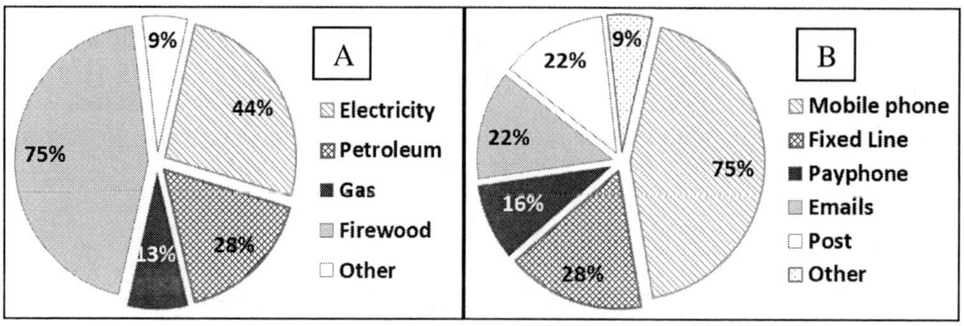

Figure 4. Energy and communications issues: A) Energy and B) Communication.

For communication, 75% of the respondents had access to mobile phones. 28% responded they had access to a fixed line phone and 22% responded they had access to emails and letters through the post. 16% responded they had access to payphones. Some responded that they relied on word of mouth, messengers from village to village, VHF radios and satellite phones while in very remote locations. Questions related to building the human capital and the skills capacity for South Africa will be the subject of further publication. During the initial study, 69% of the respondents agreed or strongly agreed that the focus should be placed on science and engineering. However, the interviews highlighted that there is a desire for a holistic agenda for education focusing on sustainable development. When asked if they thought the government should allocate more resources to promote infrastructure delivery, 90% of respondents agreed. Interviews revealed that the government needs to do more in promoting the country's image as a safe infrastructure investment area and to fight corruption.

The respondents were asked to rank twelve categories of public infrastructure services in what they considered to be of the most importance. Figure 5 shows that health facilities were ranked to be of highest importance by 72% of the respondents, followed by water and sanitation with almost 63% ranking it as most important while the regional road network was third with 58%. Education facilities were fourth with almost 55% of the respondents ranking it as most important and local roads and bridges were fifth 47% ranking them as most important.

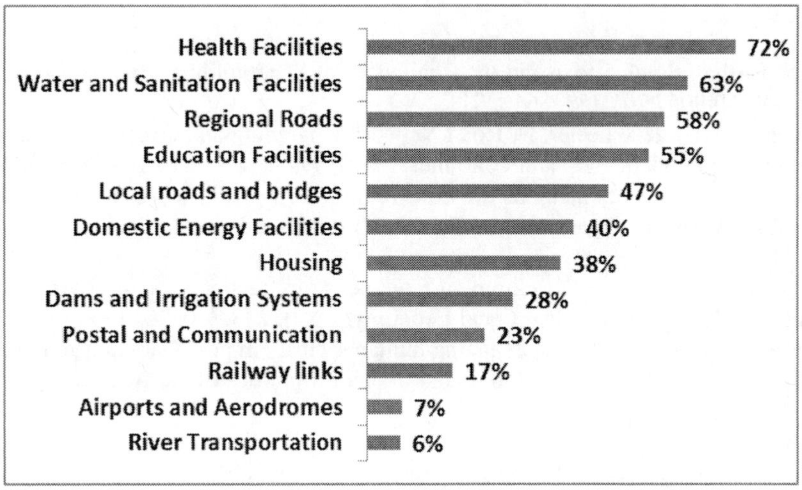

Figure 5. Infrastructure needs priorities.

4 CONCLUSIONS

The research has revealed that although South Sudan is a relatively new country, its strength lies in its resources (including water, agriculture, and young people) which are underutilized but with good management and strong leadership these can be further exploited for long term sustainability and growth. It is anticipated that opportunities for development and wealth creation will come through development projects from overseas investors, sponsorships, NGOs and neighboring sub Saharan nations, to work in partnership with the South Sudanese government in creating jobs, skills transfer, training and development opportunities for improved employability for a largely unskilled workforce.

A major weakness for South Sudan is its educational system. The government has started to address this with a planned educational reform program to public education starting from primary level through to adult education programs. This has its challenges including long-term affordability, maintaining and distributing the fiscal budget fairly on public education during periods of uncertainty.

The research shows that whilst ‘hard’ infrastructure is necessary for development, there is also a need for the government to prioritize over ‘soft’ infrastructural needs. It was originally hypothesized that hard infrastructure would be the priority for economic growth and success. However Figure 5 demonstrates that there is still a long way to go before this can be realized. The South Sudanese people will need to have their basic needs satisfied and quality of life before they can develop further trust and confidence in the government programs and initiatives.

It is therefore recommended that the government focuses its attention in the short term, to concentrate on improving basic skills and literacy which can also be blended with adult learning programs and on the job training (e.g., apprenticeships) to help up-skill the workforce. However, more long term educational reforms combined with entrepreneurial initiatives will go a long way in improving specific Higher Educational programs tailored to rebuilding the country, such as encouraging more take up in civil engineering, corporate social responsibility and governance programs that can be developed working in close nexus with businesses and in collaboration with overseas partners.

References

Belloni, R. The Birth of South Sudan and the Challenges of Statebuilding. *Ethnopolitics*, *10*(3-4), 411–429. doi:10.1080/17449057.2011.593364, 2011.

Buys, P., Deichmann, U., & Wheeler, D. Road Network Upgrading and Overland Trade Expansion in Sub-Saharan Africa. Journal of African Economies, 19(3), 399–432. doi:10.1093/jae/ejq006, 2010.

Fellows, R., & Liu, A. Research methods for Construction (3rd ed). Wiley-Blackwell: Chichester, 2009.

Government of South Sudan. Government of the Republic of South Sudan South Sudan Development Plan, 2013.

Grant, R., & Thompson, D. The Development Complex, rural economy and urban-spatial and economic development in Juba, South Sudan. Local Economy, 28(2), 218-230, 2013.

Jooma, M. Feeding the peace: Challenges facing human security in post-Garang South Sudan, 2011.

Salman, S. M. The new state of South Sudan and the hydro-politics of the Nile Basin. Water International, 36(2), 154-166, 2011.

The World Bank. Economic Overview. Retrieved from http://www.worldbank.org/en/country/southsudan/overview, 2016.

Thompson, A. A., & Strickland A. J. Strategic Management Concepts and Cases. (10th Ed). Erwin Professional Publishers, 2004.

Yoshino, Y. Industrial clusters and micro and small enterprises in Africa: from survival to growth. World Bank: Washington D. C, 2011

Integrated Solutions for Infrastructure Development
Edited by Lau, H. H., Tang, F. E., Ng, C. K., and Singh, A.
Copyright © 2016 ISEC Press

ASSESSING THE STRENGTH AND WEAKNESS OF THE GHANAIAN CONSTRUCTION INDUSTRY

BERNARD MARTIN ARTHUR-AIDOO, NAZEEM ANSARY and CLINTON AIGBAVBOA

Dept of Construction Management and Quantity Surveying,
University of Johannesburg, South Africa

Ghana as a developing nation located in the sub-Saharan Africa is characterised by its favourable political atmosphere and stable economy to do business. The country needs infrastructure development in order to satisfy both the economic and social development requirements of its economy. The construction industry, however, is a vital sector of the economy that addresses the socio-economic development needs of every country. This paper, therefore, seeks to assess the strength and weakness of the industry. The study as solely qualitative supported by an extensive review of literature via a desk-top study was embarked which captured the use of existing journal publications, government white papers, and gazette among others. The study revealed that the construction industry in Ghana is an important sector which creates employment which tends to reduce poverty to a larger extent. Also, the collaboration of the construction industry with other sectors do as well increases the gross domestic product of the economy and brings about entire growth in the economy. Similarly, the study revealed that the industry is posed with constraints which are categorised as weakness. These include access to credit facilities, a long span of control, employee management and corruption among others. The study, therefore, recommends the establishment of a construction industry development board that in Ghana to supersede all the activities of the industry in order to promote, maintain a high standard construction industry to the Ghanaian economy and society.

Keywords: Employment, Developing-nation, Infrastructure, Productivity.

1 INTRODUCTION

A developing nation like Ghana situated in the sub-Saharan part of Africa is characterised by its favourable political atmosphere and stable economy to do business. As a developing nation, infrastructure needs is imperative. As a result, the country needs infrastructure development in order to satisfy both the economic and social development requirements of its economy such as roads, houses, buildings to serve as schools, hospitals, factories and offices among others. According to Hillebrandt (2000), the construction industry of every country is a vital sector of the economy that addresses the socio-economic development needs of the country. Behm, (2008) also supported that construction is a large, dynamic, and complex industry sector that plays a significant role in the in most economies. The construction industry does not function in isolation rather; some sectors act as an auxiliary in enhancing the operations of the industry. Industries such as the automobile, pharmaceutical petrol chemical among others play crucial roles within the construction industry. The construction industry is not only a critical component of a nation's economy but also a fundamental factor in the quality of people's lives and the ability of the government to achieve many of its policy aims (Bosher *et al.* 2007). Despite significant merits that are gained from the construction industry, in Ghana, the sector is posed with constraints that limit the sector from functioning entirely. In addition, these constraints further act as barriers to the growth of the construction sector in Ghanaian economy as a whole. Nevertheless, when the construction industry which has been argued as the growth indicator of an economy is not booming, it there will be a severe effect on the entire economy. It is therefore against this backdrop this study has been embarked to explore the strength and weakness of the Ghanaian construction industry.

2 METHODOLOGY

This study was solely a qualitative approach.In order to achieve the aim of the study, an extensive review of the literature on Ghanaian construction sector between the periods of 2010 and 2015 was embarked via a desk-top study was embarked which captured the use of existing journal publications, government white papers, and gazette among others. The literature review also captured the construction industry of both developed and developing economies.

3 REVIEW OF LITERATURE

3.1 Characteristics of the Ghanaian Construction Industry

The construction industry in Ghana has similar characteristics as those in other developing countries. The industry plays an essential role in the socio-economic development of a country. The activities of the industry have immense significance to the achievement of national social and economic development goals of providing infrastructure, sanctuary, and employment. The industry deals with all economic activities directed to the creation, renovation, repair or extension of fixed assets in the form of buildings, land improvements of an engineering nature (Khan 2008). Besides, the construction industry generates substantial employment and provides a growth impetus to other sectors through backwards and forward linkages (Ganesan 2000). The industry is dominated by a huge number of small and medium-sized firms that operate mainly in the formal and informal sectors. These categories of firms are widely spread throughout all regional and district capitals of the nation. The construction industry in Ghana has longed depended on middle-aged craft people. As captured by the review of the literature the benefits that an economy gains from the activities of SMEs thus include recruiting large volumes of the population for employment amongst others. The sector is regulated by government ministries, Ministry of Roads and Highways (MRH) and Ministry of Water Resources Works and Housing (MWRWH) in the absence of the construction developed board (cidb) as practised in the advanced economies. According to Ofori (2008), there has been adhoc scheme to help the construction industry in Ghana to develop. Ribeiro and Fernandes (2010) acknowledged that construction industry is organisational complex and highly fragmented with a predominance of small and medium size firms. According to Egan (2002), the issues of organisational complexity and fragmentation of the construction industry are well captured in the Egan (1998), and Latham (1994) reports as the primary attributes of poor performance in the construction sector. In addition, Ofori and Toor (2012) argued that the construction industry is usually described as being one of the riskiest businesses such as lack of job continuity owing to the project-based nature of the industry. Further, Briscoe *et al.* (2001) emphasised that in the construction sector, the network of supply chain members can often be incredibly complex, particularly on larger projects where the number of separate supplying organisations will run into hundreds of folds.

The industry is also characterised as operating in an environment that is nomadic in nature and is affected by the adverse effect of the weather.

3.2 Strength of the Ghanaian Construction Industry

The industry of every country constitutes a substantial part of the economy. Therefore, it is essential that vital construction activity cares for the healthy growth of the economy. Osei (2013) likewise remarked that construction sector of most economies holds immense potential for stimulating growth, boosting project exports and generating employment. Studies have proved that construction contributes between 5 and 10 percent of gross domestic product (GDP) in all countries, employs up to 10 percent of the working population, and is responsible for about half of the gross fixed capital formation (Ofori 2012). Similarly, the sector's activities and its output is an integral part of a country's national economy and industrial development. The sector is often seen as a driver of economic growth especially in developing countries such as Ghana. The industry can mobilise and effectively utilise local human and material resources in the development and maintenance of housing and infrastructure to promote local employment and improve economic efficiency (Anaman and Amponsah 2007). Ofori (2012) further stressed that investments estimated in housing alone account for 2 to 8 percent of GNP; between 10 and 30 percent of gross capital formation; between 20 and 50 percent of accumulated wealth; and between 10 and 40 percent of household expenditure in the Ghanaian economy. In addition to the sector producing structures that add to productivity and quality of life, it has the unique characteristics of being labour intensive. It is evident that construction industry is the pivotal element in infrastructure development. Good infrastructure such as roads construction helps to lower productivity cost hence contributing to economic growth. The industry adds value to the country's developmental agenda through successfully undertaken such projects. Also, the industry loses credibility, trust and reputation with the populaces if projects do not address the needs and expectations of the people. Similarly, the sectors rural developmental projects which are usually embarked by indigenous small and medium-sized construction firms further trickle-down economic benefits to the country and the citizens.

3.3 Weaknesses of the Ghanaian Construction Industry

Although the construction sector as emphasized in literature, plays a vital function among other sectors in terms of the socio-economic development of a nation, there are however challenges that the industry encounters and are considered as its weakness. Challenges in the Ghanaian construction industry as highlighted by Pearce (2003) stressed on the relevance of understanding the status of the construction sector. Ofori (2012) asserted that studies have shown that the construction industries of developing countries, including that in Ghana, face many problems. Ofori, (2012) further categorized the reasons for such challenges into three namely as a result of economic weaknesses of these countries, non-recognition of the construction industry by governments of these countries and finally the inherent underdevelopment of the construction industries in these countries. This becomes difficult for the sector to deal with their weaknesses and to request for assistance, or to contribute to the efforts by the central government in accomplishing its developmental mandate. The problems facing the industry in Ghana are similar to those industries in other developing countries. The industry is faced with the following constraints; considerable maintenance problems and backlog in the road network requiring attention; the need for effective reporting and management information systems; severe management gap owing to the drift of young engineers to other organizations. The industry is also burdened with problems such as low productivity, lengthy pre-contract award procedures, corruption, and delays resulting in time and cost overruns and unsatisfactory quality of work (Ahiaga-Dagbui *et al.* 2011). Corruption has often been touted as a problem in developing countries regarding bribing government officials to win projects (Olken 2009, Olken 2007). The industry is faced with an inconsistent electrical grid, overburdened public water distribution system, poor public sanitation, overcrowded living conditions, and failing infrastructure make both the industry's future success and present state trying to sustain. If not addressed soon would further cause problems resulting in an urban slump which will have long-term detrimental impacts on Ghana's future (Osei 2013). Currency instability is another finance function, which has serious implications for several of the previously explored problem areas. In view of the large proportion of projects and programs which originate from external donors, any changes in exchange rates and prices of goods and services in the course of project implementation may necessitate an upward revision of some project cost elements. Worker training programs within the industry are facing a severe decline in the apprentice system.

4 FINDINGS AND DISCUSSION

Ghanaian context which has similar characteristics as construction industry in other developing economies. The findings also buttressed that the construction sector generates substantial employment and provides a growth impetus to other sectors through backwards and forward linkages as emphasised in the study of Ganesan, 2000. Further small and medium-size firms dominate the industry and operate mainly in either the formal and informal sectors. These SME firms are however spread throughout all the locality of the nation. The Ghana construction industry is acknowledged as an organisational complexity and highly fragmented. Also, the strength of the sector which was captured in the study includes the sector performing as a driver of economic growth particularly with its impacts on the provision of housing and other infrastructure which promotes local employment and improves economic efficiency as confirmed by (Anaman and Amponsah 2007). The results showed that the construction industry adds value to the country's developmental agenda through successfully undertaken such projects. Undoubtedly, the industry loses credibility, trust and reputation with society if executed projects do not address the needs and expectations of stakeholders. Similarly, once the sector embarks on developmental projects, jobs are created and hence there is the minimization of poverty and a direct improvement of the entire GDP of the nation. The study in addition also established some direct and related weaknesses of the industry which has an apparent effect on the sector. The industry is held back with problems such as low productivity, prolonged pre-contract award procedures, corruption, and delays resulting in time and cost overruns and inadequate quality of work as emphasised in the study by (Ahiaga-Dagbui *et al.* 2011). Further, corruption has often been touted as a problem in developing countries regarding bribing government officials to win projects (Olken 2009, Olken 2007). Additionally, the construction industry is also faced with the weakness of currency instability which has serious implications on the cost of estimated imported materials and the overall project budget. More so, the Ghanaian construction industry is characterised by the following weakness including limited access to financial credit challenges in labour management, poor health and safety practices and large span of control in communication among others.

5 CONCLUSION

The import of this study is to explore better and understand the strength and weakness of the Ghanaian of the construction industry. In order to discover the strength and weakness, a holistic overview of the construction industry in Ghana was mapped. This was done in terms of its merits to the economy by providing socio-economic development, provision of employment and job creation as well a reduction of poverty among the populace. Also, conclusions drawn from the literature indicates that construction sector in Ghana contributes between 5 and 10 percent of gross domestic product (GDP) in all and employs up to 10 percent of the working population (Ofori 2012). Further, the conclusion from the study indicates the construction industry is seen as a driver of economic growth. This is done by the construction sector mobilizing local labour and materials for the use of industry. More so, the study ascertained that the sector is healthy with the provision of good infrastructure such as roads construction helps to lower productivity cost hence contributing to economic growth. As well, the construction industry brings about economic stability, brings about Employment and job creation, and brings about a reduction in poverty, Increase in Gross domestic product. Although the construction industry as emphasised in Literature, plays a vital function among other sectors in terms of the socio-economic development of a nation, there are however challenges that the industry encounters and are considered as its weakness. Constraints which are classified as weakness includes limited access to financial credit challenges in labour management, poor health and safety practices and large span of control in communication among others. Though the Ghanaian construction industry is performing and not fully meeting the economic development of the populace, there is still the need for improvement. The study, therefore, concludes that there is the need for the establishment of the construction industry development board in Ghana to supersede all the activities of the industry in order to promote, maintain a high standard construction industry to the Ghanaian economy and society.

References

Ahiaga-Dagbui, D. D., Fugar, F. D. K., McCarter, J. W. & Adinyira, E, Potential Risks to International Joint Ventures In Developing Economies: The Ghanaian Construction Industry Experience. in: Uwakweh B.O (Ed) Proceedings of the CIBW 107 conference on innovation and sustainable construction in developing countries: 1-3 Hanoi Vietnam, November 2011.

Anaman K. A. and Amponsah C, Analysis of the causality links between the growth of the construction industry and the growth of the macroeconomy in Ghana, Institute of Economic Affairs,Accra, Ghana, May, 2007.

Behm, M., Construction Sector, Journal of Safety Research 39 (2008) 175–178, 2008.

Bosher L.S., Dainty A.R.J., Carrillo P.M., Glass J., and Price A.D.F., 'Integrating disaster risk management into construction: A UK perspective', Building Research and Information, 35 (2):163-177 , 2007.

Briscoe, G. H.; Dainty, A. R. J. and Millett, S., Construction supply chain partnerships: skills, knowledge and attitudinal requirements', European Journal of Purchasing & Supply Management, Vol.7, 2001, pp. 243 – 255, 2001.

Egan, Rethinking Construction, A Report on Construction Excellence in the UK.1998.

Ganesan, S., Employment, Technology and Construction Development. Ashgate, UK, 2000.

Government of the Republic of Ghana, Highway Network Master Plan, 2001-2020: Draft Final Report, Executive Summary. Accra, 2000.

Hillebrandt, P.M. Economic Theory and the Construction Industry, 2nd Edition. Macmillan, Basingstoke, 2000.

Khan, R. A., Role of Construction Sector in Economic Growth: Empirical Evidence from Pakistan Economy. 1st International Conference on Construction in Developing Countries (ICCIDC–I), Advancing and Integrating Construction Education, Research & Practice, Karachi, Pakistan, 2008.

Latham, M., Joint Review of Procurement and Contractual Arrangements in the United Kingdom Construction Industry. Final Report, 1994.

Olken, B. Monitoring Corruption: Evidence from a Field Experiment in Indonesia, NBER Working Paper 11753, 2009.

Ofori, G., Developing the Construction Industry in Ghana: the case of a central agency, 2008.

Ofori, G., and Toor, S. R., Leadership vs. Management: How they are different, and why, Journal of Leadership and Management in Engineering, 8(2), 61-71, 2008.

Osei, V., The Construction Industry and its linkages to the Ghanaian economy-polices to improve the sector's performance. International Journal of Development and Economic Sustainability, Vol. 1, No.1, March 2013, pp 56 -72, 2013.

Pearce, D.W. Environment and business: socially responsible but privately profitable? In J.Hirst (ed), The Challenge of Change: Fifty Years of Business Economics, London: Profile Books. 54-65, 2003,

Ribeiro, F. L., & Fernandes, M. T. Exploring agile methods in construction small and medium enterprises: a case study. Journal of Enterprise Information Management, 23(2), 161-180, 2010.

Integrated Solutions for Infrastructure Development
Edited by Lau, H. H., Tang, F. E., Ng, C. K., and Singh, A.
Copyright © 2016 ISEC Press

MANAGING MULTICULTURAL VIRTUAL ENGINEERING TEAMS

RANIA BUSADA and SAMER EZELDIN

Dept of Construction Engineering, The American University in Cairo, Cairo, Egypt

With the global trend towards internationalization, there is a need to understand the parameters that determine the success of the multicultural and multi-located teams. This research is intended to explore the performance of multicultural and multi-located teams operating in Egypt and the various managerial approaches required to enhance these teams' work efficiency. The research employed a qualitative-quantitative experiment on multicultural and multi-located teams in one of the foreign firms operating in Egypt. The groups' performance was analyzed using Porter's Model and a framework was proposed. The proposed framework was verified through three phase experiment. The final framework developed included organizational culture, mechanism for staff recruitment, communication, leadership styles and decision making approaches. This framework is intended to assist in enhancing the operation of the multicultural and multi-located design engineering teams in Egypt.

Keywords: Multiculturalism, Team management, Cultural diversity, Leadership.

1 NEED FOR THE RESEARCH

Various studies indicated that the multicultural engineering teams have a significant impact on productivity like the work of Ng and Tung (1998). Other researchers like Watson *et al.* (1993) noticed that multicultural team's performance can be 15% higher than that of the homogenous team only if the former is well managed. Managing multicultural teams enforce various parameters on the management like (i) handling geographic distances and dispersion of teams, (ii) managing cultural diversity, (iii) understanding the differences and conflicts, (iv) developing team cohesiveness, (v) maintaining communication richness, and (vi) dealing with coordination and control issues. Another approach to manage the multicultural groups is to benefit from the diversity through allocating every individual in the right position utilizing their skills like problem solving, decision making, marketing expertise etc. (Cox 1991). Most of the studies conducted focused on the European, American and the Asian market and few focused on the MENA region. The aim of this research is to analyze the management techniques required for the multicultural and multi-located teams operating in Egypt and to propose a framework to manage these teams effectively.

2 RESEARCH METHODOLOGY

The research methodology was to conduct a qualitative experiment on Firm A, a Saudi based consultant that outsources its GCC (Gulf cooperation council) work to its branch located in Egypt. The data gathered was analyzed and the parameters generated were ranked and examined using the Porter's (Diamond) Model, analytical hierarchy order (AH) and Relative Importance

Variable (RIV) values. These parameters were validated using external validation approach by conducting a quantitative survey and a framework was proposed which was verified through three-phase experiment.

2.1 Process of the Qualitative Experiment

A total of six groups were formed (table 1): two based in Egypt, two in Saudi Arabia and two virtual groups who worked online. This research was conducted assuming that certain parameters - age, gender, and employee's year of experiences- have negligible influence on the multicultural teams' performance.

Table 1. Data of the groups participating in the quantitative experiment.

Group	Nationality	Position at Firm A	Group 's Type		Communication Approach
A	Egyptian Egyptian Egyptian	Design Team Leader Structural Engineer Coordinator	Monocultural Team	Traditional Team	Face-to-Face
B	Egyptian Egyptian Lebanese	Design Architect Senior Structure BIM Coordinator	Multicultural Team	Traditional Team	Face-to-Face
C	Egyptian Lebanese Syrian	Design Architect BIM Coordinator MEP Engineer	Multicultural Team	Virtual Team	Online
D	Philippines Philippines Philippines	Design Architect Structural Engineer Technical Coordinator	Monocultural Team	Traditional Team	Face-to-Face
E	Egyptian Lebanese Philippines	Architect Engineer Structure Engineer Coordinator	Multicultural Team	Virtual Team	Online
F	Jordanian Lebanese Philippines	Coordinator Structural Engineer Design Architect	Multicultural Team	Traditional Team	Face-to-Face

In order to study the performance of these multicultural multi-located teams, the following experiment was conducted:

(1) The design brief (including the client's requirement, the project objective, mandatory technical specifications and the area requirement) was given to the architect. The architect worked on the conceptual plans and the engineer (structural/MEP) functioned on the regulations, codes and technical specifications. Concurrently, the coordinator was in charge of compiling the work together and verifying the output in accordance with the design brief.

(2) The time required to finalize a decent conceptual approach was assumed to be five hours. However, according to Tuckman (1965) insufficient time should be allocated while testing a hypothesis in order to analyze the prioritization of the group. Therefore, a total of four hours was allocated for the completion of the project.

(3) All team members were asked to work on the final presentation to encourage: (i) cohesive output, (ii) promote decision making techniques, and (iii) clarify the leadership style used among the team members.

2.1.1 Findings of the qualitative experiment

The experiment conducted was observed and the findings are as following:

(1) Organizational culture: Groups (D and F) located at the company's headquarter showed a clear understanding of the company's organizational culture, communication, and work methodology. The output of the groups (A and B) located in Egypt showed that the company's organizational culture is not well-defined and the team members are either working with previous managerial approach or influenced by their cultural diversity. The virtual teams (C and E) showed different responses with unclear output and inconsistent communication. Among the six groups, the multicultural and multi-located teams' performance was the least impressive.

(2) Communication: The main challenges that encountered the multicultural teams were: (i) the inability to find a platform for exchanging and documenting information and (ii) unsuccessful attempt to translate the client's requirement to the team members located overseas. Another issue was the presence of grapevine communication which increased the informal negative communication between the employees and the outsiders.

(3) Multicultural teams: The rise of the cultural differences along with the regular technical discrepancies caused tension among the team members. This issue was highly notable among the virtual teams (Group C and E).

(4) Decision-Making: The teams located at the company's headquarter were used to having the upper hand in the decision making process because of their direct contact with the client. This discouraged the managerial skills and decision making of the engineers at the Egyptian branch which added up to the tension between the multi-located teams and explained the inefficient performance of the teams located at this branch.

3 PROPOSED FRAMEWORK AND VERIFICATION CASE STUDY

The group performance was analyzed using the four conditions (Factor, Demand: Internal and External, Related Industry and Firm's strategy) of the Porter (Diamond) Model. Parameters for the four conditions were generated and compared using AH and RIV value which were validated using a quantitative experiment (survey). The survey targeted 62 Engineers working in multicultural firms in Egypt with more than 10 years experience. The proposed framework included the parameters that (i) scored above the condition's average score and (ii) graded four or five by 50% of the responses. These parameters with the descripted managerial approach form the proposed framework which is to be tested in section 3.1. The proposed framework is:

(1) Organizational style: In the Egyptian Design firms, the organizational culture is wrongly classified as both mean-oriented and employee-oriented which confuses the employees as to whether to be risk takers or not. Other indicators are the open system, easy working discipline, and external driven factors.

(2) Mechanism of staff recruitment: The technical experience is the main criteria for selection.

(3) Leadership style: The Egyptian engineer performs well under a participative-supportive and bureaucratic leadership style.

(4) Decision making approach: Both command and consent are powerful decision making approaches among the multicultural Egyptian design teams.

(5) Communication: Although various time breaks are scheduled to be included during various project's phases, the actual status usually includes an unorganized communication scheme between the disciplines. Among the virtual teams, the communication scheme is harder due to the absences of a common platform for documentation and collaboration.

3.1 Verification of the Proposed Framework

The proposed framework was verified using inspection method (case study). The proposed framework was tested on the same team members who were involved in the qualitative experiment.

During the first phase of the experiment, multicultural teams had low performance as compared to monocultural teams. However by the second phase of the experiment, the multicultural teams were able to identify approaches to overcome their cultural differences. Group F was able to use the diversity to resolve various discrepancies which is compatible with the research conducted by Comu *et al.* (2011) who showed that the multicultural teams are able to introduce various tactics to resolve discrepancies. Virtual Teams like group C which encountered major challenges due to cultural diversity during the first phase were able to elevate their performance after introducing job responsibilities and allocating the leadership task to a specific team member. The monocultural teams, group A and group D, had consistent outputs throughout the two phases of the experiment and no improvement was noticed in their ways of resolving discrepancies and means of communication. A third phase of the experiment was conducted after two days.

The performance of the six groups was evaluated and scored based on the key parameters of the proposed framework (except the mechanism for staff recruitment since the groups were formed by the researcher). Each of the five parameters was scored out of 50 by the technical manager of the company who evaluated the teams based on the workability of the project, the final output, the collaboration throughout the allocated four hours, organizational culture, leadership style, communication approaches and decision making techniques. The standard deviation, mean and p value were calculated using R program.

Although initially the monocultural teams had higher performance, the second and third phase showed that the multicultural teams were higher in terms of their efficiency and consistency with company's organizational culture. On the leadership level, according to Hofstede (1980) studies, the Egyptian society is a highly emotional society and thereby using leadership techniques like participative and supportive is encouraged. Groups like A, D and F used the participative leadership approach during the first phase and the second phase. The groups that had difficulties applying this leadership approach were B (multicultural team) and C (multicultural and multi-location team). In order to know whether the participative-supportive leadership approach is the most suitable style for the team members, a short questionnaire was developed which showed a high acceptance of the participative-supportive major roles with R value of 0.93. On the other hand, the virtual group showed relatively low acceptance to the participative-supportive approach and high acceptance to the bureaucratic leadership style with R value of 0.91.

Table 2. The statistical analyze showing the performances of the six groups.

Groups	Project 1	Project 2	Project 3	SD	T-test	P value	Mean
Group A	190	180	150	20.81	14.42	0.0047	173.3
Group B	200	220	240	20	19.05	0.0027	220
Group C	100	130	170	35.11	6.57	0.0223	133.3
Group D	240	240	210	17.32	23	0.0018	230
Group E	150	190	210	30.55	10.39	0.009	183.3
Group F	180	220	240	30.55	12.09	0.0067	213.3

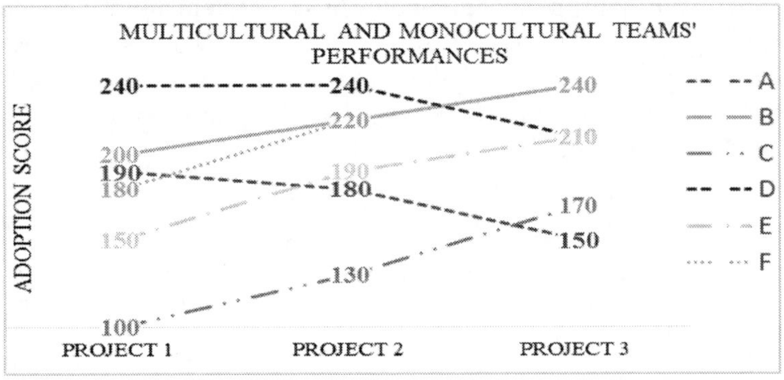

Figure 1. Performance of the Multi and Mono cultural teams throughout the three phase experiment.

4 FRAMEWORK DEVELOPED

Based on the three phases of the experiment conducted above, the final framework includes the following five parameters with a more specified managerial approach, adjusted based on the enhancement in the performance of the teams:

(1) Organizational culture: The organizational culture should facilitate the team development activities and define approaches to enhance the productivity of the multicultural teams. Open system, employee oriented methodology, clear hierarchy, detailed job responsibility- including the managerial and technical responsibilities- should be clarified. A follow-up orientation should be created for new employees to familiarize them with the company's regulations.

(2) Mechanism for staff recruitment: Specific criteria should be specified while composing the team, including individual profiling, technical experience, value of money candidate, and the ability of candidate to adapt with the company's culture.

(3) Leadership style: The leadership style to be followed can be a combination of participative and supportive styles to ensure the multicultural teams' efficiency. Virtual teams need a more bureaucratic leadership style due to the absence of face to face communication. The leaders of the industry should also consider the task's culture while exercising their duties. While working with multicultural teams, the individuals should be encouraged to lead a phase of the project. The tasks are to be allocated based on the team member's initiatives which are to be determined from the individual's profiling and culture.

(4) Decision making approach: Both command and consent are powerful decision making approaches. In case of command style, it is advised that the decision should be taken by involving all team members either through consent or majority voting.

(5) Communication: Various means of communications should be introduced for multicultural and multi-located teams which include forms, regular meetings, online discussions, and decision-making sessions. Grapevine communication is to be suppressed through communicating the difficulties and constraints faced through a fixed form system.

5 CONCLUSIONS

The qualitative experiment confirmed the hypothesis of implementing an amended managerial approach while managing multicultural teams. The framework developed is verified and validated to ensure its generalization on the Egyptian design engineering firms. The main parameters are: Organizational culture, mechanism for staff recruitment, leadership style, decision making, and communication techniques.

References

Comu, S., Taylor, J., and Messner, J., Two-Dimensional Globalizing Index for Engineering and Construction Companies, *American Society of Civil Engineers*, (2011), 10.1061/(ASCE)ME.1943-5479.0000234, 04014031.

Cox, T. H. (1991) Managing Cultural Diversity Implications for Organizational Competiveness, *Academy of Management*, 5(3), pp. 45-56, August, 1991.

Hofstede, G (1980) Culture's Consequences: International differences in work related values. Sage, Beverly Hill, CA.

Ng, E.S.W. and Tung, R.L., Ethno-cultural Diversity and Organizational Effectiveness: A Field Study, *International Journal of Human Resource Management*, Vol.9,No.6 ,pp.980-995, 1998.

Tuckman, Bruce W. "Developmental Sequence in Small Groups." *American Psychological Association* 63.6 (1965): 384-99. Web.

Watson, W. E., Kumar, K., & Michaelsen, L. K., Cultural diversity's impact on interaction process and performance: Comparing homogeneous and diverse task groups. *Academy of Management Journal*, 36: 590–602, 1993.

Integrated Solutions for Infrastructure Development
Edited by Lau, H. H., Tang, F. E., Ng, C. K., and Singh, A.
Copyright © 2016 ISEC Press

INVESTIGATION INTO WH&S ISSUES WITHIN THE RESIDENTIAL CONSTRUCTION INDUSTRIES MIGRANT WORKFORCE

SWAPAN SAHA and ANDREW MANTOUFEH

School of Computing, Engineering and Mathematics, University of Western Sydney, Penrith, Australia.

The migrant workforce is a prominent feature in the Australian construction industry and is equivalent to approximately 24% of the industry. Out of these migrant workers, 11.8% comes from English speaking countries and 12.2% from non-English speaking background (NESB) countries. The aim of this study is to investigate the issues surrounding migrant workers within the Australian residential construction industry. A number of recent studies have shown links between the increasing migrant dominated workforce and a subsequent deterioration of safety standards and quality levels of construction trade work. The main purpose of the study is to further explore this issue, in an attempt to expand on previous research and to identify the reasons behind these trends. The methodology of the research undertaken included a case study of a NSW residential construction companies. The case study applied the theory of triangulation through a range of structured formal interviews. These interviews were carried out on a range of people occupying various roles within the construction industry including site supervisors, WH&S officers and top level management. Research findings have shown considerable challenges existing today amongst the migrant workforce. Communication barriers and cultural differences are shown to have created significant challenges with regards to the safety of migrant workers and the quality assurance related to various construction trades.

Keywords: Constriction safety, Migrant workers, Australia, Quality assurance.

1 INTRODUCTION AND BACKGROUND

In 2014-15, direct construction industry output contributed 7.8% to Australia's Gross Domestic Product (GDP). This makes the construction industry Australia's third largest contributor to GDP in terms of the volume of its output. Only the financial and insurance services (8.7%) and mining (8.8%) sectors contribute larger shares of direct output to GDP. The construction industry employed 1.05 million employees (around 9% of all workers) in May 2015 (Australian Industry Group 2015). In 2009-10, the construction industry reported one of the highest OH &S incident rates in comparison to any other industry and ranked as the fourth highest contributor of serious incidence claims among all industries. Safe Work Australia's workers compensation data for 2009–10 show there were 13,640 serious worker's compensation claims in the construction industry in that period. This figure accounts for 11% of all serious claims and equates to 37.4 serious claims per day. This is equivalent to 19.6 serious claims per 1000 employees requiring one or more weeks off work because of work-related injury or disease. In terms of hours worked, the frequency rate in the Construction industry was 9.8 serious claims per million hours worked, 29% higher compared to the average across all other industries frequency rate of 7.6 (Safe Work Australia 2012- construction fact sheet). Work related injuries, illnesses and deaths impose costs on employers, workers and the community. These include both direct costs and indirect costs. Direct costs include items such as workers compensation premiums paid by employers or payments to injured workers from workers compensation jurisdictions. Indirect costs include items such as lost productivity, loss of current and future earnings, lost potential output and the cost of providing social welfare programs for injured workers. The level of costs varies with the severity of the injury or disease (Johns 2012). While measures of direct costs are understood and reasonably simple to measure, these costs cover only a fraction of the total cost of work related injury and disease. According to Safe Work Australia (2012), workers compensation incurred an economic loss of approximately 57.5 billion for the 2005-06 financial years within Australia. More recently work related injury and illness were estimated to cost $60.6 billion in the 2008–09 financial years. The Australian construction industry has sought to contain and reduce costs. This has resulted in construction corporations to focus on project management and contract out the actual physical work of construction. The outcome has been an extended division of labour, and one aspect of this has been that many subcontractors are hiring migrant workers in order to source skill requirements and contain or reduce labour costs (ILO 2012). Past literature has established a link between the migrant workforce and the construction industry's high accident rates (Trajkovski and Loosemore 2006, Loosemore and Andonakis 2007, Salleh *et al.* 2012). The migrant workforce is a prominent feature of the Australian construction industry, equivalent to approximately 24% of the industry with 11.8% originating from main English speaking countries and 12.2% from non-English speaking background (NESB) countries (Australian Bureau of Statistics 2012- Labour Force). However, it has been identified there is also a significant number of illegal immigrants working in the industry that the above figure does not account for (Wallace 2011).

It is suspected that a combination of communicative and cultural barriers, amongst non-English speaking migrant workers, has been detrimental to safety and quality standards within the residential construction industry. Recent literature has shown links between an increasing migrant dominated workforce and a subsequent deterioration of safety standards and quality levels (Rosewarne *et al.* 2012). This study further explores this issue, in order to expand on previous research and to identify the reasons behind these trends. The need is created for an investigation into work health and safety issues within the Australian residential construction industry's migrant workforce.

2 OCCUPATIONAL HEALTH AND SAFETY IN THE AUSTRALIAN CONSTRUCTION INDUSTRY

The construction industry has one of the highest OH & fatalities in comparison to any other industry and is ranked as the third highest contributor of fatalities among all industries as per Figure 1 (Safe Work Australia 2014). Transport, postal and warehousing industry recorded the highest proportion of fatalities in 2014 with the construction industry accounting 15% of all workers fatalities.

In 2012 workers statistics relate to 73% excluding self-employed workers within the construction industry who are eligible to lodge a worker's compensation claim (Australian Bureau of Statistics 2012- Labour Force).

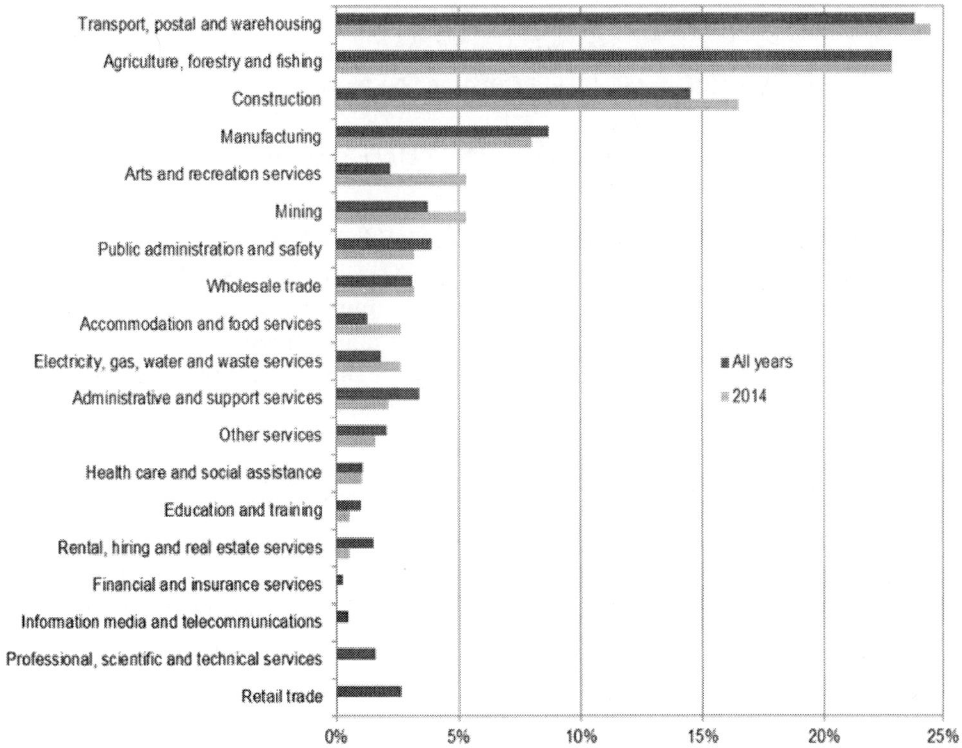

Figure 1. Worker fatalities: proportion by industry of employer, all years (2003 to 2014 combined) and 2014 (Source: Safe Work Australia 2014).

Over the 2003–13 period 77% of the fatalities in the construction industry involved employees with the remainder self-employed workers. The 311 deaths of employees in the construction industry over the 2003–13 period equates to a fatality rate of 4.20 fatalities per 100 000 employees. This is 37% higher than the fatality rate for self-employed workers of 3.06 fatalities per 100 000 self-employed workers over the 2003–13 period. Every four years, The Australian Bureau of Statistics (ABS) surveys workers on work-related injuries sustained in the previous year. The last Work-related Injury Survey (WRIS) conducted by the ABS was for the 2013–14 period. Unfortunately the results from the survey are not yet publicly available for analysis.

The last WRIS survey available is for the 2009-2010 period. The survey reported that 57000 construction workers had suffered an injury while at work. This means approximately 156 construction workers were injured each day or 5.9% of the total construction workforce (Safe Work Australia, 2016). Main injuries incurred were: i) muscular stress a result of manual handling or repetitive movement(35% of claims) ii) falls, trips and slips (26% of claims) and iii) being hit by moving objects (16% of claims). Furthermore, research on an international measure show statistics on incidence rates between different countries vary in range and the construction industry is recognised as one of the most dangerous industries to work in. This could be due to different rates of economic growth and developments. The types of developments of a particular country may have an influence on the incidence rates, as every project encounters its own individual risks, OH&S issues and requirements. The countries standards and OH&S requirements may be different and therefore no universal OH&S method exists. This contributes to different methods of prevention, construction sequences and rates of injury and fatalities in the construction industry worldwide (ILO 2012). Therefore, those who come from other countries to work in the Australian construction industry may adopt the standards of their home land country which could lead to differences in safety and construction works.

3 MULTICULTURALISM/DIVERSITY IN THE AUSTRALIAN CONSTRUCTION INDUSTRY

Multiculturalism is an increasingly prominent feature of the Australian construction industry. It is defined as communities that contain a number of different cultures. When different cultures interact it provides opportunities for the cultures to communicate with each other to create multiculturalism. Australia is a multicultural country and implements a multicultural policy which promotes equality, fairness and inclusion for all. It supports a variety of cultural, religious and linguistic diversity, shared values and cultural traditions with the law and free from discrimination (Department of Immigration and Citizenship 2011-Australia's Multicultural Policy). Therefore, steps must be taken to ensure safer working environments for all. According to the government of south Australia multiculturalism is defined through policies and practices that identify and respond to ethnic diversity of communities where members of the community may maintain distinctive cultural heritages, participate and use their skills to benefit the economic, social and cultural life of the community and live and work together harmoniously. The Australian multicultural policy suggests Australia is very diverse and multicultural. The statement below summarises this: "Australia is a multicultural nation. In all, since 1945, seven million people have migrated to Australia. Today, one in four of Australia's 22 million people were born overseas, 44 per cent were born overseas or have a parent who was and four million speak a language other than English. We speak over 260 languages and identify with more than 270 ancestries. Australia is and will remain a multicultural society." (Department of Immigration and Citizenship 2011-Australia's Multicultural Policy).

4 RESEARCH METHODOLOGY

A case study method has been completed on a NSW residential construction company in order to grasp on past research and the current situation. Furthermore, a multiple perspective observation method is undertaken through interviews on various construction personnel within the residential construction industry. The main purpose of the research is to gain first-hand knowledge and experience in relation to WH&S on a typical residential construction company. The research will also enable data collection and insight into the current situation to fulfil the research topic and satisfy the objectives of the research aims. Results has been analysed and tabulated in order to see the current trends and overview of the residential construction industry. This would show the first-hand experience of various construction personnel including site supervisors, WH&S officer, top level management and the experiences they come across when dealing with the migrant workforce. Emphasis has been placed on work subcontractor trades that do not mandate the completion of an apprenticeship including the trades of painting, plastering, tiling and final clean. An examination is conducted on issues construction personnel face with WH&S in consultation with migrant workers over the mentioned trades. Through this study, improvements in the way WH&S strategies are conducted to migrant workers has been be assessed and suggested, and implementation procedures to correct and eliminate the current problematic areas will be identified through formed suggestions and recommendations.

5 FINDINGS AND CONCLUSION

The findings from the case study have reiterated that the residential construction industry is a part of a multicultural workforce. In this study numbers of participants were interviewed including site supervisors, top level managers and WH&S officers. Findings suggest that due to the workforce diversity, a number of problems are apparent related to WH&S issues. The case established that communication is the major problem and this has a number of implications within the industry. The established language barrier results in problems associated with communicating safety requirements, simple work instructions, conducting site inductions and training. Further problems included the workers resorting to cultural norms by means of not conforming to safety requirements through the use of PPE, safe tools and resorting to unsafe work practices. Many reports of miscommunication, non-compliant work and incidents have been made apparent. Construction personnel record confusion, stress and frustration when working with Non English Speaking (NES) workers however advantages include large groups completing works fast, available to work anytime with speed. Various construction personnel do not have efficient means of dealing with the migrant workforce and they resort to the use of third party communication and documentation to aid the language barrier. Furthermore, findings established the use of new technology to aid in communication and language barriers (e.g., Apple iPhone translator application and Google translate). The result findings demonstrate deteriorated safety and quality assurance within specific areas of the residential construction industry, particularly in trades such as painting, tiling and plastering.

References

Australia Industry Group (2015), *Australia's Construction Industry: Profile and Outlook July 2015*, Retrieved from http://cdn.aigroup.com.au/Economic_Indicators/Construction_Survey/2015/Construction_industry_profile_and_Outlook.pdf on August 18, 2016.

Australian Bureau of Statistics (2012), Labor Force, Australia, Detailed, Quarterly, May, cat. no. 6291.0.55.003, Retrieved from http://www.abs.gov.au/AUSSTATS/abs@.nsf/allprimarymainfeatures/9C28A5AD6C1BFA1FCA257A77001B3F99?opendocument on August 8, 2012.

Department of Immigration and Citizenship (2011), *The people of Australia- Australia's Multicultural Policy*, Retrieved from http://www.immi.gov.au/media/publications/multicultural/pdf_doc/people- of-australia-multicultural- policy-booklet.pdf on August 4, 2012.

International Labour Organisation (ILO) (2012), The enormous burden of poor working conditions, Retrieved from http://www.ilo.org/moscow/areas-of-work/occupational-safety-and-health/WCMS_249278/lang--en/index.htm.

Johns, B. (2012), The Cost of Incidents to Australia's Building Contractors, thesis, Penrith, University of Western Sydney.

Loosemore, M & Andonakis, N. (2007), 'Barriers to implementing OHS reform the experiences of small subcontractors in the Australian construction industry', *International Journal of Project Management*, vol. 25, pp. 579-588.

Rosewarne, S., Shin, J. S., McGrath-Champ, S., & Toner, P., (2012), 'The globalisation of the construction workforce, *A report on the impact on the Australian building and construction industry'*, Sydney University, pp. 1-17.

Safe Work Australia (2012), Construction Fact Sheet, Jan 2011, Retrieved from http://www.safeworkaustralia.gov.au/sites/SWA/AboutSafeWorkAustralia/WhatWeDo/Publications/Pages/FS2010ConstructionInformation Sheet.aspx on August 2, 2012.

Safe Work Australia (2012), Workers Compensation Data, Retrieved from http://www.safeworkaustralia.gov.au/sites/swa/workerscompensation/data/pages/wcdata.aspx on September 8, 2012.

Safe Work Australia (2014), Work-related Traumatic Injury Fatalities Australia 2014, Retrieved from http://www.safeworkaustralia.gov.au/sites/swa/about/publications/pages/work-related-traumatic-injury-fatalities-australia-2014 on August 18, 2016.

Safe Work Australia (2016), Work-related injuries and fatalities in construction, Australia, 2003-2013, Retrieved from http://www.safeworkaustralia.gov.au/sites/swa/about/publications/pages/fatalities-in-construction on May 16, 2016.

Salleh, N. A. B., Nordin, N. B. M., and Rashid, A. K. B. A., (2012), 'The language problem among foreign workers in the Malaysian construction industry'. *International Journal of Business and Social Science*, vol. 3, no. 11, pp. 97-99.

Trajkovski, S, & Loosemore, M. (2006), 'Safety implications of low English proficiency among migrant construction site operatives', *International Journal of Project Management*, vol. 24, pp. 446-452.

Wallace, N. (2011), Sydney Morning Herald , 'Illegal workers rife in construction industry, union claims', Retrieved from http://www.smh.com.au/nsw/illegal-workers-rife-in-construction-industry-union-claims-20110724-1hvgo.html#ixzz2K0Jt8Dah on February 4, 2013.

Integrated Solutions for Infrastructure Development
Edited by Lau, H. H., Tang, F. E., Ng, C. K., and Singh, A.
Copyright © 2016 ISEC Press

BUILDING CONTROL AND REGULATORY BALANCE: PERSPECTIVES OF BUILDING PROFESSIONALS

JEFF CLEMENT SAMASONI and JAMES OLABODE BAMIDELE ROTIMI

Construction Management Programme, School of Engineering, Computer and Mathematical Sciences, Auckland University of Technology, Auckland, New Zealand

The New Zealand government proposed a new control regime to streamline the process of building approval by introducing risk-based inspections for certain types of building works. With this regulatory balance could be proportioned to stakeholders involved in the building process. A review of the Building Act in 2010 had indicated that a more balanced approach to building control is required to more appropriately allocate responsibility, accountability and liability between construction stakeholders. A larger study on which the current study is based, evaluates stakeholders perception of the newly introduced risk-based building inspection scheme and on regulatory balance issues. Building inspectors expressed concern in the shift in balance, as current regulatory inspections had a high proportion rate of failures involving competent building practitioners. However the scheme provides building regulators a tool to accelerate building processes without compromising cost and quality.

Keywords: Building regulations, Government, Housing, Liability, Risk, Inspections.

1 INTRODUCTION

In August 2009 the New Zealand government announced a review of the Building Act 2004 and identified issues that there were problems with ensuring responsibility sat in the right place and undue reliance on building consent authorities (BCA) (Williamson 2010a). The building consent authority post the leaky home saga in the late 1990s have been risk-averse with the consenting process requiring more documentation and checks and the addition of more inspections which has been applied to every project that is lodged with council. An example of this process would be a housing developer firm providing the same documentation on it stock standard house design which has been through the BCA numerous time but still require further information depending on which BCA officer was processing or inspecting it (May and Wood 2003). The added pressures of the Auckland house shortages (Auckland Council 2013) and the Christchurch rebuild (MBIE 2013) has forced the government to take action. The government's announcement particular targets BCA in relinquishing its control and oversight of building work and proportionally moving risk to other professional e.g., architects, engineers and licensed building practitioners who are in the best position to assess it. The rebalancing of the building consent process will be a major change in liability shift as BCA input into certain projects would be dramatically reduced and professionals picking up the shortfall.

The government focus on achieving quality homes, through building activities that is business-enabling and within an efficient regulatory framework. In other words, construction productivity that is conducted by competent personal which is value for money while being streamline through a user friendly business model. It also targeted consumers' confidence that they are able to make informed decisions in carrying out transactions in the building and housing market (DBH 2009).

A report by the Building and Construction Sector Productivity Taskforce (DBH 2009) also supports these findings in that productivity within the construction industry is relatively lower than those in other sectors of the New Zealand economy and part of its recommendation was developing quicker regulatory and consenting processes to reduce the barriers and roadblocks that urban development's face.

2 LITERATURE REVIEW

2.1 Ensuring Responsibility Sit in the Right Place

It is a well-known fact in New Zealand that responsibility and accountability for ensuring building quality and reducing the number of defects is not well aligned to the ability of each key construction stakeholder to effectively make an impact. The lessons from the leaky home saga in which the government and homeowner were usually "the last man standing" in picking up the repair bill indicated that responsibility and accountability were not fairly distributed to those best to manage risk. The government financial aid package (FAP) for leaky home owners is an example of the distribution of liability through repair funds where 25% is summoned to central government, 25% from local government (Subject to that they signed the final compliance certificate) and 50% by the owner. It is evident through this process that other key stake holders (designers, builders and product manufactures) are not part of this claim as many are known to shut down their business which have been subject to a claim and restart a new one which is a major problem in the New Zealand construction industry.

The roles of the different stakeholder and building consent authorities are not always clearly defined or understood (Massey 1999). The current regime there is strong expectation on the role of building consent authorities in reviewing and inspecting consented plans to protect consumers from defective building work, even when the risk and consequences of failure are low (May and Burby 1998). This has reverted councils into a risk-averse approach because the "duty of care" *Invercargill City Council v Hamlin*[1], imposed by the courts on local authorities in respect of residential homeowners because of their statutory responsibility to issue building consent, carry out inspections and issue code compliance certificates. Many stakeholders which had been implicated in negligence cases relating to leaky buildings (May 2003) avoided payment of the proportion of damages due to creating limited-life companies. As a result of these parties unable to contribute to their part of the cost these were deferred to those who could which were the local authority and their rate payers. The government review in 2009 emphasis that dependence on building consent authorities is out of balance with their ability to influence building quality through documentation and inspection, and their capacity to do so without considerable cost. It has the consequences of imposing higher than necessary costs. Themes that were constant through the review were excessive requests by building consent authorities for documentation and plans which contributed to slow processing of consents. A large number of inspections during the course of construction, excessive in the case of simple buildings and a reluctance by building

[1]*is a cited case in New Zealand regarding council liability for negligent inspection, as well the issue in tort when the start period for the statute of limitations for a latent defect begins*

consent authorities to approve new or alternative building designs, products or processes (DBH 2009).

2.2 Moving to a More Balanced Approach to Building Control

Another of the government's objectives of the proposal would see BCA targeting their oversight at buildings and building work where the risks and consequences of failure are most significant. Some of the options are:

- Building professionals, trades people and owner-builders would be able to undertake more of the lowest-risk work without needing building consents.
- A more streamlined process for low-risk (simple residential) buildings when the work is undertaken or overseen by licensed building practitioners.
- In the commercial sector, a more streamlined process for complex commercial building work that acknowledges the commercial risk-management and quality-assurance processes already in use.
- Importantly, retaining the current building consent system for more complex and less conventional residential building, and some commercial buildings, because of the higher risk and greater consequences of failure (Williamson 2010b).

2.3 Risk-based inspection

There is no single accepted or correct definition of risk. Rothstein *et al.* (2006) perceives risk-based regulation as allocating resources in proportion to risks to society (such as health, safety or environmental risks), in which the consequences of it happening and the overall impact are considered, in order to establish appropriate levels of control. Imrie and Street (2009) elude to the development of building controls as an example of the modernist conception of risk, whereby state directives set out minimum standards of building performance, based on a series of written rules.

Current regulations are increasingly seen as barriers to market liberalization and as a result there is an international trend towards reducing government responsibility for building quality and many countries are attempting to simplify their building control regimes, often through a combination of deregulation and the shifting of responsibility and in some cases balance to the private sector (Yau 2009). This is such the case where one large regulator the environmental agency for England and Wales, has responded to such pressures by implementing risk-based approaches to regulations (Gouldson *et al.* 2009).

Risked based inspections is an established practice in a number of areas of United Kingdom health, safety and environmental enforcement, with examples found in the enforcement of food safety, Hobbs *et al.* (2002) fire safety, Ramachandran (1999), and occupational health and safety, Tombs and Whyte (2013), In each case the frequency of inspection and associated enforcement actions are based on a combination of the risk posed by the activity and the standard of management.

The Department of Communities and local government (DCLG) in the UK has also introduced a risk-based inspection tool to their building control departments (DCLG 2012a). The risk-based inspection systems include a minimum number of phased inspections for all buildings, they typically give priority to buildings with high risk, such as environmental ones and optimize the process. Having fewer inspections for less risky buildings lowers costs without compromising safety, increasing flexibility and enabling inspectors to move away from random and phased inspections (DCLG 2012b).

Risked based inspections in New Zealand will be in the form of allocating inspections to those building projects that are at a high risk of failure and reducing or exempting inspection for those with lower risk of failure.

3 THE RESEARCH

3.1 The Research Objectives

The literature review has provided some background analysis into the movement of balance between key stakeholders in the building controls process in New Zealand, and looks at those key professional parties that will be affected. It also explores the concept of implementing a process where those who design and construct buildings and building work will need to stand by their work with the reduced over sight of regulatory checks therefore taking on more responsibility and accountability if anything should go wrong.

The introduction of risk-based inspection also provided building consent authorities a tool to focus their resources on the more complex projects and eliminate the need for simple building designs and those of low risk failures. The review concluded that there is a need to examine the reduction of regulatory involvement into the building consent process and the utilization of risk-based inspection (Williamson 2010) as a BCA tool in New Zealand. Thus the following research objectives are suggested which are also in line with the government recommendations:

(1) To promote a regulatory system that is administered in an efficient and cost effective manner

(2) To ascertain problems with ensuring responsibility sits in the right place

(3) To establish an undue reliance on building consent authorities

3.2 Research Methodology

This study is undertaken to provide an understanding of proposed changes to the building consent process in New Zealand and its effects with its key stakeholders.

The main form of data collection was in the form of semi-structured survey which went out to building designers and building control practitioners in New Zealand. Deliberate sampling was targeted at areas which had the largest volume of work through their Building Consent Authorities throughout New Zealand (Auckland, Hamilton, Wellington and Christchurch) for this not only covered the majority of the population but also covered the geographic makeup of the country. Building designer were also deliberate targeted which were members of the New Zealand Institute of Architects (NZIA) as majority of their members were registered architects. Interviewee diversification is essential to the quality of data sourced in quantitative and qualitative research, therefore all participants that had be engaged for the research will have practical understanding of the building consent process and be competent in their field of professionalism. All surveyed questionnaires responses are analysed through SPSS and open ended question through the NVivo program.

4 FINDINGS

Table 1 presents results from building professionals in which survey question asked "Does there need to be a review on the current number of inspection for a new residential dwelling". The findings in table 1 indicate that building professionals (70.25%) overwhelming agree that the current regulatory inspection required less regulatory intervention which aligns to the UK Department of Communities and local government initial intention for regulatory change to their building controls scheme for reduced inspections.

Table 1. Building professional's perspective on review of current regulatory building inspections.

Responses	Percentage %	Count	Total
Yes	70.25	88	
No	18.85	22	
Not sure	10.9	16	
			126

Table 2 presents results from building professionals in which survey question asked "Is the New Zealand construction industry ready for current regulatory inspections to be relaxed in favour of a risk-based assessment approach". The findings in table 2 indicate that nearly 60% of building professionals do not think that the relaxation of regulatory inspections in favour of a risk-based assessment approach is a positive move as more stringent regulatory governance of construction stakeholders need to be in place before responsibility is shifted.

Table 2. Building professional's perspective on a risk-based assessment approach.

Responses	Percentage %	Count	Total
Yes	24.5	34	
No	58.85	72	
Not sure	16.7	20	
			126

5 CONCLUSION

The research seeks a better understanding of the perceptions of building professionals towards regulatory balance and the implementation of risk-based inspection as a building consent authority tool. The stigma of the leaky home is still evident in the post consenting processes of building control authorities in New Zealand and the intentions of the 2004 building act were warranted but in the same token has refrained innovation and productivity. The government and construction working group through the Building Amendment Act 2010 and 2012 has recognized these bottlenecks and have proposed new objectives to get the construction industry producing to its full potential. The findings show that a majority of building professionals in New Zealand agree that there are too many regulatory inspection notification in the current system. This affirms the fact that post the leaky building crisis that building consent authorities have been over zealous in confirming regulatory compliance during consenting and construction phases.

Majority of building professionals also disagreed that the relaxation of regulatory inspections in favor of risk-based inspections is a good idea. Further commentary suggest that New Zealand construction industry is not mature enough and until the government regulated an insurance bond for licensing building practitioners who produce non-substantial work a repeat of the leaky home saga would be imminent. The conclusion of the programme study suggests that the current building regulatory inspection regime definitely requires a review but more education and understanding of risk-based inspection to all key stakeholders is needed if this is to be used as a building control authority tool.

References

Auckland Council, The Auckland Plan 2013, Housing actions plan. New Zealand, 2013.

Department of Building and Housing (DBH). Report of the Building and Sector Productivity Taskforce, New Zealand, 2009.

Department for Communities and Local Government (DCLGa). Proposed changes to the building control system: Consultation stage impact assessment. Department for Communities and Local Government. London, 2012.

Department of Communities and local government (DCLGb). Risk assessment decision making tool for building control bodies. Department for Communities and Local Government. London. 2012.

Gouldson, A., Morton, A. and Pollard, S. J. T. Better environmental regulation - contributions from risk-based decision-making. *Science of the Total Environment* 407(19): 5283-5288, 2009.

Hobbs, J., Fearne, A., and Spriggs, J. Incentive Structures for Food Safety and Quality Assurance: An International Comparison. *Food Control*, 13(2): 77-81. http://doi.org/10.1016/S0956-7135(01)00103-7, 2002.

Imrie, R. and Street, E. Risk, regulation and the practices of Architects. *Urban Studies* 46(12): 2555-2576. http://doi.org/10.1177/0042098009344231, 2009.

Invercargill City Council v Hamlin [1994] 3 NZLR 513, [1996] 1 NZLR 513 (1994).

Massey, W. Guide to the Building Act. New Zealand, 1999.

May, P. J. Performance-based regulation and regulatory regimes: The saga of leaky buildings. *Law & Policy* 25(4): 381-401, 2003.

May, P. J. and Burby, R. J. Making sense out of regulatory enforcement. *Law and Policy* 20(2), 1998

May, P. J. and Wood, R. S. At the regulatory front lines: Inspector's enforcement styles and regulatory compliance. *Journal of Public Administration Research and Theory* 13: 117-139, 2003.

Ministry of Business, Innovation & Employment (MBIE). Housing pressures in Christchurch: A summary of the evidence 2013. Wellington, New Zealand, 2013.

Ramachandran, G. Fire safety management and risk assessment. *Facilities* 17(9/10): 363-376, 1999.

Rothstein, H., Irving, P., Walden, T. and Yearsley, R. The risks of risk-based regulations: Insights from the environmental policy domain. *Elsevier*, 2006.

Tombs, S. and Whyte, D. Transcending the deregulation debate? Regulation, risk, and the enforcement of health and safety law in the UK, (July 2012): 61-79. http://doi.org/10.1111/j.1748-5991.2012.01164.x, 2013

Williamson, M. Building Act Review 5: Delivering Building regulation. Office of the Minister of Building and Construction. New Zealand, 2010a.

Williamson, M. Building Act Review 3: Stepped Consenting. Office of the Minister of Building and Construction. New Zealand, 2010b.

Yau, Y. On the proposed private certification of building works in Hong Kong. *International Journal of Law in the Built Environment* 1(3): 221-233, 2009.

Integrated Solutions for Infrastructure Development
Edited by Lau, H. H., Tang, F. E., Ng, C. K., and Singh, A.
Copyright © 2016 ISEC Press

THE INTERACTION BETWEEN LEADERSHIP STYLES AND MANAGEMENT LEVEL, AND THEIR IMPACT ON PROJECT SUCCESS

HECTOR MARTIN and KISA EDWARDS

Dept of Civil & Environmental Engineering, University of the West Indies, St. Augustine, Trinidad and Tobago

Leadership is the catalyst which ignites collaboration and participation towards achieving project success. Recent studies have acknowledged that leadership affect construction project success indirectly. However, how this success derived through leadership is directly related to management seniority remains unexplored. It is hypothesized that the higher a person advances in an organization the need to become more effective in achieving project success increases. To test this hypothesis and to improve project success rates, leadership at different managerial levels is explored through a case study using Kurt Lewin's and Multifactor Leadership (ML) questionnaires. The democratic and transformational styles of leadership were the most efficient in achieving project success. However, analysis of variance (ANOVA) revealed there is no significant relationship between project success and leadership style but there exists a strong association between management level and leadership style and a significant relationship between management level and project success. This suggest there is a maturity in leadership style as management level progresses, as such, a person should become more effective in guaranteeing project success based on how far they have progressed in the management structure of their organization.

Keywords: Critical success factors, Project management, Organization, Construction.

1 INTRODUCTION

Within project management based government organizations effective leadership continues to be a long standing problem which is evidenced by both project cost and time overruns as well as an abundance of incomplete projects. Leadership is the individual influence on a group to move towards and ultimately reaching a common goal (Northouse 2010). Leadership is concerned with what people are thinking and feeling and how they are to be interconnected to the environment, to the entity and to the job (Nicholls 1987). It therefore serves as a catalyst which ignites collaboration and participation towards project success. The link between leadership and success has been drawn by Mills (2005) who noted that the absence of leadership causes organizations to be stagnant. Leadership plays an important role within management in maximizing efficiency and achieving the organizations goals and is crucial in implementing decisions successfully (Mills 2005). The outcome of good leadership is amplified by the project manager's understanding of alternative leadership approaches and his ability to exercise them (Gehring 2007).

Though leadership has been described as one of the most important aspects of management Weihrich *et al.* (2008), it is rarely directly included as a project success factor. However, many articles support the theory that leadership influences the performance of project (Jiang 2014). Though leadership styles have been widely researched, very minimal work has been done regarding the impact of leadership styles on project success within the construction management field (Turner and Muller 2005). Further, context is not pervasive, as such, leadership studies in different environment could prove useful in its understanding.

The aim of this research is to advance knowledge of leadership styles particularly within the context of its effect on or interaction with project completion and management level. Although project success is often measured by considering cost and time overruns, this study seeks to use a more general unit of measurement, project completion. Project completion for the purposes of this study will refer to buildings where all scheduled works have been completed and as such the building is fit for occupation.

2 HOW LEADERSHIP STYLES INFLUENCES OUTCOMES

Lewin *et al.* (1939) identified three leadership styles formulated around decision making. The three leadership styles which were identified are autocratic (authoritarian), democratic (participative) and laissez-faire (delegative). The authoritarian leader does not consult any team members when making a decision (Dessler and Starke 2004). This style has the advantage of quick decision making in times of stress and solid deadlines for completion of task set by the leader. However, the downside of utilizing such a style is that it stifles creativity with respect to problem solving which ultimately decreases the performance of the group; it also leads to resentment among team members and towards the leader which may result in rebellion or high staff turnover (Lewin *et al.* 1939). The democratic type of leader customarily consults his team members and considers their suggestions although the final decision lies with the leader (Dessler and Starke 2004). The pros and cons of using this participative leadership style include higher quality of work produced although the quantity of work is decreased, and commitment to the goals as there is a sense of ownership and a valued feeling of being a part of the team (Lewin *et al.* 1939). The laissez-faire style attracts leaders that are comfortable with having minimum input in the decision making although they may still be responsible for the outcome of the decision made (Dessler and Starke 2004). This leadership style allows for the development of critical thinking skills as well as group problem solving (Dessler and Starke 2004). However, in the absence of highly qualified and effective participants the utilization of this style leads to lack of control in productivity, increased costs, and failure to meet deadlines (Lewin *et al.* 1939).

The theory of the transformational and transactional leadership was first published in 1978 by J.M. Burns and was later expanded upon by Bass (1985). Burns (1978) defines transformational leaders as having the ability to direct positive changes in a team by motivating the members to be effective and efficient. This leader is vision oriented and as such focuses on the long term goals. One of the cons of this leadership style is that it requires the leader to be surrounded by reliable, qualified personnel to take care of the minor details or the project might fail. The transformational leader expresses his personal beliefs and value system and takes special interest in the development of the team members individually which results in an overall more cohesive team with performance achievements above expectations (Bass 1985). Studies conducted by Fiol *et al.* (1999) and Lowe *et al.* (1996) concluded that leaders which utilize the transformational style are seen as more effective by subordinates and superiors. Alternatively, transactional leaders offer rewards to others in return for their adherence (Sims and Yun 2009). Transactional leaders focus on management tasks and do not identify a shared vision for the team (Giltinane 2013). These leaders are not focused on encouraging change but in ensuring that things remain the same (Odumeru and Ifeanyi 2013). This leadership style is present in construction contractual agreements; monetary rewards for work done, and as such it represents the traditional view on leadership (Thite 2001). The downsides of utilizing this leadership style is that it prevents both the leader and the follower from achieving their full potential as it does not allow for creativity, it is impersonal and focuses on the organization rather than the individual and as such results in high staff turnover (Bass 1985). Rules, processes and standards are crucial to the transactional leadership style and as such it is effective when the problems are simple and clearly defined (Bass 1985, Higgs and Dulewicz 2004). The converse of these relationships with respect to transformational leadership style holds (Turner and Muller 2005).

These two perspectives presented on leadership styles Lewin *et al.* (1939) and Bass (1985) indicates the leadership styles should be adjusted to the circumstance. Whether or not there is a dominant style associated with successful senior or junior managers remains unknown (Giritli 2008). This is explored in this paper.

3 METHOD

The target population for this case study was personnel directly involved in the leadership of construction projects and as such the leadership styles they exhibited and how this directly affects project success at each managerial level was investigated. The size of the target population is twenty-seven (27) and consisting of site supervisors, project managers, senior project managers, and divisional managers. Use of a standardize instrument should produce standardize and reliable results (Fiery 2008). The questionnaire consisted of thirty-six (36) leadership questions combined from Kurt Lewin Leadership styles as used by Northouse (2010) and the Multifactor Leadership Questionnaire MLQ developed by Bass and Avolio (1989). The number of projects completed in the last five (5) years as well as the democratic characteristics of the respondents were captured. The responses to the questions were measured on a five point Likert scale varying from strongly disagree (1) to strongly agree (5). Project performance efficiency score were determined in increment of 25 from 0 to 100. Where 0-25 represented low and 76-100 excellent. Where 0-25 suggest 25% of projects are usually completed using the particular style.

Statistical Package for the Social Sciences (SPSS) Software was used to analyze the collected data. One of the assumptions of the One-way ANOVA test is that homogeneity of variances exists across all groups and this was tested using Levene's Statistic. In this instance the null hypothesis assumes that the variances are equal. $P > 0.05$ was used for the retention of the null hypothesis and for indicating satisfaction of the homogeneity condition. The Chi-square test was used to determine if any relationship existed between the management level and the leadership style exhibited. The strength of this association was then further tested by using the Cramer's V test. Morgan *et al.* (2013) states that a Cramer's V value closer to 1 or -1 is indicative of a strong association, while 0 represents no relation.

4 RESULTS AND DISCUSSION

4.1 Leadership Styles and Project Completion

The results confirmed that all five leadership styles were present within the construction project management company. The results identify the most dominant leadership style as the democratic style which accounted for 33% of the sample population followed by transformational leadership which accounted for 26% of the population. The least popular leadership style was the laissez-faire which only accounted for a total of 8 % of the sample population. The autocratic and transactional leadership style accounted for 11% and 22% of the respondents respectively.

Turner and Muller (2005) advocated the direct measurement of leadership style and project success. The results indicate that the democratic leadership group has the highest mean project success score of 66 followed by the transformational leadership style with a mean score of 65. The autocratic leadership style had the lowest mean success score of 40. The One-way ANOVA tests whether or not a relationship exists between the leadership styles and project completion. The results showed that $p > 0.05$ indicating there is no significant relationship existing between the project completion scores and the leadership styles of the project personnel, as such, no further statistical testing was done.

4.2 Leadership Style and Efficiency

On average each leadership style efficiency was calculated by the amount of project successfully completed using a particular style. While democratic was the most dominant, in relation to the efficiency greater than 75%, there was a tie between the democratic and transformational leadership styles. This finding is supported by Blaskovics (2014) who found that the scores relating to organizational success and transformational score was above fifty percent. The results also showed that both the democratic and transformational leadership styles did not have any scores falling within the poor project success category which in this instance mirrors the conclusion that they are the most effective leadership styles. These styles can be used interchangeably as they have similar characteristics such as they both encourage the leader to consider the opinions of the followers and involving them as part of the decision making process, ultimately motivating them to ensure that the project succeeds. In addition to the relevant technical competencies when hiring new staff, the company should evaluate person's leadership style and hire personnel who display these characteristic leadership styles.

4.3 Management Level and Leadership Style

The results showed that 33%, 30% and 37% of the sample is accounted for by lower, intermediate and upper management levels respectively. Using the likelihood ratio and $p \leq 0.05$ Chi-square test indicated that there is strong association between leadership style and management level the test results showed a Crammer's V value of 0.611. The results also show that all of the autocratic leaders fall into the lower management levels while the transformational leadership style was the only one which appeared at all levels of management. This result is supported by Buble *et al.* (2014) which concluded that the autocratic leadership style often falls into the lower management levels. In particular, it should be noted that the democratic and transformational leadership styles, which were identified as the most effective management styles were also the most frequent styles displayed at the upper management level. This further reiterates the need for the project management personnel to be trained with the characteristics of the democratic and transformational leadership style in order for them to excel and increase the company's project success rate. The distribution of the leadership styles in the three management levels may also suggest that with experience the leadership styles of individuals evolve as the autocratic leadership style was only found in the lower level management with poor success ratings.

4.4 Management Level and Project Success

The One-way ANOVA tested at a significance of $p < 0.05$ shows that there is significant difference between project success scores and management level. The partial Eta squared showed that 63.3% of the between subjects' variance is accounted for by management level suggesting that management level has a strong effect on the project success scores. The power value statistic is more than 0.8 therefore the sample size is large enough to detect the effects being analyzed. Post hoc tests were then performed to establish the nature of these interactions and it was discovered that from the lower to the intermediate management level the mean project success scores increased and from the intermediate to the upper management level the mean project success scores further increased; therefore, as the management level increases so too does the mean project success score. This relationship may signify that the more experienced personnel achieve greater success with project completion. A high level of interaction among all management levels is recommended, this can be used to impart the knowledge based on experience from the upper management level to the two lower levels. Formal interactive sessions should be held among all the management levels so that the upper management can have a collective idea of the needs of the less experienced staff in terms of knowledge and guidance. A pairing of the upper management level with staff of the lower and intermediate management levels to enable them to better impart their knowledge.

5 CONCLUSION

A good leader should be able to utilize leadership styles complementary with the varying nature of projects they are presented with. This research showed no significant association between the style of leadership and the project success rate within the case study company. However, the democratic and transformational leadership styles were more prominent among successful project leaders. This may be due to external factors such as age, size of project and years of management experience which were not taken into consideration for this research. Another factor which was not considered for this study is the effect the upper management has on the lower and intermediate management levels in relation to decision making for projects. The results achieved from this research concurs with Morris (1988) who pointed out that although leadership style isn't a critical success factor it influences all if not most of the project success factor and as such indirectly affects the project success.

As the management level increases so too does the project success scores. While there is difficulty in finding previous research which analyses this relationship, it is unlikely that the respondents at the upper management level have attained these senior positions without themselves showing their ability to perform. Therefore, it maybe that with age and experience the leader has managed to refine his project approach to produce consistent project success ratings. This trend of increased project success climbing the management level suggests that the project managers within this company are rewarded for any consistent project success scores with promotions. While this is perfectly fine, when one gets to the upper management level they should have the maturity based on knowledge gained along the journey to the top to understand that the company cannot succeed, flourish or even remain operational for an extensive period of time if this knowledge is not passed on to persons on the lower levels of management. However, for this to be achieved their leadership skills would have to be sufficiently matured to display somewhat of a transformational or democratic leadership style to elicit that drive to achieve consistent project success.

There is a significant lack of literature pertaining to the relationship between management level and project success and since the results concluded that there is a significant relationship between the two, it is recommended that further research be done to investigate the nature of this relationship and the factors which influence it as it can contribute to the overall success of organizations.

References

Bass, B. M., and Avolio, B. J., *Manual for the Multifactor Leadership Questionnaire*, Consulting Psychologists Press, CA, 1989.

Bass, B. M., *Leadership and Performance Beyond Expectation*. Free Press, New York, 1985.

Blaskovics, B., Impact of leadership Styles on Project Success - The Case of a Multinational Company. *Dynamic Relationships Management Journal,* 3(2), 21-36, 2014.

Buble, M., Juras, A. and Matić, I., The Relationship between Manager's Leadership Styles and Motivation. *Management*, 14(1), 161-193, 2014.

Burns, J.M., *Leadership.* Harper & Row, New York, 1978.

Dessler, G., and Starke, F. A., *Management: Principles and Practices for Tomorrow's Leaders*. Pearson Education Canada Inc., Toronto, Ontario, 2004

Fiery, B. M., *Contextual Influences Affecting Transformational Leadership in Northwest Virginia Multiple Nonprofit Hospital Systems.* PhD Dissertation, University of Phoenix, 2008.

Fiol, C. M., Harris, D. and House, R. J., Charismatic Leadership: Strategies for Effecting Social Change. *Leadership Quarterly*, 10(3), 449–482, 1999.

Gehring, D. R. Applying Traits Theory of Leadership to Project Management, *Project Management Journal*, 38(1), 44-54, 2007.

Giltinane, C. L., Leadership Styles and Theories, *Nursing standard*, 27(41), 35-39, 2013.

Giritli, H, and Civan, I., Personality Study of Construction Professionals in the Turkish Construction Industry, *Journal of Construction Engineering and Management*, 134 (8), 630–634, 2008.

Higgs, M. J., and Dulewicz, S. V., Design of a New Instrument to Assess Leadership Dimensions and Styles. *Selection and Development Review*, 20(2), 7–12, 2004.

Jiang, J., The Study of the Relationship between Leadership Style and Project Success, *American Journal of Trade and Policy*, 1(1), 51-55, 2014.

Lewin, K., Lippitt, R. and White, R. K., Patterns of Aggressive Behavior in Experimentally Created Social Climates, *Journal of Social Psychology*, 10(2), 271-301, 1939.

Lowe, K.B., Kroek, G.K. and Sivasubramaniam, N. Effectiveness correlates of transformational and transactional leadership: a meta-analytic review of the MLQ literature. *Leadership Quarterly*, 7(3) 385–425, 1996.

Mills, D. Q., *Leadership: How to Lead, How to Live*. Mind Edge Press, Waltham, MA. 2005.

Morgan, G. A., Nancy L. L., Gene W. G., and Barrett, K. C. *IBM SPSS for Introductory Statistics: Use and Interpretation,* Routledge, New York, 2013.

Morris, P. W. G. *Managing Project Interfaces. In Project Management Handbook,* Cleland, D. I. and King, W. R. (ed.), 479-512, Project Management Institute, New York, 1988.

Nicholls, J., Leadership in Organisations: Meta, Macro and Micro. *European Journal of Management* 6(1), 16-25,1987.

Northouse, P., *Leadership: Theory and Practice,* Sage Publications, Los Angeles, 2010.

Odumeru, J. A, and Ifeanyi, G.O. Transformational vs. Transactional Leadership Theories: Evidence in Literature. *International Review of Management and Business Research*, 2(2), 355-361, 2013.

Sims, H.P., and Yun, S. When Should a Leader Be Directive or Empowering? How to Develop Your Own Situational Theory of Leadership, *Business Horizons,* 52 (2), 149-158, 2009.

Thite, M. Leadership Styles in Information Technology Project, *International Journal of Project Management*, 18(4), 235–24, 2001.

Turner, J.R., and Muller, R., The Project Manager's Leadership Style as a Success Factor on Projects: A Literature Review, *Project Management Journal*, 36(2), 49-61, 2005.

Weihrich, H., Cannice, M.V. and Koontz, H. *Management.* 12th, Mc Graw Hill, New Delhi, 2008.

Integrated Solutions for Infrastructure Development
Edited by Lau, H. H., Tang, F. E., Ng, C. K., and Singh, A.
Copyright © 2016 ISEC Press

RISK ASSESSMENT ON DESIGN AND BUILD PROJECTS – A COMPARISON OF CONTRACTORS, CONSULTANTS AND OWNERS PERCEPTIONS

SABIHAH SAAIDIN[1], INTAN ROHANI ENDUT[1], SITI AKMAR ABU SAMAH[2], AHMAD RUSLAN MOHD RIZDUAN[1], and NUR NABIHAH ABD RAZAK[1]

[1]*Faculty of Civil Engineering, Universiti Teknologi MARA, Shah Alam, Selangor*
[2]*Academy of Language Studies, Universiti Teknologi MARA, Shah Alam, Selangor*

Design and Build (DB) has been classified as the most risky project due to the complexities and uniqueness of the project itself. Appropriate identification, analyzing, controlling and monitoring for the project risk are required to minimize the risk in DB projects. It is essential that all parties play an important role in minimizing risk inherent in design and build projects. Therefore, this study has been carried out to conduct risk assessment on design and build projects by the comparison of contractors, consultants and owners perceptions. To achieve the objective, the survey questionnaire was distributed among contractor, consultant and owner, that were directly involved in design and build projects. A total of 128 useable questionnaires were received and analyzed using mean ranking, Kruskal-Wallis and Mann-Whitney Test. The result of these analyses shows that the contractors, consultants and owners have the same insight in risk factors presented namely: "client financial capability" and "inadequate cash flow by contractor". There are no significant differences in between all parties except "bureaucracy in government agencies" and "inflation". This study helps to increase understanding of contractors and owners on the importance of implementing risk management process at early stage of the projects.

Keywords: Construction, Procurement, Project risk, Stakeholder perceptions, Risk factors.

1 INTRODUCTION

Design and Build (DB) contract is an innovative procurement method whereby a single contractor is responsible for the design and construction to deliver a construction project to a client's satisfaction (Akintoye and Fitzgerald 1995). Although the DB's are expected to be faster and cheaper project delivery as compared to the others, not all the projects involved in DB procurement are accomplished on time (Oztas and Okmen 2004). Design and build projects are also exposed to the risks due to the unique features and complexity of the construction activities. In addition, this type of contract transfers more risk to the contractor than any other construction contract (Seng and Yusof 2006).

Literally, quite often the construction projects have failed to achieve the objectives of cost, time and quality due to the occurrence of unexpected events which the contractor has failed to identify, analyze and manage risk properly (Akintoye and Macleod 1997). Likewise, risk is being less concerned and almost non-existing in construction projects (Seng and Yusof 2006). However, it is essential that all parties play an important role in minimizing risk in DB project

(Akintoye and Macleod 1997). Therefore, this study has been carried out to conduct risk assessment on design and build projects with regard to contractors, consultants and owners perceptions.

2 RISK IN DESIGN AND BUILD PROJECTS

Design and build procurement method is a risky system of project delivery for both owners and contractors. It is due to the combination of design activity, on site supervision and participation in the actual construction project by different parties. Design and build procurement method is prone to several risks either borne by the contractor, owner or shared by both parties. Several empirical studies relevant to the risk factors influencing DB project were reviewed.

Banik and Hannan (2003) state that the allocation of risk comes from the owner and contractor or shared by both parties. A total of eight DB construction project risks was allocated to the owner of which site access/right of way, differing site condition, delayed payment on contract, unidentified utilities, permits and ordinances, change in work, Government Acts and Regulation and tax rate change and Acts of God. Furthermore, nine risks were identified among the twenty six risks that were allocated to the DB contractors. These were namely defective materials, quality of work, safety and accidents, contractor competence, defective design, labor disputes, establishment of a project cost, actual quantities of work and defensive engineering. Meanwhile, the risk factors shared by both the owner and contractor is contract delay resolution, indemnification and hold harmless, financial failure any party, change order negotiations and third party delay. Further research by Kartam and Kartam, (2001) supported the results apportionment of construction risk that also comes from owner and contractor or shared by both parties.

Financial capability is consider most important to an owner and contractor in DB projects. However, insufficient of financial capability will contribute to project risky. That is only the big scale companies can afford the financial cost and gain competitiveness at the bidding stage (Cheng 1995).

Oztas and Okmen (2004) have also indicated that the lack of experience and knowledge of owners and contractors contributed to risk in DB project. Therefore, the owner has a large role to play when deciding to use DB procurement method in minimizing the risk during construction project.

3 RESEARCH METHODOLOGY

Extensive literature review was undertaken to identify sources and types of risks in DB projects. Based on the literature, a research questionnaire was designed to elicit information from respondents such as owner, consultant, and contractor who have been involved in DB projects in Malaysia. A comprehensive list of 64 risks was identified and used in the questionnaire that was designed to get the perception of the construction practice in DB projects and only 10 risky factors were selected. The first section of the questionnaire solicited demographic information about the respondents. The second section consists of three parts of 64 risk factors. The two parts were related to the probability and impact of each risk factor in design and build project on a five-point Likert scale. The scale for risk probability ranged from 1 (low) to 5 (high). Meanwhile, the scale for risk impact ranged from 1 (insignificant) to 5 (catastrophic). A total of one hundred twenty eight (128) questionnaires were received from the owners, consultants and contractors. The results were analyzed by using Mean Ranking, Kruskal Wallis and Mann-Whitney Test.

4 DATA ANALYSIS AND DISCUSSION

4.1 Rank and Comparison of Risk Factors on Owner, Consultant and Contractor Perceptions

Based on the overall results, the risk factors in design and build project were shown in Table 1, all 64 numbers of risk factors analyzed and ranked accordingly to fulfill the objective of this study. The results only show the most ten risky factors on owner, consultant and contractor perceptions as indicated below.

Inadequate cash flow by contractor and client financial capability was given the first and second ranking on owner and consultant on their perceptions, instead of contractor's perceptions which has given client financial capability as the first ranking and followed by inadequate cash flow by contractor as the second ranking. It should be noted that financial capability is the most important in construction project as considered either by owner or contractor. Failure to that will affect the cost and time overruns. For example, Dada and Jagboro (2007) identified in their study, finance as one of the main risk factors that contributed to the construction project. These results opined by Kartam and Kartam (2001) financial capability was also at a higher rank of the survey risk in Kuwaiti. Enshassi *et al.* (2006) affirms the opinion further by agreeing that contractors could fail financially due to the following reasons: dependence on banks and paying high rates, lack of capital, and lack of experience in the line of work, cash flow management, low margin of profit due to competition, lack of experience in contracts and award contracts to the lowest price.

Insufficient time for completion date was given third ranking on the owner perception, while consultant and contractor perception was ranked seventh and eighth, which means the least ranked. Insufficient time for completion date may result from inadequate program scheduling planning, innovative design or contractors' lack of knowledge in planning construction programs. To minimize this risk, an informative program schedule should be worked out during the design stage together with the owner and contractors or project manager who is skillful in program coordination. The ability to manage construction programs may lead to the key criteria in appointing DB contractors (Ahmed *et al.* 1999).

Lack of payment (delayed progress payment by owner to the contractor) was ranked fourth and fifth on owner, consultant and contractor perceptions, respectively. This finding is consistent with literature in developing countries where the construction industry has always been closely related to the national economy (Frimpong *et al.* 2003). Adams (2008) reveals that lack of payment is the most important factor that brings to the project risks. Frimpong *et al.* (2003) have identified bureaucracy in Ghanian government department as a contributory factor, coupled with the nature of the funding of projects which could either be through domestic savings or foreign funding. Unfortunately, for most construction organizations, the government is the main owner of the construction industry. There is so much bureaucracy in the government agencies that it takes ages for certificates to be issued for payments. Tuuli *et al.* (2007) observe that there is no form of compensation to contractors due to lack of payment.

Incompetent sub-contractors were fifth, third and tenth ranked on owner, consultant and contractor perceptions. This means that on consultant perspective the sub-contractor is one of the most important parties that will contribute to the risk in DB project. Meanwhile, for the contractor perspective it is the least risky. Responsibilities of sub-contractors normally allocate their manpower and resources to the project and supervise by contractor. Therefore, the competency and knowledge should be regarded as one of the key criteria for appointing sub-contractors to meet owner and contractor expectation (Ahmed *et al.* 1999).

Lack of design/build knowledge/experience/competency was ranked sixth by owner and consultant and contractor was ranked third. Meanwhile, contractor's lack of staffs knowledge/experience was ranked eight by owner and consultant and ranked fifth by the contractor. Lack of project manager competency and authority was ranked ninth, tenth and seventh by owner, consultant and contractor perceptions. Lack of design/build knowledge / experience / competency was ranked third by contractor. It is important to note that the contractors should have sufficient knowledge, experience and competency because the contractor is a single entity taking the total responsibility of DB project.

Mistake during construction was ranked seventh by the owner and ninth and sixth by the consultant and contractor, respectively. Responsibility and experience of designer to determine the design correctly can help to illuminate the black box and minimize the mistake during construction to the contractors/subcontractors involved in the project. Inadequate or insufficient drawings information should be avoided by establishing an efficient communication scheme among the owner, consultant and contractor (Saaidin *et al.* 2016b).

Lack of payment (delayed progress payment by contractor to the sub-contractor) was ranked tenth by the owner and fourth and ninth by the consultant and contractor, respectively. Financial risks have an important impact on the firms and on the economy as a whole. Lack of payment can create a difficult situation to the subcontractor to carry out projects smoothly. Akintoye and Macleod (1997) have also acknowledged financial risk as having the most adverse consequences on the successful completion of the construction projects.

Table 1. Comparison of risk factors on owners, consultants, and contractors perceptions.

Risk Factors in Design and Build	**Mean**					
	Owner	**Rank**	**Consultant**	**Rank**	**Contractor**	**Rank**
Inadequate cash flow by contractor	4.000	1	4.200	1	4.260	2
Client financial capability	4.000	2	4.175	2	4.320	1
Insufficient time for completion date	4.000	3	3.950	7	4.020	8
Lack of payment (delayed progress payment by owner to the contractor)	3.895	4	4.025	5	4.120	4
Incompetent sub-contractors	3.895	5	4.100	3	3.920	10
Lack of design/build knowledge / experience / competency	3.868	6	4.025	6	4.200	3
Mistake during construction	3.868	7	3.925	9	4.080	6
Contractors lack of staffs knowledge/experience	3.816	8	3.950	8	4.120	5
Lack of project manager competency and authority	3.789	9	3.775	10	4.060	7
Lack of payment (delayed progress payment by contractor to the sub-contractor)	3.763	10	4.075	4	3.940	9

A number of discrepancies were observed between the ranking of three groups involved in the survey as discussed above. This necessitated investigating whether the overall ranking is applicable to the general population. This necessitated identifying the differences in the dataset that are statistically significant among three different parties. Therefore, Kruskal-Wallis H test was conducted to determine whether or several independent items come from the same population. The results of the test are illustrated in Table 2 on the perception of all three groups. Bureaucracy in government agencies ($p=.036<.05$) and inflation ($p=.027<.05$) were found with significant difference from each other on their perceptions. Kruskal-Wallis H test is a compilation test statistic and based on the results it cannot tell which specific groups independent variables are statistically different from each other.

Table 2. Kruskal-Wallis Test on owner, consultant, and contractor perceptions.

	Mean			Kruskal-Wallis Test	
Risk Factors	**Owner**	**Consultant**	**Contractor**	**Chi-square**	**p-Value**
Bureaucracy in government agencies	3.026	3.550	3.400	6.624	0.036*
Inflation	2.974	3.450	3.460	7.239	0.027*

*Indicates that the p-value is less than 0.05

Furthermore, there were follow-up pair-wise comparisons using Mann-Whitney U test as in Table 3. The results show that the pair consultant-contractor has no significant difference on their perceptions with regard to bureaucracy in government agencies ($U = 901, z = -.843, p>.05, r = .399$) and inflation ($U = 991, z = -.078, p>.05, r = .938$); so the assumption is satisfied. However, the pair owner-contractor has significant difference when it comes to their perceptions with regard to inflation ($U = 678, z = -2.408, p<.05, r = .016$) and for bureaucracy in government agencies ($U = 745, z = -1.845, p>.05, r = .065$) have not shown significant difference in their perceptions. This means that the inflation has higher risk factor on contractor perceptions to DB project than owner perceptions. As this inflation risk is usually unavoidable, the price of construction materials is always changing in response to the inflation and the relation between supply and demand in the construction material market. According to Ahmed *et al.* (1999), inflation should best be shared among the parties between the owner and contractor by including in contract clauses which provide a clause to pay for fluctuation in labor wages. One fair way to deal with inflation is client should add the price fluctuation to the contingency premium in the contract document (Saaidin *et al.* 2016a).

The result of the pair owner-consultant has also shown a significant difference on their perception with regard to bureaucracy in government agencies ($U = 522, z = -2.497, p<.05, r =.013$) and inflation ($U = 544, z = -2.278, p<.05, r = .023$). Bureaucracy in government agencies and inflation risk has higher risk factor on consultant perceptions to DB project than owner perceptions. This risk is normally out of control of the project stakeholders. Bureaucracy in government agencies are always being complained by owner, consultant and contractor. To minimize the risk in bureaucracy, the stakeholders should create a friendly environment by communicating with each other as much as possible and to always adopt the strategies of maintaining close relationship with local government officers.

Table 3. Mann-Whitney Test on owner, consultant, and contractor perceptions.

	Perceptions		
Risk Factors	**Owner Consultant**	**Owner Contractor**	**Consultant Contractor**
Bureaucracy in government agencies	Significant difference	No Significant difference	No Significant difference
Inflation	Significant difference	Significant difference	No Significant difference

5 CONCLUSION

This paper has presented the perception of owners, consultants and contractors on risk assessment in Malaysian DB project. The most top 10 risk factors were summarized in Table 1 and should be considered among the owners, consultants and contractors. Namely, the owners should possess the ability to clearly define the project scope to prevent insufficient time, have sufficient financial capability, adequate competent and experience staff for the projects, and manage the contract and communication effectively. Meanwhile, for the contractors, they are required to have sufficient financial resources and DB project experience and corporate management capability. As key stakeholders in DB project, both owners and contractors should possess certain competencies to ensure the success of DB projects. The stakeholder's should work cooperatively to address the potential risk effectively from the feasibility phase onwards.

References

Adams, F. K., Risk perception and Bayesian Analysis of International Construction Contract Risks: The case of Payment Delays in a Developing Economy. *International Journal of Project Management,* 26, 138-48, 2008.

Ahmed, S. M., Ahmad, R. and Saram, D. D., Risk Management Trends in the Hong Kong Construction Industry: A comparison of Contractor and Owners Perception, *Engineering, Construction nd Architectural Management,* 6(3), 225-234, 1999.

Akintoye, A. S., and Macleod M. J., Risk Analysis and Management Construction, *International Journal of Project Management,* 15(1), 31-38, 1997.

Akintoye, A., and Fitzgerald, E., Design and Build: A Survey of Architects' Views, *Engineering, Construction and Architectural Management,* 1, 27-44, 1995.

Banik, G. C. and Hannan, F., Specialty Contractors' Perspectives on Risk Importance and Allocation of Design-Build Contracts. *Journal of Construction,* 2003.

Cheng, R. T. L., Design & Build – Contractor's Role. *Design and Build Project – International Experiences International Congress on Construction, Singapore,* 1995.

Dada, J. O. and Jagboro, G. O., An Evaluation of the Impact of Risk on Project Cost Overrun in the Nigerian Construction Industry. *Journal of Financial Management of Property and Construction,* 12(1), 37-44, 2007.

Enshassi, A., Hallaq, K. and Mohamed, S., Causes of Contractors' Business Failure in Developing Countries:The Case of Palestine, *Journal of Construction in Developing Countries,* 11(2), 1-14, 2006.

Frimpong, Y. Oluwoye, J. and Crawford, L., Causes of Delay and Cost Overruns in Construction of Groundwater Projects in Developing Countries: Ghana as A Case Study. *International Journal of Project Management,* 21, 321-26, 2003.

Kartam, N. A., and Kartam, S. A., Risk and its Management in the Kuwaiti Construction Industry: A contractors' Perspective, *International Journal of Project Management,* 19, 325-335, 2001.

Oztas. A. and Okmen. O., Risk analysis in fixed price design build construction projects. *Building and Environment.* 39,229-237, 2004.

Saaidin, S., Endut, I. R., Abu Samah, S. A., M. Ridzuan, A. R. and Abd Razak, N. N., Risk Variable on Contractor's Tender Figure in Malaysia. *Journal Teknologi,* 78(5-2), 85-89, 2016a.

Saaidin, S., Endut, I. R., Abu Samah, S. A. and M. Ridzuan, A. R., The Current Practice of Design and Build Procurement Process in Malaysia. *Social and Management Research Journal,* 13(2), 1-10, 2016b.

Seng. N. W. & Md. Yusof. A., The success factors of design and build procurement method: A literature visit. *Proceedings of the 6th Asia Pacific Structural Engineering and Construction Conference. APSEC.* Kuala Lumpur, Malaysia, 2006.

Tuuli, M. M, Baiden, B. K. and Badu, E., Assessment and Enforcement of Liquidated damages in Construction Contracts in Ghana. Structural Survey, 25(3/4), 204-19, 2007.

Integrated Solutions for Infrastructure Development
Edited by Lau, H. H., Tang, F. E., Ng, C. K., and Singh, A.
Copyright © 2016 ISEC Press

THE STABILITY OF SLOPE USING CELL – BASED SMOOTHED FINITE ELEMENT METHOD AND SECOND ORDER CONE PROGRAMMING

HAI LE NGUYEN, HAI THAN NGUYEN and THIEN VO MINH

Faculty of Civil Engineering, University of Tech, Ho Chi Minh National University, Vietnam

In this paper, the numerical limit analysis procedure, associating the cell-based smoothed finite element method (CS-FEM) with the (second-order cone) primal-dual interior point algorithm, for cohesive-frictional materials problem is described. The soil is modeled as a cohesionless frictional Mohr-Coulomb material with the associated flow rule. Kinematically admissible velocity fields are established using CS-FEM. The underlying non-smooth optimization problem is formulated as a problem of minimizing a sum of Euclidean norms, ensuring that the resulting optimization problem can be solved by an efficient second order cone programming algorithm. The core purpose of this study is to evaluate collapse loads as well as failure mechanisms of footings on slope which will be obtained directly from solving the optimization problems. In this study, the properties of soil and the width of footing and distance from footing to the edge of the slope are considered. Several numerical examples of slope stability are given to show the performance of the proposed method.

Keywords: Limit analysis, Cohesive-frictional, CS-FEM, SOCP, Optimization, Upper bound.

1 INTRODUCTION

In limit analysis, upper bound and lower bound solution will give the bracket consisting of exactly collapse load. However, the statically admissible stress field is difficult to establish rigorously comparison with the kinematically admissible velocity field. Thus, the upper bound solution normally employed to estimate critical state of structures, especially in geomechanics.

In upper bound limit analysis problems, kinematically admissible velocity fields need be approximated by using a computational method. Once the displacement fields are approximated and the upper bound theorem of plasticity theory applied, limit analysis becomes a problem of optimization involving and can be solved using linear or non-linear programming techniques. This paper proposes a new numerical procedure for kinematic limit analysis problems-the cell-based smoothed element method (CS-FEM) with second-order cone programming-to find out acceptable solutions for some problems in geomechanics such as bearing capacity and slope stability problems (Canh 2009, Canh 2010, Liu 2008, Hung 2009).

2 UPPER BOUND LIMIT ANALYSIS FORMULATION

Consider a rigid-perfectly plastic body of area $\Omega \in R^2$ with boundary Γ, which is subjected to body forces $\boldsymbol{f}$ and to surfaces tractions $\boldsymbol{g}$ on the free portion Γ_t of Γ. The constrained boundary Γ_u

is fixed and $\Gamma_u \cup \Gamma_t = \Gamma$, $\Gamma_u \cap \Gamma_t = \varnothing$. Let $\dot{\boldsymbol{u}} = [\dot{u} \quad \dot{v}]^T$ be plastic velocity or flow fields that belong to a space U of kinematically admissible velocity fields. Where $\dot{u}$ and $\dot{v}$ are the velocity components in the x and y directions respectively. The strain rates can be expressed by relations:

$$\dot{\varepsilon} = \left[\dot{\varepsilon}_{xx} \ \dot{\varepsilon}_{yy} \ \dot{\gamma}_{xy}\right]^T = \nabla \dot{\boldsymbol{u}} = \begin{bmatrix} \frac{\partial}{\partial x} & 0 & \frac{\partial}{\partial y} \\ 0 & \frac{\partial}{\partial y} & \frac{\partial}{\partial x} \end{bmatrix}^T \dot{\boldsymbol{u}} \tag{1}$$

The external work rate associated with a virtual plastic flow $\dot{\boldsymbol{u}}$ is expressed in the form as:

$$W_{ext}(\dot{\boldsymbol{u}}) = \int_{\Omega} \boldsymbol{f}^T \dot{\boldsymbol{u}} d\Omega + \int_{\Gamma_t} \boldsymbol{g}^T \dot{\boldsymbol{u}} d\Gamma \tag{2}$$

The internal plastic dissipation of the two-dimensional domain Ω can be written as: $W_{int}(\dot{\varepsilon}) = \int_{\Omega} D(\dot{\varepsilon}) d\Omega$, where the plastic dissipation $D(\dot{\varepsilon})$ is defined by $D(\dot{\varepsilon}) = \max_{\psi(\sigma) \le 0} \boldsymbol{\sigma}.\dot{\boldsymbol{\varepsilon}} \equiv \boldsymbol{\sigma}_{\varepsilon}.\dot{\boldsymbol{\varepsilon}}$, with σ represents the admissible stresses contained within the convex yield surface $\psi(\sigma)$ represents the stresses on the yield surface associated to any strain rates $\dot{\boldsymbol{\varepsilon}}$ through the plasticity condition.

The kinematic theorem of plasticity states that the structure will collapse if and only if there exists a kinematically admissible displacement field $\dot{\boldsymbol{u}} \in U$, such that:

$$W_{int}(\dot{\varepsilon}) < \lambda^+ W_{ext}(\dot{\boldsymbol{u}}) + W_{ext}^0(\dot{\boldsymbol{u}}) \tag{3}$$

where λ^+ is the collapse load multiplier, $W_{ext}^0(\dot{\boldsymbol{u}})$ is the work of any additional loads $\boldsymbol{f}_0$, $\boldsymbol{g}_0$ not subjected to the multiplier.

If defining $C = \{\dot{\boldsymbol{u}} \in U | W_{ext}(\dot{\boldsymbol{u}}) = 1\}$, the collapse load multiplier λ^+ can be determined by the following mathematical programming:

$$\lambda^+ = \min_{\dot{\boldsymbol{u}} \in C} \int_{\Omega} D(\dot{\varepsilon}) d\Omega - W_{ext}^0(\dot{\boldsymbol{u}}) \tag{4}$$

3 CELL-BASED SMOOTHED FINITE ELEMENT METHOD (CS-FEM)

In CS-FEM, the problem domain is discretized into elements as in FEM, such as $\Omega = \Omega^1 \cup \Omega^2 \cup ..... \cup \Omega^{nel}$ and $\Omega^i \cap \Omega^j = \varnothing, i \neq j$ the displacement fields are approximated for each element as:

$$u^h(x) = \sum_{I=1}^{n} N_I(x) d_I \tag{5}$$

where n is the number of node per element and $d_I = [u_I v_I]^T$ is the nodal displacement vector.

Elements are then subdivided into several smoothing cells, such as shown in Figure 1, and smoothing operations are performed for each smoothing cell (SC).

A strain smoothing formulation is given by Liu (2006).

$$\begin{aligned} \tilde{\varepsilon}^h(x_C) &= \int_{\Omega^e c} \varepsilon^h(x) \varphi(x, x - x_c) d\Omega \\ &= \int_{\Omega^e c} \nabla u^h(x) \varphi(x, x - x_c) d\Omega \end{aligned} \tag{6}$$

where $\tilde{\varepsilon}^h$ is the smoothed value of strains ε^h for smoothing cell $\Omega^e{}_C$, and φ is a distribution function or a smoothing function that has to satisfy the following properties (Liu 2006): $\varphi > 0$ *and* $\int_{\Omega^e{}_C} \varphi d\Omega = 1$. For simplicity, the smoothing function φ is assumed to be a piecewise constant:

$$\varphi(x, x - x_C) = \begin{cases} 1/A_C, & x \in \Omega^e{}_C \\ 0 \quad, & x \notin \Omega^e{}_C \end{cases} \tag{7}$$

with A_C is the area of the smoothing cell $\Omega^e{}_C$.

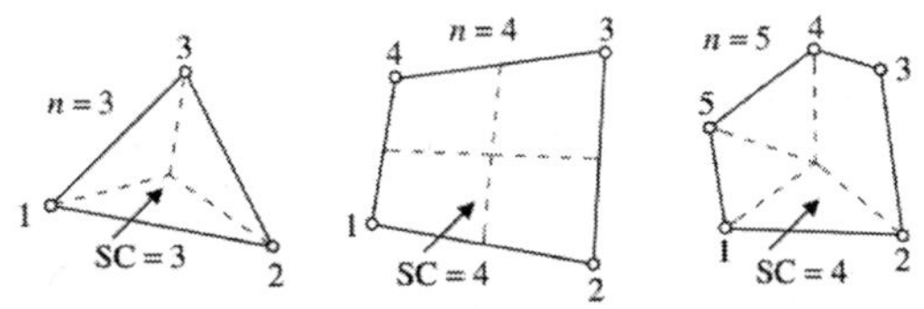

Figure 1. Smoothing cells for various element types: triangular, quadrilateral and polygonal.

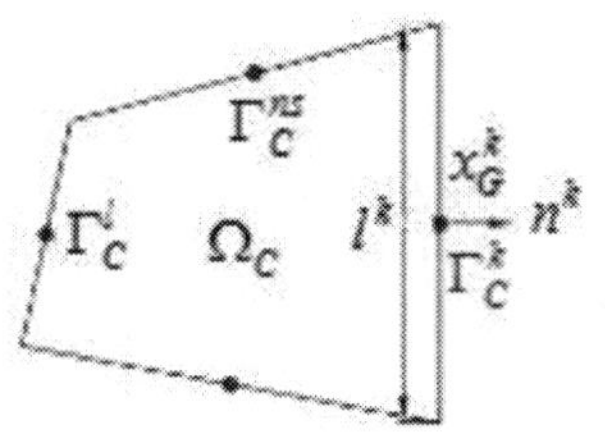

Figure 2. Geometry de finition of a smoothing cell.

Substituting Eq. (6) into Eq. (4), and applying the divergence theorem:

$$\tilde{\varepsilon}^h(x_C) = \frac{1}{A_C} \int_{\Omega^e{}_C} \nabla u^h(x) d\Omega = \oint_{\Gamma_C} n(x) u^h(x) d\Gamma \tag{8}$$

where Γ_c is the boundary of $\Omega^e{}_c$ and n is a matrix with components of the outward surface normal.
The smooth version of the strain rates can be expressed as:

$$\dot{\tilde{\boldsymbol{\varepsilon}}}^h(\boldsymbol{x}_C) = \tilde{\boldsymbol{B}} \dot{\boldsymbol{d}} \tag{9}$$

where $$\dot{\mathbf{d}}^T = [\dot{u}_1, \dot{v}_1, ... \dot{u}_n, \dot{v}_n] \tag{10}$$

$$\tilde{\mathbf{B}} = \begin{bmatrix} \tilde{N}_{1,x}(x_C) & 0 & \dots & \tilde{N}_{n,x}(x_C) & 0 \\ 0 & \tilde{N}_{1,y}(x_C) & \dots & 0 & \tilde{N}_{n,y}(x_C) \\ \tilde{N}_{1,y}(x_C) & \tilde{N}_{1,x}(x_C) & & \tilde{N}_{n,y}(x_C) & \tilde{N}_{n,x}(x_C) \end{bmatrix} \tag{11}$$

with $$\tilde{N}_{I,\alpha}(x_C) = \frac{1}{A_C} \int_{\Omega^e{}_C} N_I(x) n_\alpha(x) d\Gamma = \frac{1}{A_C} \sum_{k=1}^{ns} N_I(x_G{}^k) n_\alpha{}^k l^k, I = 1, 2, ..., n \tag{12}$$

where $\tilde{N}_{I,\alpha}$ is the smoothed version of shape function derivative $N_{I,\alpha}$; n_s is the number of edges of a smoothing cell Ω_c as shown in Figure 2; x_G^k is the Gauss point of Γ_C^k boundary segment which has length l_x and outward surface normal n^k.

4 CS-FEM FORMULATION FOR PLANE STRAIN WITH MOHR-COULOMB YIELD CRITERION

In this study, the Mohr-Coulomb failure criterion is used:

$$\psi(\sigma)=\sqrt{(\sigma_{xx}-\sigma_{yy})^2+4\tau_{xy}^{\ 2}}+(\sigma_{xx}+\sigma_{yy})\sin\varphi-2c\cos\varphi \tag{13}$$

The plastic strains are assumed to obey the normality rule $\dot{\varepsilon}=\dot{\mu}\frac{\partial\psi}{\partial\sigma}$, where the plastic multiplier $\dot{\mu}$ is non-negative.

Hence, the power of dissipation can be formulated as a function of strain rates for each domain i as $D(\dot{\varepsilon})=cA_it_i\cos\varphi$.

The upper bound limit analysis for plane strain using the smoothed strains can be formed as:

$$\lambda^+=\min\sum_{i=1}^{n_{SD}xnel} cA_it_i\cos\varphi \tag{14}$$

$$s.t\begin{cases} W_{ext}(\dot{\boldsymbol{u}}^h)=1 & \\ \dot{\boldsymbol{u}}^h=0 \quad on\ \Gamma_u & \\ \dot{\tilde{\varepsilon}}_{xx}^h+\dot{\tilde{\varepsilon}}_{yy}^h=t_i\sin\varphi & i=1,2,\ldots,nel*n_{SD} \\ \|\boldsymbol{\rho}\|_i\le t_i & i=1,2,\ldots,nel*n_{SD} \end{cases} \tag{15}$$

where n_{SD} is the smoothing cell and *nel* is the number of element in the whole investigated domain. And the fourth constraint in problem (15), resulting optimization problem is cast in the form of a second – order cone programming (SOCP) problem so that a large-scale problem can be solved efficiently (Canh 2009, Canh 2010, Mosek programming).

5 UNDRAINED STABILITY OF FOOTINGS ON SLOPE USING

In this section, a strip footing with a width of B is rested on slope, L is the distance from footing to the edge of the slope, as shown in Figure 3. The ultimate bearing capacity depends on L, the soil unit weight which has the effect on the overall stability of the slope. This is not similar to the surface foundation resting on level ground where soil unit weight has no influence on the ultimate bearing capacity. The ultimate bearing capacity of the considered problem can be stated as:

$$\frac{p}{\gamma B}=f\left(\beta,\frac{L}{B},\frac{H}{B},\frac{c_u}{\gamma B},\frac{q}{\gamma B}\right) \tag{16}$$

with p is the average limit pressure acting on the footing and q is the distributed load.

In this paper, the ratio $H/B = 3$ which ensures that the destruction occurs above the toe of the footing is used. In order to analyse the influence of the footing distance to rest, an angle slope $\beta=90^0$ is considered and $c_u/\gamma B=5$, $q/\gamma B=0$ is established in the limit analysis problem, the footing distance to rest $L/B=0\div4$ is considered so as to compare with result of Shiau (2011) using finite element limit analysis formulations of the upper bound theorem.

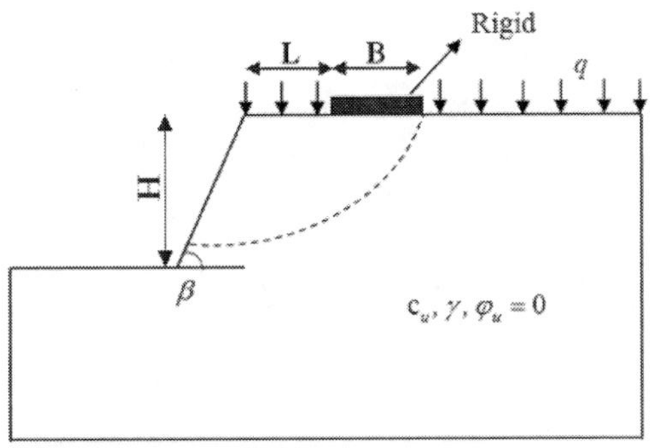

Figure 3. Footing strip resting on slope.

Analysis model using 16000 elements and the result obtained is reported in Table 1.

Table 1. Ultimate bearing capacity (p/γB) for (cu/γB = 5 and q/γB = 0).

	Footing distance to rest L/B				
	0	1	2	3	4
CS-FEM	9.5639	15.85	19.21	22.36	24.98
Shiau	9.5	16.12	19.64	22.73	25.35

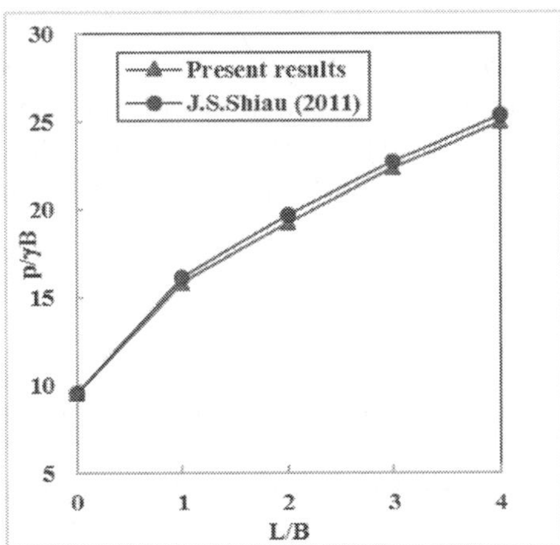

Figure 4. Ultimate bearing capacity (p/γB) with c_u/γB = 5 and q/γB = 0.

As shown at Figure 4, the present result is lower (better) than result of Shiau (2011).

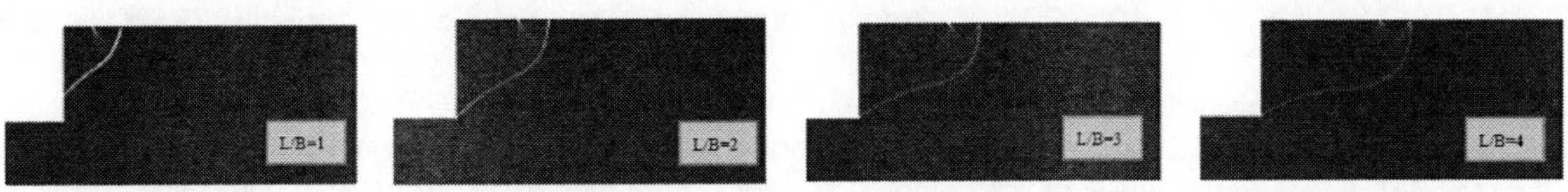

Figure 5. Plastic dissipation distribution with various *L/B* for: $\beta = 90^o$, *cu/γB = 5, q/γB = 0.*

5.1 Effect of $c_u/\gamma B$

In the case L/B = 0 with 31600 elements, various ratio of cu/γB from 0.566 to 25 was chosen, a normalized bearing capacity obtained will be shown in Table 2. Although the result of CS-FEM is slightly higher than those of Kusakabe (1981) that using numerical method based on the upper bound theorem of perfect plastic solids, it's better than the result of Shiau (2011) as well as Narita and Yamaguchi's (1990) result based on the limit equilibrium method.

Table 2. Bearing capacity with different cu/γB.

c_u/gB	L/B =0				
	Narita and Yamaguchi (limit equlibrium)	**Kusakabe (upper bound)**	**Shiau (lower bound)**	**Shiau (upper bound)**	**CS-FEM**
25	107	102	97.5	104.33	103.6923
5	21.1	20.2	19.61	20.69	20.5269
1	3.94	3.84	3.73	3.93	3.8578
0.75	-	-	2.59	2.85	2.7687
0.714	-	-	2.36	2.68	2.5403
0.6	-	-	No feasible solution	1.87	1.6116
0.556	-	-	No feasible solution	1.34	1.1137

6 CONCLUSION

A novel procedure for performing upper bound limit analysis using CS-FEM and SOCP has been described. A key advantage of applying the CS-FEM to limit analysis problems is that the size of optimization problem is reduced. Moreover, numerical examples show that when the underlying optimization is cast in the form of a standard SOCP large-scale engineering problems can be solved with a minimal computational cost, and gives acceptable upper bounds for both drained and undrained analysis.

References

Canh, L. V., Mattheu, G., and Harm, A., Limit analysis of plates using the EFG method and second-order cone progreamming, *International Journal for Numerical Methods in Engineering*, 78, 1532–1552, February 2009.

Canh, L. V., Hung, N. X., Askes, H., Bordas, S., and H., N. V., A cell-based smoothed finite element method for kinematic limit analysis, *International Journal for Numerical Methods in Engineering,* 83(12), 1651-1674, September 2010.

Hung, N. X., Liu, G. R., Trung, N. T., and C., N. T., An edge-based smoothed finite element method for analysis of two-dimentional piezoelectric structures, *Smart Materials and Structures,* 18(6), May 2009.

Kusakabe, O., Kimura, T., and Yamaguchi, H. (1981)., Bearing capacity of slopes under strip loads on the top surface, *Soils and Foundations*, 21(4), 29–40, December 1981.

Liu, G. R., Trung, N. T., and Lam, K. Y., An edge-based smoothed finite element method (ES-FEM) for static, free and forced vibration analyses of solids, *Journal of Sound and Vibration*, Elsevier, 320(4-5), 1100–1130, October 2008.

Liu, G. R., Trung, N. T., and Lam, K. Y., Theoretical aspects of the smoothed finite element method (SFEM), International Journal for Numerical Methods in Engineering, 71, 902-930, December 2006.

Mosek. The MOSEK optimization toolbox for MATLAB manual. http://www.mosek.com.

Narita, K., Yamaguchi, H., Bearing capacity analysis of foudations on slopes by use of log-spiral sliding surfaces, *Soils and Foundations*, 30(3), 144–152, September 1990.

Shiau, J. S., Merifield, R. S., Lyamin, A. V., and Sloan, S. W., Undrained Stability of Footings on Slopes, *International Journal of Geomechanics,* 11(5), 381-390, October 2011.

Integrated Solutions for Infrastructure Development
Edited by Lau, H. H., Tang, F. E., Ng, C. K., and Singh, A.
Copyright © 2016 ISEC Press

ANALYTICAL STUDY ON PILED-RAFT FOUNDATION CONSIDERING NONLINEAR BEHAVIOR OF GIBSON SOIL

HE HUANG and HARUYUKI YAMAMOTO

Graduate School for IDEC of Hiroshima University, Hiroshima, Japan

Randolph (1994) was proposed simplified estimated equations to carried out the proportion of load between the components of piled-raft foundation. These equations are based on elastic theory. Therefore, Authors focus on the point of applicability for nonlinear behavior of soil. This paper is a part of study on behavior of piled-raft foundation under considering the nonlinear behavior of ground. Authors already discussed the effect of nonlinear behavior of soil on the equations by Randolph. In addition, as the extension of research, authors also discussed the applicability of these equations for irregular raft shape. However, foundations were rarely set in an ideal homogeneous single soil layer in our previous studies. Therefore, the applicability of these equations are discussed in case of inhomogeneous soil (Gibson soil) in this paper.

Keywords: Settlement stiffness, Loading share ratio, Gibson soil, Nonlinear analysis, Piled-raft foundation.

1 INTRODUCTION

The piled-raft foundation transfers load to the ground by the raft and the piles together and it was proposed in the 1970s. The key point that arises in the design of piled-raft foundation focus on the relative proportion of load carried out by raft and the piles separately. Then, this question is solved by Randolph (1994) and he pointed out methods which could carried out the proportion of load between the components. In the same paper, Randolph also proposed a method to estimate the settlement stiffness of piled-raft foundation. However, these methods were proposed based on elastic theory with homogeneous soil. The utilization of these methods is challenging because soil always exhibit nonlinear behavior even at low working loads in reality. Therefore, authors discuss the effect of nonlinear behavior of soil on the equations by Randolph. In addition, as the extension of research, authors also discussed the applicability of these equations to the raft with irregular shape. However, piles are rarely set in an ideal homogeneous single soil layer. Then, the following question arises. How about the applicability of these equations if inhomogeneous of soil is taken into account? Consequently, authors will discuss the applicability of these equations for inhomogeneous soil which has increasing stiffness with depth, and is known as the Gibson soil.

It is not realistic to do full scale tests for this research because it's costly. On the other hand, Numerical analysis satisfies our requirements and it could simulate complicated model and ground condition as we want. 3-D FEM (Finite Element Method) was adopted in this study. Soil has the characteristic of elastic-plastic material with Drucker-Prager failure criterion. Piled-Raft foundation is always used on soft ground to improve differential settlement. Then, we speculate that these equations which have a good feasibility on soft ground. However, it will be a thrill if they also give as a good applicability on hard ground and super-soft ground. Therefore, three different ground conditions are simulated in this study. More specifically, super-soft ground (case (A)), soft ground (case (B)) and hard-ground (case (C)).

2 GIBSON SOIL AND ESTIMATIMATED EQUATIONS

2.1 Gibson Soil

Gibson (1967) discusses an elastic ground layer in which the stiffness increases linearly with depth. Such a ground is described as nonhomogeneous and is known as Gibson soil, as shown in Figure1 (left side).

$$E=E_0+\lambda z \tag{1}$$

where, E_0 is surface stiffness of soil and λ is the rate of increase of the stiffness modulus with depth z. For the purpose of analysis, the ground system will be idealized as constant Possion`s ratio, its surface stiffness and adhesion are zero as shown Figure 1 (right side), the maximum stiffness is at the point reached rigid soil layer assumed as 2E and 2C.

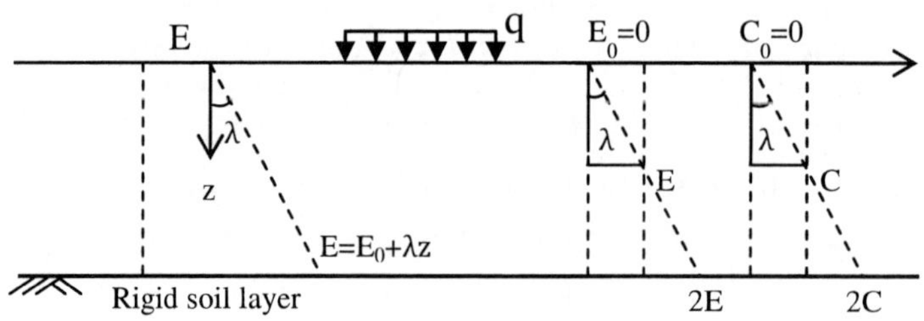

Figure 1. Gibson soil.

2.2 Estimated Equations

Estimated equations of settlement stiffness and loading share ratio of piled raft foundation proposed by Clancy and Randolph are shows as follows (Clancy and Randolph, 1996). Settlement stiffness is estimated by Equation (2).

$$\frac{k_{pr}}{k_r}=\frac{1+(1-2\alpha_{rp})(k_r/k_p)}{\{1-\alpha_{rp}{}^2(k_r/k_p)\}(k_r/k_p)} \tag{2}$$

where, k_{pr}: settlement stiffness of piled-raft foundation; k_p: settlement stiffness of friction pile group foundation; k_r: settlement stiffness of raft foundation; α_{rp}: influence coefficient of pile group for the raft foundation. Pile group and raft foundation loading share ratio is estimated by Equation (3).

$$\frac{P_r}{P_p}=\frac{(1-\alpha_{rp})(k_r/k_p)}{1-\alpha_{rp}(k_r/k_p)} \tag{3}$$

where, P_p: shared load of friction pile group; P_r: shared load of the raft foundation.

$$\alpha_{rp}=1-\frac{\ln(n)}{\ln(2r_m/d_p)} \tag{4}$$

where, n is the ratio of equivalent circular raft diameter to pile diameter d_p, and $r_m = 2.5\rho L_p(1 - \upsilon_s)$. L_p is the embedded length of a pile. ρ is a soil inhomogeneity factor varying from unity for homogeneous conditions to 0.5 in case of the stiffness is proportional to depth (Randolph, 1994).

3 ANALYTICAL MODEL

Three basic analysis models were set as below. Model ①: Raft foundation, Model ②: Pile group, Model ③: Piled-raft foundation as shown Figure 2.

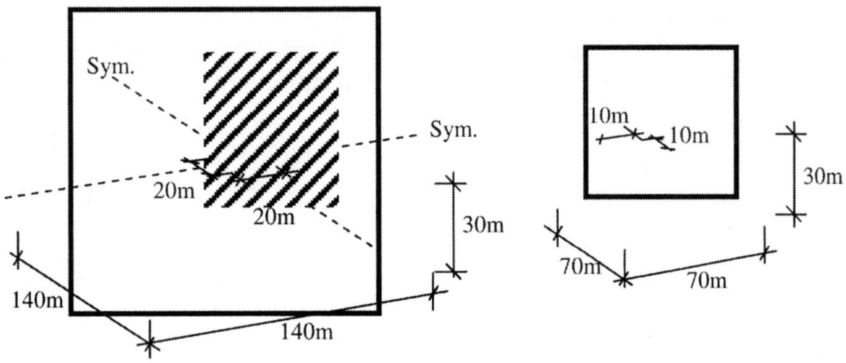

Figure 2. Analytical model.

Piles are assumed as elastic material made of concrete (diameter d=1m, length l=15m, 21m, 28m, Young's modulus E_c=2.1 × 10^7kN/m^2) and raft foundation is assumed to be rigid. The Drucker-Prager failure criterion with associated flow rule was used for the analytical model of ground. Poisson's ratio υ_s of ground was assumed 0.3. 1/4 model was analyzed by using the symmetry condition as shown in Figure 2. Figure 3 shows the FEM mesh of model. The soil properties of the analytical model are assumed by three parameters of elastic modulus E_s, the internal friction angle φ and adhesion c. Assuming a normal consolidation sand ground with N value of about 20, the internal friction angle φ of the ground was set by the equation of Osaki (AIJ, 2001). For the elastic modulus, and set the approximate value based on the empirical equation E = 1.4N (Mpa) (AIJ, 2001). For the adhesion c, it is the value of the order corresponding to the soft ground. Ground parameters are set as shown in Table1.

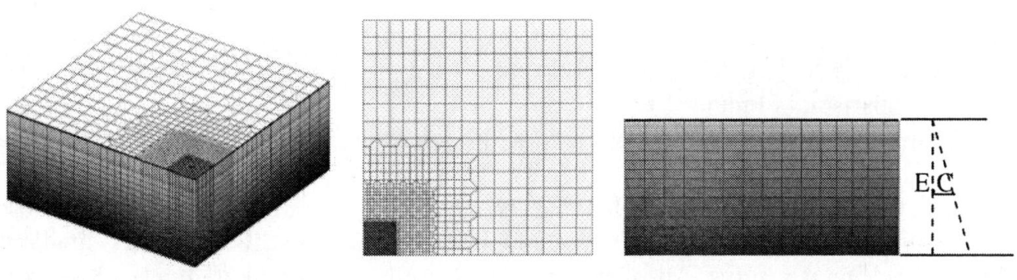

Figure 3. Mesh of analytical model.

Table 1. Ground Parameters.

	(A)	(B)	(C)
Es (kN/m^2)	10000	30000	150000
φ(deg.)	30	35	40
c (kN/m^2)	5	10	20

4 ANALYTICAL CASE

In consideration that the distribution range of k_r/k_p is very important when setting the analytical cases, the placements of pile are set as shown in Table 2. Number of pile is increased to be installed in the order of model (0) to (3), and piles length are increased on model (4) and (5). In other words, for the same ground conditions and loading conditions, k_p is increase and the value of k_r/k_p decreases from (0) to (5). Value of α_{rp} is determined by the placement of these piles. Total number of analytical case is 18 from (A)-(0) to (D)-(5).

Table 2. Pile arrangements.

	(0)	(1)	(2)	(3)	(4)	(5)
Pile spacing; s (m)	10	9	6	4.5	4.5	4.5
s/d	10	9	6	4.5	4.5	4.5
NxN	2x2	3x3	4x4	5x5	5x5	5x5
Pile length; L (m)	15 m	15 m	15 m	15 m	21 m	28 m
α_{rp}	0.26	0.38	0.47	0.54	0.58	0.61

5 ANALYTICAL RESULTS

5.1 Estimated Equation for the Settlement Stiffness

The secant settlement stiffness mentioned above (k_p, k_r, k_{pr}) are calculated to discuss the applicability of these estimated equations. The analytical results and the approximate solution from estimated equation (2) are compared by the relationships between the ratio k_r / k_p, and the ratio k_{pr} / k_r as shown in Figure 5.

As shown in Table 2, the range of α_{rp} in this analytical model is 0.26 to 0.61, two curves of approximate solution by the estimated equation (2) are also drawn (α_{rp} = 0.26 and α_{rp} = 0.61) in Figure 5 for comparison. Analytical results of first step (loading increments is very small) is seemed close to the initial elastic state, and it was considered as the first considering point (larger marked point in Figure 5 and Figure 6). Then analytical results were examined for every increment step of a settlement ratio; 0.01 (settlement/ pile diameter), and discussion range of settlement ratio was unified to 0.1 in this paper. Analytical results for homogeneous soil was shown as a comparison in Figure (5)-(a).

Pile group settlement stiffness k_p increase with increasing number of pile, therefore the value of k_r/k_p are reduced with in turns of case (0), (1), (2), (3) as the same with homogeneous soil case. Value of k_r/k_p is over 1 in case (A)-(1) with homogeneous soil, but it is below 1 for Gibson soil because its stiffness increases linearly with depth, then its surface stiffness is very small and lead to lacking of bearing capacity. All of the analytical results are distributed very close to the two curves obtained by the estimated equation (2) in Figure 5. It can be seen that good approximate results have been obtained regardless of elastic or non-linear regions.

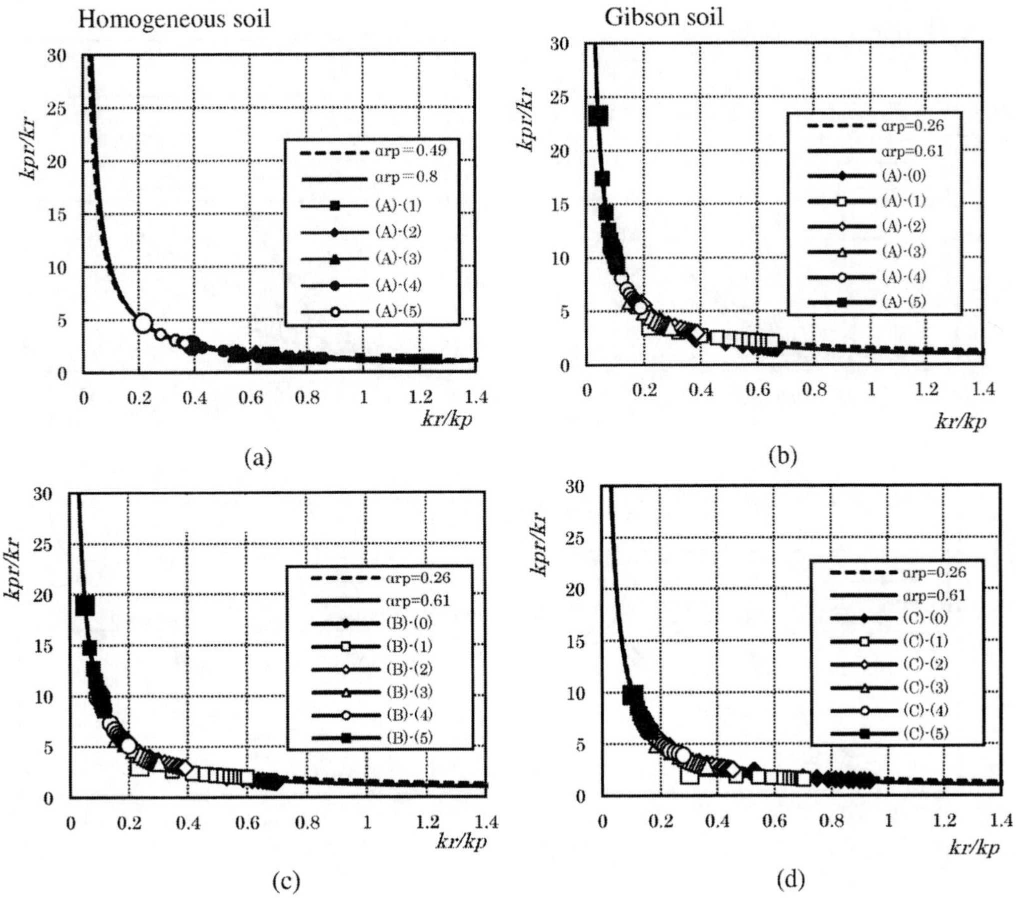

Figure 5. Compartion of analytical results and approximate solutions from estimated equation (2).

5.2 Estimated Equation for Loading Share Ratio

The analytical results and the approximate solutions from estimated equation (3) are compared by the relationships between the ratio k_r/k_p and the ratio of Pr/Pp as shown in Figure 6 with different ground condition. As described above, the estimated equation each analytical case, the analytical results are distributed close to the two curves obtained by the estimated equation (3) regardless of non-linear region or not. However, it seems large relative difference in case (0) of Gibson soil. Case (0) is an extreme model with number of pile for each ground conditions, it has many influential factors of difference. Furthermore, in fact it is necessary to examine the applicability of Piled-raft foundation.

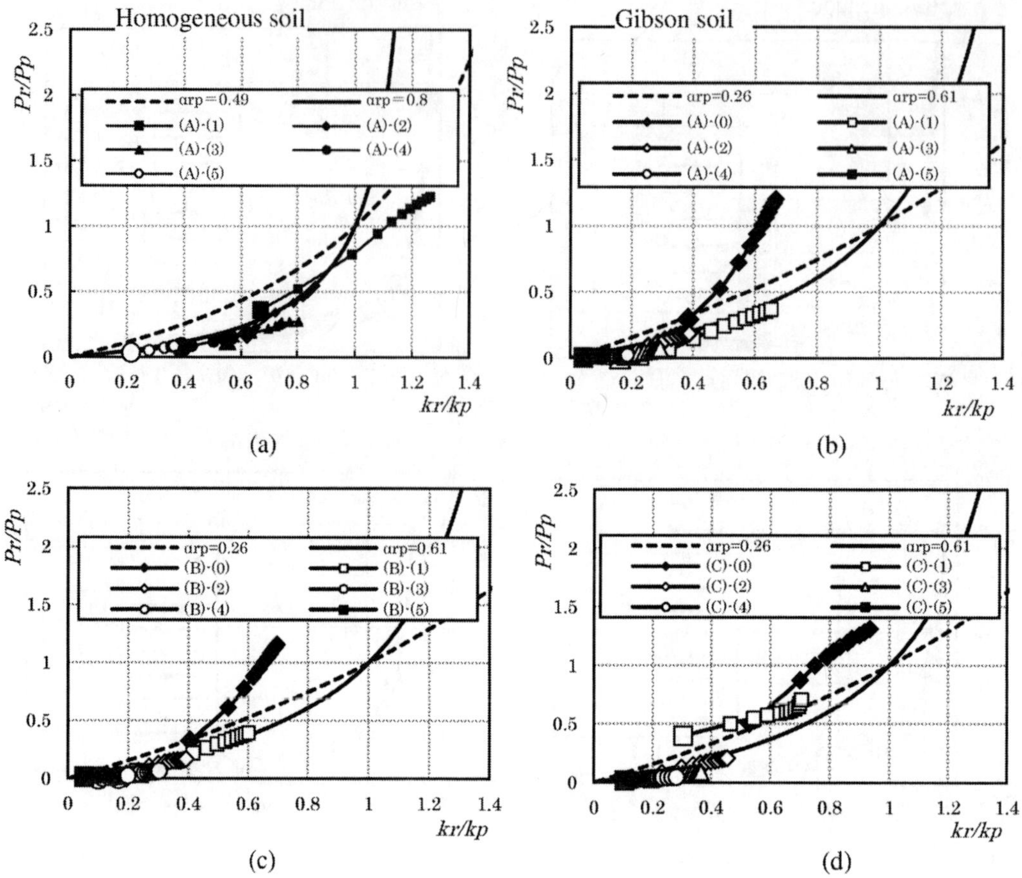

Figure 6. Compartion of analytical results and approximate solutions from estimated equation (3).

6 CONCLUSIONS

On the equations which estimate the settlement stiffness of the piled-raft foundation and the loading share ratio, it finds that most of analytical results are well in agreement with the results obtained by the estimating equation in each case under Gibson soil. There is almost no influence by a nonlinear behavior and the applicability of this estimating equation is very well also in a nonlinear domain (within the settlement ratio = 0.1). However, it shows some difference in the extreme case with fewest pile (case (0)) for the equation of the loading share ratio. It is necessary to examine the estimating equation for extreme case in the further.

References

Architectural Institute of Japan, Recommendation for Design of Building Foundations, 113-146, 2001. (In Japanese)

Clancy,P. and Randolph,M.F.: Simple Design Tools for Piled Raft Foundations, Geotechnique 46, No.2, pp.313-328,1996.

Gibson,R.E.:Some Results Concerning Displacement and Stresses in a Non-homogeneous Elastic Half-space, Geotechnique 17, No.1,pp.58-67,1967

Randolph, M.F.: Design Methods for Pile Groups and Piled Rafts. 13th ICSMFE, New Delhi, India, pp.61-82,1994.

Integrated Solutions for Infrastructure Development
Edited by Lau, H. H., Tang, F. E., Ng, C. K., and Singh, A.
Copyright © 2016 ISEC Press

APPLICABILITY OF A NEW SURFACE PREPARATION METHOD FOR STEEL STRUCTURE USING LASER

HIROSHI OGAMI[1], KATASHI FUJII[2], YUKIO MANABE[3], KOHEI OTA[4], and ASUO YONEKURA[5]

[1]*Hiroshima University, Hiroshima, Japan*
[2]*Professor, Hiroshima University, Hiroshima, Japan*
[3]*Center for Collaborative Research Community Cooperation, Hiroshima, Japan*
[4]*Hiroshima Prefectural Technology Research Institute, Hiroshima, Japan*
[5]*Yonekura institute of Technology for Social Infrastructure Co., Ltd, Hiroshima, Japan*

Surface preparation is very important in re-painting of steel structures so as to extend the effective term of corrosion prevention. Though grinding or blasting have been widely used to remove rust and old coating film on steel surface, both these methods have difficulty in completion of rust-removal and cause some problems such as dust scattering, noise, etc. In order to solve these problems, this paper presents the laser cleaning method which instantaneously sublimes/evaporates the rust on the surface of steel structure. The authors investigate the effects and the applicability of laser cleaning using the specimens made by accelerated corrosion method. The test results confirmed that the laser cleaning can remove the rust almost much as using sand blast, and the salt on the surface of steel can also be evaporated as good as the rust. Moreover, this method seems can be applied on wet surface condition because the moisture and water on the surface of steel can also be evaporated.

Keywords: Evaporate, Remove rust, Remove salt, Wet condition, Fiber laser.

1 INTRODUCTION

In recent years, maintenance of the steel structure is regarded as important. For the long life of steel structure, re-painting is carried out periodically. As listed in Table 1, the durability of the anti-corrosive is greatly influenced by the quality of surface preparation[1]. Therefore, blasting which can completely remove rust and old coating becomes the basic of surface preparation. However, blasting has some problems such as the need for dust scattering prevention and noise control, and the increasing in cost processing of large amount of grinding material and old coating film. As a solution to the problems above, laser cleaning method has been noticed[2,3]. This method instantaneously sublimes/evaporates rust by irradiating high energy density laser.

---

[1] Japan Bridge Association Inc.
[2] Laser cleaning of steel for paint removal, G.X. Chen, T.J. Tan, Y.S. Choo, M.H. Hong, Appl Phys A Materials Science & processing, 2010.
[3] Corrosion Deterioration of Painted Steels with Different Surface Preparations of Steel Substrates, Yoshihito Itoh, In-tae Kim, Shigenobu Kainuma, Yoshihisa Kadota, Japan Society of Civil Engineers No.766/I-68, 291-397, 2004.

Moreover, it can be expected that not only rust and old coating film but also salt can be simultaneously removed from the surface of material.

In this paper, primary tests were conducted for verification of the applicability of laser cleaning technology. At the same time, in consideration of the implementation in marine structures and steel structures under the rainy weather, the authors also examined whether or not can be performed immediately painting after the surface preparation to the surface which was in wet condition.

Table 1. Influence of various factors to the lifetime of coating.[1)]

Factor	Contribution ratio (%)
Surface preparation	52.7
Number of coating times	16.1
Types of coating material	7.9
Other factors	23.3

Note 1: The test environment is the actual bridge of countryside
Note 2: Test paint is lead-based anticorrosive plus ready-mixed paint

2 SPECIMENS AND METHOD OF THE TEST

The specimens were prepared by carrying out the brine bubbling accelerated corrosion test[4] as listed in Table 2.

Figure 1 illustrates the rust condition of the specimen. Laser cleaning was conducted by laser marking device to the specimens. Summary and appearance of the laser marking device are shown in Table 3 and Figure 2. Laser irradiation range is rectangular 40mm by 20mm, and the initial setting of laser at the time of cleaning is the moving speed of 1.0mm/s and the heat input of about 1.3j/mm^2. In addition, each specimen has changed the frequency for each No.1=30kHz, No.2=50kHz, No.3=100kHz. We irradiated with a laser several times at the same range of each specimen, and the test was terminated when the rust on the surface cannot be confirmed by eyes.

After the test, we performed component analysis of the steel surface using a scanning electron microscope to confirm the effect of removing the rust and salt. In addition, the surface condition at the time of laser cleaning was not the dry condition which is required by the dry blasting etc, but it was wet condition such as marine structures and steel structures under the rainy weather. If this method can remove the rust equal to blasting in wet condition, it can be expected to be greatly reduced the limitation by the environmental condition.

Figure 1. The rust condition of the specimen.

[4] Guidelines for Evaluation of Durability and Load-carrying Capacity for Steel Structures under Marine Environment, Japan Society of Civil Engineers, March 2009.

Table 2. The test conditions.

Item	Details
Steel grade	SS400
Surface condition	Nothing is coated
Salinity	About 5%
Temperature	50℃
Air quantity	2.5 l/mm
Test period	About 2 month

Table 3. Summary of the laser marking device.

Oscillation method	Yb-fiber laser
Output	30 W (Processing end 26.6W)
Wavelength	1060-1080 nm
Focus diameter	About 40 μm
Power density at the focal point	7.1×10^8 W/cm^2 note:30kHz oscillation mode

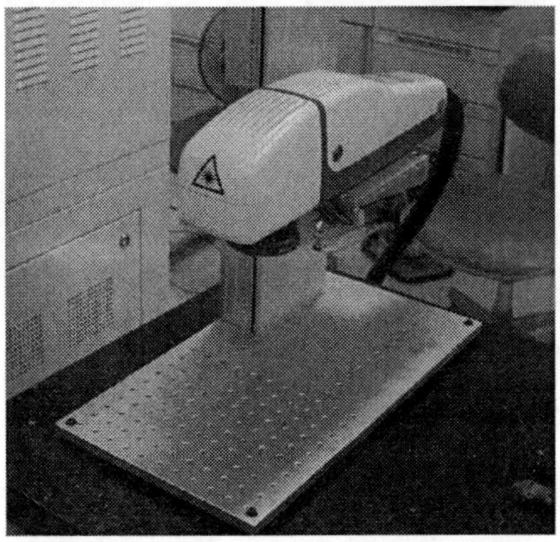

Figure 2. Appearance of the laser marking device.

3 RESULT

3.1 Consideration about Appearance of Test Specimens

Figure 3 shows the analysis results of steel surface by the scanning electron microscope. In this test, each laser irradiation number of times is No.1 = 7 times, No.2 = 5 times and No.3 = 6 times. It can be seen that the rust is roughly removed in all specimens (Figure 3). The surface appearance of No.1 was different from the other ones; it has light yellow color because of the impact of the oxidation reaction at the time of laser irradiation. On the other hand, the surface of No.3 came to burn a little. Consequently, the surface of No.2 has become the best-looking by Laser cleaning in this test.

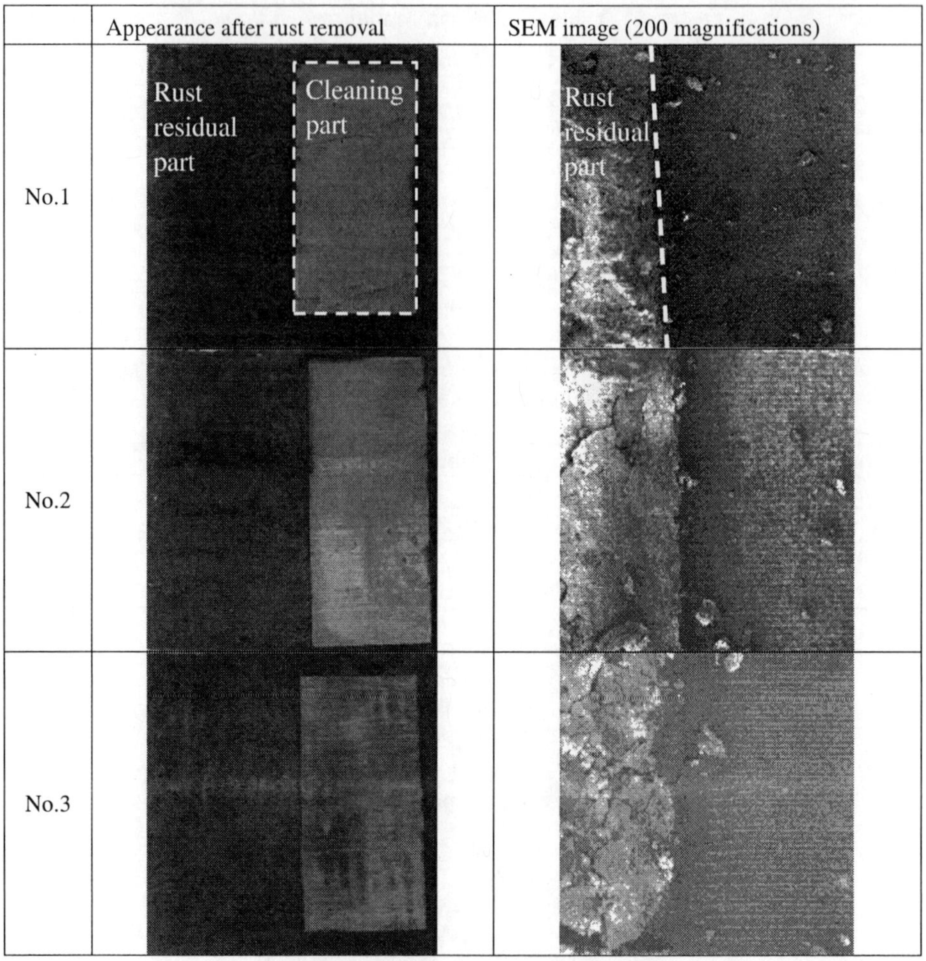

Figure 3. The analysis results of steel surface.

Thus, it is found that there are differences in appearance of the steel surface by the initial setting of the laser. However, it was revealed that regardless of the initial setting of the laser, the surface preparation with laser was approximately equal to the blasting in visual confirmation. In previous study, it has been said that there exist initial thresholds during laser cleaning[5]. Therefore, we need to investigate a more appropriate initial setting by performing the laser cleaning with many more parameters from now on. As a result, the water on steel surface was thoroughly evaporated.

3.2 The Effect of Salt Removal by Laser Cleaning

Figure 4 shows the results of the energy dispersive x-ray analysis using the scanning electron microscope. As shown in the Figure 4, chlorine molecular (Cl) is represented by gray color. Compared to the rust residual part, it can be seen that the chlorine molecular (Cl) i.e. the salt is

[5] Parameters and surface performance of laser removal of rust layer on A3 steel, Zemin Wang, Xiaoyan Zeng, Weiling Huang, Surface and Coatings Technology 166, 2003

removed in cleaning part. This also suggests from Table 4 that shows the comparison of the weight percent of sodium molecular (Na) and chlorine molecule (Cl) in specimens No.2. Table 4 indicates that both molecules can be eliminated by this method more than 98% in the measurement range. And also it can be realized that the rust is eliminated at the same time from Table 5 that shows the weight percent of Oxygen molecule (O) and iron atom (Fe) in specimens No.2.

According to the above, it is thought that the simultaneous removal the rust and salt of steel structure is possible by laser cleaning. In addition, it can be expected to apply the laser cleaning in marine structures and the steel structures under the rainy weather.

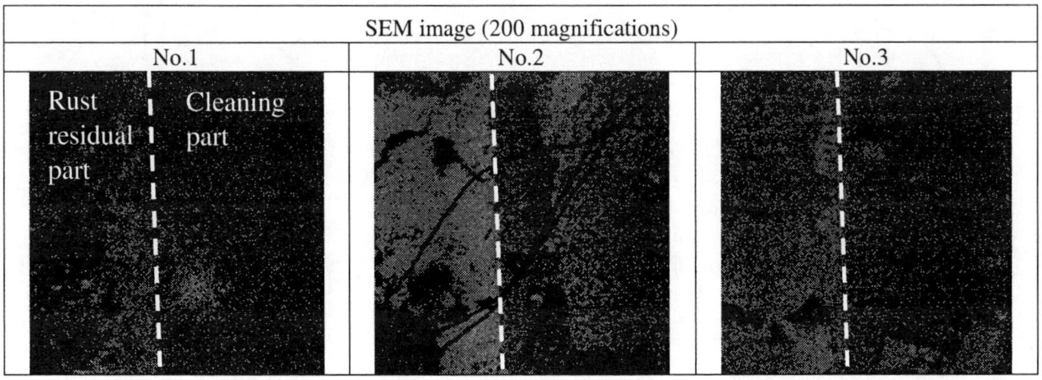

Figure 4. The result of steel surface analysis (chlorine molecular Cl).

Table 4. The weight percent ratio of Na and Cl in the specimen No.2.

	Rust part (%)	Cleaning part 1 (%)	Cleaning part 2 (%)
Na	15.95	0.31	0.10
Cl	10.19	0.07	0.05

Table 5. The weight percent ratio of O and Fe in the specimen No.2.

	Rust part (%)	Cleaning part 1 (%)	Cleaning part 2 (%)
O	15.58	4.83	5.62
Fe	52.3	88.0	74.6

3 CONCLUSION

Surface preparation using a laser has equal effect as blasting in the case of rust removal. Regarding the setting condition of laser in this study, in order to completely remove the rust, it is required a laser irradiation of about 5 times in the same range.

Based on the analysis result using the scanning electron microscope, when using a laser for surface preparation, not only the rust but also the salt is removable at the same time.

In this paper, the steel surface was in wet condition such as marine structures and steel structures under the rainy weather. In the case of blasting, it is difficult to perform the painting just after surface preparation because of difficulty of water removal. However, the construction time can be minimized because the laser cleaning is able to evaporate water.

References

Chen, G. X., Kwee, T. J., Tan, K. P., Choo, Y. S., & Hong, M. H. (2010). Laser cleaning of steel for paint removal. *Applied Physics A*, *101*(2), 249-253.

In-Tae, K. I. M., KAINUMA, S., & KADOTA, Y. (2004). Corrosion deterioration of painted steels with different surface preparations of steel substrates. *Doboku Gakkai Ronbunshu*, *2004*(766), 291-307. Japan Bridge Association Inc., Q and A of the steel bridges, September, 2006.

Wang, Z., Zeng, X., & Huang, W. (2003). Parameters and surface performance of laser removal of rust layer on A3 steel. Surface and Coatings Technology, 166, 2003.

Integrated Solutions for Infrastructure Development
Edited by Lau, H. H., Tang, F. E., Ng, C. K., and Singh, A.
Copyright © 2016 ISEC Press

ESTIMATING LOAD-DISPLACEMENT BEHAVIOR OF SOFT CLAYS

MURAT TONAROGLU[1], CEM AKGUNER[2] and MURAT E. SELCUK[1]

[1]*Dept of Civil Engineering, Yildiz Technical University, Istanbul, Turkey*
[2]*Dept of Civil Engineering, TED University, Ankara, Turkey*

Prediction of long term settlement of soft soils below civil engineering structures is an important issue in geotechnical engineering. The data from laboratory consolidation tests are used to estimate the ultimate settlement and problems arise in predicting settlement-time behavior in the field. In order to overcome this difficulty some empirical models or adjustments have been proposed based on field settlement measurements, even if they are available at least for the early stages of loading. In this study, laboratory test results and field settlement measurements obtained for the long term settlement of clayey layers underlying the Alibey Dam in Istanbul, Turkey are used in conjunction with a model proposed by Edil and Mochtar (1984) for peat-like soils. The soil parameters of the proposed model have independently been obtained using both laboratory test data and field measurements, and model predictions are compared with actual recorded settlements. It is observed that the field measurements could be predicted more closely if the model parameters are obtained from field measurements, but predictions based on laboratory consolidation and creep test results also provide satisfactory results following the initial stages of loading.

Keywords: Consolidation, Strain, Strain rate, Long-term settlement, Peat, Field measurement.

1 INTRODUCTION

Over the years, the prediction of long term in-situ settlement-time behavior of clay layers underlying civil engineering structures has been a common topic for many researchers. Similarly, in this study, a method proposed by Edil and Mochtar (1984) for peat-like soils, which is a simplified approach to Gibson-Lo (1961) model, is utilized to model the long-term compressive behavior of organic clayey soils comprising the foundation of Alibey Dam in Istanbul, Turkey. This unique dam was constructed in approximately fifteen years using a staged construction approach. Furthermore, it was monitored by the installed instrumentation for more than twenty-five years (Berilgen *et al.* 2005, Berilgen *et al.* 2006, Ozcoban *et al.* 2007). Both laboratory incremental loading consolidation and creep test results and recorded field settlements at different stages of embankment construction of the dam are used in conjunction with a prediction approach proposed by Edil and Mochtar (1984) for long term field settlement of clayey with organic content. Computed values are compared with the actual recorded values.

2 MODEL FOR SETTLEMENT-TIME BEHAVIOR

The model proposed in the study by Edil and Mochtar (1984), which is similar to that by Edil and Dhowian (1979), is used to predict the long-term compression of peat-like soils. In this model, the strains due to the long-term compression of soils are obtained using the following equation;

$$\varepsilon(t) = \Delta\sigma\left[a + b(1 - e^{-(\lambda/b)t})\right] \tag{1}$$

here, $\varepsilon(t), \Delta\sigma, a, b, \frac{\lambda}{b}$ are the time-dependent vertical displacement, stress increment, primary and secondary compressibility parameter, and the rate factor for the secondary compression respectively. These empirical parameters are obtained by first drawing the $(t - \log \dot{\varepsilon}_v)$ curve to form a linear set of points at the time interval corresponding to the portion of secondary compression of the total settlement. Secondly, the slope of the best straight line is drawn through the linear portion and the intersection point of the straight line and the vertical y-axis is calculated to determine the needed parameters:

$$\text{The slope, m} = -0.434\lambda / b \tag{2}$$

$$\text{The intersection} = \log(\Delta\sigma\lambda) \tag{3}$$

$$a = \frac{\varepsilon(t_k)}{\Delta\sigma} - b + be^{-(\lambda/b)t_k}\Delta\sigma\left[a + b(1 - e^{-(\lambda/b)t})\right] \tag{4}$$

where, the parameter $\Delta\sigma$ in Eq. 3 is the difference between the value of stress at any time and stress at the ultimate time during a stage. The parameter t_k in Eq. 4 represents the time at which the last strain readings were taken at a specific loading stage.

3 APPLICATION OF MODEL

The suggested model has been applied to results from multiple stage consolidation and creep tests with corresponding $\log \dot{\varepsilon}_v - t$ curves for each loading stage to obtain the model parameters. However, in multiple staged loading tests the loading is usually terminated before long-term compressions are completed and the obtained data is not sufficient for a long-term prediction model. Therefore, a best-fit curve is drawn through available experimental measurements to obtain a continuous compression curve and then these settlement-time curves are used to obtain $\log \dot{\varepsilon}_v - t$ curve. It is worthwhile to note that a linear behavior is observed in these curves after the initial stages of loading. As mentioned above, if the slope of the linear section is denoted with *m* and the intersecting point of the extension of linear portion with the vertical axis with *n*, Eqs. 2,3 and 4 can be used to obtain model parameters and strain values at any time can be determined using Eq. 1. Comparison of the compression-time curves obtained by using the above relationships with the laboratory test readings are shown in Figure 1 for a load of 250 kPa and the model parameters used in analysis are listed in Table 1.

Although the model by Edil and Mochtar (1984) is not able to predict the early stages of the compression-time behavior observed in laboratory tests, the match with laboratory measurements is very well for the points after approximately 300 minutes.

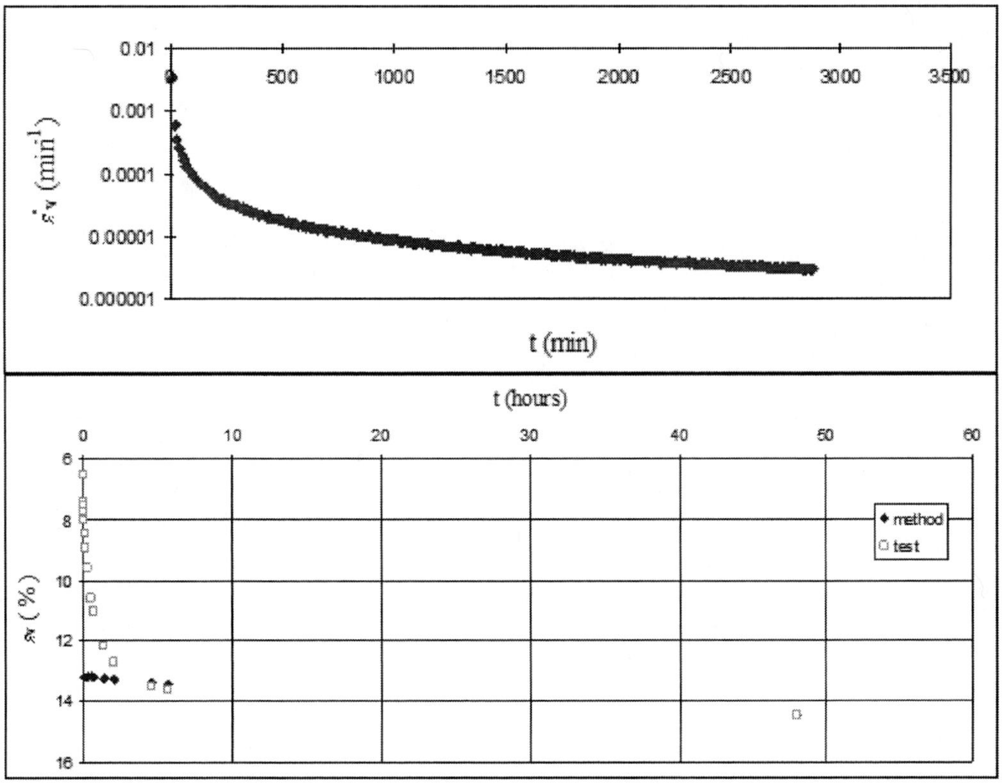

Figure 1. Consolidation data for staged loading and comparison the suggested method.

Table 1. The values of model parameters obtained from a creep test.

Model parameters	λ / b (min^{-1})	b (kPa^{-1})	a (kPa^{-1})
$\sigma'_v = 250kPa$	4x10^{-4}	1.2x10^{-4}	1x10^{-3}

In the second step of the study, settlements of soft clayey underlying Alibey Dam foundation layers are attempted to be predicted utilizing the same methodology. The embankment for the Alibey Dam in Istanbul, Turkey was constructed in stages with extended periods of waiting during loading stages in order to avoid bearing capacity problems and excessive displacements. A comprehensive in-situ measurement scheme was employed to monitor the field behavior during and after construction. The in-situ settlement readings at the dam cross-section during two of the stages (2 and 3) of construction are compared with the settlements calculated by the method (Figure 2). The model parameters obtained from in-situ measurements for all six stages of construction are presented in Table 2.

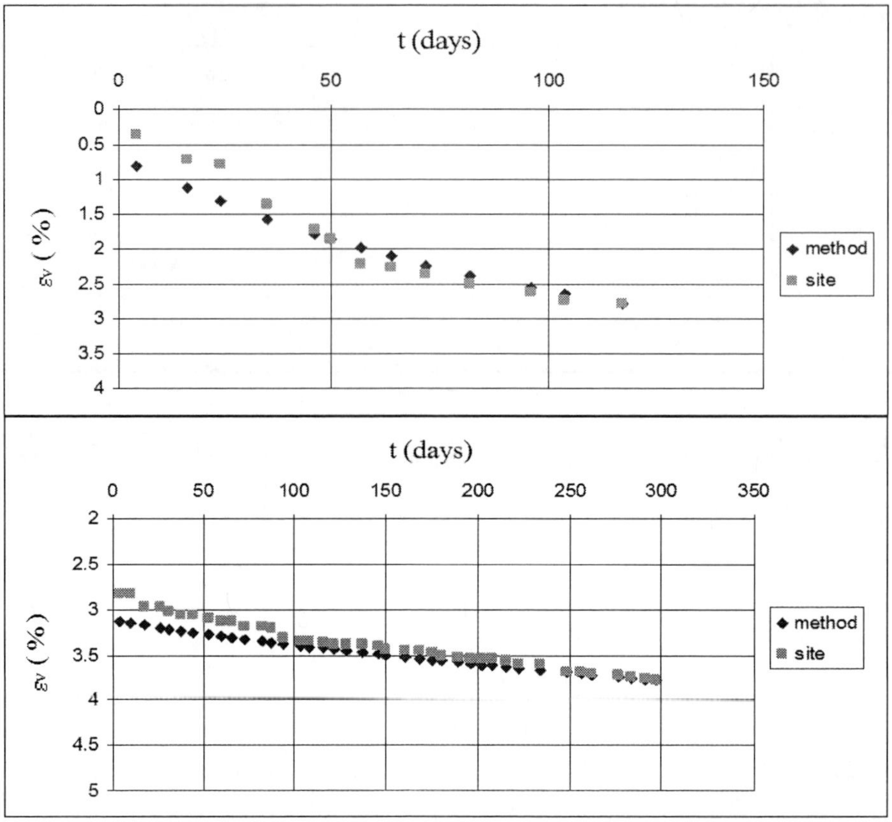

Figure 2. Comparison of computed and recorded strain curves for several loading stages.

Table 2. The values of model parameters obtained from in-situ measurements for six stages.

Model parameters	Stage 1	Stage 2	Stage 3	Stage 4	Stage 5	Stage 6
λ / b ($\min^{-1}$)	$1x10^{-5}$	$6.2x10^{-6}$	$1.3x10^{-6}$	$2x10^{-6}$	$4x10^{-7}$	$1.5x10^{-7}$
b (kPa^{-1})	$9x10^{-7}$	$5.8x10^{-4}$	$2.7x10^{-4}$	$1x10^{-4}$	$3x10^{-4}$	$2x10^{-4}$
a (kPa^{-1})	$7x10^{-6}$	$1.3x10^{-4}$	$5.7x10^{-4}$	$1x10^{-3}$	$2x10^{-3}$	$2x10^{-3}$

Starting with the second stage of loading there is an increasing degree of agreement between the computed and measured settlement. The computed and recorded values are observed to be particularly close at final stages of loading at each loading stage. Thus, it can be stated that for the case of model parameters being obtained from in-situ measurements, the model is very successful in predicting the long-term settlements. In the last step of this study, the method has been utilized with the model parameters obtained from laboratory consolidation tests selected based on the loading conditions in the field. The model parameters obtained from laboratory consolidation tests carried out in loading increments compatible with the field embankment construction program are given in Table 3. The computed and measured settlements under the dam for a period of more than twenty years at the considered section are shown in Figure 3. Settlements computed with parameters obtained from both laboratory tests and field measurements are provided on the same figure together with the measurements. Considering the complicated field soil conditions and staged embankment construction extending over 15 years, it can be said that a reasonable degree of agreement is achieved between the predicted and recorded settlements starting with the second stage of the loading.

Table 3. The values of model parameters obtained from laboratory consolidation tests.

Model parameters	Stage 1	Stage 2	Stage 3	Stage 4	Stage 5	Stage 6
λ / b ($\min^{-1}$)	$3x10^{-4}$	$3x10^{-4}$	$3x10^{-4}$	$3x10^{-4}$	$3x10^{-4}$	$3x10^{-4}$
b (kPa^{-1})	$2x10^{-4}$	$2x10^{-4}$	$1x10^{-4}$	$1x10^{-4}$	$2x10^{-4}$	$2x10^{-4}$
a (kPa^{-1})	$2x10^{-4}$	$2x10^{-4}$	$9x10^{-4}$	$9x10^{-4}$	$1.8x10^{-3}$	$1.8x10^{-3}$

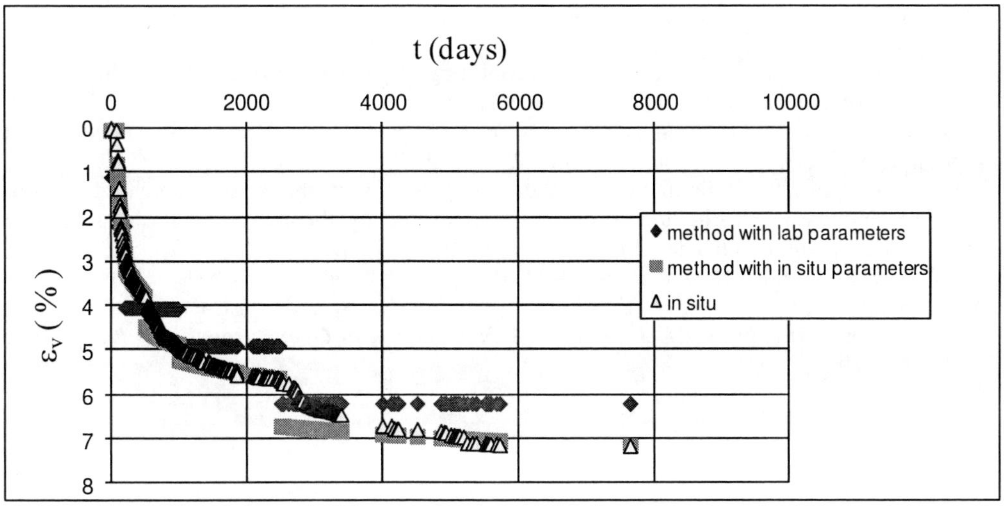

Figure 3. The comparison of all strain values obtained using methods with actual measurements.

Based on the findings of this study, the use of the method with parameters obtained from short-term in-situ readings is quite successful in predicting the field settlements. However, when the model parameters are obtained from laboratory tests, the compatibility between computed and recorded values are not as good especially at the initial stages of loading steps.

4 CONCLUSIONS

In this paper, findings of a study in which the prediction of long term settlements of soft clayey soils underlying an earth dam embankment constructed through a staged fill approach are presented. A soil compression model proposed by Gibson-Lo (1961) and modified for peat-type soils by Edil and Mochtar (1984) is used to predict the long-term settlements below an earth dam constructed over a period of about fifteen years. Comparisons with the field settlement measurements indicate that if model parameters are obtained from laboratory tests performed by taking into account the loading and stress conditions in the actual construction, although at the initial stages of loading the agreement is not very good, there is an increasing degree of agreement with increased settlements, and ultimate values especially are predicted quite accurately. On the other hand, if the model parameters are obtained from in-situ measurements ability to predict the settlement- time behavior improves considerably. In general, it may be stated that the proposed methodology may be utilized as a good tool to predict the long-term field behavior from laboratory test data and/or field settlement readings recorded for a limited time.

Acknowledgement

The authors would like to thank Prof. Dr. Kutay Özaydın and Prof. Dr. Tuncer Edil for their support, encouragement and kindness.

References

Berilgen, S.A., Berilgen, M.M. and Özaydın, K., Determination of Consolidation Behaviour of Haliç Dredged Material Using a Seepage Induced Consolidation Test, Management of Environmental Quality: An International Journal, 16-6, 681-690, 2005.

Berilgen, S.A., İpekoğlu, P. and Berilgen, M.M. and Özaydın, K., Assessment of Consolidation Behavior of Golden Horn Marine Dredged Material, *Marine Georesources and Geotechnology*, Taylor and Francis Group, 24, 1-16, 2006.

Gibson, R. E. and Lo, K. Y., A Theory of Soils Exhibiting Secondary Compression, Publication No. 41, Norwegian Geotechnical Institute, Oslo, 1-16, 1961.

Edil, T. B. and Dhowian, A. W., Analysis of Long Term Compression of Peat, *Geotechnical Engineering Southeast Asian Society of Soil Engineering*, 159-178, 1979.

Edil, T. B. and Mochtar, N. E., Prediction of Peat Settlement, Proceedings, Sedimentation Consolidation Models Predictions and Validation, symposium sponsored by the ASCE Geotechnical Engineering Division in conjunction with The ASCE Convention in San Francisco, CA, 1 Oct. 1984, R. N. Young and F. C. Townsend, Eds., New York, 411-424, 1984

Ozcoban, S., Berilgen, M.M., Kılıç, H., Edil, T.B., Özaydın, İ.K., Staged construction and settlement of a dam founded on soft clay, *ASCE Journal of Geotechnical and Geoenvironmental Engineering,* Vol. 133 (8), 1003-1016, 2007.

Integrated Solutions for Infrastructure Development
Edited by Lau, H. H., Tang, F. E., Ng, C. K., and Singh, A.
Copyright © 2016 ISEC Press

PULL-OUT PROPERTIES OF GYPSUM BOARD ANCHORS

YOSHINORI KITSUTAKA[1] and FUMIYA IKEDO[2]

[1]*Dept of Architecture and Building Engineering, Tokyo Metropolitan University, Tokyo, Japan*
[2]*Kajima Co., Ltd., Tokyo, Japan*

In this study, pull-out properties of various anchors embedded in gypsum board were investigated. Tensile tests for anchors embedded in 200mm square size gypsum board were conducted to measure the load - load displacement curves. Strength of gypsum board was changed for three conditions and twelve kinds of anchors were selected which were ordinary used for gypsum board anchoring. The loading conditions were a monotonous loading and a repeating loading controlled by a servo-controlled hydraulic loading system to achieve accurate measurement. The fracture energy for each anchors were estimated by the analysis of consumed energy calculated by the load - load displacement curve. The effect of the strength of gypsum board and the types of anchors on the pull-out properties of gypsum board anchors was cleared. A numerical model to predict the load-unload curve of pull-out deformation of gypsum board anchors caused by such as the earthquake load was proposed and the validity on the model was proved.

Keywords: Gypsum board, Anchor, Tensile test, Pull-out property, Repeating loading, Load-unload curve.

1 INTRODUCTION

Gypsum boards have been used as an interior material for numerous buildings, with various interior fixtures and anti-seismic devices being fitted to them. However, screws directly driven into gypsum boards are prone to pullout, due to insufficient load-bearing capacity of the boards, making it difficult to safely secure such fixtures and devices to the boards. The use of gypsum board anchors (hereafter simply referred to as "anchors") is effective in securing such devices on the boards and complementing the load-bearing capacity. While anchors come in wide ranges of shapes and materials, few studies have compared their pull-out properties (Takahashi and Okamoto 2013) with no quantitative indices having been provided for their selection. This paper reports on a study in which tension tests were conducted on various types of anchors by monotonic and cyclic loading to investigate the characteristics of each type and proposes a method of estimating a unidirectional history curve, which is necessary for evaluating the load-bearing capacity under cyclic seismic loads.

This paper reports on a study in which tension tests were conducted on various types of anchors by monotonic and cyclic loading to investigate the characteristics of each type and proposes a method of estimating a unidirectional history curve, which is necessary for evaluating the load-bearing capacity under cyclic seismic loads.

2 EXPERIMENT OVERVIEW

Figure 1 shows the anchors used in the tests. The symbol JN represents a self-drilling screw type; JU, a self-drilling drive type; SO, a drilling insert type; and SN, a drilling screw type.

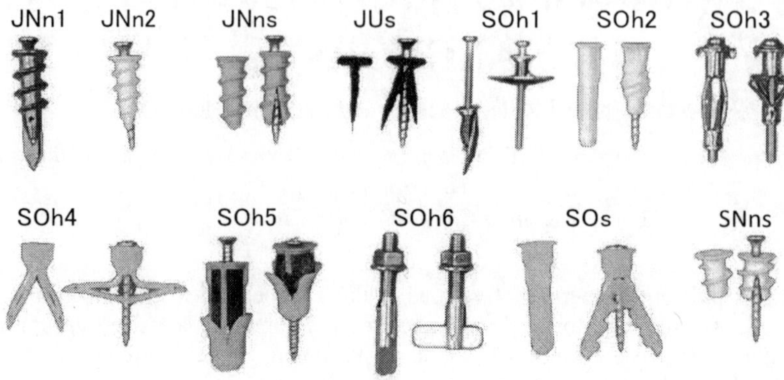

Figure 1. Anchor bolts use in this study.

The gypsum boards were regular gypsum boards 12.5 mm in thickness. An anchor was fixed to the center of each gypsum board 200 mm by 200 mm in size, which was fixed to a steel reaction board, and subjected to unidirectional cyclic tensile displacement using a displacement-controlled fatigue testing machine shown in Figure 2.

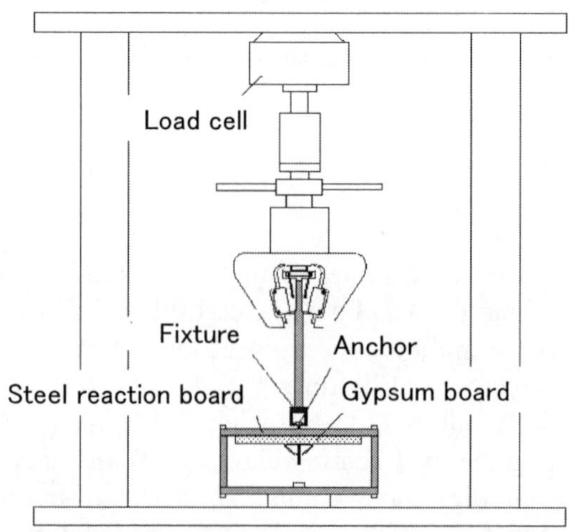

Figure 2. Overview of tensile test.

The displacement rate was 1 mm/min under repeating conditions shown in Figure 3 conforming to the test method specified in the standard test method for metal fastenings and fasteners "Test method of metal fastenings and fasteners for wood frame construction, Chapter 4" (HOWTEC 2015). Monotonic loading tests were also conducted for comparison.

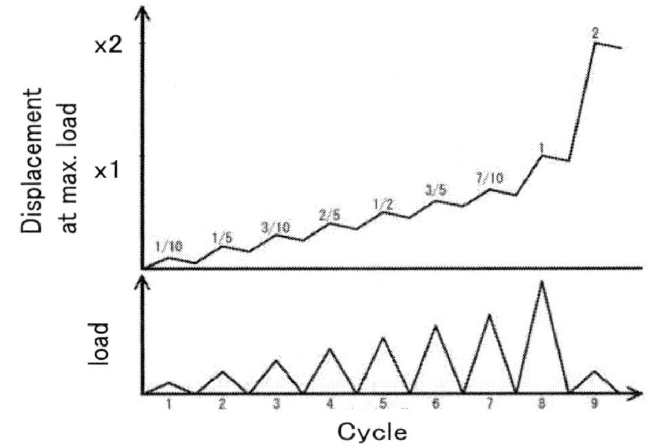

Figure 3. Displacement repeating conditions.

3 ESTIMATION OF LOAD-UNLOAD CURVES BY CYCLIC LOADING TESTS

Figure 4(a) shows a conceptual drawing of a load-displacement curve and unloading-reloading curves at arbitrary deformations and energy changes (Kitsutaka 1997). The unloading curves at C and D are linearly approximated to CA and DB. The elastic strain energy (Ue) at C and D is expressed as the area of ΔCAF and ΔDBE, respectively. The energy externally given to the specimen from C to D (dUw) is expressed as the area of □CFED, and the energy consumed for pullout (dUf) is dUw – dUe. Therefore, dUf = □CFED – (ΔDBE - ΔCAF) = □CAED - ΔDBE =□CABD, which is the gray part of Figure 4(a). As to the accumulated change in the energy to the anchor pull-out length (a), the energy given from the outside (Uw) is the sum of the energy consumed for pullout (Uf) and the elastic strain energy (Ue). This relationship is illustrated as Figure 4(b). Figure 4(b) is determined by determining dUf, dUe, and dUw from the cyclic loading test results of each anchor. Ue is determined from P at an unloading point on the load-displacement curve, where estimation is desired, and the pull-out length (a) in Figure 4(b). Linearly approximated unloading-reloading curves are then evaluated, and history curves under various cyclic tensile loads can be estimated (Figure 5).

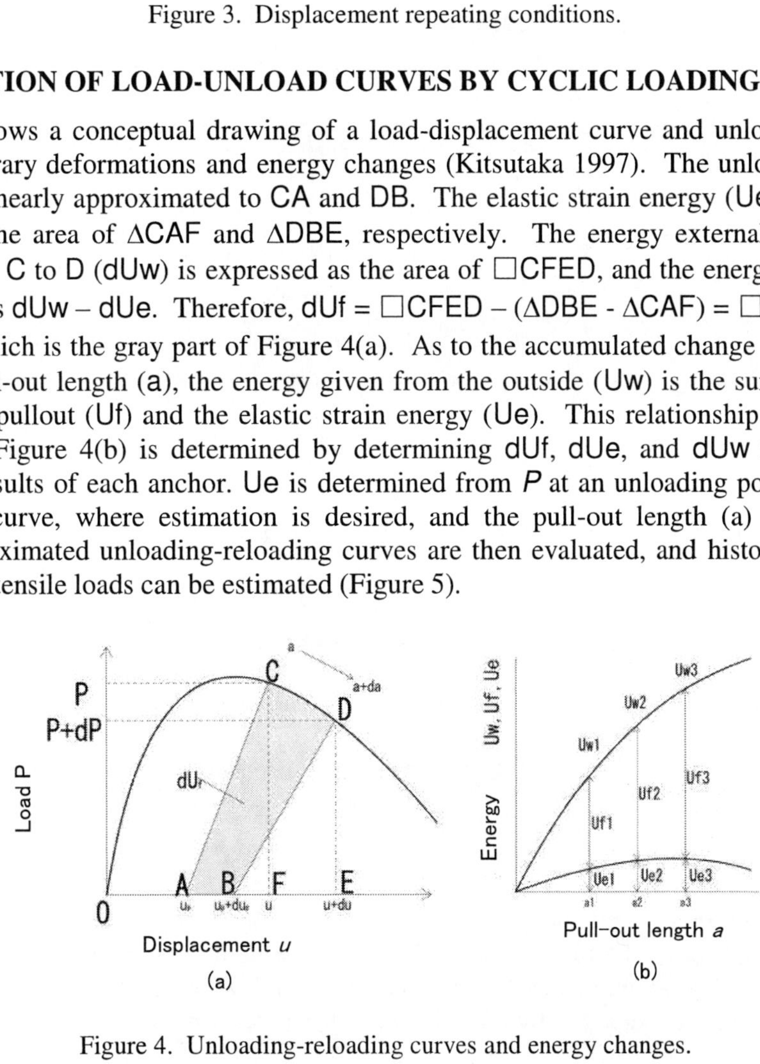

Figure 4. Unloading-reloading curves and energy changes.

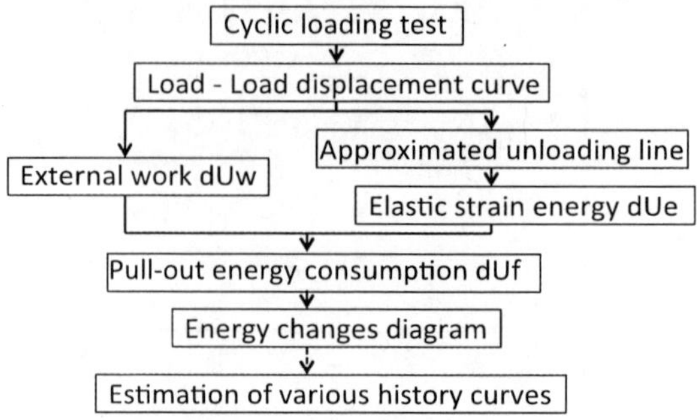

Figure 5. Flowchart of estimating historic curves.

4 TEST RESULTS AND DISCUSSION

Figure 6 shows the test results. The pull-out properties widely vary depending on the anchor type.

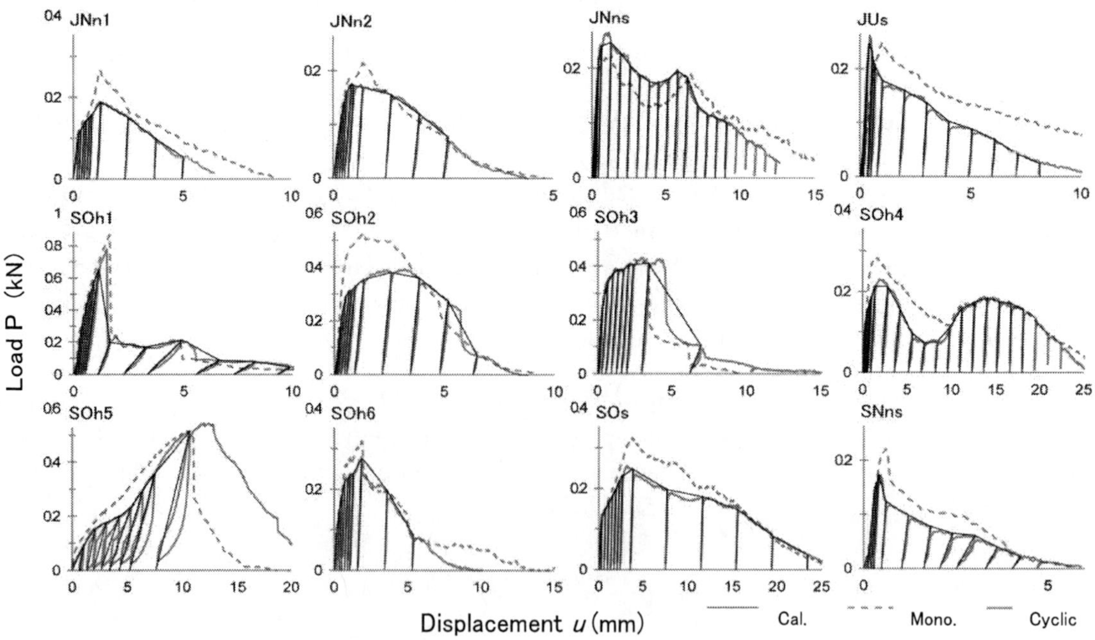

Figure 6. Load – load displacement curves.

4.1 Self-drilling Anchors

Specimens JNn1 and JNn2 lead to small maximum loads and small displacements to 0 kN. JNns lead to a small maximum load and a post-peak increase in the load at around a displacement of 5 mm, due to friction fixing by the expanded end. Also, reductions in the load are relatively slow. JUs lead to a small maximum load, with slow reductions after reaching the peak. Also, self-drilling anchors are characterized by the relatively small effect of the gypsum board type on the pull-out capacity.

4.2 Anchors Requiring a Pilot Hole

As to SOh1, failure of the anchor was observed in the regular-hard gypsum board. This led to rapid increases and reductions in the load, attaining the largest pull-out capacity of all anchors. SOh2 shows a relatively large maximum load, with anchor failure in the regular-hard gypsum board. However, the load did not increase after the peak, as the gypsum board did not break. SOh3 leads to anchor failure associated with rapid increases and reductions in the load. SOh4 leads to anchor failure, but the reductions in the load are very slow, presumably due to the shape of the anchor after failure resembling the type of friction fixing by an expanded end. SOh5, which was made of a soft anchor material, shows increases in the load slower than the other anchors, but the post-peak reductions are rapid. The maximum loads of anchors that grip the board from behind other than SOh4 are very large when being placed into regular-hard gypsum boards. The increases and reductions in the load on SOs are both very slow. SNns shows a small maximum load, with the displacement to 0 kN being small. It was therefore confirmed that the load-displacement curves are clearly characterized by the methods of fixing.

4.3 Differences among Fixing Methods

Figure 7 shows a conceptual drawing illustrating the differences in the pull-out resistance and characteristics among the fixing methods. Anchors fixed by screw friction show small maximum loads at pullout and subsequent slow reductions in proportion to the displacement. These anchors are mostly self-drilling, being impossible to drive into high strength gypsum boards. The type of gypsum board scarcely affected the pull-out capacity. Anchors fixed by friction of expanded ends show moderate curves of increases and reductions in the load, with the type of gypsum board slightly affecting the pull-out capacity. As to anchors that grip the board from behind, the effect of the gypsum board type was significantly observed on the pull-out capacity. These anchors are characterized by the rapid increases in the load to large maximum loads followed by rapid reductions with the failure of the gypsum board. The anchors that grip the board from behind also include specimens in which the anchor fractured, causing re-increases in the load after reductions. SOh4 is a characteristic example in which the expanded end re-resisted the tensile forces after pull-out.

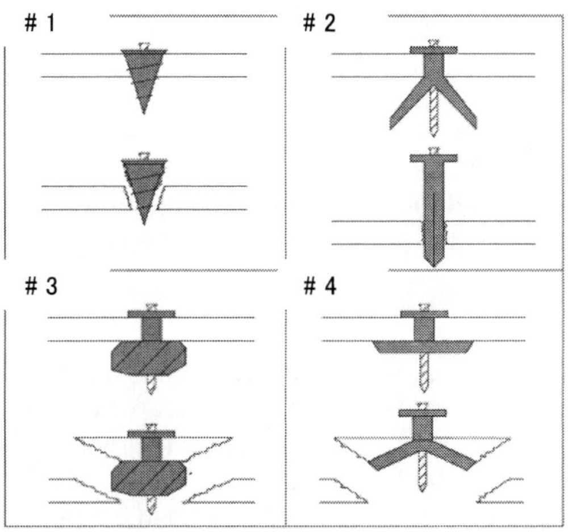

Figure 7. Conceptual drawing illustrating the pull-out fracture.

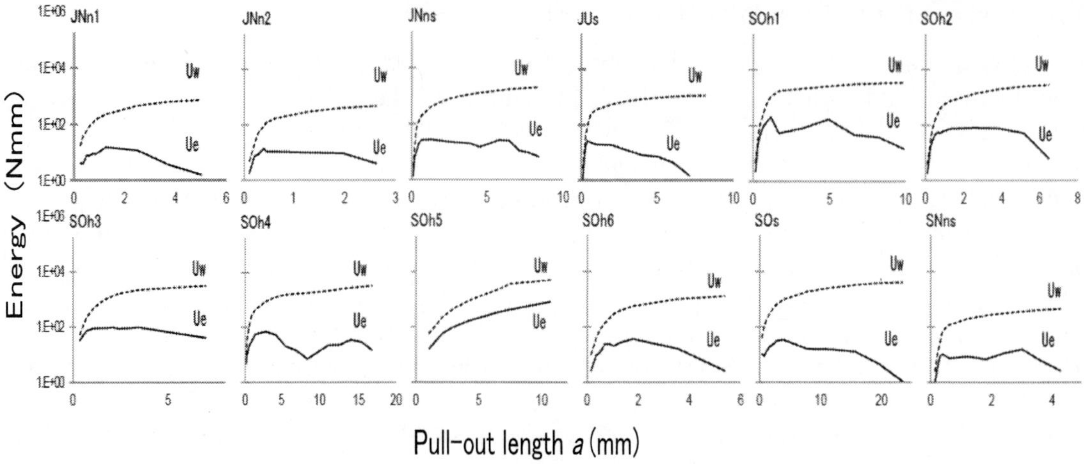

Figure 8. Energy changes in pull-out test.

4.4 Results of Cyclic Loading Tests

In Figure 8, the load-displacement curves under cyclic tensile loads as well as those under monotonic loading were shown. The envelope curves of anchors under monotonic and cyclic loading nearly agree with each other. Figure 8 shows the results of determining changes in the energy on the assumption that the unloading-reloading curves obtained from cyclic tension tests are straight lines. The unloading-reloading curve of each anchor assumed from the energy changes shown in Figure 8 following the estimation flow shown in Figure 7 is also superimposed in Figure 8. The estimations nearly agree with the test results, proving the validity of the estimation method.

5 CONCLUSIONS

Major conclusions of this study are as follows:

(1) Tension tests were conducted on various types of gypsum board anchors by applying monotonic and cyclic loads to grasp the characteristics of each type.

(2) Differences in the pull-out resistance and characteristics among the fixing methods of anchors were cleared.

(3) A method of estimating unidirectional unloading-reloading curves under cyclic loads, such as seismic loads was proposed and the results of estimation by the proposed method agreed with the test results, verifying the validity of the estimation method.

References

Kitsutaka, Y., Fracture Parameters by Poly-linear Tension Softening Analysis, *Journal of Engineering Mechanics,* ASCE, Vol.123, No.5, pp.444-450, (1997).

Standard Test Method of HOWTEC, Test method of metal fastenings and fasteners for wood frame construction, *Japan Housing & Wood Technology Center,* (2015).

Takahashi, H. and Okamoto, H., et al., Pull-out Strength of Tilted Screw from Furring Channel, *Summaries of Technical Papers of Annual Meeting, Architectural Institute of Japan,* A-1, 1109-1110, (2013).

Integrated Solutions for Infrastructure Development
Edited by Lau, H. H., Tang, F. E., Ng, C. K., and Singh, A.
Copyright © 2016 ISEC Press

EFFECT OF SURFACE PREPARATION GRADE ON REPAIR BY MEANS OF COVER PLATES AND ADHESIVE

RYUICHI YOKOTA[1], KATASHI FUJII[1], HISAKAZU HORII[2] and YUYA HIDEKUMA[3]

[1]*Hiroshima University, Hiroshima, Japan*
[2]*Konishi Co., Ltd. Saitama, Japan*
[3]*Nippon Steel & Sumikin Materials Co., Ltd. Tokyo, Japan*

There are a lot of corroded steel bridges, which have to be prolonged its life in Japan. Corroded members in these bridges generally have local corrosion damages, and are often repaired by covering on the corroded area with steel plates in order to rehabilitate the strength. Recently, adhesive such as epoxy resin is often used for bonding the cover plates made of steel or CFRP, even though high-tensile bolts or welding have been adopted conventionally. In this case, surface preparation is the most important issue to bond the cover plates completely. However, complete removal of rust and wastes are quite difficult in the fieldwork of surface preparation. This paper presents recovery effect of local corrosion damage by means of covering method used by adhesive when the rust is not completely removed. Axial tensile tests were conducted for rectangular steel plates of 1,100mm x 100mm x 12mm with 400mm length of corrosion-surface-unevenness on the both surfaces, and the defects of surface preparation are made by putting tape paper at the damaged area. Then various sizes of the repair plates are bonded to consider the influence to repairing. From the test results, the ultimate strength decreased 7-10% compared with non-rust state when the defect area rate was about 30%.

Keywords: Corrosion, Adhesive repair, Steel prate, CFRP, Partial repairing.

1 INTRODUCTION

Currently, aging of infrastructure constructed in 1970s has become a social problem in Japan. The damage of a steel bridge is roughly categorized into corrosion and fatigue, where corroded steel has been conventionally repaired by cover plates used high-tensile bolt or welding. On the other hand, adhesive is recently being expected for bonding cover plates such as steel plate and CFRP (Carbon Fiber Reinforced Plastics), and CFRP has been sometimes used in real structures.

In this case, it is necessary to remove completely rust and dusts in corroded area and non-corroded area around the corroded area, in order to bond cover plates firmly. However, when the bonded area is too large, such as corroded area occurred most of the flange in a plate girder, it may not be so easy to bond adequately to the area. Therefore, in such a case, it can be expected to repair partially against only the dangerous corroded area so as to bear the design load. Furthermore, it is also not easy to remove the rust perfectly by electric tools such as a grinder. In this paper, we investigate the effects of the repairing with partial cover plate and the effects of incomplete surface preparation in order to create a basic data for establishment of design method in practice.

2 EXPERIMENT[1] TO CHANGE ADHESIVE LENGTH

We performed a static tensile testing of the specimens, which are subjected to simulate corrosion based on the Corroded Surface Generation Model[1)]. In this model, thinning due to corrosion is reproduced by scraping surface of undamaged steel plate by endmill. Thus, it is able to evaluate the repair effect quantitatively by using a number of specimens with the same decrease.

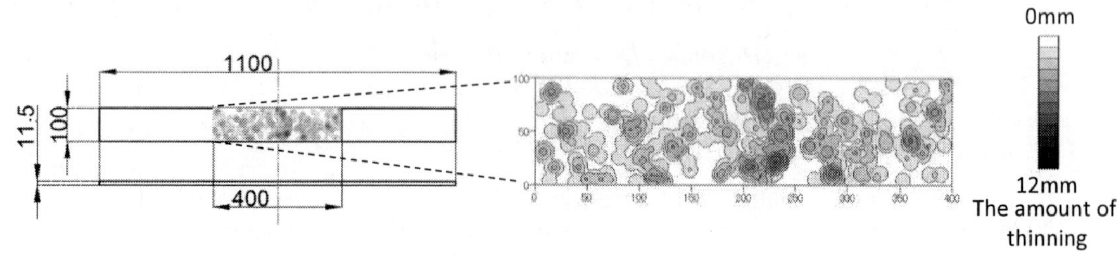

Figure 1. The details of decreased base steel.

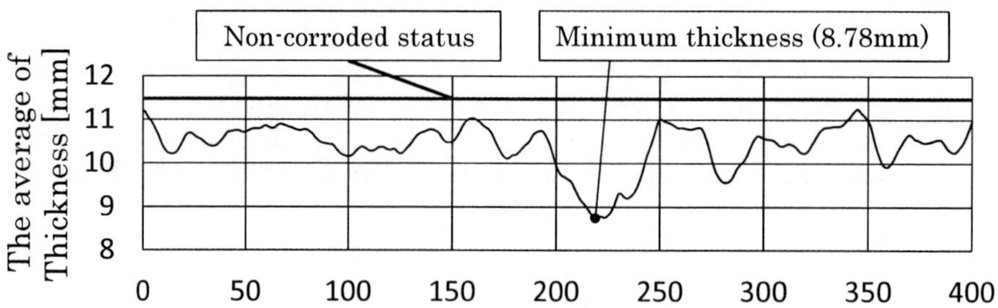

Figure 2. Profile of reduced average thickness.

Table 1. The detail of base steel, repair steel, CFRP, and adhesive.

	Thickness [mm]	Yield stress [N/mm²]	Tensile strength [N/mm²]	Young's modulus [kN/mm²]
Base steel	11.5	296	441	205
Repair steel plate	2.3*	326	436	203
Repair CFRP plate	0.143x4*		1900	640

* Per side.

Name	Application		Compression shear strength [N/mm²]
E258	Bonding steel		27.5
FP-N9	Bonding CFRP	Primer	18.2
FB-E9S		Unevenness modifying	16.7
FR-E9P		impregnating	15.4

Figure 1 shows the details of the specimens, and the designed average thickness of the cross-section is shown as Figure 2. It was verified that the reduced thickness of each specimen was nearly same as designed thickness by measuring with a laser measuring instrument. The average of reduced cross-sectional area at the 219mm position is minimum with value of 8.78mm,

[1] Fujii, K., Kaita, T., Nakamura, H., and Ario, I., A model generating surface irregularities with consideration of corrosion progress in aging, Japanese Society of Civil Engineers, 50A, 657-665, 2004.

therefore it is decided to adhere repair plates around this position. In this section, both methods (steel plate and CFRP) were implemented by five types of repair length (50mm, 100mm, 200mm, 300mm, and 450mm). In the case of adhesive repairing, repair effect is not obtained at the end of the adhered area; nevertheless the adhered area (length) is defined repair area (length) in this study for convenience.

Table 1 shows the details of steel plate, repair plate, and adhesive. In the case of CFRP repair, the product of the cross-sectional area and Young's modulus is assured in adhered four layers per side[2)].

3 RESULTS

3.1 Results of Method to Adhere Steel Plates

Figure 3 shows the maximum load of each specimen before the effect of repair plates is lost. It clarifies that the maximum load increases when the adhesive length is longer. S-5 specimen having 450mm in repair length is adhered to all of damaged area and a part of undamaged area; therefore, the strength of S-5 specimen has increased higher than that of undamaged status. On the other hand, in both of the specimens having 200mm and 300mm repair length, these strength have increased up to 98-99% compared with that of the undamaged status.

Figure 4 shows the strain of the base steel when the specimens was subjected by the loading 300kN; whereas, 300kN is the load that base steel without repairing will yield partially. At the 219mm position, which is the minimum cross-sectional area position, it can be seen that the strain of all the specimens except for N-1 and S-1 is almost equal. It is known that there is no stress concentration in cross section having certain distance or more from the end of adhering-plate area[3)].

From Figures 3 and 4, it can be said that it is possible to obtain sufficient partial repair effect if steel plate is adhered in the range of 200mm or more in this case.

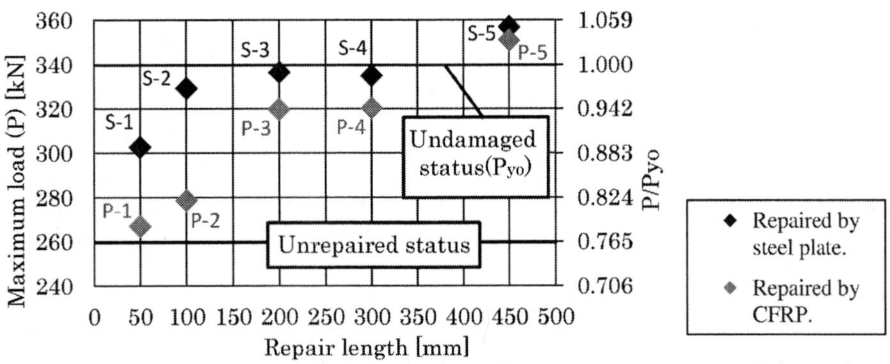

Figure 3. Maximum load.

---

[2] Nippon Expressway Research Institute Company, Guide to design and construct repairing of steel structures using CFRP, 2013.

[3] Aoki, Y., Banno, R., Ishikawa, T., Kawano, H., and Adachi, Y., Bending behavior of thickness-reduced steel member repaired with patch plate on one side, Japanese Society of Civil Engineers, 59A, 647-656, 2013.

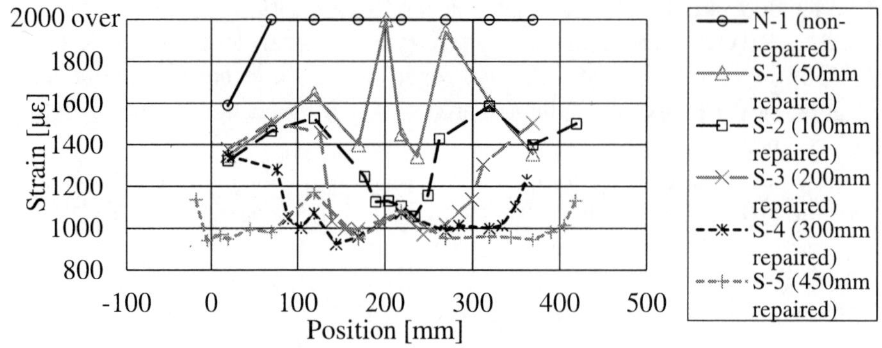

Figure 4. Strain of base steel (Repaired by steel plate).

3.2 Result of Method to Adhere CFRP

From figure 3, the strength of P-5 specimen having 450mm in repair length has increased higher than that of undamaged status, similar to the specimen adhered the same length of steel plate. On the other hand, both of the strengths of the specimens having 200mm and 300mm repair length have increased up to 94% that of the undamaged status. Figure 5 shows the strain of the base steel loaded 300kN. It is clearly that the strain of all the specimens except for N-1, P-1, and P-2 is almost equal at the 219mm position.

From Figures 3 and 5, it can be said that it is also possible to obtain sufficient strength repair effect if CFRP is adhered in the range of 200mm or more in this case.

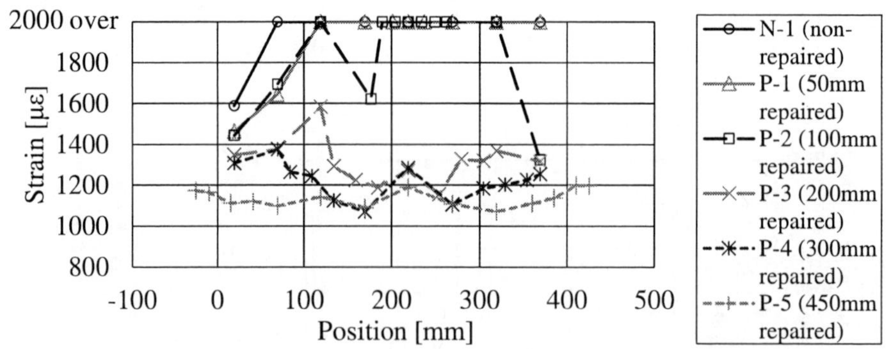

Figure 5. Strain of base steel.

4 EXPERIMENT TO CHANGE SURFACE PREPARATION GRADE

In the following section, we consider the effects of remained rust. Based on the results of the previous section, repair length is kept constant at 200mm in this section. The amount of reduced thickness is the same as previous specimens (Figures 1 and 2). Surface preparation grade is defined by the eq. (1).

$$P_{sp} = \frac{A_0 - A_c}{A_0} \times 100 \tag{1}$$

P_{sp} : Surface preparation grade [%]. A_0 : Adhere area [mm^2]. A_c : corroded area in adhere area [mm^2]

Figure 6 shows the example of artificial simulated rust made of tape paper. Both methods (steel plate and CFRP) were implemented by four types of surface preparation grade.

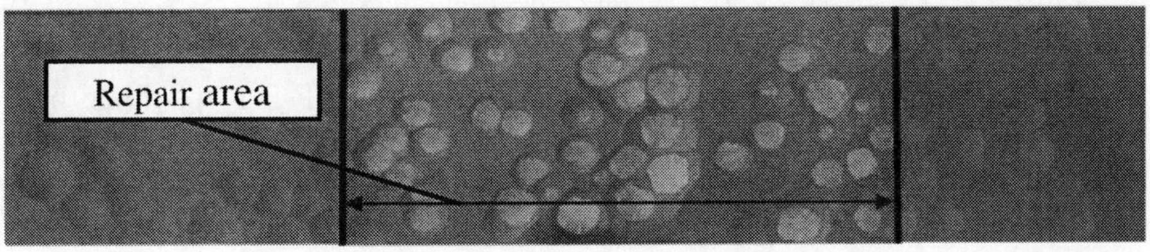

Figure 6. Artificial simulated rust.

5 RESULTS

5.1 Results of Method to Adhere Steel Plates

Figure 7 shows the maximum load of each specimen before repair plate was peeled off. For S-3 specimen, the surface preparation grade of which is 100%, its strength withstood up to 99% of the load compared with undamaged specimen yielding. However, other specimens repaired by steel plate withstood up to only 88-96% of the load compared with undamaged specimen yielding. Thus, strength recovery effect is obviously reduced by the influence of the remained rust.

Figure 8 shows the strain of the base steel loaded 300kN. S-3-1 specimen's strain is larger than that of other specimens, even though S-3-2,3,4 and S-3 specimens do not have big difference. Thus, it is suggested that strength recovery effect will be maintained unless steel plate is peeled off.

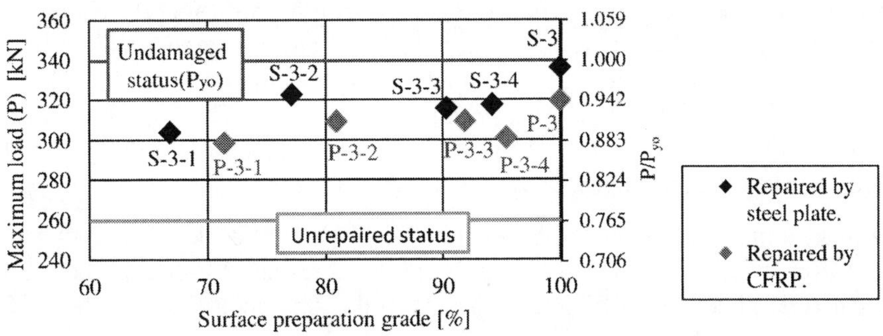

Figure 7. Maximum load.

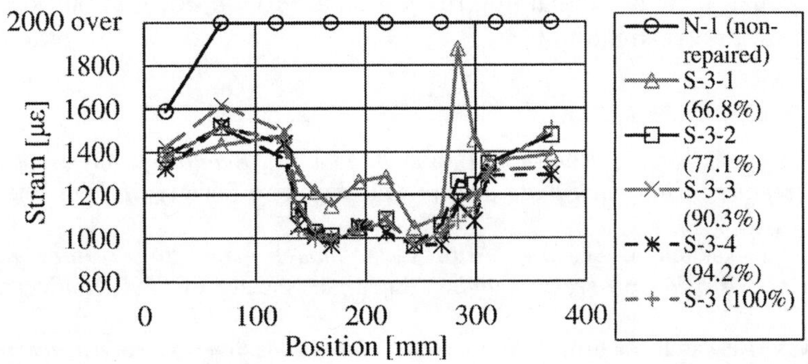

Figure 8. Strain of base steel (Repaired by steel plate).

5.2 Result of Method to Adhere CFRP

According to Figure 7, similar to specimens using adhered steel plate, P-3-1,2,3,4 specimens only withstand up to 86-91% of the load compared with undamaged specimen yielding. On the other hand, P-3 specimen withstands up to 94%.

Figure 9 shows the strain of the base steel loaded 300kN. P-3-1 specimen had already be peeled off CFRP, as a result it had no repair effect when 300kN was loaded to that. However, other specimens did not have big difference. Thus, it is suggested that strength recovery effect will be also maintained unless CFRP is peeled off.

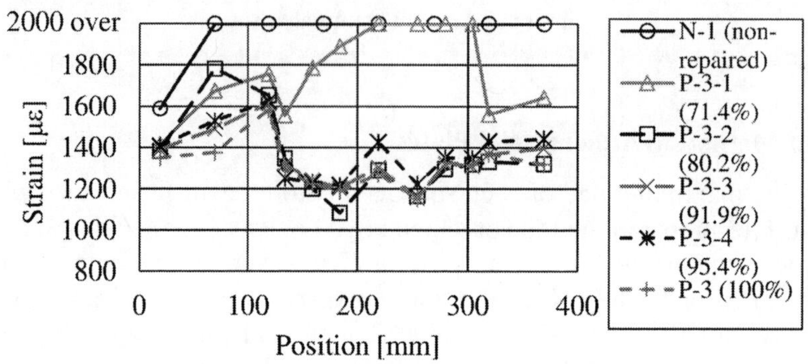

Figure 9. Strain of base steel (Repaired by CFRP).

6 CONCLUSION

1. The ultimate load increased when the repair length was longer. However, specimens having 200mm or 300mm repair length had the almost same repair effect each other. Furthermore, the strain of the base steel in the specimens loaded 300kN having 200-450mm repair length were also nearly equal each other. Thus, the reduction of strain of the base steel and the effect of the partial strength recovery can be expected if there are 200mm or more repair length.
2. The maximum load before the repair effect were lost decreased when the surface preparation grade was worse. The repair effect decreased 9 points in the specimens using adhered steel plate, and decreased 7 points in the specimens using adhered CFRP. Even though, unless steel plate or CFRP peeled off, the repair effect did not decreased.
3. It would be the future tasks to establish a new maintenance manual by clarifying the effect of repair area and remaining rust on repair effect. For that purpose, it is needed to conduct further investigation.

References

Aoki, Y., Banno, R., Ishikawa, T., Kawano, H., and Adachi, Y., *Bending behavior of thickness-reduced steel member repaired with patch plate on one side*, Japanese Society of Civil Engineers, 59A, 647-656, 2013.

Fujii, K., Kaita, T., Nakamura, H., and Ario, I., *A model generating surface irregularities with consideration of corrosion progress in aging*, Japanese Society of Civil Engineers, 50A, 657-665, 2004.

Nippon Expressway Research Institute Company, *Guide to design and construct repairing of steel structures using CFRP*, 2013.

Integrated Solutions for Infrastructure Development
Edited by Lau, H. H., Tang, F. E., Ng, C. K., and Singh, A.
Copyright © 2016 ISEC Press

STUDY ON RETROFIT EFFECTS FOR HORIZONTAL PLANE OF TRADITIONAL WOODEN HOUSES BASED ON FULL-SCALE TEST

MITSUHIRO MIYAMOTO and HARUKA OKUHIRO

Dept of Safety Systems Construction Eng., Kagawa University, Takamatsu, Japan

In the present study, few studies have focused on the horizontal plane of traditional wooden houses in Japan. This study aims to examine the retrofit effects for the horizontal plane of traditional wooden houses based on full-scale tests. The first part of this paper is devoted to the experimental study performed to determine the structural behavior and characteristics of full-scale roof specimens. A horizontal shear test was conducted to obtain the fracture mode and relationship between the applied load and deformation angle. The second part deals with a static pushover analysis of the full-scale roof specimens. The results between the experimental test and the static pushover analysis are presented and discussed. The analysis model used for the static pushover analysis is proposed; the results were in good agreement with the tests.

Keywords: Roof plane, Structural plywood, Structural characteristics, Static pushover analysis.

1 INTRODUCTION

Many large earthquakes have occurred in Japan since the Southern Hyōgo Prefecture earthquake in 1995. Therefore, ensuring the seismic resilience of traditional wooden houses has become very important for their preservation. A great number of studies in Japan focused on the vertical plane of traditional wooden houses (e.g., seismic walls and connections); however, only a few have focused on the horizontal plane of the houses (e.g., floor and roof plane).

This paper aims to examine the retrofit effects for the horizontal plane of traditional wooden houses based on full-scale tests. The first part of this paper is devoted to the experimental study conducted to determine the structural behavior and characteristics of full-scale roof specimens. The test parameter is chosen to examine the retrofit effects for the roof plane. A horizontal shear test was performed to obtain the fracture mode and relationship between the applied load and deformation angle. The second part deals with a static pushover analysis of the full-scale roof specimens. The results between the experimental test and the static pushover analysis are presented and discussed in this paper.

2 FULL-SCALE TESTS

In the study, full-scale horizontal shear tests of roof specimens were conducted to examine the retrofit effects for the roof plane. The relationships between the lateral load and the deformation angle as well as the collapse mechanism were clarified. The test parameters were chosen to help examine the retrofit effect on the seismic performance of a roof plane. One specimen was used for each parameter.

2.1 Specimen Outline

A total of four specimens were used for the full-scale horizontal shear tests (Figure 1 and Table 1). The specimen lengths of the longitudinal and transverse directions were 2730 mm and 2730 mm, respectively. The beam dimensions to the longitudinal and transverse directions, vertical roof strut and purlin, brace for out-of-plane, rafter, roof board, structural plywood, and brace were 120 mm × 240 mm, 120 mm × 180 mm, 120 mm × 120 mm, 120 mm × 30 mm, 45 mm × 60 mm, 150 mm × 15 mm, 910 mm × 9 mm, and 13 mm × 104 mm, respectively. All wooden members were made of cedar from Kochi Prefecture, Japan. The specimen roof pitch was four. The beam was connected to each other with a cogged joint and a steel bolt. The vertical roof strut was connected to the beam or purlin using only a short tenon with height, width, and thickness of 50 mm, 90 mm, and 30 mm, respectively. The short tenon on specimen RA-4 was reinforced with steel plate to avoid the pull-out fracture of the connection. The other connections were connected with screws.

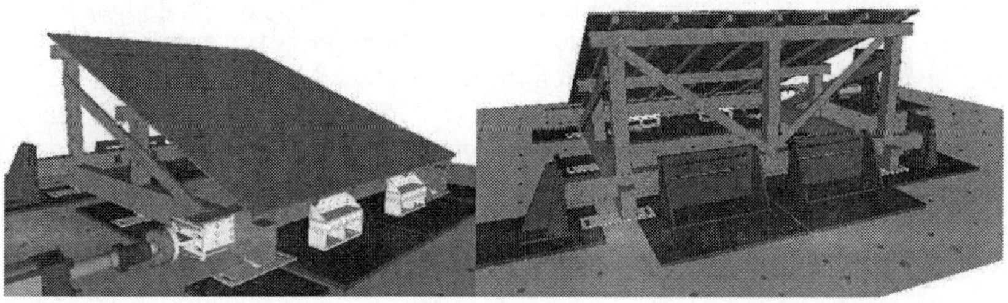

(a) Front side (b) Back side

Figure 1. Specimen RA-4.

Figure 2. Photograph of the loading instrument.

Figure 3. Photograph of the bearing.

Table 1. Specimen specification.

Specimen	Rafter	Roof board	Structural plywood	Brace
RA-1	O			
RA-2	O	O		
RA-3	O	O	O	
RA-4	O	O	O	O

2.2 Test Method

Figure 2 shows a photograph of the loading instrument. One side of the specimen beam was fixed with anchor bolts. The other side was subjected to lateral loads through a servo actuator with six bearings under the specimen beam (Figure 3). The specimen was subjected to cyclic lateral loads, with the real shear deformation angle γ_0 gradually increased symmetrically from 1/600, 1/450, 1/300, 1/200, 1/150, 1/100, 1/75, 1/50, and 1/30 to 1/15 rad. One cyclic loading was applied. The specimen was finally deformed to 1/10 rad at one end. The plus and minus directions are defined in this experiment (Figure 2). The lateral loads applied to the specimen were measured using a load cell attached to the edge of the servo actuator.

2.3 Results and Discussion

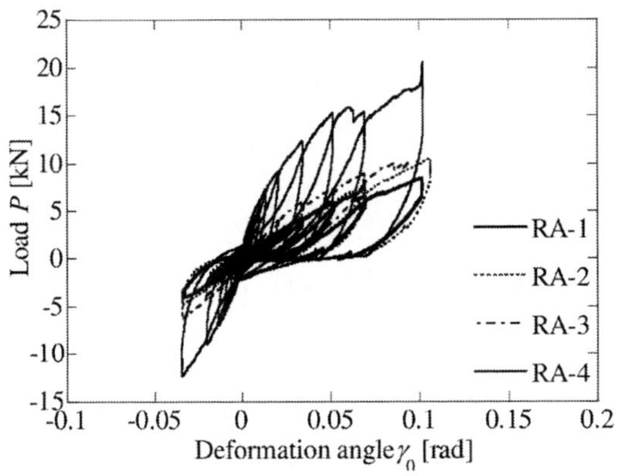

Figure 4. Relationship between the load and the deformation angle.

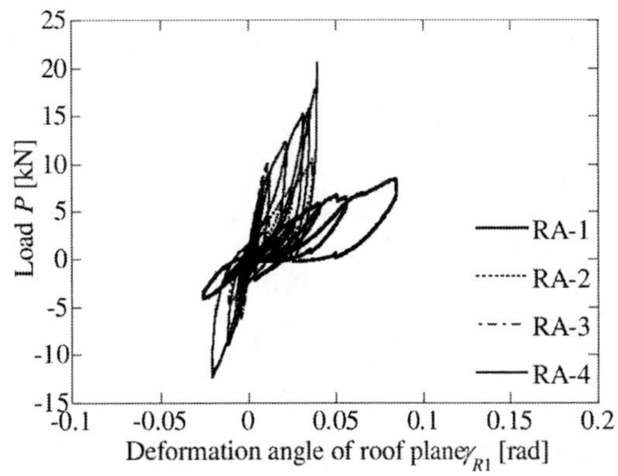

Figure 5. Relationship between the load and the deformation angle of the roof plane.

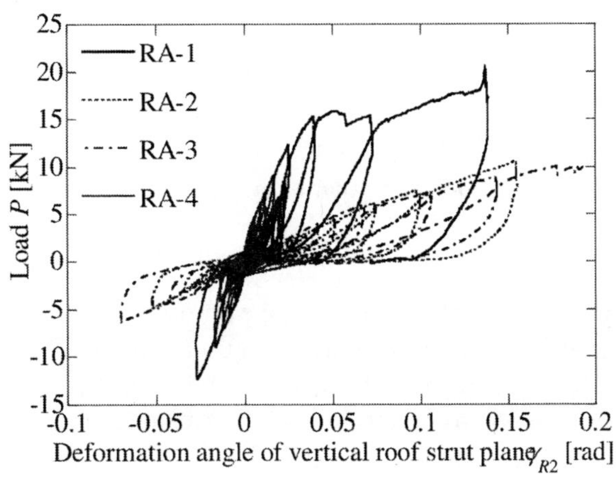

Figure 6. Relationship between the load and the deformation angle of the vertical roof strut plane.

Figure 7. Buckling of brace (RA-4).

Figure 4 shows the relationship between the load and the deformation angle. Figure 5 illustrates the relationship between the load and the deformation angle of the roof plane. Figure 6 presents the relationship between the load and the deformation angle of the vertical roof strut plane of each specimen. Figure 7 shows the fracture mode of the RA-4 specimen. The loads on specimens RA-1–3 in both directions are almost similar; hence, the retrofit effect of the structural plywood without brace on the vertical roof strut plane is very small. However, a large difference is found between the deformation angles of the other planes. The deformation angle of the roof plane on specimen RA-1 is larger than that of the vertical roof strut plane. Meanwhile, the deformation angle of the roof plane on specimens RA-2 and 3 is smaller than that of the vertical roof strut plane because of the retrofit effect of the structural plywood. The maximum load on specimen RA-4 is approximately twice as large as that on the other specimens because of the retrofit effect of brace on the vertical roof strut plane in addition to the structural plywood on the roof plane. The load on specimen RA-4 becomes almost constant after 1/20 rad because of the buckling of the brace (Figure 7). The initial stiffness on all specimen during the second cycle is much smaller than during the first cycle because of the embedment deformation at each connection.

3 SIMULATION ANALYSIS

The simulation analysis was executed using a 3D model to estimate the relationship between the load and the deformation angle of each specimen.

3.1 Analysis Model Outline

Figure 8 shows the analysis model used for the static pushover analysis. The cogged joint and the connection of the short tenon were expressed as a rotation spring. The rood board and structural plywood were expressed as a brace. The connections between the beam and the rafter and the rafter and the purlin were assumed to be pin joints. The hysteresis model of each rotation spring and roof board brace is a bi-linear model. The initial stiffness and yield moment were calculated from the previous study (Architectural Institute of Japan, 2006, Isoda et al., 2007, Architectural Institute of Japan, 2009, Takino et al., 2014). The mechanical property of cedar was estimated from the results of non-destructive tests for each wooden member. The hysteresis model of the structural plywood brace was a tri-linear model (Japan Housing and Wood Technology Center, 2008). The unit multiplier of the structural plywood was assumed to be 2.5.

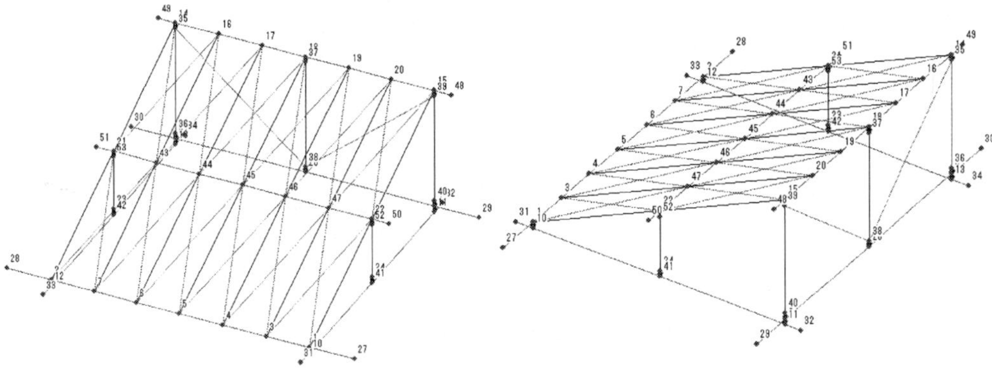

(a) Front side (b) Back side

Figure 8. Analysis model of specimen RA-4

3.2 Comparison of Test and Analysis Results

The static pushover analysis was performed using a general purpose analysis software, SNAP (Ver. 6). Figure 9 shows the comparison of the relationship between the load and the deformation angle of each specimen as obtained from the test and analysis results. The analysis results were in good agreement with the tests for each specimen. The initial stiffness on specimens RA-1 and 2 obtained by the analysis was larger than that obtained by the tests because of the gap at the connections or between wooden boards, such as the roof board and the structural plywood. The load on specimens RA-4 after 1/50 rad obtained from the analysis was larger than that from the tests because the brace edge was expressed as a pin joint without considering the brace's embedment deformation.

4 CONCLUSIONS

We performed horizontal shear tests to obtain the fracture mode and relationship between the applied load and deformation angle. Furthermore, the static pushover analysis of the full-scale roof specimens was executed. The results between the experimental test and the static pushover analysis were compared. The main findings of this study are as follows:

1) The retrofit effect of the structural plywood without brace on the vertical roof strut plane was very small. However, the deformation angle of the roof plane was smaller than that of the vertical roof strut plane because of the retrofit effect of the structural plywood.

2) The maximum load was approximately twice larger than that of the other specimens because of the retrofit effect of the brace on the vertical roof strut plane in addition to the structural plywood on the roof plane. The load became almost constant after 1/20 rad because of the buckling of the brace.

3) The analysis model used for the static pushover analysis was proposed. The analysis results were in good agreement with the tests for all specimens.

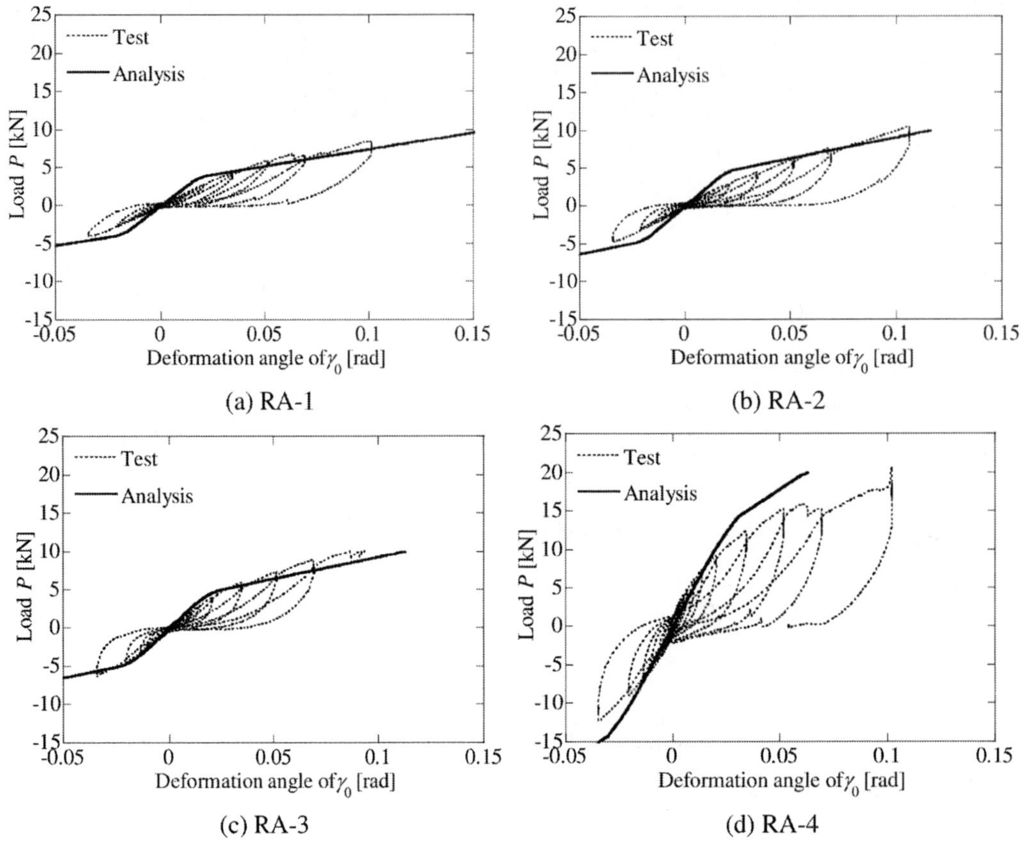

(a) RA-1 (b) RA-2

(c) RA-3 (d) RA-4

Figure 9. Comparison of test and analysis results.

Acknowledgments

This research was supported by the Maeda Engineering Foundation and the Forest Technology Center of Kochi Prefecture. We gratefully acknowledge this support.

References

Architectural Institute of Japan (edited), *Design Manual for Engineered Timber Joints*, Maruzen, Japan, 2009.

Architectural Institute of Japan (edited), *Standard for Structural Design of Timber Structures*, Maruzen, Japan, 2006.

Isoda, H., and Kawai, N., Hysteresis Model of Walls on Japanese Conventional Construction: Study on seismic behavior of wooden construction, *AIJ Journal of Structural and Construction Engineering*, Architectural Institute of Japan, 72(616), 157-163, June, 2007.

Japan Housing and Wood Technology Center (edited), *The Allowable Stress Design of Conventional Wooden Houses*, Japan Housing and Wood Technology Center, Japan, 2008.

Takino, A., Azumi, Y., Okamoto, S., Nakagawa, T., Shibutani, T., and Murakami, M., Formula to Predict the Shear Performance of Floor Construction of Traditional Wooden Structures Considering the Increasing Strength Based on Friction, *AIJ Journal of Structural and Construction Engineering*, Architectural Institute of Japan, 79(703), 1329-1336, September, 2014.

Integrated Solutions for Infrastructure Development
Edited by Lau, H. H., Tang, F. E., Ng, C. K., and Singh, A.
Copyright © 2016 ISEC Press

EVALUATION OF THE OPTIMAL ENGINEERING SOLUTION FOR SWIMMING POOL CONSTRUCTION

RUTA MINIOTAITE

Dept of Civil Engineering Technologies, Kaunas University of Technology, Kaunas, Lithuania

The article describes the evaluation of the optimal engineering solution for monolithic reinforced concrete swimming pool (one of the most durable swimming pools). There are three alternatives for constructing an indoor swimming pool. The three swimming pool construction alternatives differ by flashing and finishing technologies. Three alternative options are analysed in order to find the optimal building technology alternative for the construction of indoor swimming pool: pairwise comparison method (only subjective significance is determined), the cross-entropy method, and expert method. The most rational (optimal) technological solution for the construction of the swimming pool was selected by means of multiple-criteria decision analysis. The closeness to the ideal point method and efficiency value method were applied. The optimal alternative was found by means of weighted sums. The swimming pool construction price in EUR/m^2 was the most important evaluation criterion in the selection of the optimal swimming pool building technology alternative. The results obtained by means of two multiple-criteria analysis methods were different; therefore the weighted sum model was applied. The best alternative for the construction of indoor swimming pool was as follows: concrete reinforced with 300 mm bars, waterproofing tape BT2025S and 1.5 mm thick reinforced waterproofing finishing film.

Keywords: Swimming pool, Technological solution, Multi-criteria evaluation, Consistency of methods.

1 INTRODUCTION

The swimming pool is a water facility of a certain capacity, which is suitable for swimming, procedures of water, hygiene, disease prevention, treatment or rehabilitation purposes. As per environmental conditions, a pool can be indoor or outdoor. The indoor pool can be used all year round and the usage does not depend on the weather conditions. The indoor swimming pools do not require so much care. An indoor pool requires extra costs for the installation of ventilation and maintaining the same level of humidity in the room. By the purpose of using, all indoor and outdoor pools are divided into private pools and public pools. The installation and equipment of the pool depends on the particular purpose.

The finishing of a swimming pool can be:

- Tiles or mosaics (ceramics of glass) – the advantage of mosaics and tiles is that the interior designer can decorate the pool in an artistic manner. However, this decoration of pools has a disadvantage because it is difficult to ensure the pool waterproofing. Special tiles suitable for swimming pools must be used;

- The waterproofing membrane is 1.5 mm thick reinforced membrane especially designed for swimming pools. It absolutely ensures the waterproofing of the pool and it is a good-looking decoration as well. Pool is covered with this membrane are pleasant, soft and warm to walk. It is anti-bacterial, extremely strong and is also UV resistant. Almost 80% of all pools are covered with this membrane in Lithuania. The proposed colour range of membranes is wide enough and it provides great opportunities for the design of the pool bottom to match the interior design or landscape.

This article describes the optimal evaluation of engineering solutions of a monolithic reinforced concrete pool, which planned to be installed in a multifunctional sports and entertainment centre. This kind of pools is one of the strongest and most durable pools.

2 OPTIONS OF SWIMMING POOL INSTALLATION

There are three alternatives for constructing an indoor swimming pool. Monolithic reinforced concrete is used for building the water pool in all three alternatives. The water pool is reinforced with metal bars that are tied together. A high swelling bentonite waterproofing tape BT 2025S is used to seal reinforced concrete construction joints. The structural solutions of pool are shown in Table 1.

Table 1. Structural solutions of pool.

Alternative	Structural drawing	Structural layer, materials
A1	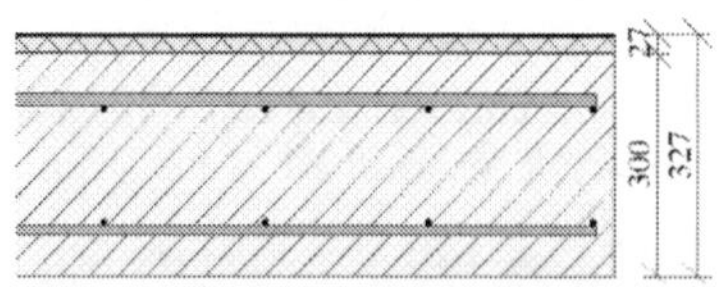	Reinforced concrete – 300 mm Waterproofing tape BT 2025S Waterproof Xypex Concentrate Waterproof Xypex Modified Cement / adhesive mixture – 20 mm Mosaic tiles – 7 mm
A2	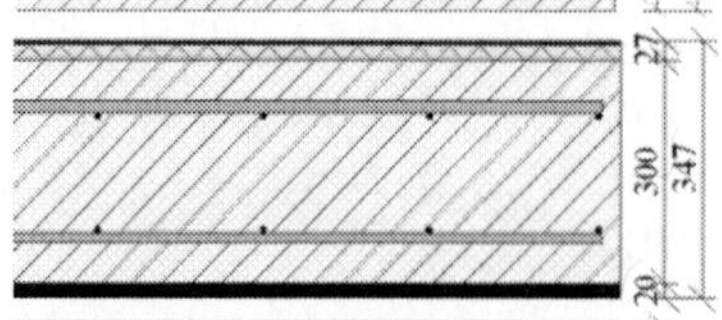	Bituminous waterproofing – 2 layers Reinforced concrete – 300 mm Waterproofing tape BT 2025S Cement / adhesive mixture – 20 mm Mosaic tiles – 7 mm
A3	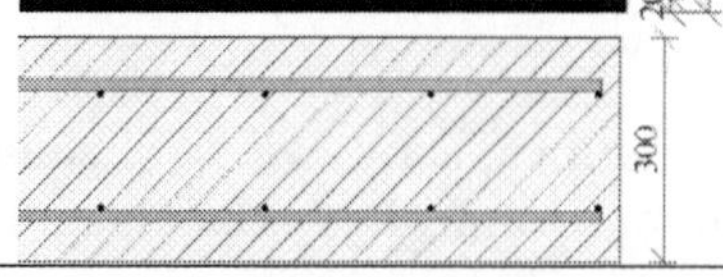	Reinforced concrete – 300 mm Waterproofing tape BT 2025S 1,5 mm thick reinforced waterproofing finishing film

3 EVALUATION OF ASSESSMENT CRITERIONS OF POOL CONSTRUCTION TECHNOLOGY AND SIGNIFICANCE

The criterions of assessment are divided into two groups: 1) technical-economic criteria; 2) qualitative characteristics (Sarka *et al.* 1999, Rapceviciene 2010).

Each criterion has a numeric value, the unit of measurement or its own significance. An importance of the criteria indicates how one criterion is more important compared to another criterion. An importance of the criterion can be one of three types: subjective, theoretical, complex.

The criterions of significance assessment can be determined in several ways: the method of pairwise comparison; the method of entropy; the expert method.

3.1 The Method of Pairwise Comparison

The line of priorities and significances of the optimal variation of the pool construction technology are set, when the method of pairwise comparison is chosen and the Kendall concordance coefficient is checked. The essence of this method is that all possible options are compared with each other in pairs. If it is said that the variant x_i is better than the variant x_j it means that the first one variant got "1 point" and the second one variant got "0 points".

Evaluation criterions: X_1- cost of pool installation EUR/m^2; X_2 – the duration of works (man-hours); X_3 – durability, in points; X_4 – aesthetics, in points. According to the data of experts, the total matrix is formed. The total matrix of the pairwise comparison is shown in Table 2.

Table 2. The total matrix of the pairwise comparison.

Reasons	X_1	X_2	X_3	X_4	S_i	q_i	The priority order
X_1	-	8	8	8	24	0,3692	4
X_2	2	-	3	8	13	0,2000	1
X_3	5	6	-	7	18	0,2769	2
X_4	1	6	3	-	10	0,1538	3
				Sum	65	1,0000	

An importance of the criterions is determined by calculating the sum of every each line of i version (Eq. (1)):

$$S_i = \sum_{k=1}^{n} x_{ik}^{*} \tag{1}$$

As the higher value of S_i is, then the variation is more effective or criterion is more significant. The priority order: $q_1 > q_3 > q_2 > q_4$; where q_i – the subjective significance of criterion.

In this case, all sizes of q_i are not less than 5%. This means that all data are overwrite in the total and transformed matrix of the pairwise comparison. There are no eliminated data. Carry out the transformation of the matrix. When the calculations are done, it is necessary to check the concordance coefficient and whether the opinions of experts match. Calculation of the necessary data: 1) calculation of the possible combinations according to the experts; 2) possible combinations according to the evaluation criterions; 3) the total amounts are calculated; 4) calculation of the concordance coefficient.

When the method of pairwise comparison (W = 0,719) is used, the concordance coefficient is higher W > 0,6. All opinions of experts match each other; as a result the data can be used for the further calculations of optimization.

3.2 The Method of Entropy

The entropy – this is a measure of random size of uncertainty. The entropy is used to determine the theoretical and complex significance of the evaluation criterion.

According to the estimates of calculated comparative variations, the initial data sheet is prepared. Table 3 provides initial data of alternative design solutions.

In order to compare all engineering solutions, this matrix needs to be normalized, because the matrix is made by the different criterions, which have different mathematical units. When a few different criterions of evaluation are used, it is recommended to normalize the matrix with this formula (Eq. (2)):

$$\bar{P}_{ij} = 1 - \frac{\left| x^*_j - x_{ij} \right|}{x^*_j} \quad (i = \overline{1,m} ; j = \overline{1,n}) \tag{2}$$

where x_{ij} is a maximum (minimum) value of the criterion. The normalized matrix is obtained where all elements are dimensionless quantities.

If all criteria are equally important or there are no determined or known subjective or experts' evaluation of the significance, the theoretical significance of the criterion is determined. If the subjective values of significance of criterion are known, then the complex significance of criterion can be determined as well. Table 4 provides the significance of criterion.

Table 3. Initial data of alternative design solutions.

Criteria / Alternative decisions	X_1 Work cost (EUR/m²)	X_2 The duration of works (man-hours)	X_3 Durability (points)	X_4 Aesthetics (points)
A1	163,23	3,07	10	10
A2	134,42	2,46	8	10
A3	94,5	1,52	6	7
Optimum	MIN	MIN	MAX	MAX

Table 4. The significance of criterion.

Criteria	X_1 Work cost (EUR/m²)	X_2 The duration of works (man-hours)	X_3 Durability (points)	X_4 Aesthetics (points)
The theoretical significance	0,3382	0,2124	0,1457	0,3038
The subjective significance	0,34	0,21	0,15	0,30
The complex significance	0,419	0,165	0,178	0,338

The priority order: $q_1 > q_4 > q_2 > q_3$.

A comparison the significance of the theoretical and complex criterions shows that the cost of the work is the most important criterion (X_1) and the duration of works is the least significant criterion (X_2).

4 EVALUATION OF MULTI-CRITERION OPTIMAL ENGINEERING DESIGN OF THE POOL

According to methods of multi-criterion, the optimal technological solution of the swimming pool will be selected. Two methods will be analyzed: the method of proximity to an ideal point and the method of efficiency value.

4.1 The Method of Proximity to an Ideal Point

The essence of this method is that the optimal alternative has to the minimum distance from the ideal solution and the maximum distance from the worst solution. The main principle of the method of proximity to an ideal point: forming the generalized complex criterion, based on comparison of options deviation from the so-called ideal criterion, which consists of the best options under consideration criterion. This method takes into account the fact that each version of the decision criterion function of utility can be increased or decreased monotonically. The algorithm, when the method of proximity to an ideal point is used and the complex significance of criterion is evaluated: the matrix of alternative architectural solutions is made; the normalization of the matrix; a weighted matrix is obtained.

The optimal variation of the indoor pool construction technology, when the method of proximity to an ideal point is used and the significance of criterion are evaluated.

The matrix is made of different criterions and these criterions are made of different mathematical units. In order to compare all engineering solutions, this matrix must be normalized. The best (the most rational) architectural solution will become the one, which K_{bit} value will be max ($K_{bit,i}$=max). Using the values we form the priority sequence utility degree establishment. We compare the value of the variant examined with the value of ideal variant.

4.2 The Method of Efficiency Value

When the method of efficiency value is chosen, the following criterion is used: subjective, complex and theoretical significance. According to the individual design solutions, these criterions are not equally significant, so the evaluation takes into account the importance of the criterion to each other. The highest utility value is selected as the best option.

Method evaluation algorithm:

A set of evaluation criteria is formed; →a relative weight to each criterion is assigned; →values of criteria are found; →the efficiency value from 0 to 10 points is assigned; →the numerical value of efficiency is determined; →the total efficiency value of alternative solutions is determined.

The best alternative is *A3*, because in accordance with three criterions, the sum of utility of these options is the biggest and it is 25%.

5 RESULTS OF THE OPTIMAL DESIGN EVALUATION OF POOL

Several methods were used for the chosen optimal design evaluation of pool; a few different variations were received. Table 5 provides a summary of the results; according to the results, the points of priority are given (the most useful - 3, the least useful - 1). After the summing of scores, an optimal solution was found. The resume of results shows that the third option got the maximum number of points, and this option is selected as optimal.

Table 5. The total resume of results.

Method	Significance	Alternative		
		A1	A2	A3
The method of proximity to an ideal point	Theoretical	100,00	100,00	99,96
	Subjective	71,16	89,56	100,00
	Complex	31,23	62,75	100,00
The method of the efficiency value	Theoretical	5,46	6,98	8,10
	Subjective	5,47	6,98	8,11
	Complex	6,33	7,86	8,84
The method of proximity to an ideal point	Theoretical	3	3	2
	Subjective	1	2	3
	Complex	1	2	3
The method of the efficiency value	Theoretical	1	2	3
	Subjective	1	2	3
	Complex	1	2	3
	Sum	8	13	17

6 CONCLUSIONS

(1) The swimming pool construction price in EUR/m^2 was the most important evaluation criterion in the selection of the optimal swimming pool building technology alternative.

(2) When the optimal design evaluation of pool was proved by two multi-criterion methods of evaluation, the final results were different, so the total evaluation was carried out. In order to find out the optimal technological solution for pool construction, three different technological solutions were analysed.

(3) The best alternative is *A3*, i.e. the price of square meter is 95 EUR. The best alternative for the construction of indoor swimming pool was as follows: concrete reinforced with 300 mm bars, waterproofing tape BT2025S and 1.5 mm thick reinforced waterproofing finishing film. 7 points were given for the aesthetics of this alternative, and 6 points for durability.

References

Rapceviciene, D., Evaluation of multi residential house renovation efficiency: *Science – Future of Lithuania,* 2(2), 83-89, 2010.

Sarka, V., Ustinovicius, E.K., Zavadskas, E.K., Synthesis of the projects applying the compromised compensating models in construction, *Civil engineering,* V, 6, Technika, Vilnius, 374-385, 1999.

Integrated Solutions for Infrastructure Development
Edited by Lau, H. H., Tang, F. E., Ng, C. K., and Singh, A.
Copyright © 2016 ISEC Press

SHEAR STRENGTH OF PERFORATED PLATE GIRDERS WITH INCLINED STIFFENERS

M. R. AZMI, M. Y. M. YATIM and W. H. WAN BADARUZZAMAN

Dept of Civil & Structural Engineering, Universiti Kebangsaan Malaysia, Selangor, Malaysia

A series of experimental tests on perforated steel plate girders with inclined stiffeners is presented herein. Ten simply supported girders of practical size were tested for failure under a single concentrated load applied at the center of gravity of the section. Circular web openings of various diameters were introduced at the middle panels of the girders. Effects of different inclination angles of intermediate stiffeners on the ultimate load behavior are the main focus of the present study. Five different angles, measured from the bottom flange, were considered in the tests, viz., 90°, 75°, 60°, 45°, and 30°. Variations of strength, failure characteristic, and load-deflection response have been observed. The ultimate strength was found to increase considerably from 21% to 92% as the angle of inclined stiffeners reduced from 90° to 30°.

Keywords: Plate girder, Inclined stiffener, Web opening, Ultimate load behavior, Post-buckling strength.

1 INTRODUCTION

The use of plate girders in civil engineering construction has found increasing popularity in railroad bridges and industrial buildings. Studies on plated structures were carried out by many scholars since decades ago in order to establish the design philosophy of plated structures and other forms of thin-walled structures (Basler and Thurlimann 1960, Rockey and Skaloud 1971, Porter *et al.* 1975, Lee and Yoo 1998, Horne and Grayson 1983, Alinia and Moosavi 2008). It is common to allow ducts and services pass through the plate girders by providing openings in the web. (Hagen *et al.* 2009) and (Hagen and Larsen 2009), for instance, carried out some analyses and proposed simple design calculations to determine the shear capacity of steel girders having large web openings. In addition, (Shanmugam *et al.* 2002) investigate the ultimate load behavior of plate girders containing central web openings of rectangular and circular shapes. This form of structural member results in reduced floor height, systematic installation of pipes or ducts, and cost-effectiveness, but at the same time, causes penalty on the shear strength of the girders depending on the parameter of the openings.

It has long been known that vertical stiffeners, except those at load points, do not carry any load and serve only to prevent the web from buckling. By placing the intermediate stiffeners diagonally across each panel, a trussed girder is formed and the stiffeners thereby carrying a portion of the load in addition to performing in an efficient manner, their given tasks of preventing buckling. Inclined stiffeners would also have the advantage of limiting the shear factor without requiring additional longitudinal stiffeners (Guarnieri 1985). However, the use of inclined stiffeners causes unequal diagonal length and unequal subdivisions of web panels at the top compression and bottom tension flanges thus, results in a complex behavior of the panels and complicates the analysis. Also, far too little attention has been paid to study inclined stiffeners in the plated structures as found in the literature.

Therefore, there is a necessity to get a clear picture of the behavior of such girders through physical observations in order to enhance the understanding of the elastic and inelastic responses under loading. An experimental investigation on plate girders having different angles of inclined stiffeners has, therefore, been undertaken. Details of the girders and experimental procedures are described in this paper. Attention was focused on the variations of ultimate capacity, load-deflection behavior, and failure characteristics. This paper will demonstrate the potential application of inclined stiffeners in plate girders containing web openings in which a definite increase in the efficiency of plate girders can be obtained over the conventional ones.

2 EXPERIMENTAL PROGRAM

Tests were conducted on ten perforated steel plate girder specimens of practical size containing inclined stiffeners. The specimens are basically same in dimensions such as span length, L = 2410 mm, depth of web panel, d = 500 mm, thickness of web, t_w = 2 mm, width of flange, b = 100 mm, thickness of flange, t_f = 10 mm, and the web slenderness ratio, d/t_w = 250. The differences between all specimens are the angle of inclination of the intermediate web stiffeners which varies from 90° to 30° in the increment of 15° and the diameter of web openings, d_0 which varies from $0.2d_0$ to $0.5d_0$. The test specimens are noted in the text as PG-90-Cr100, PG-90-Cr200, PG-75-Cr150, PG-75-Cr250, PG-60-Cr100, PG-60-Cr200, PG-45-Cr150, PG-45-Cr250, PG-30-Cr150, and PG-30-Cr250 in which the notations 90, 75, 60, 45, and 30 refer to the angles of inclination of the stiffeners. Notation Cr100, for example, indicates 100 mm diameter of circular opening. A typical test girder is shown in Figure 1.

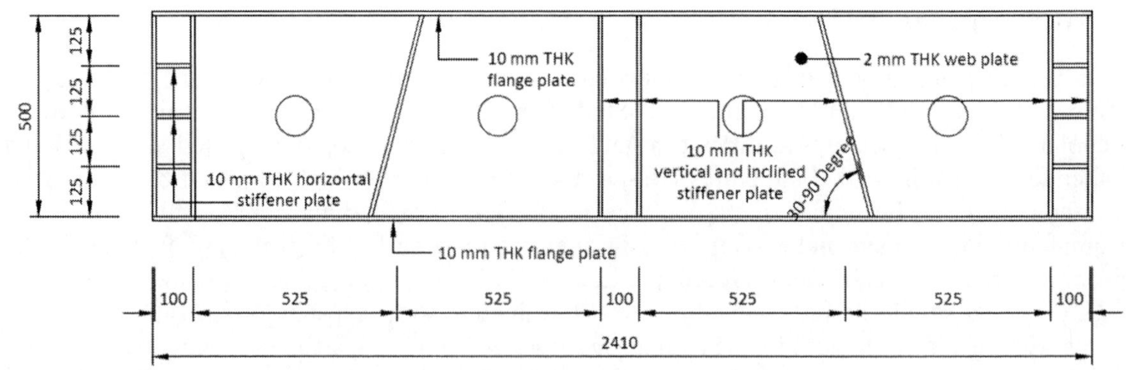

Figure 1. Typical details of the test girder.

Mild steel plates of grade S275 were chosen in the fabrication of the plate girder specimens. The simply supported girder was placed over the strong support system mounted to the test frame. It is very important to ensure the correct positioning of the test specimen in such a way that the midpoint of the girder is in line with the centre line of the hydraulic jack. Seven LVDT transducers were placed at seven selected points below the bottom flange of the girder to measure the vertical deflection during the testing program. Experimental data such as increment of loads and deflections were recorded using an appropriate data acquisition system. Figure 2 shows a typical test set-up.

Prior to testing, all specimens were loaded about 10 percent of the expected ultimate load and then unloaded in order to ensure the specimens are properly seated on the support system. This procedure also helps to check the functionality of the strain gauges and transducers used in the test. Then, the application of load will be increased gradually by a predetermined increment to failure. After the ultimate load was reached, the application of load was continued in order to investigate the post-buckling behavior. Buckling shapes due to tension field action were visible for each specimen during the test. The ultimate load and mode of failure for all specimens were recorded. The applied shear load was then removed when it dropped to a certain level beyond failure. In this study, the test procedures are similar for all girders.

Figure 2. Typical view of the test set-up.

3 TEST RESULTS AND DISCUSSION

3.1 Ultimate Load Capacity

All of the girders were tested for failure under shear load applied at the mid-span. The ultimate loads obtained from the experiments, P_u are tabulated in Table 1. The test specimens were designed so that they could be used to determine the influence of inclined stiffeners and the effect of openings in the web panels. Figure 5(a) compares the behavior of girders of PG-75-Cr150 and PG-75-Cr250 which have the same inclination angle of intermediate stiffeners but different circular opening size. The results show a substantial decrease in the ultimate strength about 27% when the circular opening size increases. Figure 5(b) shows the behavior of girders of PG-75-Cr250, PG-45-Cr250, and PG-30-Cr250 which have the same opening size but different inclination angle of intermediate stiffeners. It is evident from the results that the ultimate load of the girder with the same size of web opening is increased by 92% when the angle of intermediate stiffeners is reduced from 75° (i.e., PG-75-Cr250) to 30° (i.e., PG-30-Cr250). Same findings were found where the ultimate load capacity of girder PG-60-Cr100 increased about 21% compared to girder PG-90-Cr100 and the ultimate load capacity of girder PG-60-Cr200 increased about 25% compared to girder PG-90-Cr200. These indicate that the intermediate inclined stiffeners carry some of the applied load through the distribution of forces as in truss members. However, it can also be observed that the ultimate loads of girders with the same angle of stiffeners inclination decrease when the web opening is larger.

3.2 Load-Deflection Behavior

Figure 5(a – b) shows the comparisons of load-deflection response of the girders by considering the influences of the inclination angle of stiffeners and web opening size. Both curves represent mid-span deflection plotted against the applied load. The curves primarily show the bending behavior of the girders. A sudden drop of the applied load can be observed in almost girders after reaching the maximum capacity, indicating the yielding of web material. It is evident from the figures that the two curves correlate well especially at the initial stages of loading.

Table 1. Ultimate loads.

Specimen	Ultimate Load, P_u (kN)
PG-90-Cr100	187.7
PG-90-Cr200	126.4
PG-75-Cr150	169.5
PG-75-Cr250	124.4
PG-60-Cr100	226.2
PG-60-Cr200	157.5
PG-45-Cr150	221.7
PG-45-Cr250	180.3
PG-30-Cr150	304.4
PG-30-Cr250	238.7

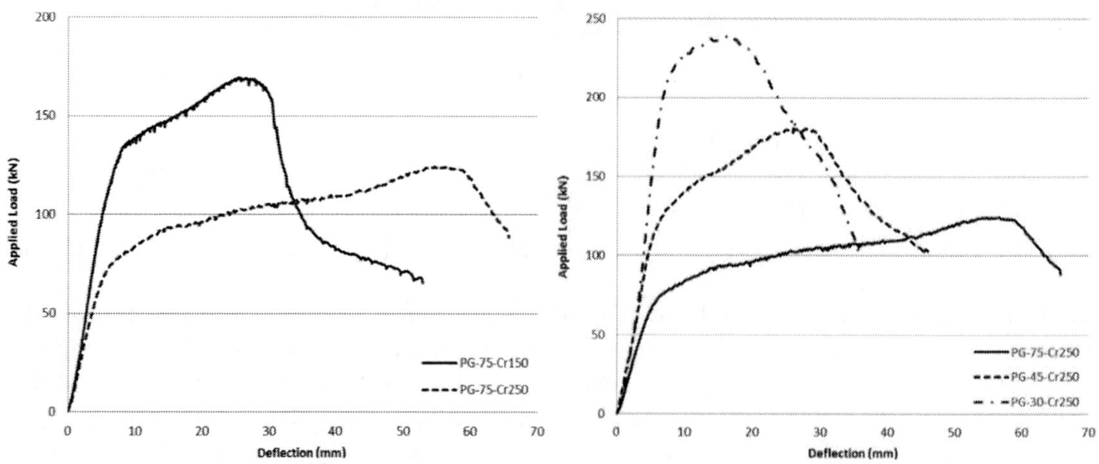

Figure 5. Comparisons of load-deflection behavior.

3.3 Failure Characteristic

All of the girders exhibited almost identical behavior at failure load. After reaching the elastic buckling load, all web panels started to develop diagonal-shaped bucklings as shown in Figure 6. Further increment in load beyond this stage was resisted by the web tension field action which, from then on, caused an increase in out-of-plane deformation of the diagonal buckling. Upon attaining the ultimate load, plastic hinges were formed in the compression and tension flanges, and the deformation of the girders was obvious, particularly those having large web openings.

Beyond this point, the girder swayed to one side as presented in Figure 7 and a gradual drop in the applied force occurred. This indicates that the girder has totally lost its capacity and the web material started yielding. All of the girders appeared to have this similar ductile failure characteristic.

Figure 6. Typical deformation of webs at critical buckling load.

Figure 7. Typical deformation of girder at failure.

4 CONCLUSIONS

Ten perforated steel plate girders with different inclination angles of intermediate stiffeners were tested for failure under shear load at mid-span. The following conclusions were made based on the experimental results. It is clear from the experimental results, with the same size of web openings, the ultimate strength increased when the angles of intermediate stiffeners were reduced. This finding shows the significant contribution of inclined stiffeners in carrying some percentages of forces. Appreciable variations may be observed in the load-deflection responses. The curves show a sudden drop in the applied force soon after attaining the ultimate point. Corresponding curves obtained from the non-linear finite element analysis are superimposed for comparison, only to find that the two curves correlate with each other well especially at the initial stages of loading. Overall, this study has provided some insights regarding the ultimate load behavior of plate girders having inclined stiffeners and therefore, further detailed analytical works and design recommendations are strongly suggested.

Acknowledgements

The study was carried out with the generous financial assistances by; the Fundamental Research Grant Scheme (FRGS/2/2014/TK02/UKM/03/1) and Geran Galakan Penyelidik Muda (GGPM-2015-015). The authors acknowledged the facilities provided by the Department of Civil and Structural Engineering, Universiti Kebangsaan Malaysia.

References

Alinia, M. M., and Moosavi, S. H., "A Parametric Study on the Longitudinal Stiffeners of Web Panels", *Thin-Walled Structures*, 46, 1213-1223, 2008.

Basler, K., and Thurlimann, B., "Carrying Capacity of Plate Girders", in Preliminary Publication, 6th Congress of International Association for Bridge and Structural Engineering *(IABSE)*, 16, 1960.

Guarnieri G., "Collapse of plate girders with inclined stiffeners", *Journal of Structural Engineering ASCE*, 111(2), 378-399, 1985.

Hagen N., Larsen P., Aalberg A., "Shear capacity of steel plate girders with large web openings, Part I: Modeling and simulations". *Journal of Constructional Steel Research*, 65:142–50, 2009.

Hagen N., Larsen P., "Shear capacity of steel plate girders with large web openings, Part II : Design guidelines" . *Journal of Constructional Steel Research,* 65:151–8, 2009.

Horne, M. R., and Grayson, W. R., "The Ultimate Load Behavior of Longitudinally Stiffened Web Panels Subjected to Shear Stress", Proceedings of the Institution of Civil Engineers, 75(2), 175-203, 1983.

Lee, S. C., and Yoo, C. H., "Strength of Plate Girder Web Panels Under Pure Shear", *Journal of Structural Engineering ASCE*, 124(2), 184-194, 1998.

Porter, D. M., Rockey, K. C., and Evans, H. R., "The Collapse Behavior of Plate Girders Loaded in Shear", *The Structural Engineer* 53(8), 313-325, 1975.

Rockey, K. C., and Skaloud, M., "The Ultimate Load Behavior of Plate Girders Loaded in Shear", International Association for Bridge and Structural Engineering (IABSE) Colloquium: Design of plate box girders for ultimate strength, 1-19, 1971.

Shanmugam N., Lian V., Thevendran V., "Finite element modeling of plate girders with web openings". *Thin-Walled Structure*, 40:443–64, 2002.

Integrated Solutions for Infrastructure Development
Edited by Lau, H. H., Tang, F. E., Ng, C. K., and Singh, A.
Copyright © 2016 ISEC Press

NONLINEAR ANALYSIS OF JAPANESE TRADITIONAL WOODEN FRAMES WITH FITTING-TYPE JOINT

NAOTO IDATE, YUKIMASA YAMADA and NORIKO TAKIYAMA

Division of Architecture and Urban Studies, Tokyo Metropolitan University, Tokyo, Japan

This paper reports our research project on the seismic performance of traditional wooden frames with large-section beams. First, based on past cyclic loading tests on four frames, a simple analysis model of a fitting-type joint was constructed using beam elements. Next, past experimental results were simulated by the model. Finally, a sensitivity analysis was performed to study about the effect of height arrangement of beams on the performance of the whole frame. The major findings of this study are as follows: (a) Some material tests are conducted for wooden test pieces cut down from all column of specimens. The compressive spring properties are decided. (b) Analysis models are constructed, and the restoring forces of frame specimens are simulated precisely. (c) The relation between height arranging beams and the change in restoring force is considered by sensitivity analysis.

Keywords: Traditional wooden structure, Fitting-type joint, Large-section beam, Material test, Nonlinear analysis, Sensitivity analysis.

1 INTRODUCTION

Numerous earthquakes have occurred in Japan. The 1995 Hyogoken-Nanbu Earthquake (M7.3) damaged many structures. Subsequently, there have been many reports of wooden structures collapsing as a result of large earthquakes. Damage to traditional wooden structures often occurs around the joints. Therefore, it is essential to conduct studies on the seismic resistance of joints.

In a seismic evaluation method based on strength-limit calculation, a shear force having a one-to-one correspondence with a load-bearing element is defined, and the shear force of construction only adds to the restoring forces of the element (Editorial 2008). However, there are various joint shapes and types of materials within a single load-bearing element, and the calculation method does not depend on them.

Therefore, we aimed to understand the seismic performance of wooden frames with uneven, large-section beams, and to study the influence of beams on the behavior of the whole frame. In a past study, we conducted a cyclic loading test on four test frames (Takiyama 2015a). Moreover, we constructed a simple analysis model of column-beam joint by using beam elements and compressive springs. Then, the initial stiffness was simulated by linear analysis (Takiyama 2015b). In this study, we simulated skeleton curves of restoring forces by a nonlinear analysis that used material characteristics verified by material tests.

2 OUTLINE OF ANALYSIS

First, the material characteristics are verified by material tests. Next, a model is proposed to simulate the result of past cyclic loading test on four test frames by nonlinear analysis. Here, material nonlinearity and geometric nonlinearity are considered.

2.1 Specimen Design

We designed four specimens based on a typical traditional wooden frame, as shown in Figure 1 and conducted a cyclic loading test based on past studies (Takiyama 2015a, b). Specimen F is the standard frame. Specimen S1 is a frame with a large-section beam. Specimens S2 and S3 are two-span frames with beams. The uneven beams of specimen S3 are exaggerated. After the cyclic loading test, bending test and longitudinal compressive test are conducted on test pieces from all columns based on Japanese Industrial Standard "JIS Z 2101." The material strength of the columns are indicated in Table 1.

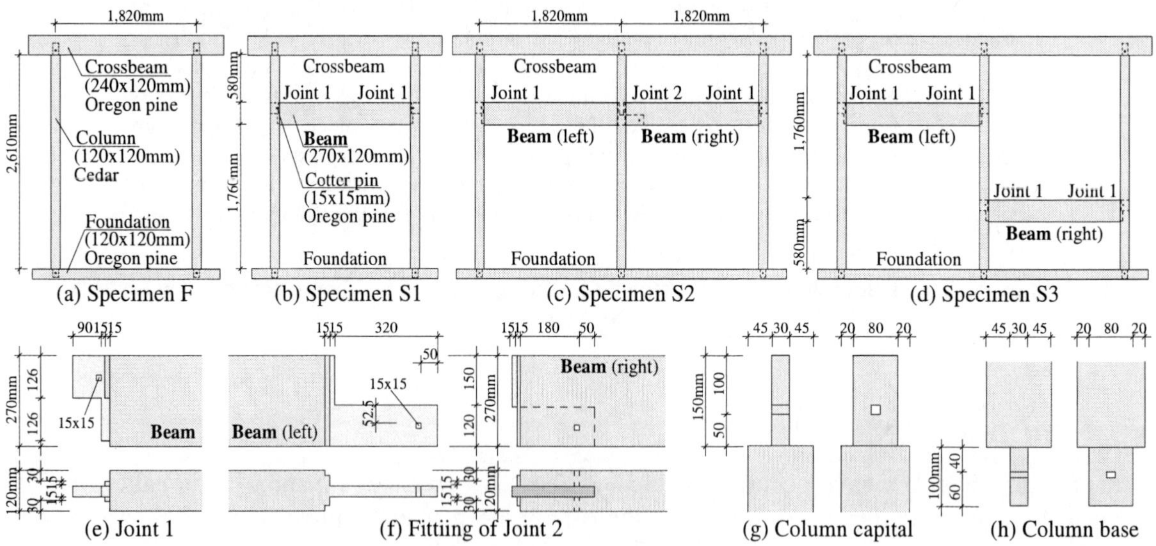

Figure 1. Details of specimens.

Table 1. Material Strength of columns.

Specimen	Column	Strength (N/mm^2)		Young's Modulus (kN/mm^2)	
		Bending	Compressive	Bending	Longitudinal Compressive
F	Left	71.004	56.755	12.908	10.166
	Right	74.683	55.966	13.858	11.346
S1	Left	54.377	38.927	8.820	8.373
	Right	52.906	41.124	8.806	7.456
S2	Left	66.443	45.442	11.208	9.440
	Middle	67.914	51.200	12.297	11.953
	Right	52.906	41.646	10.006	9.466
S3	Left	48.345	42.286	8.938	8.192
	Middle	56.952	38.682	10.376	8.515
	Right	42.238	50.379	6.509	7.107

2.2 Elements of the Analysis Model

The spring has a bilinear restoring force. The following assumptions were made:

a) The analysis is conducted for large deformations. Deformations are concentrated on the edge of members. After calculation, the damaged member is checked for stress such as breakage or shear fracture parallel to the grain.

b) The resistance elements are the compressive strain resistance of the fiber orthogonality, frictional resistance, and shear resistance of the cotters.

c) The compressive strain occurs only in the direction of the fiber orthogonality. Triangular displacement compression occurs in the tenon (Architectural 2009). Embedment resistance on the tenon is a triangular displacement embedment and the embedded length is half of the tenon length.

d) For the frictional resistance, only the dynamic frictional force is considered. The frictions that occur are the encroaching and sliding frictions.

e) Two symmetrical double shears act on the cotter.

f) Only compressive strain acts on the springs; the spring properties were obtained from the literature (Architectural 2009).

g) All of the members were represented by wire rods on the materials axis.

h) The second stiffness is 1/10 times the first stiffness.

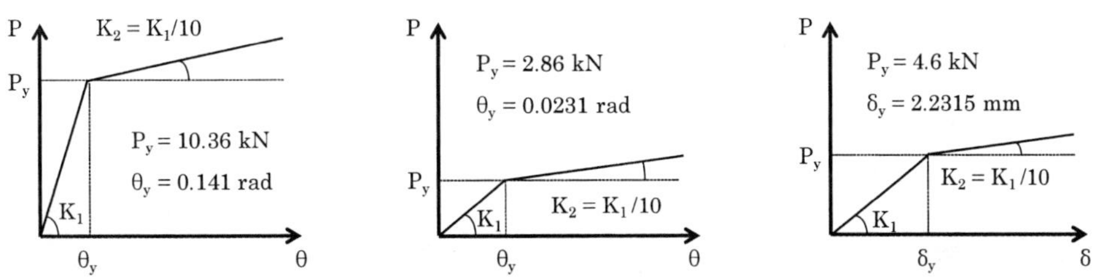

(a) Embedment resistance (Tenon of column base) (b) Embedment resistance (Ground sill) (c) Shear force of cotter pin (joint of ground sill)

Figure 2. Examples of restoring force of resistance element.

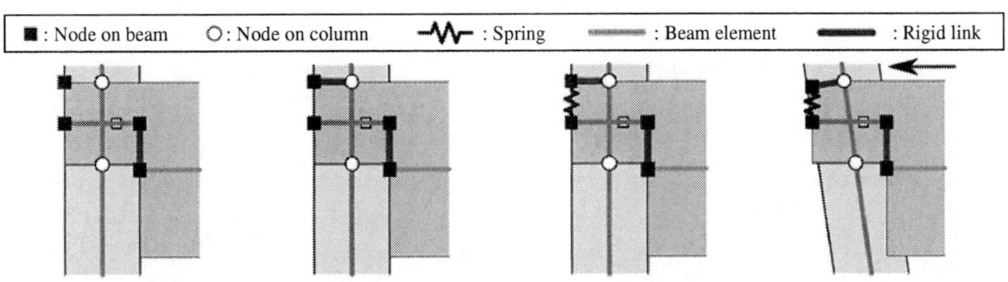

(a) (b) (c) (d)

Figure 3. Example of modeling resistance for column-beam joint in positive loading.

First, the nodes were determined for the columns, beams, crossbeams, and base. The nodes on the base were fixed, while others had three degrees of freedom. Next, the columns and horizontal members were replaced with beam elements that have material properties and a cross-section. The two nodes were connected by a rigid link and a spring; the rigid link is a rigid element.

The material properties of the columns are decided based on Table 1. For other materials, these are decided from mechanical stress grade (Architectural 2006).

From the above, some examples of restoring force of resistance element on the column base are shown in Figure 2.

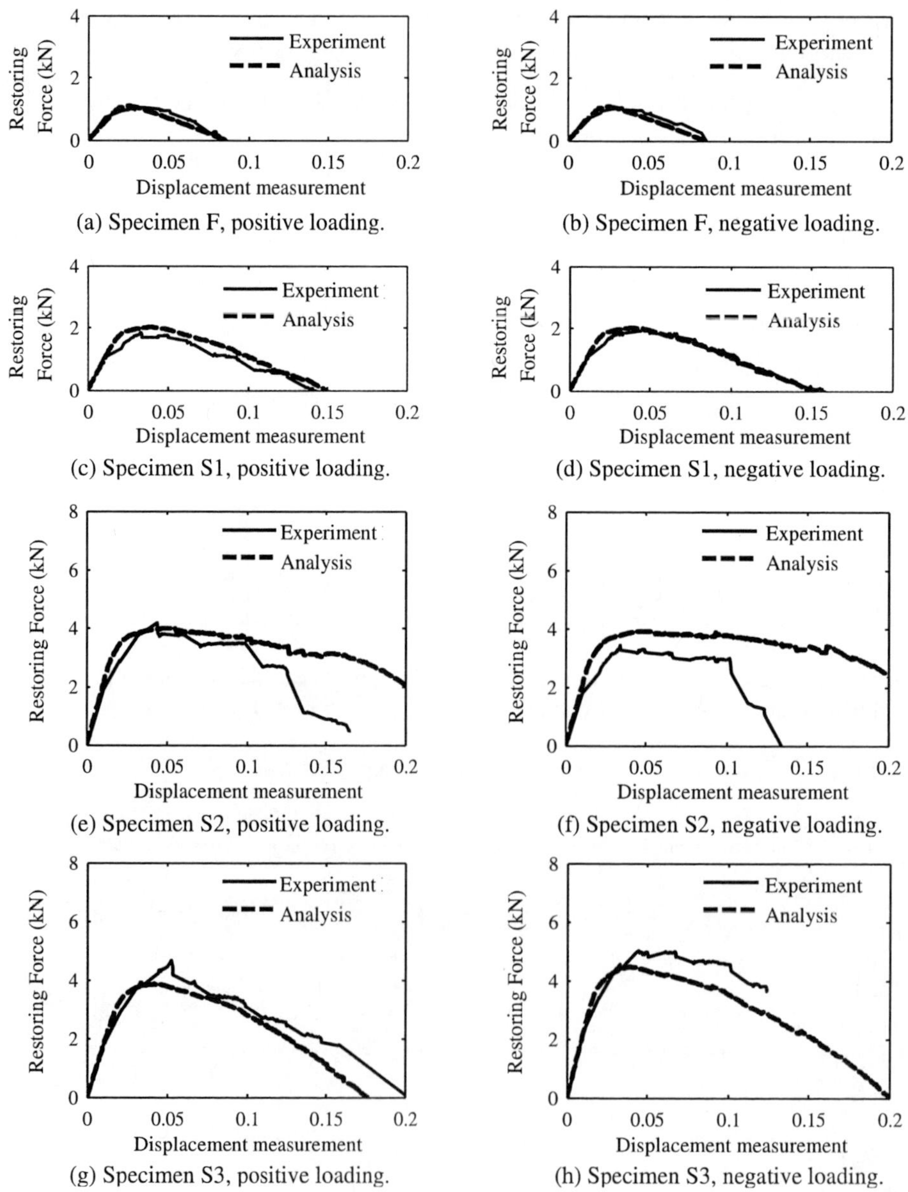

Figure 4. Simulation of skeleton curve on experimental result.

2.3 Modeling Procedure of Resistance on Joint

As an example of the procedure for modeling the resistance occurring at a column-beam joint in the positive direction, Figure 3 shows the compressive strain resistance on the side of the column.

a) In Figure 3(a), the nodes are determined on the columns and beams and are connected by beam elements.

b) For a column, a rigid link is established from the column toward the point assumed to be subject to compressive strain; for a beam, the point-changing section of the member is shown in Figure 3(b).

c) In Figure 3(c), the nodes subject to compressive strain are connected by springs representing the compressive strain resistance.

d) When the joint moves, as shown in Figure 3(d), the compressive strain force is represented by the spring.

3 SIMULATED RESULT

Figure 4 compares the values of restoring force determined from the experiment and those from nonlinear analysis. The results of the analysis are almost equal to the test results for every specimen. Because it cannot check breaking member enough in program, the restoring force in negative loading is separated after 1/10 rad on specimen S2.

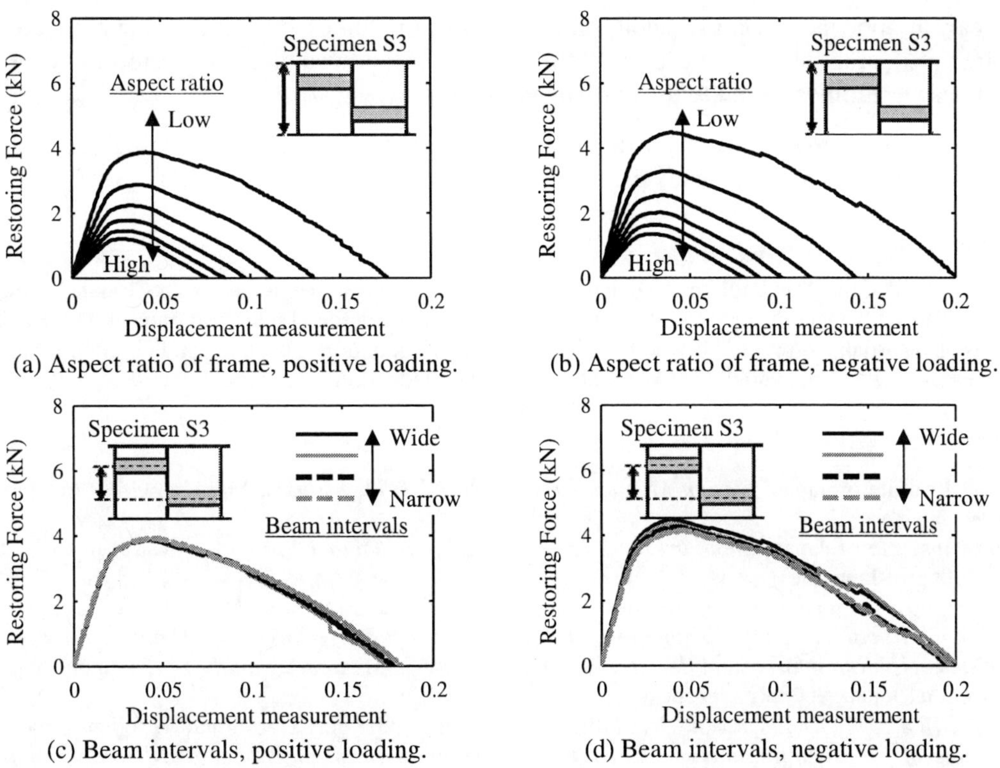

(a) Aspect ratio of frame, positive loading.
(b) Aspect ratio of frame, negative loading.
(c) Beam intervals, positive loading.
(d) Beam intervals, negative loading.

Figure 5. Sensitivity analysis of changing beam arrangement.

4 SENSITIVITY ANALYSIS OF BEAM ARRANGEMENT

We conducted a sensitivity analysis to study the effect of height arrangement of beams on the behavior of the whole frame. The results are shown in Figure 5. In the case of changed aspect ratio of frame for specimen S3, while keeping the beam height arrangement constant, it is observed that the restoring force decreases as the aspect ratio increases for both positive and negative loadings, as shown in Figure 5(a) and (b).

In the case of decreased interval of beam height arrangement for specimen S3, the restoring force increases a little in positive loading, but decreases a little in negative loading. In addition, the difference of restoring forces depends on the loading direction, as in Figure 5(c) and (d).

5 CONCLUSIONS

We simulated past cyclic loading tests on four traditional wooden frames with large-section beams to determine the seismic performance of the frames, while also studying the influence of the beams on the overall behavior of the frame. First, a simple analysis model of a fitting-type joint was constructed using beam elements. Next, this model was used to simulate the experimental results. Finally, a sensitivity analysis was conducted to study about the effect beam height arrangement on the behavior of the whole frame.

The major findings of this study were as follows:

a) Some material tests are conducted for wooden test pieces cut down from all columns of specimens. The compressive spring properties are decided.

b) Analysis models are constructed, and the restoring forces of frame specimens are simulated precisely. The results of the simulation showed that the restoring force values were almost equal to the experimental results.

c) The relation between height arranging beams and the change in restoring force is considered by sensitivity analysis.

Acknowledgments

This study was part of the “Technology of Urban Architecture Rooted in Regional Asian Climate” project, supported by TMU advanced research under the “Asian Human Resources Fund” program of TMG. This research was also partially supported by the Housing Research Foundation “Jusoken” and Grants-in-Aid for Scientific Research (A) (No.15H02275). We gratefully acknowledged this support.

References

Architectural Institute of Japan, *Design Manual for Engineered Timber Joints*, Maruzen Co., Ltd., Japan, 2009. (in Japanese)

Architectural Institute of Japan, *Standard for Structural Design of Timber Structures*, Maruzen Co., Ltd., Japan, 2006. (in Japanese)

Editorial committee for manual of seismic design for wooden frame structures, *Manual of seismic design for wooden structures taking advantage of traditional structural techniques - Methods for seismic design and seismic reinforcement design based on response-limit capacity analysis*, Gakugei Shuppan Sha Co., Ltd., Japan, 2008. (in Japanese)

Takiyama, N., Idate, N., and Yamada, Y., Fitting-type Joint Model for Traditional Wooden Structure, *Proceeding of The 2nd International Conference on Preservation, Maintenance and Rehabilitation of Historical Buildings and Structures (REHAB2015)*, Porto, Portugal, July, 2015

Takiyama, N., and Idate, N., Seismic Behavior of Frames with Uneven Large Section Beams of Traditional Wooden Structure, *Proceeding of The Eighth International Structural Engineering and Construction Conference*, Sydney, Australia, Nov., 2015

Integrated Solutions for Infrastructure Development
Edited by Lau, H. H., Tang, F. E., Ng, C. K., and Singh, A.
Copyright © 2016 ISEC Press

AN APPLICATION OF HYPERPLASTICITY TO DETERMINE THE BEHAVIOR OF A PLANE STRESS CANTILEVER BEAM

HAI THAN NGUYEN and LAM NGUYEN SY

Faculty of Civil Engineering, Ho Chi Minh City University of Technology, Vietnam

Hyperplasticity is an approach to plasticity theory based on thermodynamic principles. By using this concept, the entire constitutive model behavior can be determined from two scalar potential functions, one is the energy function and the other is the yield function or the dissipation function. In this paper, the hyperplasticity model for homogeneous and isotropic material is derive and implemented in finite element method (FEM) to solve a particular problem in structure analysis, the behavior of a cantilever beam. The Gibbs free energy function and the Von Mises yield function are chosen to derive the elasto-plastic response of structures. Besides, kinematic hardening with single yield surface model is employed to describe the behavior when yield occur. The result shows that the application of hyperplasticity to FEM can be another proper way for non-linear analysis. Moreover, it reveals the ability to develop more sophisticated model using multiple yield surfaces in cases of cyclic loading.

Keywords: Elasto-plastic, Non-linear analysis, Kinematic hardening, Thermodynamic, Yield surface, Flow rule.

1 INTRODUCTION

The principle root of hyperplasticity approach is originally proposed by Ziegler (1977). This work shows that constitutive model of a deformable solid can be defined by using only two scalar functions, the energy function and the dissipation function. Collins and Houlsby (1997) developed an approach to plasticity theory base on thermodynamic for geotechnical materials. Houlsby and Puzrin (2000a and 2000b) continued this work by developing this approach into a consistent and rigorous thermomechanical framework base on generalize thermodynamic for the rate-independent dissipative materials, which can be applied for a wide range of engineering materials.

There are several advantages of hyperplasticity model. One significant feature of this approach is that the derivation of constitutive behavior is established in strict accordance with the first and the second law of thermodynamics; therefore, the hyperplasticity approach guarantees that incremental derivation of plasticity will automatically obey the laws of thermodynamics. Another advanced is the ability to capture behavior under cyclic loading. This is because the past history of loading-unloading process is always reflected clearly through the state of the internal variables.

Various researches have been done by applying the hyperplasticity approach. Nguyen-Sy (2005) employed this concept for the modelling of circular shallow offshore foundation.

Likitlersuang (2005) developed a hyperplasticity model for clay. Nguyen (2005) and Einav (2007) applied this framework for the coupled damage and plasticity models.

In this paper, the hyperplasticity model for a rate-independent kinematic hardening plastic material is introduced and applied to derive behavior of a cantilever metal beam, which is considered as a homogeneous and isotropic material. A comparison between the conventional approach and the hyperplasticity concept is also conducted. Aim of this work is to introduce another rigorous way to simulate the structure behavior, which has enough ability to derive the realistic work from the viewpoint of thermodynamic.

2 THE HYPERPLASTICITY THEORY

The specific formulation of this concept from classical thermodynamic for fluid to solid can be found in Houlsby and Puzrin (2000a and 2000b). This section introduces a brief summary of this concept for small strain continuum mechanic.

The formulation starting by the introduction of an internal state variable in tensor form (α_{ij}). In this paper, the Gibbs free energy function, which take the form $g = g(\sigma_{ij}, \alpha_{ij}, \theta)$, is used. The main goal of the definition of the internal variables is to supply the tools for the study of the dissipation of energy. Formulation from first and second laws lead to the form of dissipation function $d = d(\sigma_{ij}, \alpha_{ij}, \theta, \dot{\alpha}_{ij}) \geq 0$. Tensors $\overline{\chi}_{ij} = -\partial g / \partial \alpha_{ij}$ and $\chi_{ij} = \partial d / \partial \dot{\alpha}_{ij}$ are the generalized stress and the dissipative generalized stress respectively. The fundamental of constitute hypothesis is $\overline{\chi}_{ij} = \chi_{ij}$.

Collins and Houlsby (1997) shows that rate-independent material, which the dissipation is homogeneous first order function, is a degenerate case of Legendre transformation. Therefore, the transformation of and derivative dissipation function give $\dot{\alpha}_{ij} = \lambda \dfrac{\partial y}{\partial \chi_{ij}}$. The role of y is identified with the role of yield surface in conventional approach, but the difference is that it respects to generalized stress χ_{ij} instead of true stress σ_{ij}. Besides, α_{ij} is now identified with the plastic strain.

3 HYPERPLASTICITY FOR SINGLE YIELD SURFACE KINEMATIC HARDENING

In this section, the behavior of model using Von Mises yield criteria and kinematic hardening will be derive. Thermal effect is not considered in this paper; therefore, the dependence on the temperature will not be taken into account.

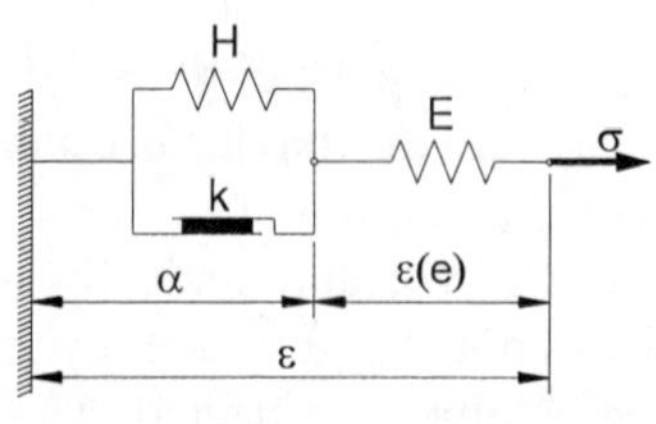

Figure 1. Friction slip and spring model for kinematic hardening.

The mechanical behavior of kinematic hardening with single yield surface in one-dimensional model is expected to be as shown in Figure 1. The hardening parameter H can be seen as the modulus of a spring when yield occur.

The first function specified is the Gibbs free energy function. As Puzrin and Houlsby (2001), the Gibbs free energy function for multi-dimensional model takes the form:

$$g = -\frac{1}{18K}\sigma_{ll}\sigma_{kk} - \frac{1}{4G}\sigma'_{ij}\sigma'_{ij} + \frac{H}{2}\alpha_{ij}\alpha_{ij} - \sigma_{ij}\alpha_{ij} \tag{1}$$

Where K is the bulk modulus, G is the shear modulus and H is the hardening parameter of model. From Eq. (1) we can derive:

$$\varepsilon_{ij} = -\frac{\partial g(\sigma_{ij},\alpha_{ij})}{\partial \sigma_{ij}} = \frac{1}{9K}\sigma_{kk}\delta_{ij} + \frac{1}{2G}\sigma'_{ij} + \alpha_{ij} \tag{2}$$

$$-\frac{\partial^2 g}{\partial \sigma_{ij}\partial \sigma_{kl}} = \frac{1}{3K}I^V_{ijkl} + \frac{1}{2G}I^D_{ijkl} = C_{ijkl} \tag{3}$$

In which I^V_{ijkl}, I^D_{ijkl} are tensors that $\sigma_{kk} = I^V_{ijkl}\sigma_{ij}$, $\sigma'_{ij} = I^D_{ijkl}\sigma_{ij}$ and $C_{ijkl} = D^{-1}_{ijkl}$ is the elastic tensor.

The second function is the Von Mises yield function, which takes the form:

$$y^g = \chi'_{ij}\chi'_{ij} - 2k^2 \tag{4}$$

Where k is the yield stress in simple shear. From this we can derive as follow:

$$\dot{\alpha}_{ij} = \lambda\frac{\partial y^g}{\partial \chi_{ij}} = 2\lambda I^D_{ijkl}\chi'_{kl} = 2\lambda\chi'_{ij} \tag{5}$$

The multiplier λ can be calculated from Houlsby and Puzrin (2001) and the result $\lambda = \frac{\chi'_{ij}\dot{\sigma}_{ij}}{4Hk^2}$ with note that λ is a non-negative multiplier whenever $y = 0$.

In this paper, the increment of stress and strain tensor are not decoupled into deviatoric and volumetric parts, which is different from the formulation as Puzrin and Houlsby (2001). The incremental response is as follow:

$$\dot{\varepsilon}_{ij} = -\frac{\partial^2 g}{\partial \sigma_{ij}\partial \sigma_{kl}}\dot{\sigma}_{kl} + \frac{\partial^2 g}{\partial \sigma_{ij}\partial \alpha_{kl}}\dot{\alpha}_{kl} = C_{ijkl}\dot{\sigma}_{kl} + \frac{\chi'_{ij}\dot{\sigma}_{ij}}{4Hk^2}2\chi'_{kl} = \left(C_{ijkl} + \frac{\chi'_{ij}\chi'_{kl}}{2Hk^2}\right)\dot{\sigma}_{kl} \tag{6}$$

The vector notation is needed in order to apply to FEM. In vector notation, the stress component is written as $\sigma = \{\sigma_x \quad \sigma_y \quad \sigma_z \quad \tau_{yz} \quad \tau_{zx} \quad \tau_{xy}\}^T$. The notation is similar for $\varepsilon, \alpha, \chi$. The incremental stress-strain response from Eq. (6) can be rewritten as:

$$\dot{\varepsilon} = C\dot{\sigma} + \frac{(\chi')^T P\dot{\sigma}}{2Hk^2}\chi' = \left(C + \frac{\chi'(\chi')^T P}{2Hk^2}\right)\dot{\sigma} \tag{7}$$

$$\dot{\sigma} = D_{ep}\dot{\varepsilon} = \left(C + \frac{\chi'(\chi')^T P}{2Hk^2}\right)^{-1}\dot{\varepsilon} \tag{8}$$

Where $C=D^{-1}$ is the inverse of elastic matrix and $P = diag[1 \quad 1 \quad 1 \quad 2 \quad 2 \quad 2]$ is diagonal scaling matrix for the result in vector notation equal to product in tensor notation.

The form of the incremental equation is similar in comparison with conventional approach. However, it is derived by internal state variable and generalized stress instead of true stress.

4 NUMERICAL EXAMPLE

This section explains the result when apply hyperplasticity theory to derive the behavior of a cantilever beam under cyclic load in plane stress problem. See Figure 2 for definition. It should be noted that this is just a theoretical example, so the parameters are assumed base on the properties of steel to view the stress-strain response and the ultimate strength is not considered in this example. A more realistic model needs data from experiments, which requires further researches.

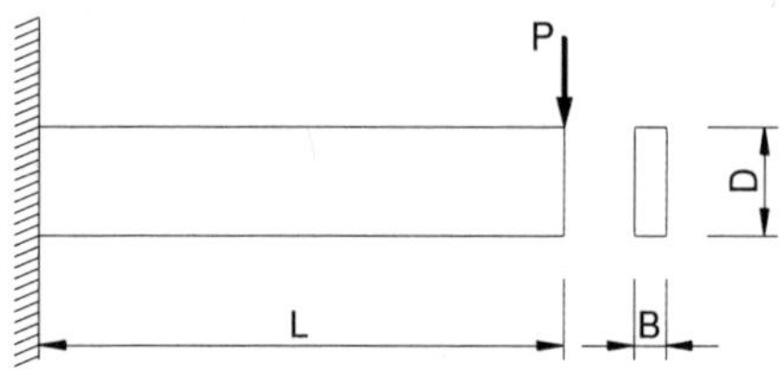

Figure 2. Problem definition.

Geometric parameters: L = 1,000 mm, D = 200 mm, B = 50 mm. Material properties: E = 210,000 MPa, $\nu = 0.3$, H = 30,000 MPa and yield stress σ_P = 240 MPa. There are four loading processes as shown in Table 1.

Table 1. Loading processes.

Step	Process			
	1	2	3	4
1	From 40 kN to 176 kN	From 40 kN to 176 kN	From 40 kN to 176 kN	From 40 kN to 176 kN
2	From 176 kN to 56 kN	From 176 kN to -84 kN	From 176 kN to -184 kN	From 176 kN to -184 kN
3	From 56 kN to 216 kN	From -84 kN to 216 kN	From -184 kN to 216 kN	From -184 kN to 188 kN
4				From 188 kN to -200 kN

4.1 The Increment Response of Applied Load and Displacement

In order to check the reasonable of new approach, we have made codes based on the approach of Owen and Hinton (1980) and use the same input data (process 1) to compare the results of load-displacement response at center of free-end section between two approaches as show in Figure 3.

The result shows that the shape of the responses of hyperplasticity model are similar in comparison to the conventional approach. However, there are little difference in values, especially when the yield behavior occur. This is because the meaning of hardening parameter. Owen and Hinton (1980) used the work hardening rule whereas the strain hardening is employed in our work. Despite of the differences, the results show that hyperplasticity is a promising approach. Moreover, the advanced of the new approach is that it can simulate the behavior of cyclic load in three other processes (2, 3 and 4), which cannot be done by the conventional approach. The responses of center point of free-end section in four processes are shown below.

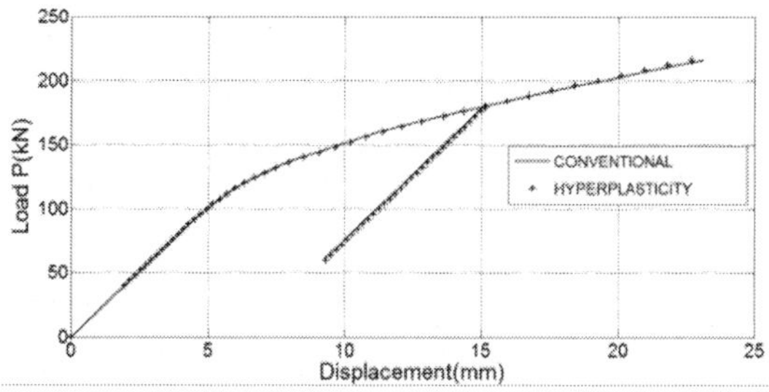

Figure 3. Load-displacement response in comparison with conventional approach.

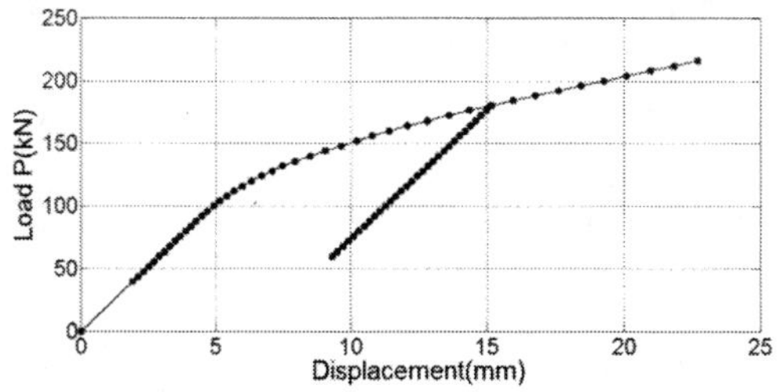

Figure 4. Load-displacement response for process 1.

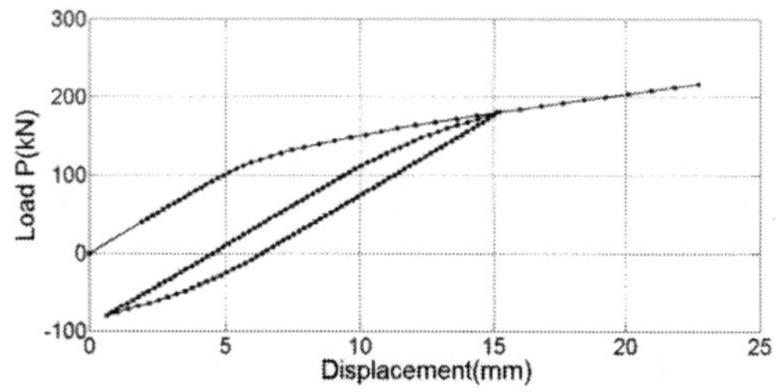

Figure 5. Load-displacement response for process 2.

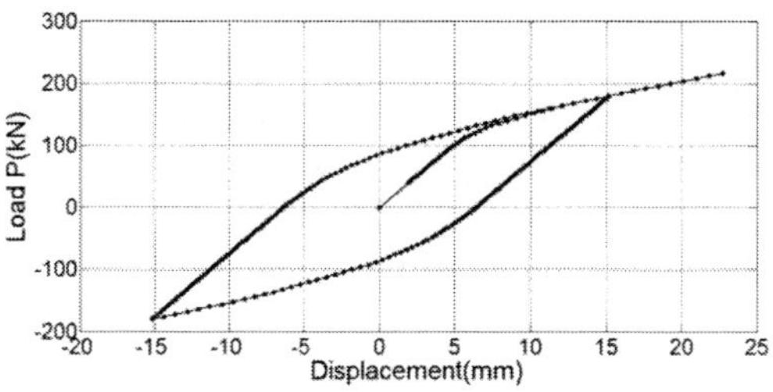

Figure 6. Load-displacement response for process 3.

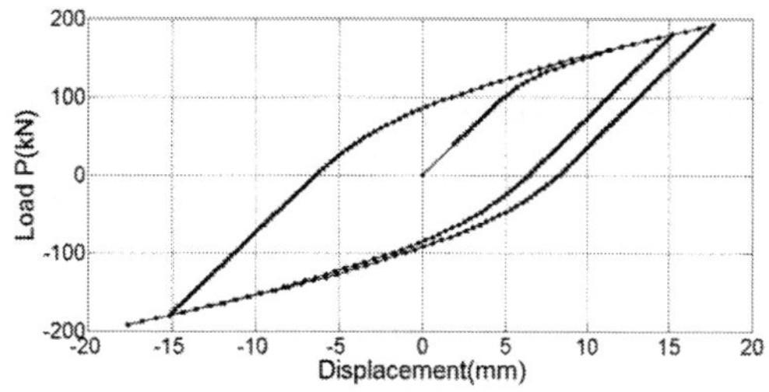

Figure 7. Load-displacement response for process 4.

When the unloading occurs in the first process, the stress state point moves within the yield surface, but it does not touch the yield surface again, so the behavior is elastic and the reloading is elastic too as shown in Figure 4. The second process, see Figure 5, shows a further step in which the stress point touch the yield surface again during unloading. In the third process, which can be seen from Figure 6, the range of unloading step is expanded in order to get over the state at the first time. The fourth process adds another unloading period as shown in Figure 7. It can be seen from the results that if the cyclic processes continue, the load-displacement response will obey the Mashing rule as discussed in Nguyen-Sy (2005).

4.2 The Stress Distribution of the Cantilever Beam

The stress (σ_x) distribution during process 4 is presented at four reference stages: (1) After the first loading increment, when yield behavior has not appeared, (2) at the end of load step 1, (3) at the end of load step 2 and (4) at the end of load step 3.

As shown in Figure 8, the stress distribution looks reasonable. At the first stage, the tensile stresses distribute on top and compressive stresses at bottom of section (the stress on top and at bottom of fixed end section are 126.78 MPa and -126.42 MPa respectively). During the loading period, the distribution areas of tension and compression enlarge to the center of section and to the free end of the cantilever beam as shown at second reference time (maximum stress is 459.25 MPa and minimum stress is -458.14 MPa). The third reference stage is similar to the second but inverse in value due to the change of loading direction (stress on top of fixed end section is -458.70 MPa whereas at bottom is 457.54 MPa). When the reloading is conducted, the distribution at fourth stage is similar to second stage (497.17 MPa and -495.96 MPa on top and at bottom respectively).

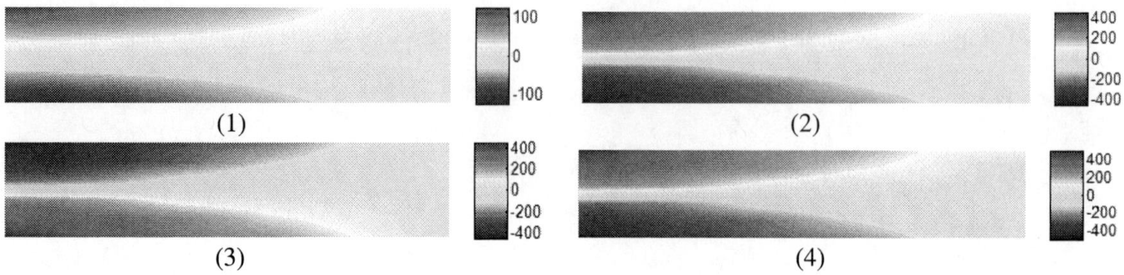

Figure 8. The distribution of σ_x (MPa) during process 4 at four reference stage.

5 CONCLUSION

Applying hyperplasticity theory to nonlinear behavior can be a proper way to simulate the realistic behavior of structure engineering. The presented model can be used to derive the response of structures in various ways of loading process. For a more realistic result, parameter of this approach should be calibrated with experimental data. Besides that, the approach for single yield surface shows the ability to go further to the concept of multiple yield surface, which is more sophisticated model. Hyperplasticty theory can also apply for behavior of various models in structure analysis such as slender beam, pin-jointed structures etc. by appropriate potential functions, which require further studies.

Reference

Collins, I. F., and Houlsby, G. T., Application of thermomechanical principles to the modelling of geotechnical materials. *Proceedings of the Royal Society of London A: Mathematical, Physical and Engineering Sciences*. Vol. 453. No. 1964. The Royal Society, 1997.

Einav, I., Houlsby, G. T., and Nguyen, G. D., Coupled damage and plasticity models derived from energy and dissipation potentials. International Journal of Solids and Structures, 44(7), 2487-2508, 2007.

Houlsby, G. T., and Puzrin, A. M., An approach to plasticity based on generalised thermodynamics. *Constitutive Modelling of Granular Materials*. Springer Berlin Heidelberg, 319-331, 2000a.

Houlsby, G. T., and Puzrin, A. M., A thermomechanical framework for constitutive models for rate-independent dissipative materials. *International Journal of Plasticity* 16.9 (2000): 1017-1047, 2000b.

Nguyen, G. D., A thermodynamic approach to constitutive modelling of concrete using damage mechanics and plasticity theory. Oxford: University of Oxford, 2005.

Nguyen-Sy, L., The theoretical modelling of circular shallow foundation for offshore wind turbines. University of Oxford, 2005.

Likitlersuang, S., A hyperplasticity model for clay behaviour: an application to Bangkok clay. University of Oxford, 2003.

Owen, D. R., and Hinton, E., *Finite elements in plasticity, Pineridge Press, UK,* 1980.

Puzrin, A. M., and Houlsby, G. T., Fundamentals of kinematic hardening hyperplasticity. *International journal of solids and structures* 38.21 (2001): 3771-3794, 2001.

Ziegler, H., *An introduction to thermomechanics*, North Holland, Amsterdam, 1977.

Integrated Solutions for Infrastructure Development
Edited by Lau, H. H., Tang, F. E., Ng, C. K., and Singh, A.
Copyright © 2016 ISEC Press

NUMERICAL ANALYSIS OF AXIALLY LOADED COLD-FORMED STEEL BUILT-UP BOX SECTION

T. C. H. TING, H. H. LAU, and M. S. JOO

Dept of Civil and Construction Engineering, Curtin University, Sarawak Malaysia

Built up box section is a doubly symmetrical section in which two C-channel connected at the flange using self-drilling screw. They are generally used in the construction industry to obtain higher rigidity, when a single section is not sufficient. Despite being widely used in the construction industry, there are limited studies on built-up box section. Moreover, the current design code does not provide clear design guideline for the calculation of built-up box section. Therefore, research on the built-up box section is important. This paper presents a finite element model to predict the compressive strength and simulate the behavior of cold-formed steel built-up box section. The finite element model was developed using ABAQUS CAE/6.14. Comparison of the finite element and experimental results showed good correlation. The model well predicted the behavior of the built-up box column.

Keywords: Effective width method, Direct strength method, Finite element analysis.

1 INTRODUCTION

Cold-formed steel structural members are generally used for individual structural framing members such as purlin, joist and studs, panels and decks (Yu 2010). However, due to increasing demand, cold-formed steel are often used to stretch over large span. The major issue with large span is the deflection, lateral stability and constructability. In order to obtain higher rigidity, built-up section such as built-up box section is formed (Cheng & Schafer 2007). The built up box section used in this research is a doubly symmetrical section in which two C-channels are connected at the flanges using self-drilling screws (Figure 1).

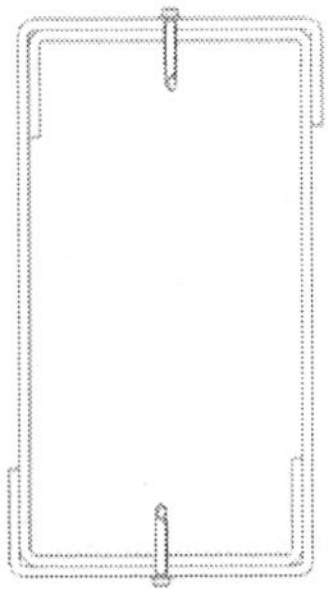

Figure 1. Built-up box section.

Finite element modeling has successfully been used in the past to predict the behavior of cold-formed steel structures. Finite element model could simulate the buckling and compressive strength of a cold-formed steel built-up column, provided that important parameters such as element type, mesh size, and surface contact of the finite element model are properly configured (Becque & Rasmussen 2009). In this research, finite element models were developed by using ABAQUS CAE/6.14. The finite element models were developed to simulate the compression test from the authors' previous publication (Ting & Lau 2010).

2 FINITE ELEMENT METHOD

Thin walled members such as cold-formed steel are modeled by using shell type element. Quadrilateral shell element (S4R) was selected to develop the finite element model. S4R account for three translational and three rotational degrees of freedom at each node of the element. The influence of shear flexibility in laminated composite shell models was taken into consideration by these elements. The element provides more flexibility to model curved and straight geometry such as the corner of the C-channel column.

In order to study the effect of the mesh size on the accuracy of the finite element analysis results, convergence study has been conducted. The mesh sizes modeled were 2mm, 5mm, 10mm and 15mm with total mesh number of 20700, 3780, 1125 and 570 respectively for a 450mm length column. Mesh size of 5mm × 5mm with a difference of 2.6% (with 2mm × 2mm as reference) is selected for the further analysis.

The Young's modulus of the steel is 20,500MPa and Poisson's ratio of 0.3 is used to simulate the elasticity and plasticity properties of the cold-formed steel. The yield strength of the cold-formed steel is 550.5MPa. The plastic behavior was analyzed by using Dynamic-Implicit method.

In the compressive test, the stub column has flat-end condition and intermediate column under pinned-end condition. In the finite element analysis, the ends of the stub columns were fixed, except for the y-axis displacement at the top end of the column. The displacement was imposed on the top end of the column. For the pinned-end condition, the rotation at weak axis (y-axis) was released. The automatic incrimination was used with a time period of 1 second for stable load incrimination.

In the compression test, the built-up box section was assembled together by using self-drilling screw. To simulate the fastener in the finite element analysis, discrete rigid 2mm thin strip was created to attach the outer flange and inner flange of the built-up box section, as shown in Figure 2 below. Discrete rigid part was selected because there is no failure in the screw connection during the compression test.

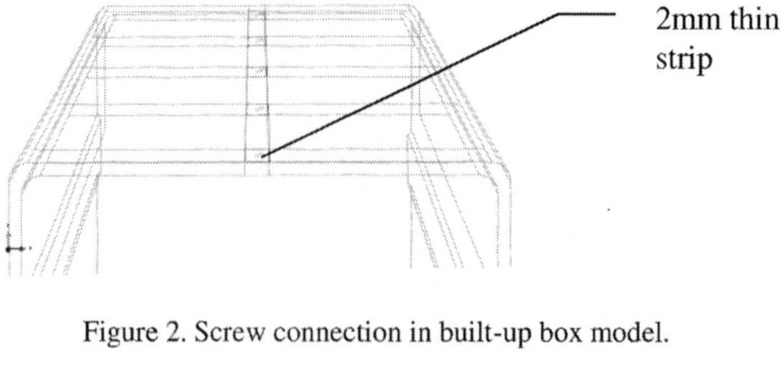

Figure 2. Screw connection in built-up box model.

3 RESULTS AND DISCUSSIONS

3.1 Load Carrying Capacity

Tables 1 and 2 show the comparison of test results to finite element results. Generally, the finite element model well predicts the compressive strength of the column. The test results for intermediate columns are higher than the finite element and design calculated results. This could be due to the end condition of the test specimens. The friction between ball bearing and the end plate causes the pinned-end condition to behave as partially fixed.

Table 1. Comparison of load carrying capacity for stub columns.

Specimens	P_{Test} (kN)	P_{FE} (kN)	$\frac{P_{Test}}{P_{FE}}$
C75-450-1	57.54	48.47	1.17
C75-450-2	55.50		1.12
C75-450-3	54.20		1.12
C75-450-4	57.74		1.19
Mean			1.15
B75-450-1	124.80	131.34	0.95
B75-450-2	127.52		0.97
B75-450-3	130.50		0.99
B75-450-4	129.70		0.99
Mean			0.98

Table 2. Comparison of load carrying capacity for intermediate columns.

Specimens	P_{Test} (kN)	P_{FE} (kN)	$\frac{P_{Test}}{P_{FE}}$
C75-1500-1	39.00	36.68	1.06
C75-1500-2	40.50		1.10
C75-1500-3	42.00		1.15
C75-1500-4	45.50		1.24
Mean			1.14
B75-1500-1	84.00	76.89	1.09
B75-1500-2	81.00		1.05
B75-1500-3	92.00		1.20
B75-1500-4	91.00		1.18
Mean			1.13

3.2 Buckling Behavior

The buckling behavior of stub columns is as shown in Figure 3. From observation, local buckling at web occurred as an initial buckling mode for the stub columns. Later, the flanges deflected when load was close to the ultimate load. After that, deflection in web and flanges increased and accompanied by a sudden drop in load when the specimen failed. The location of maximum deflection occurs near the mid-length of the column.

C75-450 *B75-450*

Figure 3. Failure mode of FE models and tested specimens of stub columns.

Figure 4 shows the buckling behavior of intermediate columns. At the initial stage of the compression test, the intermediate columns shows local buckling at web and then distortional buckling at the flanges. Significant global buckling only occurred after reaching the maximum load. This is because the high friction between the ball bearing and the end plates at both ends of the column restricted the rotation.

C75-1500 *B75-1500*

Figure 4. Failure mode of FE models and tested specimens of intermediate columns.

4 CONCLUSION

A finite element model was developed using ABAQUS CAE/6.14 to study the load carrying capacity and buckling behavior of boxed built-up stub column. The model was verified against previously published experimental results. Test data of eight C-channel and eight boxed built-up columns also show that the failure of stub columns was resulted from local and distortional buckling. Similar failure mode was observed during the tests with maximum deformation occurred near the mid height of the columns. Predictions correlate well with the test results.

Acknowledgments

The authors would like to acknowledge the the financial support under Curtin - EcoSteel Sdn Bhd Industry Research Grant and also acknowledge Mr. Chin Mei Chuo for his contribution towards this research project.

References

Becque, J., and Rasmussen, K.J.R., Numerical investigation of the interaction of local and overall buckling of stainless steel I-columns, *Journal of Structural Engineering*, American Society of Civil Engineering, 2009.

Cheng, Y. and Schafer, B.W., Simulation of Cold-Formed Steel Beams in Local and Distortional Buckling with Applications to the Direct Strength Method, *Journal of Constructional Steel Research* 63 (5): 581-590, 2007.

Ting, T.C.H., and Lau, H.H., Load Carrying Capacity of High Strength Cold-Formed Steel Built-up Box Sections, Proceeding of the 4th International Conference on Steel & Composite Structures, ICSCS, Sydney, Australia, 2010.

Yu, W.W., Cold-Formed Steel Design 4th Edition, John Wiley & Sons, Inc., 2010.

Integrated Solutions for Infrastructure Development
Edited by Lau, H. H., Tang, F. E., Ng, C. K., and Singh, A.
Copyright © 2016 ISEC Press

NUMERICAL STUDIES OF CONCRETE PLATES UNDER LOCALIZED BLAST LOADS

SOFIA W. ALISJAHBANA[1], WIRATMAN WANGSADINATA[2], and IRENE ALISJAHBANA[3]

[1]*Civil Engineering Dept, Bakrie University, Jakarta, Indonesia*
[2]*Wiratman and Associates, Consulting Engineering, Jakarta, Indonesia*
[3]*Civil Engineering Dept, Universitas Indonesia, Depok, Indonesia*

An investigation has been carried out to examine the dynamic behavior of concrete plates subjected to localized blast loading. The concrete plate is modeled as a thin plate with finite dimensions sitting on an elastic three-parameter soil foundation model. The localized blast loading is expressed by using the Dirac Delta function for different value of the load's position. The governing equation of the problem is solved using the modified Bolotin method for determining the natural frequencies and the wave numbers of the system. The orthogonal properties of Eigen functions are used to find the general solution of the problem. Special emphasis is focused on the mid-point displacements. The results obtained allow an insight into the effect of the load's position, the thickness of the concrete plate on the response of the concrete plate under localized blast loading and indicate that the load's position and the thickness of the plate can affect their overall behavior.

Keywords: Dynamic behavior, Localized blast load, Dirac delta function, Three-parameter soil foundation, Natural frequencies, Eigen function.

1 INTRODUCTION

In recent years, public buildings and structures have unexpectedly been exposed to the risk of terrorist attacks, particularly in the form of vehicle bombing or other portable detonation devices. These potential threats give rise to a challenging question of structural safety, provided that any structural part could be subjected to unpredictable loading that were not primarily designed against, in terms of both the loading type and intensity.

The dynamic analysis of orthotropic plates with fully fixed supported boundary conditions under localized blast loading was presented by Alisjahbana and Wangsadinata in 2014. Effects of various parameters such as the position of the blast loading, plate thickness and damping ratio on the maximum dynamic deflection of the plates subjected to localized blast loading were considered (Alisjahbana and Wangsadinata 2014). In most models used previously, the dynamic response of concrete plate is taken into account only by the inertia of the plate (Lu 2001). Concerning the soil, inertia is neglected in dynamic modeling of pavement structure. To extend the previous work, the present study investigates dynamic response of concrete pavement resting on Pasternak foundation under localized blast loading. The concrete pavement is modelled as a thin orthotropic plate with semi rigid boundary condition in its edges. In order to take into account its inertia, the soil is modelled as a three-parameter type (Gibigaye *et al.* 2016).

2 GOVERNING EQUATION

In this research work, a rectangular orthotropic concrete plate of thickness h sits on an elastic three-parameter soil foundation model as shown in Fig. 1 is considered. According to the classic theory of thin plates and taking into account the reduced mass of soil, the transverse deflection of the Kirchhoff plate satisfies the following differential equation (Gibigaye *et al.* 2016):

$$D_x \frac{\partial^4 w(x,y,t)}{\partial x^4} + 2B \frac{\partial^4 w(x,y,t)}{\partial x^2 \partial y^2} + D_y \frac{\partial^4 w(x,y,t)}{\partial y^4} - G_s \left(\frac{\partial^2 w(x,y,t)}{\partial x^2} + \frac{\partial^2 w(x,y,t)}{\partial y^2} \right) + \gamma h \frac{\partial w(x,y,t)}{\partial t} + (\rho h + m_0) \frac{\partial^2 w(x,y,t)}{\partial t^2} = p(x,y,t) \quad (1)$$

where, $w(x,y,t)$ is the deflection of the plate which is equal to the deflection of the plate-soil interface, D_x is the flexural rigidity of plate in the x-direction, B is the torsional rigidity of the plate, D_y is the flexural rigidity in the y-direction, G_s is the shear modulus of the foundation, γ is the logarithmic decrement of the soil, ρ is the mass density of the plate, m_0 is the inertial factor of the foundation soil.

3 IDEALISATION OF BLAST LOADING

Determination of the exact localized blast loadings is almost unrealistic considering the complicated process of the interaction of the blast wave with the target in concern (Li *et al.* 2009). In order to reduce this complex problem of blast loadings to reasonable terms, Alisjahbana and Wangsadinata have suggested a simplified blast loading function as a stepped triangular function as shown in Fig. 2 (Alisjahbana and Wangsadinata 2011):

$$P(t) = P_0\left(1 - \frac{t}{td_3}\right) \text{ for } 0 \le t \le td_1;\ P(t) = P_2\left(1 - \frac{t}{td_2}\right) \text{ for } td_1 \le t \le td_2\ ;\ P(t) = 0 \text{ for } t > td_2 \quad (2)$$

where, P_0 and P_2 are the maximum amplitude of the blast loading, td_1, td_2 and td_3 are duration of the blast loading.

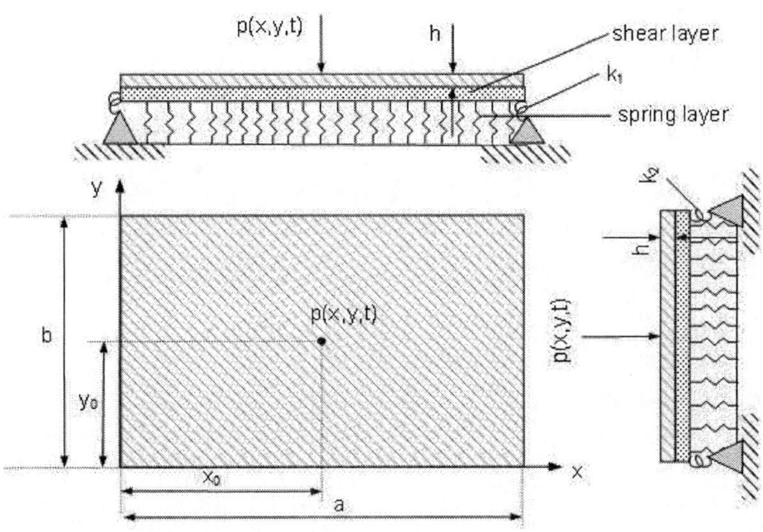

Figure 1. An elastic rectangular concrete plate resting on an elastic three-parameter soil foundation.

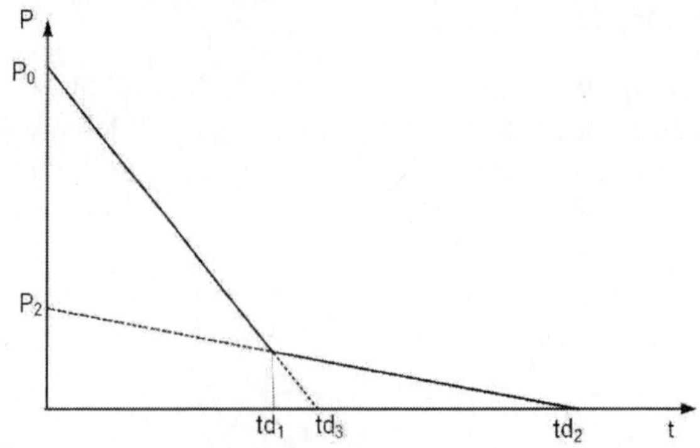

Figure 2. Model of a stepped triangular blast loading.

4 DETERMINATION OF THE EIGEN FREQUENCIES

In order to solve the governing equation (1) of the problem, the free vibrations solution of the problem is set as:

$$w(x,y,t)=\sum_{m=1}^{\infty}\sum_{n=1}^{\infty}W_{mn}(x,y)\cos(\omega_{mn}t) \tag{3}$$

ω_{mn} is the circular frequency of plate and $W_{mn}(x,y)$ is the function of position coordinates determined for the mode numbers m and n in x-direction and y-direction, respectively, which can be determined from the first and second auxiliary Levy-type problem (Alisjahbana and Wangsadinata 2014).

4.1 First Auxiliary Levy-type Problem

Based on the Modified Bolotin Method the solution of Eq. (1) for the first auxiliary problem can be expressed as (Alisjahbana and Wangsadinata, 2011)

$$W(x,y)=\sum_{m=1}^{\infty}\sum_{n=1}^{\infty}X_{mn}(x)\sin\left(\frac{q\pi y}{b}\right) \tag{4}$$

Satisfying the semi rigid boundary conditions along x=0 and x=a:

$$X_{mn}(x)=0\ ;\ -D_x\left(\frac{\partial^2 W(x,y)}{\partial x^2}+vy\frac{\partial^2 W(x,y)}{\partial y^2}\right)=k_1\frac{\partial W(x,y)}{\partial x}\quad \text{at } x=0 \text{ and } x=a \tag{5}$$

Substituting Eq. (4) into the homogeneous form of Eq. (1) results in the Eigen mode of the plate in the x-direction X_{mn}(x):

$$X_{mn}(x)=\cosh\left(\frac{\beta\pi}{ab}x\right)+\left(\frac{bpk_1(c_1-C_1)+a(F_1+F_2)s1}{k_1(bpS1-\beta s_1)}\right)\sinh\left(\frac{\beta\pi}{ab}x\right)-\cos\left(\frac{p\pi}{a}x\right)+\left(\frac{ab(F_1+F_2)S1+\beta k_1(c1-C1)}{k_1(\beta s1-bpS1)}\right)\sin\left(\frac{p\pi}{a}x\right) \tag{6}$$

where:

$$\beta = \sqrt{\frac{2.Bq^2a^2}{D_x} + p^2b^2 + \frac{G_s a^2 b^2}{\pi^2 D_x}}\ ;\ F_1 = D_x\left(\frac{\beta}{ab}\right)^2 \pi - \upsilon y D_x\left(\frac{q}{b}\right)^2 \pi\ ;\ F_2 = D_x\left(\frac{p}{a}\right)^2 \pi + \upsilon y D_x\left(\frac{q}{b}\right)^2 \pi\ ;$$

$$C_1 = \cosh\left(\frac{\pi\beta}{b}\right);\ c_1 = \cos(p\pi);\ S_1 = \sinh\left(\frac{\pi\beta}{b}\right);\ s_1 = \sin(p\pi)$$

4.2 Second Auxiliary Levy-type Problem

The solution of Eq. (1) for second auxiliary problem that satisfies the boundary conditions of

$$Y_{mn}(y) = 0;\ -D_y\left(\frac{\partial^2 W(x,y)}{\partial y^2} + v_x \frac{\partial^2 W(x,y)}{\partial x^2}\right) = k_2 \frac{\partial W(x,y)}{\partial y} \quad \text{at } y=0 \text{ and } y=b \tag{7}$$

can be expressed as

$$W(x,y) = \sum_{m=1}^{\infty}\sum_{n=1}^{\infty} Y_{mn}(y) \sin\left(\frac{p\pi x}{a}\right) \tag{8}$$

Substituting Eq. (8) into the homogeneous form of Eq. (1) results in the Eigen mode of the plate in the *y*-direction Y_{mn}(y):

$$\begin{aligned} Y_{mn}(y) = {} & \cosh\left(\frac{\theta\pi y}{ab}\right) + \frac{a(s_2 b(F_3+F_4) + qk_2(c_2 - C_2))}{k_2(S_2 qa - s_2\theta)} \sinh\left(\frac{\theta\pi y}{ab}\right) - \cos\left(\frac{q\pi y}{b}\right) \\ & + \frac{(S_2 ab(F_3+F_4) + k_2\theta(c_2 - C_2))}{k_2(\theta s_2 - S_2 qa)} \sin\left(\frac{q\pi y}{b}\right) \end{aligned} \tag{9}$$

where:

$$\theta = \sqrt{\frac{2Bp^2b^2}{D_y} + q^2a^2 + \frac{G_s a^2 b^2}{\pi^2 D_y}}\ ;\ F_3 = \left[D_y\left(\frac{\theta}{ab}\right)^2 \pi - \upsilon_x D_y\left(\frac{p}{a}\right)^2 \pi\right];$$

$$F_4 = \left[D_y\left(\frac{q}{b}\right)^2 \pi + \upsilon_x D_y\left(\frac{p}{a}\right)^2 \pi\right];C_2 = \cosh\left(\frac{\pi\theta}{a}\right);\ c_2 = \cos(q\pi);\ S_2 = \sinh\left(\frac{\pi\theta}{a}\right);\ s_1 = \sin(q\pi)$$

The unknown quantities *p* and *q* are calculated from the transcendental equation as:

$$\begin{aligned} & -2a^2b^3k_1{}^2 p\beta + 2ab^3k_1{}^2 p\beta\cos[p\pi]\cosh\left[\frac{\pi\beta}{b}\right] + \\ & + \left(a^2b^2k_1{}^2\left(b^2p^2 - \beta^2\right) + D_x{}^2\pi^2\left(b^2p^2 + \beta^2\right)^2 \sin[p\pi]\sinh\left[\frac{\pi\beta}{b}\right]\right) = 0 \end{aligned} \tag{10a}$$

$$\begin{aligned} & -2a^3b^2k_2{}^2 q\theta + 2a^3bk_2{}^2 q\theta\cos[q\pi]\cosh\left[\frac{\pi\theta}{a}\right] + a^4\left(b^2k_2{}^2q^2 + D_y\pi^2q^4\right) \\ & + a^2\left(\left(-b^2k_2{}^2 + 2D_y{}^2\pi^2q^2\right)\theta^2 + 2D_y{}^2\pi^2\theta^4\right)\sin[q\pi]\sinh\left[\frac{\pi\theta}{a}\right] = 0 \end{aligned} \tag{10b}$$

Once the value of *p* and *q* are determined from Eqs. (10a)-(10b), the Eigen modes of the system are determined as the product of Eq. (6) and Eq. (7). The natural frequency of the system can be expressed as:

$$\omega_{mn}^2 = \frac{\pi^4}{\rho h + m_0}\left[D_x\left(\frac{p}{a}\right)^4 + 2B\left(\frac{pq}{ab}\right)^2 + D_y\left(\frac{q}{b}\right)^4\right] + \frac{k_f}{\rho h + m_0} + \frac{G_s}{\rho h + m_0}\left[\left(\frac{p\pi}{a}\right)^2 + \left(\frac{q\pi}{b}\right)^2\right] \tag{11}$$

4.3 Determination of the Time Function

The time function that satisfies Eq. (1) can be expressed as:

$$\ddot{T}_{mn}(t) + 2\xi\omega_{mn}\dot{T}_{mn}(t) + \omega_{mn}^2 T_{mn}(t) = \frac{\int_0^a X_{mn}(x)dx \int_0^b Y_{mn}(y)dy}{(\rho + m_0)Q_{mn}} P(t)\delta[x - x(t)]\delta[y - y(t)] \quad (12)$$

where $\delta[.]$ is the Dirac Delta function and Q_{mn} is the normalization factor of the Eigen modes that can be expressed as:

$$Q_{mn} = \int_0^a (X_{mn})^2 dx \int_0^b (Y_{mn})^2 dy \quad (13)$$

5 NUMERICAL APPLICATIONS, RESULTS AND DISCUSSION

Using the procedure described above, an orthotropic concrete plate on the inertial type of soil subjected to a localized blast loading is analyzed. In this paper, a finite rectangular concrete plate is considered. The structural properties of the plate are the size (4 m x 3 m); the thickness of 0.12 m, the physical characteristics of the plate are ρ= 2400 kg.m^{-3}; υ_x= 0.2; $\upsilon_{y;}$ = 0.3; k_1=8.5.10^7 Nm/rad/m; k_2= 4.75.10^7 Nm/rad/m; E_x= 27.8.10^9 Pa; E_y= 30.10^9 Pa; m_0= 1261.63 kg.m^{-3}. The stepped triangular blast loading magnitude are P_0= 12.10^4 N; P_2= 3.10^4 N; td_1= 2.293.10^{-3} s; td_2=40.10^{-3} s; td_3= 3.10^{-3} s.

5.1 Variation of Deflection as a Function of Blast Loading Position

Figure 3 shows the variation of dynamic deflection as a function of blast loading position for soft soil condition (H_s= 5 m). The maximum dynamic deflection at the mid-span occurred when the position of the blast loading occurred at (a/8, b/2) and $0 \le t \le td_1$. This result indicates that the maximum dynamic deflection is influenced by the position of the blast load and the duration of the blast load.

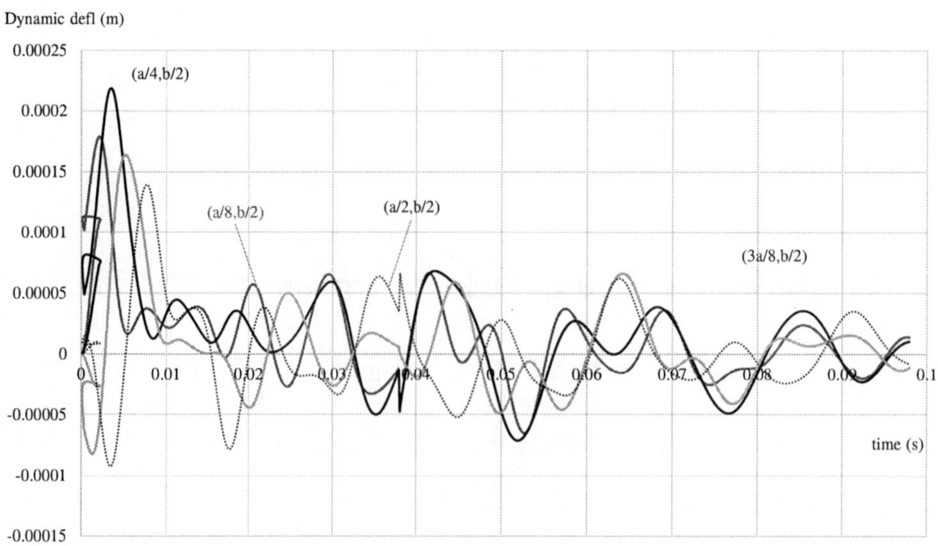

Figure 3. Dynamic deflection time history of concrete plate for different blast load's position.

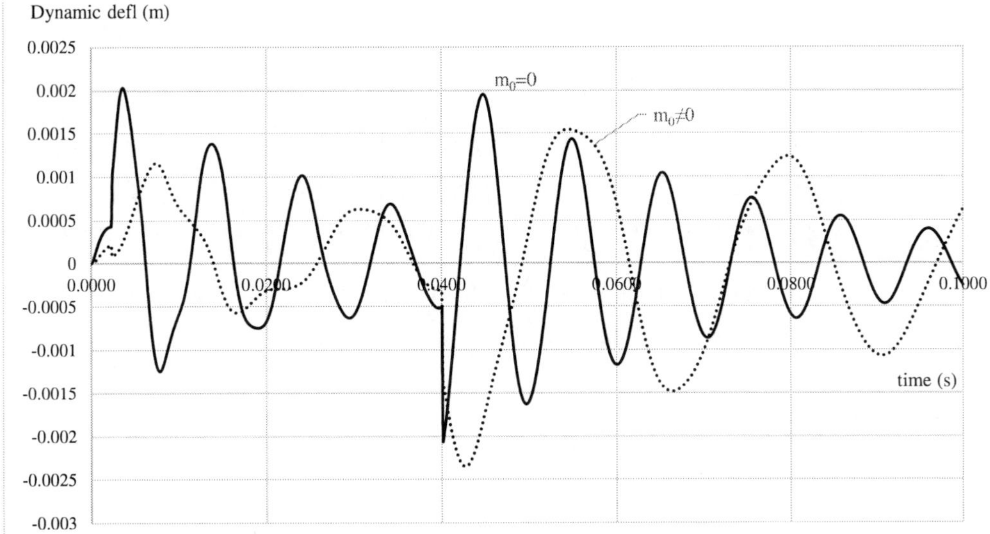

Figure 4. Dynamic deflection time history of concrete plate as the function of inertial soil (m_0).

5.2 Effect of Inertial Soil Factor on Dynamic Deflection

Figure 4 shows the variations of the deflection at the center of the mid-span of the plate for 2 types of foundation models. It can be seen that by including the inertial soil factor (m_0) into the equation of motion, the maximum dynamic deflection can be reduced by 14.6% compared with the maximum dynamic deflection of the plate on Pasternak foundation model.

6 CONCLUSION

In this paper the dynamic behavior of the orthotropic plate with semi rigid boundary condition subjected to stepped triangular blast loading was studied. The position of the blast loading, the duration of the blast load and the effect of inertial soil factor, which may affect the dynamic response of the plates subjected to blast loading was considered. The position of the blast loading is found to influence the development of maximum response. The three type parameters of foundation model will significantly reduce the maximum dynamic deflection.

References

Alisjahbana, S.W. and Wangsadinata, W., Response of Damped Orthotropic Plates Subjected to a Stepped Triangular Blast Loading, *The Twelfth East Asia-Pacific Conference on Structural Engineering and Construction*, Procedia Engineering 14, 115-120, 2011.

Alisjahbana, S.W. and Wangsadinata, W., Numerical Dynamic Analysis of Orthotropic Plates under Localized Blast Loads, *Sustainable Solutions in Structural Engineering and Construction*, 115-120, Bangkok, Thailand, 2014.

Gibigaye, M., Yabi, C. P., Alloba, I. E., Dynamic Response of a Rigid Pavement Plate Based on an Inertial Soil, International Scholarly Research Notices, Volume 2016, Article ID 4975345, 2016.

Li, B., Pan, Tso-Chien, Nair, A., A Case Study of the Effect of Cladding Panels on the Response of Reinforced Concrete Frames subjected to Distant Blast Loadings, *Nuclear Engineering and Design*, 239, 455-469, 2009.

Lu, S., Dynamic Displacement Response of Beam-type Structures to Moving Line Loads, International Journal of Solids and Structures, Vol. 38, No. 48-49, 8869-8878, 2001.

Integrated Solutions for Infrastructure Development
Edited by Lau, H. H., Tang, F. E., Ng, C. K., and Singh, A.
Copyright © 2016 ISEC Press

PERFORMANCE CONFIRMATION TEST FOR TIMBER COLUMN–GROUND SILL JOINTS REINFORCED WITH ARAMID FIBER SHEETS

AKARI YAMAGUCHI, XINYAN CHEN and NORIKO TAKIYAMA

Division of Architecture and Urban Studies, Tokyo Metropolitan University, Tokyo, Japan

High-performance aramid fiber sheets are a new class of composite materials composed of weaved polyamide fibers. This seismic performance and failure behavior of timber column–ground sill joints reinforced with aramid fiber sheets were investigated. We conducted bending tests under cyclic loading for three column–ground sill specimens. The maximum bending moments were estimated using a simple method and compared with the experimentally obtained moments. After reinforcement with aramid fiber sheets, joint strength improved but was dependent on the sheet-attaching form. Moreover, it is remarkable to break from peeling sheet. Further, the proposed sheet-attaching form (with widened crossing sheets) exhibited a higher restoring force than did the conventional form because of sheet is further away from the center of rotation.

Keywords: Restoring force, Damage, Resin, Bending test, Bending moment, Estimation.

1 INTRODUCTION

Many timber structures collapse during earthquakes because of joint damage. Therefore, the Building Standards Law in Japan was revised in 2000 to include joint specifications for timber structures (Ministry 2016). However, the seismic resistance of timber frames with steel joints can be compromised by the partial loss of member sections, such as bolt holes (Sawai 2012).

High-performance aramid fiber sheets are a new class of composite materials composed of weaved polyamide fibers. A method of reinforcing timber joints with aramid fiber sheets was recently proposed; through this method, existing timber structures can be reinforced without large-scale demolition because aramid fiber sheets can be easily glued to the joints by using an adhesive agent.

In this study, to evaluate the seismic performance and failure behavior of joints reinforced with aramid fiber sheets, we conducted bending tests under cyclic loading for three column–ground sill specimens with T-shaped joints. In addition, we proposed a new method to attach the sheets to the timber joints. In the tests, sheets were attached to the joints by using three methods: two conventional methods and the proposed method.

2 REINFORCEMENT OF TIMBER JOINTS WITH ARAMID FIBER SHEETS

The fibers in aramid fiber sheets may be woven in one or two directions, as depicted in Fig. 1. These sheets possess superior corrosion-, water-, and heat-resistance properties (Fibex 2016).

Using aramid fiber sheets in construction can improve structural strength and prevent structural degradation. Impregnating resin in aramid fiber sheets decreases their porosity and increases their plasticity; such sheets are light and soft and can be used to easily reinforce structural members without requiring additional expertise or large working spaces. These sheets have largely been used to reinforce RC structures but have recently been applied to traditional wooden structures as well.

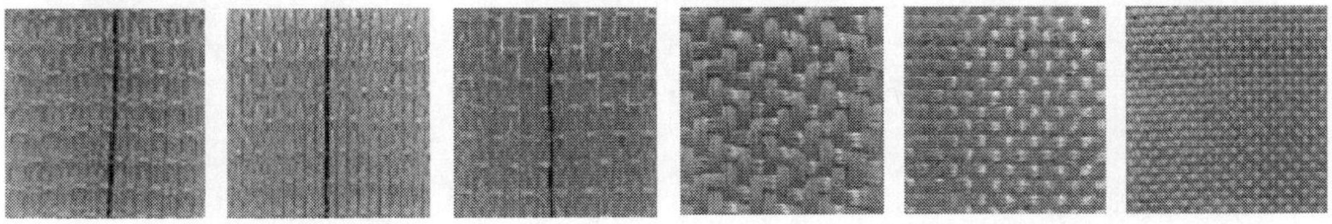

(a) One-dimensional weave (b) Bidimensional weave

Figure 1. Aramid fiber sheets.

3 CYCLIC LOADING TESTS

3.1 Specimens

Three T-shaped column–ground sill specimens were designed (Fig. 2 (a)). The columns and the ground sills (105 mm x 105 mm) were made from cedar wood, and the joints were fastened using V-shaped steel members. The cyclic tests involved three steps: (1) apply the load on the specimen, (2) remove the damaged specimen and reinforce it with aramid fiber sheets, and (3) reapply the load on the specimen.

The sheets were attached to both sides of the joints of the specimens by using two conventional methods—vertical form (specimen V-1, Fig. 2 (b)) and vertical form with an added crossing sheet (specimen C-1; Fig. 2 (c)) (Tezuka 2003) —and a newly proposal form in which the crossing sheet was widened (specimen C-2, Fig. 2 (d)). The specifications of the materials used to fabricate the specimens are listed in Table 1.

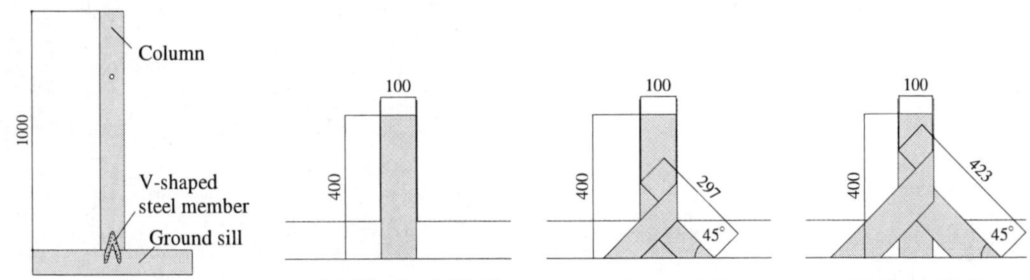

(a) Specimen (b) Vertical (V-1) (c) Cross (C-1) (d) Cross (C-2)

Figure 2. Test specimens.

Table 1. Materials used in the tests (Fibex 2016, Asahibond 2016).

Material	Item	Notes	Item	Notes
Aramid fiber Sheet	Description	FiBRA Sheet, AK-40	Manufacturer	Fibex Co., Ltd
	Tensile strength	2060 N/mm^2	Young's Modulus	118 kN/mm^2
	Design thickness	0.193 mm	Width	100 mm
Resin	Description	Asahi bond, 701	Manufacturer	Asahi Bond, Inc.
	Tensile strength	38.3 N/mm^2	Bending strength	66.0 N/mm^2

The First, the timber face is cleaned and the resin is pasted on joint (see Fig. 3 (a)). Next, the sheet is attached on resin (see Fig. 3 (b)). Then, the resin is impregnated in the sheet and repasted on the sheet (see Fig. 3 (c)). Finally, they are cured about one week.

(a) Cover the joint with resin (b) Place the sheet over the joint (c) Impregnate the sheet with resin

Figure 3. Method of attaching the aramid fiber sheet to the joints of the specimens.

3.2 Loading System

The loading system is illustrated in Fig. 4. Vertical displacement at the top of the column was measured using a displacement transducer, and the rotation angle R was determined trigonometrically from the measured vertical displacement and the known column length. Cyclic loading was applied to the specimen such that the amplitude of R increased gradually. Strain gauges were attached to both faces of the member. For the analysis, the left side of Fig. 4 is assumed to be the positive direction; the front is designated the S side and the back the N side.

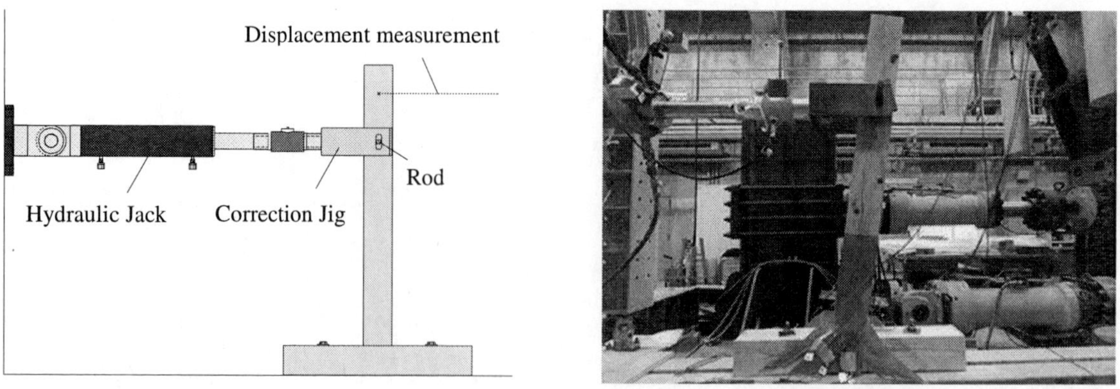

(a) Schematic diagram (b) Loading state

Figure 4. Loading system.

3.3 Damage and Restoring Force

The damaged specimens are shown in Fig. 5, and the relationship between bending moment and rotation angle is presented in Fig. 6, where black lines indicate the hysteresis loop in the unreinforced specimens and red lines the skeleton curve in the reinforced specimens.

For unreinforced V-1, the restoring force reached a maximum of 1.8 kN at +1/10 rad, and the ground sill split at the nail of the joint at +1/6 rad. After reinforcement, line formed on the compression zone of the aramid fiber reinforced plastic layer at the boundary of the joint at +1/50 rad. The edge of the sheet began peeling off at +1/30 rad. In the N side, the sheet peeled off entirely, and the restoring force reached a maximum of 1.0 kN at −1/30 rad, whereas in the S side, the line of the boundary broke at +1/20 rad.

For unreinforced C-1, the restoring force reached a maximum of 1.6 kN at +1/10 rad. The ground sill split at the nail of the joint at +1/6 rad. After reinforcement, the edge of the vertical sheet began peeling off at +1/50 rad. In the N side, the crossing sheet of ground sill start to peel off and the restoring force reached a maximum of 1.3 kN at +1/20 rad. The crossing sheet of column peels off at +1/15 rad.

For specimen C-2 with hardware, the ground sill split at the nail of hardware at +1/15 rad. and the restoring force reached a maximum of 1.5 kN at +1/10 rad. After reinforcement, the crossing sheet attached to the ground sill began peeling off at +1/30 rad, and the restoring force reached a maximum of 3.6 kN at +1/20 rad; the sheet peeled off entirely at +1/8 rad.

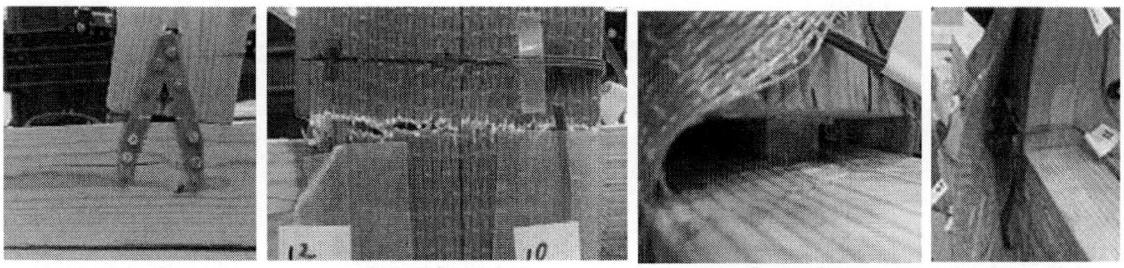

(a) Joint hardware (b) Vertical (V-1) (c) Cross (C-1) (d) Cross (C-2)

Figure 5. Damaged specimens.

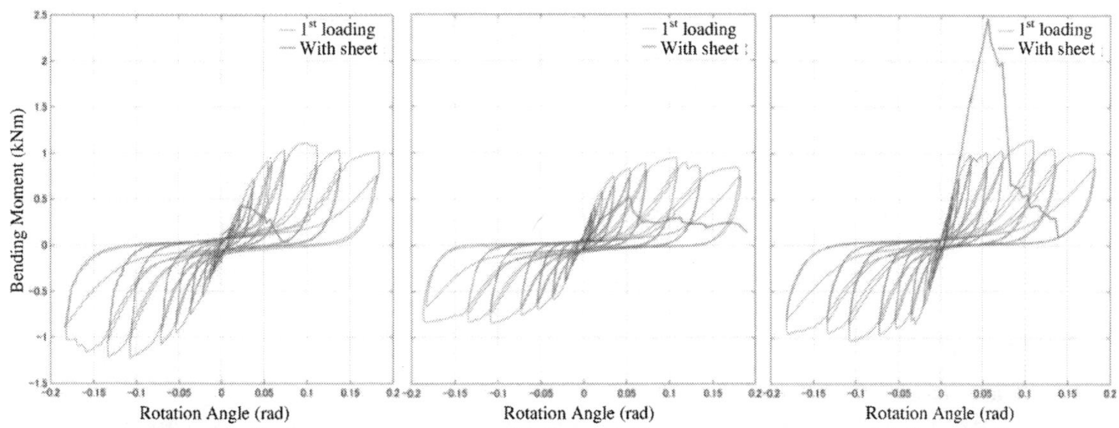

(a) Vertical (V-1) (b) Cross (C-1) (c) Cross (C-2)

Figure 6. Relationship between bending moment and rotation angle.

4 ESTIMATION OF BENDING MOMENT

The maximum bending moment after reinforcement was estimated using a simple model (Fig. 7) under the following assumptions (Architectural 2006):

(a) The center of rotation of the joint is on the edge of the column base.

(b) The reinforcing sheet is subjected to tension stress only.

(c) The sheet experiences triangular tension stress, as evidenced by the strain gauges.

(d) The maximum bending moment at a point is the sum total of the product of tension stress along the direction of the fiber and its distance from the center of rotation.

The estimated and experimentally obtained bending moments are compared in Fig. 8. The maximum bending moment of C-1 was overestimated.

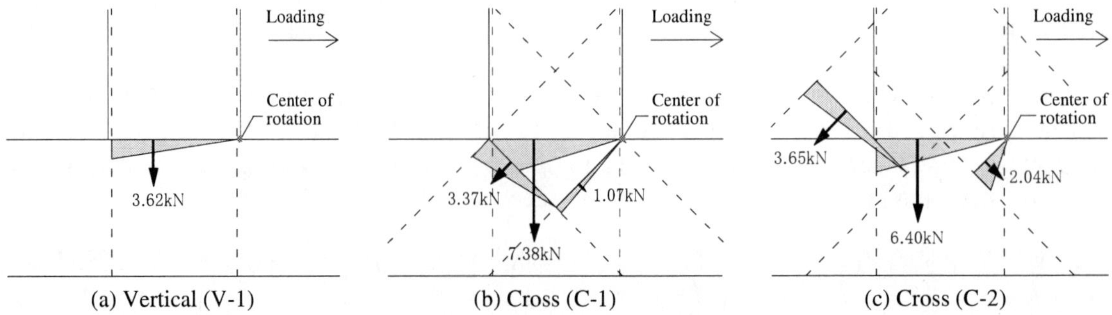

Figure 7. Simple model for estimating the maximum bending moment.

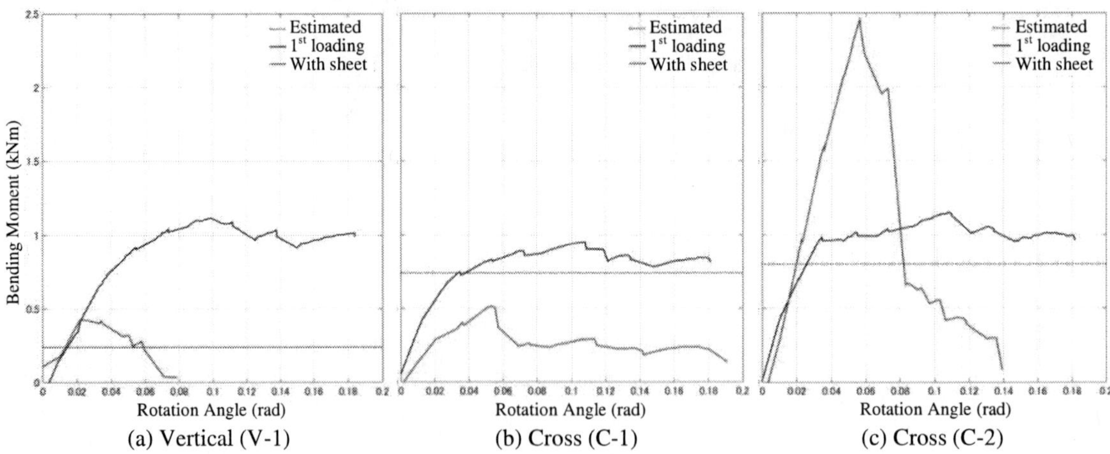

Figure 8. Estimated and experimentally obtained bending moments.

5 CONCLUSIONS

We conducted bending tests under cyclic loading for three column–ground sill specimens in order to investigate the seismic performance and failure behavior of joints reinforced with aramid fiber sheets. The maximum bending moments were estimated using a simple method and compared with the experimentally obtained moments. The major findings of this study can be summarized as follows:

(a) For joint fastened using V-shaped steel members, the ground sill split at the nail of the joint at a maximum restoring force. After reinforcement with aramid fiber sheets, joint strength improved but was dependent on the attaching form. However, the destruction occur brittleness.

(b) After reinforcement with aramid fiber sheets, it is remarkable to break from peeling sheet.

(c) The proposed sheet-attaching (with widened crossing sheets) exhibited a higher restoring force than did the conventional form because of sheet is further away from the center of rotation.

Acknowledgments

We are grateful to Fibex Co., Ltd., for providing the aramid fiber sheets and to Asahi Bond, Inc., for providing the resin used in this study. We thank the graduate students of Tokyo Metropolitan University for their assistance with the cyclic loading tests.

References

Architectural Institute of Japan, *Standard for Structural Design of Timber Structures*, Maruzen Co., Ltd., Japan, 2006. (in Japanese)

Asahibond, Inc. Retrieved from http://www.asahibond.co.jp/ on June 1, 2016. (in Japanese)

Fibex Co., Ltd. Retrieved from http://www.fibex.co.jp/index.html on June 1, 2016. (in Japanese)

Ministry of Land, Infrastructure, Transport and Tourism. Retrieved from http://www.mlit.go.jp/index.html on June 1, 2016.

Sawai, S., *Timber Structure Seismic Construction Perfect Guide*, X-Knowledge Co., Ltd., Japan, 2012. (in Japanese)

Tezuka, J., Noguchi, T., Yamashita, J., and Taniguchi, H., Experiment of Joint of Timber Structure Reinforced with Aramid Fiber Sheet and Metal Fittings, *Summaries of Technical Papers of Annual Meeting*, Architectural Institute of Japan, C-1(III), 2003. (in Japanese)

Integrated Solutions for Infrastructure Development
Edited by Lau, H. H., Tang, F. E., Ng, C. K., and Singh, A.
Copyright © 2016 ISEC Press

SHAPE OF YIELD SURFACE ESTIMATED BY USING NATURAL STRAIN UNDER PRE-DEFORMATION OF SHEAR AFTER UNIAXIAL TENSION

YASUYUKI KATO

Dept of Mechanical Engineering, College of Science & Technology
Nihon University, Tokyo, Japan.

In the present paper, the anisotropy of yield surface generated under large deformation is investigated on the basis of the Natural Strain theory, which can be treated systematically from the infinitesimal deformation to the large deformation. After applying the pre-deformation of uniaxial tension, simple shear and combined load of them to the test pieces, proportional loading tests are carried out again with changing the ratio of tension and torsion. Then, the shape of yield surface is investigated by estimating the yield stress in an arbitrary direction in the stress space. As the method for determining the yield stress, the slope of the tangent at yielding in the deviatoric stress- deviatoric strain curve is used. In order to reveal the influence of deformation history on the development of anisotropy in yield surface, the pre-deformation that the deformation history is different from previous study, namely, the pre-deformation of shear after uniaxial tension, is examined in this study. Then, the relation between the development of anisotropy in the yield surface and the deformation history of pre-deformation is revealed by comparing with the results of previous study.

Keywords: Deformation history, Yield stress, Anisotropy, Finite deformation, Elasto-plastic analysis, Ductile materials, Proof stress, Slope of tangent.

1 INTORODUCTTION

In general, as for the yield surface under the multi-axial loading conditions, the combined hardening model that the isotropic hardening and kinematic hardening coexists is used. On the other hand, in the case of the theory for representing the anisotropy of yield surface, the models by ellipse shape and convex shape are commonly used. However, in any case, almost these models are defined on the basis of infinitesimal deformation theory, and as for the yield surface under large deformation, it is not always fully elucidated.

The Natural Strain suggested in this study is associated with a certain identical line element in a body, and it is obtained by integrating an infinitesimal strain rate on an identical line element over the whole process of the deformation path. Consequently, since the shearing strain component of Natural Strain is represented by a pure angular strain, the rigid body rotation can be clearly removed. Moreover, since the additive law of strain on an identical line element can be satisfied, this strain rate can be clearly decomposed into the elastic component and the plastic component. Furthermore, the elastic component can also be decomposed into deviatoric part and volumetric part in the same manner as conventional infinitesimal deformation theory. Therefore, the Natural Strain is the effective strain representation which can be systematically treated from infinitesimal deformation to the large deformation.

In a series of our previous study (Kato and Kazama 2013), after applying the large pre-deformation of uniaxial tension or simple shear to the test pieces, proportional loading tests have been carried out again with changing the ratio of tension and shear. Then, the shape of yield surface has been estimated by investigating the slope of the tangent in those deviatoric stress and deviatoric strain curves. Consequently, it was revealed that the shape of yield surface in a stress space becomes convex at the pre-deformation side and flat at the opposite side of the pre-deformation. Thus, the existence of anisotropy in the yield surface has been confirmed. However, in these studies, the pre-deformation was limited to the uniaxial state of tension or shear. Hence, in subsequent work (Kato 2014), the pre-deformation of combined loading of tension and shear with the proportional loading state has been investigated.

However, the detailed studies concerning the effect of the deformation history of pre-deformation have not been conducted before. So, as the next step of this research, another types of pre-deformation, namely, the deformation path is different but the final deformation is the same as the pre-deformation of proportional loading of tension and shear, have been chosen as the subject of this research. Especially, in a recent study (Kato 2016), the deformation history of tension after simple shear has been examined.

In the present study, in order to elucidate the effect of deformation history on the development of anisotropy of yield surface more in detail, the pre-deformation that is reversed the order of deformation path, namely, the pre-deformation of shear after uniaxial tension, is examined.

2 THE DEFORMATION HISTORY OF PRE-DEFORMATION

In the present paper, the pre-deformation, which the deformation path is different but the final deformation state is the same as the deformation path of tension after simple shear that has been investigated in the previous study (Kato 2016), is chosen as the subject. Namely, the shear after uniaxial tension is examined as the deformation history of pre-deformation. Figure 1 (a), (b) shows the differences in the deformation history between the pre-deformation of the present paper (O→A→B) and the pre-deformation of proportional loading (O→B) by comparing the deformation gradients (the normal components D_{11} , D_{22} and the shearing component D_{12}). As for the deformation path in this study, in the first stage, the uniaxial tension is applied during O→A, and after that, in the second stage, the simple shear is applied during A→B.

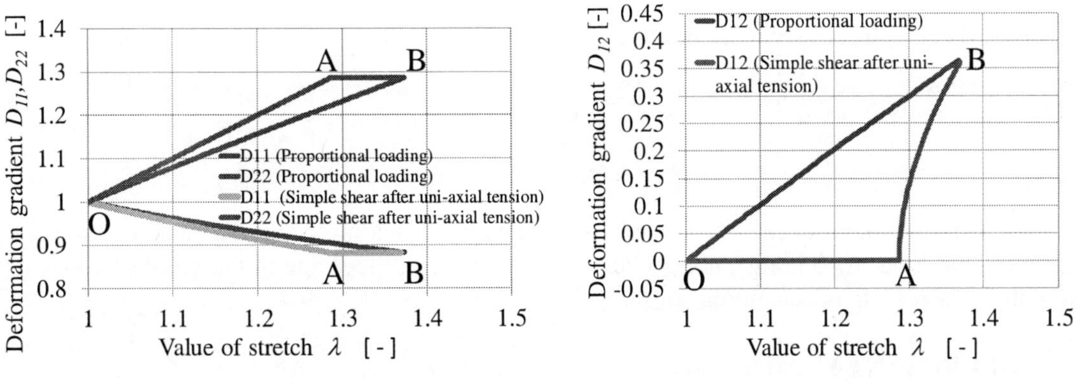

(a) Deformation gradient D_{11}, D_{22} (b) Deformation gradient D_{12}

Figure 1. Deformation path during pre-deformation (shear after uniaxial tension).

3 METHOD FOR ESTIMATION OF YIELD STRESS

Figure 2 shows the schematic diagrams of the principal deviatoric stress- deviatoric strain curve generated after applying the large pre-deformation. Since the yield stress has been already found from the final stress condition of the pre-deformation (see point B), the slope of tangent at yielding can also be specified in advance (see strait blue line at point A').

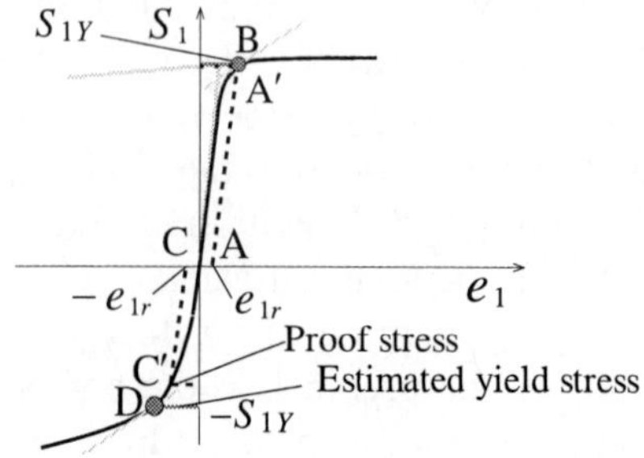

Figure 2. Principal deviatoric stress-deviatoric strain curve.

First, the experimental equation for the principal deviatoric stress, which becomes shallow curve around yielding point, is formulated as

$$S_1 = a\left(1 - exp\left(b e_1\right)\right) + c e_1 + d \tag{1}$$

where, *a*, *b*, *c*, *d* are coefficients determined by using the *Levenberg-Marquardt Method*, which is one of the non-linear least-squares method. And, in this figure, the experimental equation is described by the red solid curve. Moreover, in order to derive the slope of the tangent, the following equation is obtained by differentiating Eq.(1), i.e.,

$$\frac{dS_1}{de_1} = -ab\, exp\left(be_1\right) + c \tag{2}$$

Hence, the slope of the tangent at yielding can be specified by using Eq.(2). As one example, considering the case of the opposite side of the pre-deformation, a point D, which has the same value as the slope in pre-deformation side, is assumed to be a yield stress. On the other hand, the yield stress by proof stress, which is defined by using residual strain, is estimated smaller as shown in point C'.

Moreover, the strain hardening modulus *h* is derived as follows.

$$h = \frac{2G}{\frac{dS_1}{de_1}} - 1 \tag{3}$$

As previously mentioned, since the slope of tangent at yielding is already specified, the strain hardening modulus at yielding can also be specified. Hence, the shape of the yield surface can be estimated based on the distributions of *h*.

4 EXPERIMENTAL METHOD

In the experiments, in order to apply a large pre-deformation to the specimens, the cylindrical specimens made from high purity tough pitch copper, i.e., outer diameter 22 [mm], inner diameter

16 [mm] , the gauge length 30 [mm] and purity 99.99%, are adopted. And, the multi-axial loading test machine, which can be applied tension and torsion at the same time, was used. As for the measurements of longitudinal displacements, the displacement meters equipped to the circular jig are used, and the rotary encoders are used for the measurements of twisting angles (see Fig.3).

Figure 3. Test machine.

Next, the experiments are composed of two stages. Namely, the experiments applying the pre-deformation of shear after uniaxial tension to the specimens are firstly conducted. In the second place, the proportional loading tests of tension and torsion are conducted.

(i) Experimental condition of pre-deformation of shear after uniaxial tension

The uniaxial tension, i.e., the principal stretch m =1. 28, is firstly applied, then, the torsion, i.e., shear deformation D_{12} =0.37, is applied. However, the final deformation is the same value in all experiments, i.e., the principal stretch λ =1.38 [-].

(ii) Experimental condition of proportional loading tests for tension and shear

Attaching the tri-axial strain gages to the specimens, the experiments are conducted with the principal axis of stress being fixed in arbitrary angle, i.e., 0, 10, 25, 35, 45, 50, 55, 65, 70, 80, 90, 100, 110, 120, 135, 145, 155,170 [deg.]. Here, the principal strain is measured from the extensional strain components of the tri-axial strain gauge in consideration of the Natural Strain theory. Then, according to the method explained in the previous chapter, the value of yield stress is determined. Moreover, the shape of the yield surface in a stress space is estimated from the distribution of the yield stress.

5 EXPERIMENTAL RESULTS

Figure 4 shows the locus of stress points in each proportional loading tests, which are conducted after applying the pre-deformation to the specimens. Here, ① is the result of 0 [deg.], i.e., the case of uni-axial compression, ② is 10[deg.], ③ is 25[deg.], ④is 35[deg.], ⑤ is 45[deg.], i.e., the case of the simple shear and it is the final direction of the pre-deformation, ⑥ is 50[deg.], ⑦ is 55[deg.], ⑧ is 65[deg.], ⑨ is 70[deg.], ⑩ is 80[deg.],⑪ is 90[deg.], i.e., the case of uni-axial tension, ⑫ is 100[deg.], ⑬ is 110[deg.], ⑭ is 120[deg.], ⑮ is 135[deg.], i.e., the simple shear in the reversed direction, ⑯ is 145[deg.], ⑰ is 155[deg.] and ⑱ is 170[deg.].

As the example, the principal deviatoric stress- deviatoric strain curves in case of ⑤, ⑪ and ⑮ are described in Figure 5 (a), (b) and (c), respectively. Here, the experimental equation, i.e., Eq. (1), is described by the solid red curve in each figure. As is obvious from these figures, the stress and strain relation is a sharp curve at shear side of ⑤. However, it becomes gradually shallow curve as approaching the reverse shear side, which is the opposite direction to the final state of pre-deformation.

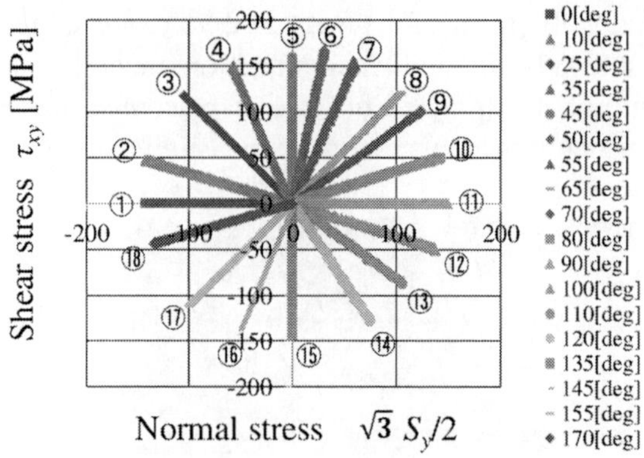

Figure 4. Stress points in each proportional loading tests.

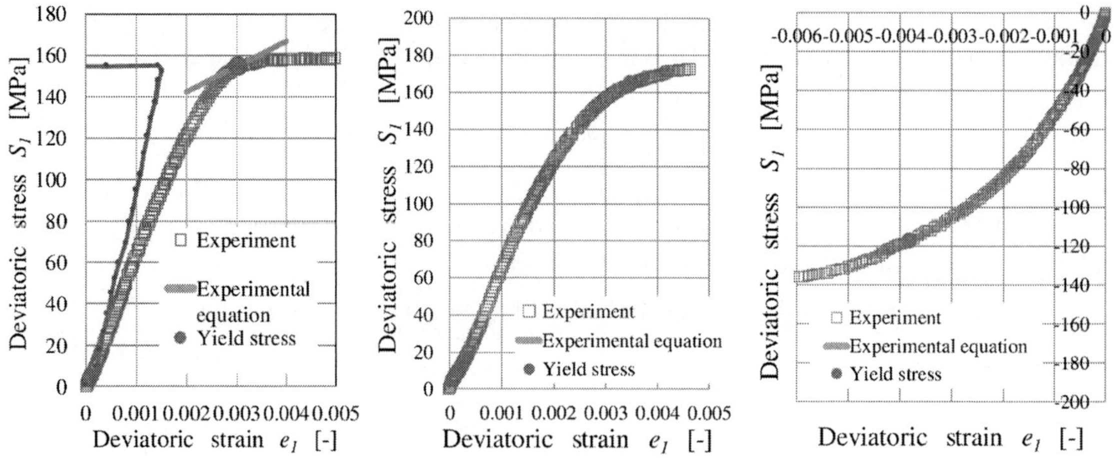

(a) 45[deg.] (b) 90[deg.] (c) 135[deg.]

Figure 5. Principal deviatoric stress – deviatoric strain diagram.

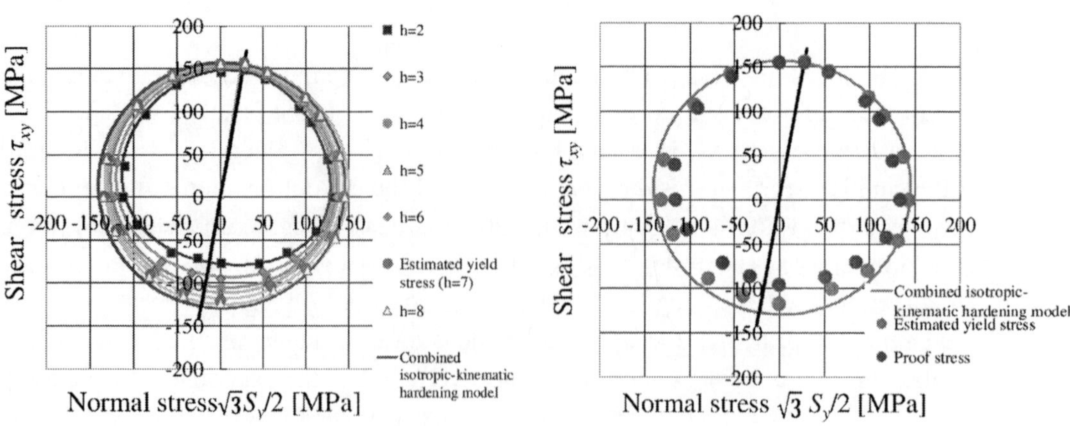

(a) Distribution of h and yield surface (b) Comparison of yield surface

Figure 6. Shape of yield surface and distributions of h (shear after uniaxial tension).

Figure 6 (a) shows the results of the estimated yield stress in all directions in a stress space. In this figure, the circular plots (●) indicate the estimated yield stress. Moreover, the distributions of the strain hardening modulus h are also described in this figure. As the loading direction approaches to the simple shear side that is the second stage of the pre-deformation, the distance of curves in h becomes narrow. However, the distance becomes larger as the loading direction approaches the revers shear side, i.e., 135[deg.]. On the other hand, Figure 7 (a), (b) shows the results of yield stress generated under pre-deformation of uniaxial tension after simple shear in the previous study. It can be confirmed that the shapes of these curves become the convex at the tension side and become flat at compression side. Therefore, the anisotropy of the yield surface is closely related to the deformation applied at the second stage of pre-deformation.

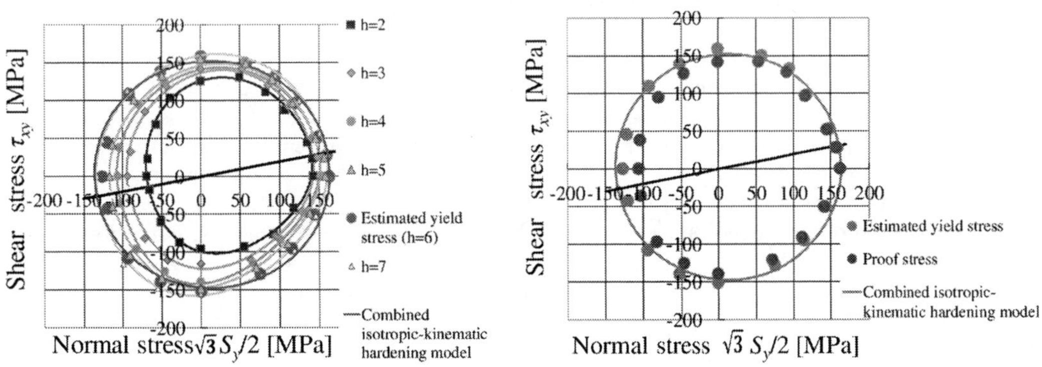

(a) Distribution of h and yield surface (b) Comparison of yield surface

Figure 7. Shape of yield surface and distributions of h (tension after simple shear).

6 CONCLUSIONS

In the present study, using the test piece applying the pre-deformation of shear after uniaxial tension, the yield stress was estimated by examining the slope of tangent on the principal deviatoric stress – deviatoric strain curve. As the result, since the shape of the yield surface becomes convex near the shear side and becomes flat at the reversed shear side, it was revealed that the anisotropy of the yield surface is formed closely related to the shear deformation applied at the second stage of pre-deformation. As for the distributions of strain hardening modulus h, it was revealed that the distance of these curves is small at the shear side, however, at the reversed shear, the distance becomes large. Moreover, the shape of estimated yield surface was compared with the shape by conventional proof stress. Consequently, both yield surfaces almost coincide at the shear deformation side, but at the opposite side, the yield surface by proof stress is small compared with the yield surface estimated in this study.

References

Kato, Y., Kazama, A., Elasto-plastic Analysis using Natural Strain: Shape of Yield Surface generated after Pre-deformation of Large Simple Shear, Advanced Materials Research Vols. 875-877, pp 518-523, 2013.

Kato, Y., Natural Strained Yield Surface Shape Estimated under Pre-deformation of Tension and Torsion, Sustainable Solutions in Structural Engineering and Construction, Chantawarangul, K., Suanpaga, W., Yazdani, S., Vimonsatit,V., Singh, A. (ed.), 153-158, ISEC Press, 2014.

Kato, Y., Yield surface shape estimated by using Natural Strain under pre-deformation of tension after simple shear, Interaction between Theory and Practice in Civil Engineering and Construction, Komurlu, R., Gurgun, A., Yazdani, S., and Singh, A.(ed.), 247-252 , ISEC Press, 2016.

Integrated Solutions for Infrastructure Development
Edited by Lau, H. H., Tang, F. E., Ng, C. K., and Singh, A.
Copyright © 2016 ISEC Press

A STUDY ON THE PUSHOVER TEST AND NUMERICAL ANALYSIS OF GFRP FRAMES

YEOU-FONG LI, BO-SHIANG WANG and JIAN-YU LAI

Dept of Civil Engineering, National Taipei University of Technology, Taipei, Taiwan, R.O.C.

This paper presents the use of glass fiber reinforced plastic (GFRP) composite material to produce a structure frame; the behaviors of the GFRP frames were analyzed by using pushover test and a numerical analysis software. Double-web FRP I-beams are used for the beams and columns of the frame, and joints made from metal and FRP. Three types of frame specimens were involved: an un-braced frame, a compression-braced frame and a tension-braced frame for each joint type. The joints were bonded to the frame components using epoxy resin but also adding bolts in the beam-column joint. The pushover test was used to investigate the mechanical behaviors and failure modes of the GFRP frames. The analysis software SAP2000 was used for the pushover analysis of the GFRP frames, and it was shown that the ultimate strength and force-displacement relationships of the analytical results were similar to that of the experimental ones.

Keywords: Composite material, Brace, Experiment, Joint.

1 INTRODUCTION

FRP material, because of its high strength, high durability and light weight, can be advantageous when used to construct prefabricated houses in the mountain areas or seaside without being affected by corrosion. This study focuses on assembled GFRP Building frame specimens with related experiments and analysis.

This study collected previous related research achievements which are summarized as follows. Ascione *et al.* (2010) proposed bolt opening of porous GFRP deck experiments; and the establishment of numerical model, the experiment results of the bolts were compared with analysis results from the numerical model. Al-Kharat and Rogers (2007) describe the use of cold-formed steel to build a frame model. The model was used to estimate the performance of inelastic structural element, to determine the frame ductility, plastic strain concentration or possible position in the component and weld seam damage results. Bao *et al.* (2011) uses 3 groups of displacement meters to measure vertical displacement of beams and then calculate the rotation angle. Davalos *et al.* (1996) for different types of types I and box section member for three points and four points bending test, using the hierarchical systems theory, the mechanics of laminated beams and finite element analysis model to predict the deformation and strain of FRP components. In addition to the above references, several other literatures provide vital information relevant to this topic but cannot be covered in this text for the sake of brevity. Literature from previous studies is helpful to understand content and experiment planning for this paper. The analysis results were compared with experimental results and finally submit conclusions.

2 EXPERIMENTAL PROCEDURES

2.1 Materials of GFRP

In this study, GFRP frames are assembled with beams and columns using pultruded GFRP double web I type components. Bracing members are pultruded GFRP rectangular section components. Metal joints and FRP joints are used to connect and configure the members into a frame after which the members are bolted in place. The bracing also uses metal pieces to connect to the frame joint. GFRP member parameters are shown in Table 1.

Table 1. GFRP component parameters.

Elastic Modulus (kN/cm^2)	Poisson's Ratio	Shear Modulus (kN/cm^2)
$E_x = 1{,}722$	$v_{xy} = 0.11$	$G_{xy} = 310$
$E_y = 551$	$v_{yz} = 0.33$	$G_{yz} = 310$
$E_z = 551$	$v_{xz} = 0.11$	$G_{xz} = 310$

2.2 Experimental Setup

The experimental setup details of single-span GFRP frame was shown in Fig. 1. A displacement controlled pushover test was done with unidirectional loading. In order to avoid external deformation of the GFRP, two lateral supports were provided along with pulley rollers attached to them to reduce the in-plane deformation from friction. Rotation was calculated according to the literature Bao *et al.* (2011). The specimen names and configurations can be seen in Table 2 and Table 3, respectively.

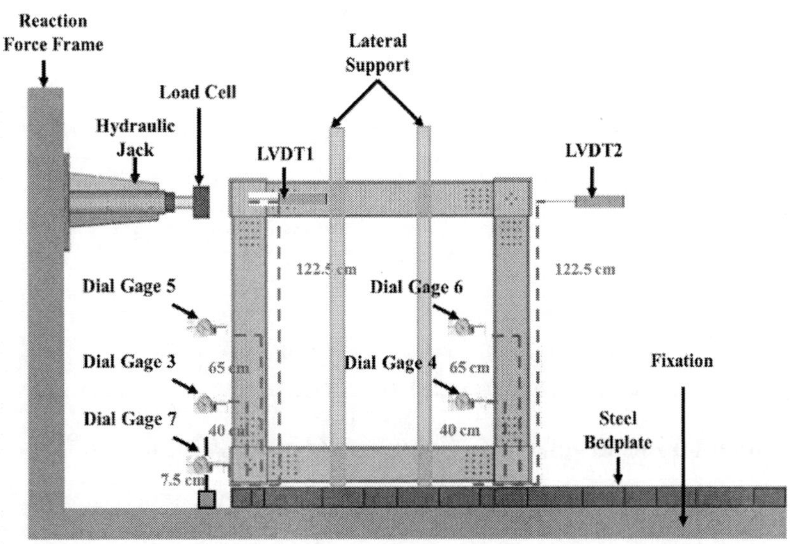

Figure 1. The schematic diagram of single-span specimen.

In this study, the single-span GFRP frame height and width are both 130 cm. The double-span GFRP frame height is 130 cm and width is 245cm. GFRP members clear height is 100 cm. The metal joints are 15 cm deep and 3.3 cm wide. GFRP components are bolted to the metal joints using M6 bolts. The bracing is attached to the joint by using a steel block and at the tip of the bracing member is inserted a small steel block which is bolted to the member to prevent shearing damage to the bracing member from the bolts.

3 TEST RESULTS AND DISCUSSIONS

The experimental results of single-span frame of metal joints and FRP joints with bolts were discussed as the following.

3.1 The Experimental Results of Single-Span Frame of Metal Joints

FP1T and FP1C had better stiffness and ultimate strength at the initial stages than the prototype specimen FP1. This proves that adding a brace can improve the overall stiffness and strength of the frame. GFRP has higher tensile than compressive strength, therefore the ultimate strength of FP1T is even higher than that of FP1C.

For specimens with bracing, the bracing member becomes the first casualty whether in tension or compression damage. Tension damage (FP1T) occurs in the form of bolt hole tearing by the bolts due to high concentration of force whilst compression damage (FP1C) is visible when the bracing member buckles. The overall experimental results were compared and shown in Table 2.

Table 2. Comparing experimental results of single-span frames with metal joints.

Specimen Name	FP1	FP1T	FP1C
Configuration	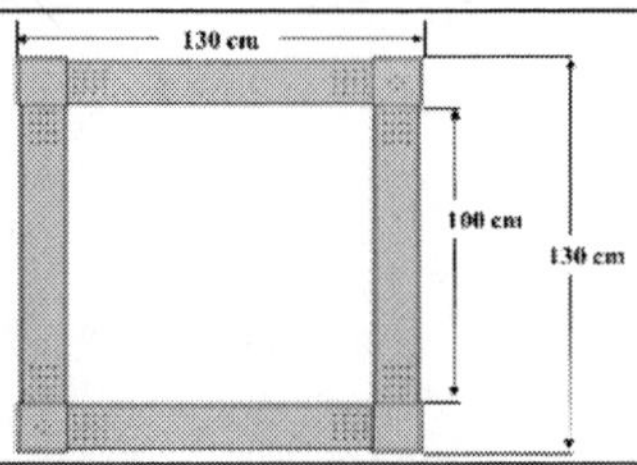	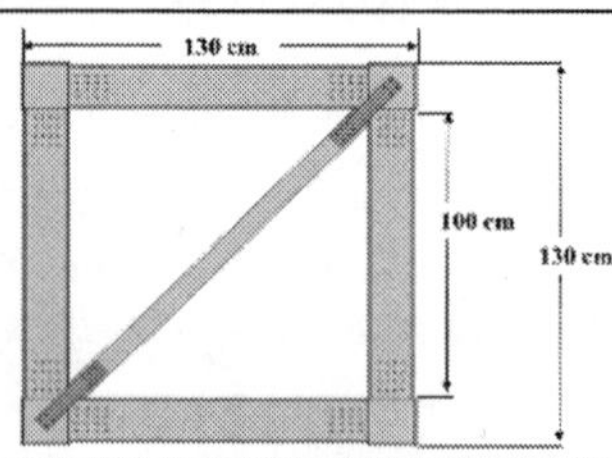	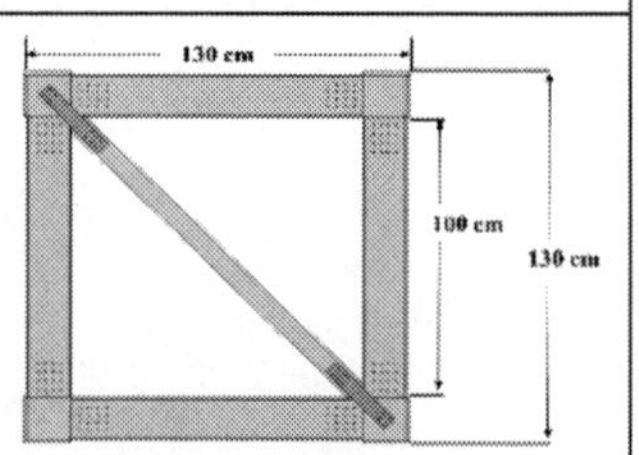
Linear Stiffness (kN/cm)	15.9	42.2	26.2
Maximum Load (kN)	33.3	56.4	36.9
Energy Dissipation (kN.cm)	154.51	216.96	192.32

3.2 The Experimental Results of Single-Span Frame of FRP Joints

FB1T and FB1C had better stiffness and ultimate strength than the prototype specimen FB1 at the initial stages. This proves that adding a brace can improve the overall stiffness and strength of the frame. The ultimate strength of FB1T is even higher than that of FB1C. The overall experimental results were compared and shown in Table 3.

Table 3. Comparing experimental results of single-span frames with FRP joints.

Specimen Name	FB1	FB1T	FB1C
Configuration	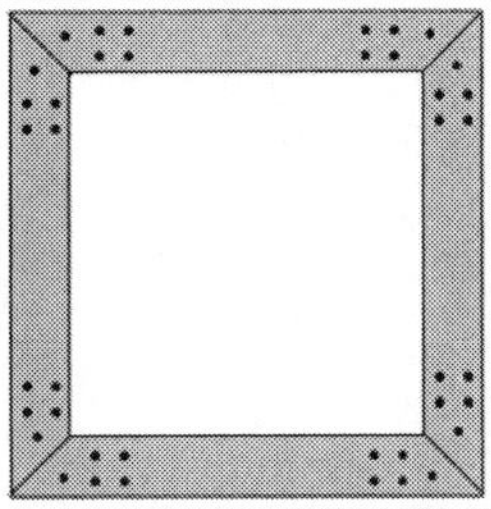	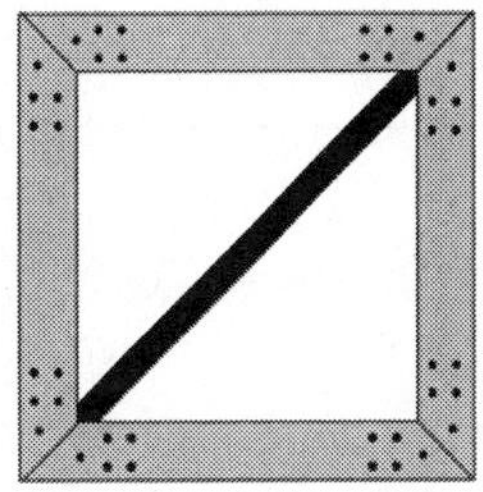	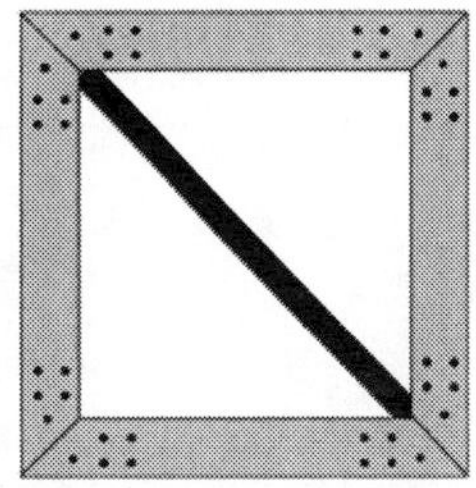
Linear Stiffness (kN/cm)	17.8	59.7	50.7
Maximum Load (kN)	17.75	30.24	24.7
Energy Dissipation (kN.cm)	88.6	161.54	117.71

4 NONLINEAR PUSHOVER ANALYSIS

Two-dimensional models of the frames were numerically analyzed using SAP2000 Version 14. Plastic hinges were manually defined for each column member, one in the upper part and one in the lower end. A plastic hinge was also defined at the middle of each bracing member.
The nonlinear nature of the problem requires the use of plastic hinges to simulate the behavior of the frame after yielding. Before yielding, the frame is in an elastic state and elastic behavior takes place over the entire length of the member but once the yielding point is reached, deformation behavior is propagated from the hinges. The Hinge-overwrite settings allow for proper simulation of plastic hinge locations. The plastic hinge parameter was set by using a single-column force-displacement relationship diagram and converting it into a moment-rotation relationship. The comparison of the experimental and numerical results of the force-displacement relationships of the GFRP frames were shown in Fig. 3.

As seen from Fig. 3, the analysis results compared with the experimental results, showing good correlation at the maximum load for all specimen, with absolute error rate less than 4 %. The analysis results can also predict experimental initial failure of the braces for specimens FP1T, FP1C, FB1T and FB1C correctly.

5 CONCLUSIONS

Based on this study, the following conclusions can be drawn:

1. Single-span of metal joints FP1T with tension bracing had a 165 % increase in initial stiffness and 69 % increase in maximum load. FP1C had 65 % increase in initial stiffness and 11 % increase in maximum load.
2. Single-span of FRP joints FB1T with tension bracing had a 235 % increase in initial stiffness and 70 % increase in maximum load. FB1C had 33 % increase in initial stiffness and 39 % increase in maximum load.
3. GFRP frames with FRP joints increase higher initial stiffness compare to those with metal joints, because the rigidity of FRP joint is higher than metal joint. GFRP frames with metal joints increase higher maximum load compare to those with FRP joints, because the strength and ductility of metal joint is higher than FRP joint.

4. Analysis results compared with the experimental results, showing correlation. This proves the use of SAP2000 can make engineers works easier as the analysis software has a high accuracy.

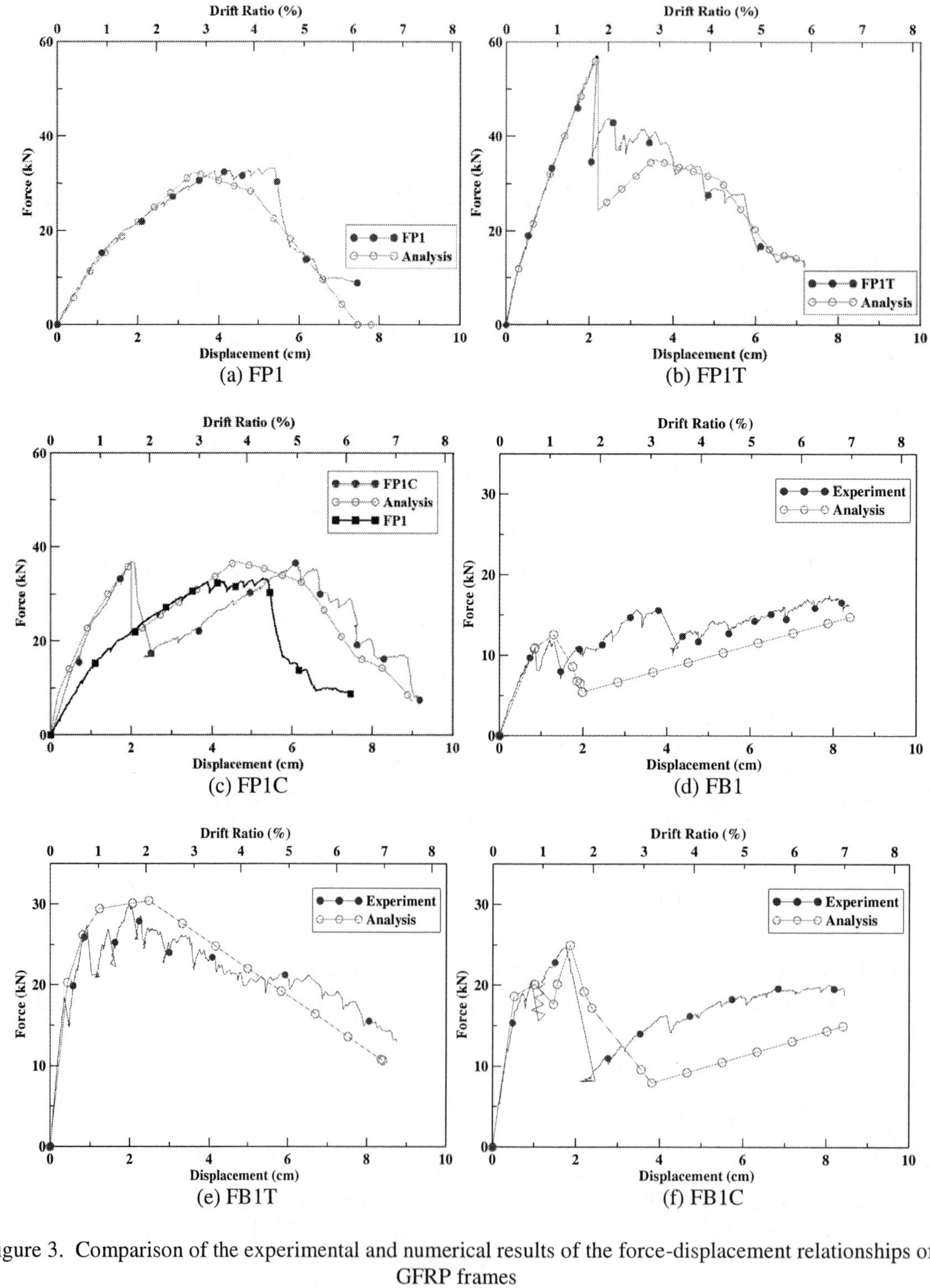

Figure 3. Comparison of the experimental and numerical results of the force-displacement relationships of GFRP frames

Acknowledgements

The authors are grateful for the financial support of the Ministry of Science and Technology of Taiwan, under contract No. NSC-100-2622-E-027-027-CC3.

References

Ascione, F., Feo, L., Maceri, F., On the Pin-Bearing Failure Load of GFRP Bolted Laminates: An Experimental Analysis on the Influence of Bolt Diameter, *Composites Part B: Engineering*, 41(6), 482-490, 2010.

Al-kharat, M., Rogers, C. A., Inelastic Performance of Cold-Formed Steel Strap Braced Walls, *Journal of Constructional Steel Research*, 63(4), 460-474, 2007.

Bao, W., Shao, Y. S., Xing, L. T., Measuring Method of Rotation Used in Research on Steel Frame Semi-Rigid Beam-to-Column Connection, *Advanced Materials Research*, 243, 1439-1442, 2011.

Davalos, J. F., Salim, H. A., Qiao, P., Lopez-anido, R., Analysis and Design of Pultruded FRP Shapes under Bending, *Composites Part B: Engineering*, 27(3), 295-305, 1996.

Integrated Solutions for Infrastructure Development
Edited by Lau, H. H., Tang, F. E., Ng, C. K., and Singh, A.
Copyright © 2016 ISEC Press

FIRE SPREADING ANALYSIS OF A GROUP OF WOODEN HOUSES IN TOWNSCAPE FORMATION DISTRICT

TOMIYA TAKATANI

Dept of Civil Engineering & Architecture, National Institute of Technology, Maizuru College, Maizuru, Kyoto Prefecture, Japan

There was a fire spreading disaster for 18 wooden houses in the historical townscape formation district in a famous spa resort area. In this paper, a fire spreading analysis of a group of wooden houses was carried out under some conditions of wind velocity and direction in order to numerically simulate a fire spreading phenomenon in the spa resort area and also investigate some fire spreading countermeasures for a group of wooden houses and also investigate an appropriate countermeasure for the fire spreading disaster of wooden houses using both fireproof wall and quasi-fireproof wall on wooden houses. As a result, it should be noted that the installation of fireproof wooden wall in every third wooden house can economically reduce a fire spreading area in comparison with the installation of quasi-fireproof wooden wall in every other wooden house.

Keywords: Fireproof wall, Quasi-fireproof wall, Fire disaster simulation.

1 INTRODUCTION

It is well-known that there exist a lot of Japanese-style three-story wooden inns in some famous spa resort areas in Japan, and they may be almost cultural properties and also may produce comfortable cultural scenery for many visiting tourists. Many tourists enjoy spa scenery and street scene with wooden three-story Japanese inn structures and many street trees on both road sides along a river in this historical townscape formation district. Therefore, local city office with these cultural properties must be forced to take prompt countermeasure for their operation and maintenance.

There was a fire spreading disaster for 18 wooden houses in the historical townscape formation district in a famous spa resort area whose burnt area is 2,730m^2. Figure 1 shows a fire fighting state and a fire extinguishing one at Kinosaki Spa resort area. This spreading disaster in the spa resort area is a serious problem for the local city, and an appropriate countermeasure for the fire spreading disaster of a group of wooden houses must be conducted by the local city office.

In this paper, a fire spreading analysis of a group of wooden houses was carried out under some conditions of wind velocity and direction in order to numerically simulate a fire spreading phenomenon in the spa resort area and also investigate some fire spreading countermeasures for a group of wooden houses and also investigate an appropriate and economical countermeasure for the fire spreading disaster of wooden houses previously mentioned using both fireproof wall and quasi-fireproof wall on wooden houses.

Figure 1. Fire spreading site at Kinosaki Spa resort area (Kobe Newspaper 2015).

2 OUTLINE OF FIRE SPREADING ANALYSIS

Many researchers have reported fire spreading theories and papers before. In this paper, a fire spreading velocity equation proposed by Hamada (1969) is briefly described in this section. Figure 2 indicates an outline sketch of fire spreading area under wind condition. Fire spreading area is assumed to be an oval equation with $l_1 + l_2$ in the major axis and $2l_3$ in the minor one.

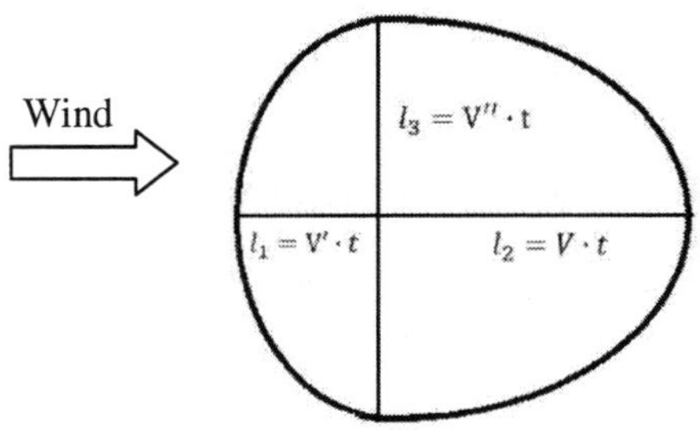

Figure 2. Shape of fire spreading area.

Based on the fire spreading area shown in Figure 2, a fire spreading velocity equation proposed by Hamada can be written in the following equations.

(1) Time from an ignition to the neighboring house's ignition, t(min)

(Leeward)

$$t_0 = \frac{3 + \frac{3}{8}a + \frac{8d}{D_i}}{1 + 0.1v} \quad \text{(First house)} \tag{1}$$

(Upwind)

$$t_i = \frac{3 + \frac{3}{8}a + \frac{8d}{D_i}}{\alpha\left(1 + 0.1v + 0.007v^2\right)}, \quad \alpha = 1.6\frac{t_0 + 14}{t_0 + 25} \tag{2}$$

(Wind side)

$$t' = \frac{3 + \frac{3}{8}a + \frac{8d}{D'}}{1 + 0.002v^2} \tag{3}$$

$$t'' = \frac{3 + \frac{3}{8}a + \frac{8d}{D''}}{1 + 0.005v^2} \tag{4}$$

(2) Spreading velocity, V(m/min)

(Leeward)

$$V = n\frac{\alpha(a+d)(1+0.1v+0.007v^2)}{3 + \frac{3}{8}a + \frac{8d}{D_i}} \tag{5}$$

(Upwind)

$$n = \frac{a'+b'}{a'+\frac{b'}{0.6}}(1-c') \tag{6}$$

(Wind side)

$$V' = n\frac{(a+d)(1+0.002v^2)}{3 + \frac{3}{8}a + \frac{8d}{D'}} \tag{7}$$

$$V'' = n\frac{(a+d)(1+0.005v^2)}{3 + \frac{3}{8}a + \frac{8d}{D''}} \tag{8}$$

(3) Spreading limit distance, D (m)

(Leeward)

$$D_0 = 1.15(5+0.5v), \quad D_i = \beta_i D_0 \tag{9}$$

$\beta_i = 1.0$ (Until 10 minutes from an ignition), $\beta_i = 1.5$ (From 10 minutes to 30 minutes)

$\beta_i = 3.0$ (From 30 minutes to 60 minutes), $\beta_i = 5.0$ (Over 60 minutes from an ignition)

(Upwind)

$$D' = 1.15(5+0.2v) \tag{10}$$

(Wind side)

$$D'' = 1.15(5+0.25v) \tag{11}$$

where, a is an average length of house structure (m), d is a distance to the neighboring house (m), v is wind velocity (m/sec), a' is a wooden structure mixing rate, b' is a fire prevention structure mixing rate, c' is a fireproof structure mixing rate.

3 FIRE SPREADING SIMULATION RESULTS

Figure 3 shows a housing map of Kinosaki Spa resort area. In this paper, several fire spreading analyses are conducted to simulate a real fire spreading disaster occurred in 2015 in Kita district shown in Figure 4. Fire spreading simulation can be conducted using a group of house structures, that is, housing information data. Fundamental housing information data with a scale level 1/2,500 can be easily downloaded from the Geographical Survey Institute in Japan, and also can be modified in accordance with the present house structure shape. Wind data with velocity and direction on that day in 2015 were measured by the Meteorological Agency in Japan, and are employed in this simulation. Wall material of wooden house structure is assumed to be "naked wooden house" without any fire proof walls, wooden house with "quasi-fireproof" walls and fireproof walls. Quasi-fireproof wall and fireproof wall materials mean that the deformation, dissolution and other damage to occur a structural strength reduction does not cause for 45 minutes and 60 minutes respectively even if walls and eaves are heated by a fire. A comparison between the installation of fireproof wooden wall in every third wooden house and the installation of quasi-fireproof wooden wall in every other wooden house is numerically investigated in this paper.

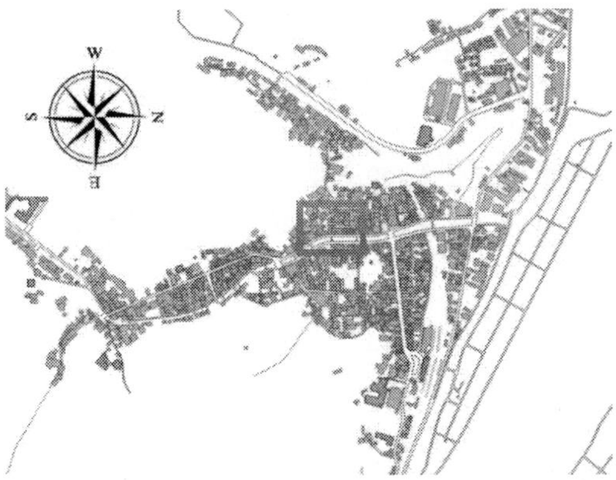

Figure 3. Housing map of Kinosaki Spa resort area.

Figure 4. Kita district.

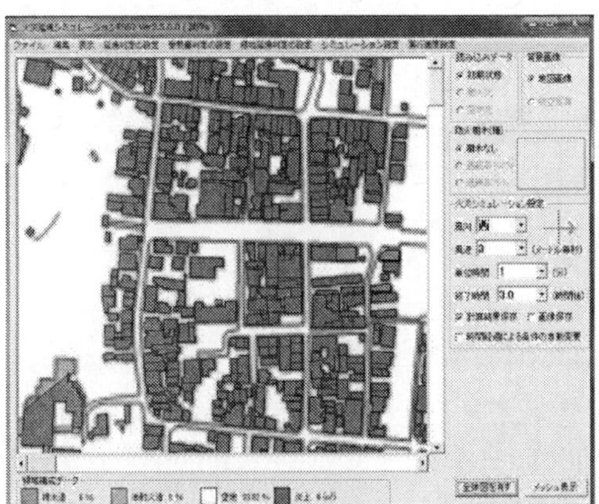

Figure 5. Fire spreading simulation screen.

Fire spreading simulation software developed at Ehime University is briefly described in this paper.

Figure 5 illustrates a simulation screen on the fire spreading simulator, and a fire outbreak point, wind direction, wind velocity, simulation time, and unit time can be easily set on this screen.

Figure 6 shows a fire spreading simulation result for a group of "naked wooden houses" without any fireproof wall materials after 5 hours from fire outbreak. 17 wooden houses were burned down in this fire spreading simulation. 18 wooden houses were burned down in the real fire spreading disaster in Kita district of Kinosaki Spa resort area occurred in 2015. It is found from this figure that fire spreading area spreads to neighboring houses along the front road and also spreads behind the house with a fire outbreak. It should be noted that the effectiveness of fire spreading analysis with time progress can be confirmed through this simulation.

A fire spreading simulation result for a group of wooden houses with quasi-fireproof walls after 5 hours from fire outbreak is indicated in Figure 7. In comparison with "naked wooden houses" without any fireproof wall materials, the installation of quasi-fireproof material on every wooden house can decrease the number of wooden house burned down and also can restrain a fire spreading behind wooden house.

Figure 8 illustrates a fire spreading simulation result in the case of the installation of quasi-fireproof wooden wall in every third wooden house after 5 hours from fire outbreak. As a matter of course, a fire spreading phenomenon to the neighboring house for the installation of quasi-fireproof wooden wall in every third wooden house is faster than that for the installation of quasi-fireproof material on every wooden house.

A fire spreading simulation result in the case of the installation of fireproof wooden wall in every other wooden house after 5 hours from fire outbreak is shown in Figure 9. It can be obvious from Figures 7 and 8 that the installation of fireproof wooden wall in every other wooden house can be significantly reduced a fire spreading area in comparison with the installation of quasi-fireproof material on every other wooden house and fire spreading to the neighboring houses not only in the lateral direction but also the backward can be restrained.

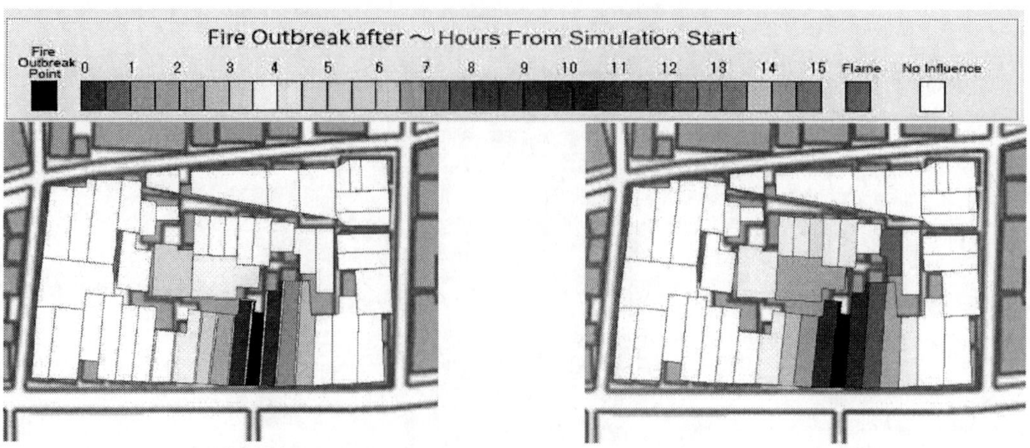

Figure 6. Fire spreading simulation result after 5 hours (Naked wooden house).

Figure 7. Fire spreading simulation result after 5 hours (Installation of quasi-fireproof wall in every wooden house)

Figure 8. Fire spreading simulation result after 5 hours (Installation of quasi-fireproof wall in every other wooden house).

Figure 9. Fire spreading simulation result after 5 hours (Installation of fireproof wall in every third wooden house).

4 CONCLUSIONS

In this paper, a fire spreading analysis of a group of wooden houses was carried out under some conditions of wind velocity and direction in order to numerically simulate a fire spreading phenomenon in the spa resort area and also investigate some fire spreading countermeasures for a group of wooden houses and also investigate an appropriate countermeasure for the fire spreading disaster of wooden houses previously mentioned using both fireproof wall and quasi-fireproof wall on wooden houses.

The summary obtained in this paper is as follows:

(1) An actual fire disaster for a group of wooden houses can be accurately simulated under some conditions of wind velocity and direction by fire spreading analysis.

(2) Fireproof wooden walls on wooden houses can restrain a fire spreading phenomenon in wooden houses in comparison with a fire spreading simulation for wooden houses without fireproof wall.

(3) Installation of fireproof wooden wall in every third wooden house can significantly reduce a fire spreading area in comparison with the installation of quasi-fireproof wooden wall in every other wooden house.

Acknowledgments

This fire spreading simulation was conducted by Mr. Daiki Muranaka, who was a graduate student at National Institute of Technology, Maizuru College. The author would like to thank him for his research support and corporation.

References

Ehime University Disaster Information Center: Fire Spreading Simulator, http://cdmir. jp/simu-lator/.
Hamada, M., Fire, Kyouritsu Publishing Co. Ltd., 1969.
Kobe Newspaper Co. Ltd., http://www.kobe-np.co.jp/news/shakai/201501/0007646059.shtml.

Integrated Solutions for Infrastructure Development
Edited by Lau, H. H., Tang, F. E., Ng, C. K., and Singh, A.
Copyright © 2016 ISEC Press

ANALYSIS OF ECONOMIC DETERMINANTS OF AFFORDABLE HOUSING PRICES

ABDULLATEEF OLANREWAJU[1], LEE LIM TAT[1], SEONG YEOW TAN[1], MINE NAOTO[1], ZAFARULLAH NIZAMANI[1], and ABDUL RASHID ABDUL AZIZ[2]

[1]*Faculty of Engineering and Green Technology, Universiti Tunku Abdul Rahman, Malaysia*
[2]*Universiti Sains Malaysia, Pulau Pinang Malaysia*

There is high possibility that the housing shortage in Malaysia will remain for a long time. However, while the existing housing stock is insufficient, constraints on income distributions have created affordability gaps by creating an imbalance between demand and supply. To increase housing supply, the government is providing homes directly as well as providing tax reliefs and subsidies to the developers, contractors and prospective home buyers. However, despite the measures the housing supply and distribution gaps are growing persistently. Therefore, there is the need to provide answers to a number of questions. It is the aim of this current study to examine the economic determinants of affordable housing. Through a cross sectional survey questionnaire, involving 71 homeowners, 13 cost determinants were established for affordable housing. The study provides insight into cost determinants of affordable housing. The results will be useful to policy makers, developers towards and the prospective home buyers.

Keywords: Subsidy, Functional cost modeling, Micro-determinants, Income, Malaysia.

1 INTRODUCTION

Housing serves as both capital and consumption goods depending on the stakeholders involved. Housing price is known to be the major priority of homebuyers. The price is often determined in the market. The literature on housing price determinants is very extensive and enormous. Literature cited various complex determinants for housing prices (Zou and Chau 2015, Gokm and Keceli 2015). The determinants include construction costs, availability of credit facility, topology, land use regulation, gestation period, weather, neighbourhoods, and types of housing, housing characteristics, the performance of the construction industry, household income, location, population and culture. Due to the complexities of the price determinants, the determinants have been grouped into different classifications by various authors and writers. For instance, while some studies focus on the housing demand side some addressed the supply side (Ball *et al.* 2010). Yet, some classified the determinants into economic, social and environmental dimensions. In fact, while some have discussed the micro-determinants, some have attended to the macro-determinants (Panagiotidis and Printzis 2015). At the micro level, housing prices were determined in terms of housing characteristics. The housing characteristics include housing size, housing quality, number rooms, toilets, and kitchens. For the households and developers, addressing the housing characteristics appears meaningful. The concerns on the increase in the housing prices and the future volatility are very high among homebuyers, especially as the housing price is outstripping inflation. The homebuyers use the micro-determinants to measure the affordability of the house. The theme of the study reported in this paper is similar to this. However, this study focused on the price determinants of the 'affordable housing'. Although some studies in Malaysia focused on demand and supply sides and even developing pricing model, no-one examines the price determinants of affordable housing. In other words, the aim of this study is to determine the cost determinants of affordable housing. The determinants of affordable housing are expected to be different from other types of housing. For instance, housing affordability is usually a function of income; house price – income ratio. The smaller this ratio becomes, the higher the possibility (i.e., affordability) of the householders to own a house. This reasoning is consistent with microeconomics theories, because the poorer a family the higher the proportion of their income to house price/rent. However, to emphasise, affordability does connote poor minimum standards or cheap house. Rather, it aims primarily to obtain best value at the desired level of expenditure. Therefore, the main responsibilities and duties of the providers (public and private) are to provide value for money for the prospective householders or homebuyers that money is economically allocated to the various building functions. In the past 20 years, Malaysia's housing prices have steadily outstripping residents' incomes. A survey conducted by the Department of Statistics, Malaysia, revealed that the Malaysian homeownership rate is decreasing by approximately 1% annually. Importantly, those in the low and medium income groups are losing their homes and renters are increasingly unable to pay rent. Debts on utility bills are also on the increase. While the household debt has increased to 89.1% of the GDP, 63% of the debt is for property and finance investment (Yong 2016). The increasing in wages and salaries of those in low and medium has done less to cushion their housing cost burden. Primarily because, as income increase costs of other goods and, services like, utilities, transportation, foods, education and hospital bills also increase in tandem. The main aim of the theory of the economic determinants of affordable housing is to explain the costs that limit the housing demand. Householders' knowledge of the determinants of housing price has a significant impact on the strategies and measures to procure and operate the house. The broader insight of this research is to stimulate an existing situation in the housing marketplace to generate information that is useful for household decision making as well as policy makers.

2 BACKROUND TO THE STUDY

"One of Malaysia's long standing development objectives is the provision of affordable housing for Malaysians in both rural and urban areas, with a focus on lower-income groups" (Government of Malaysia 2010). To achieve this objective, many schemes were introduced. The schemes include: People's Housing Program, MyHome Scheme, Civil Servants Housing Program and Perumahan Rakyat 1Malaysia. These schemes aim to address the housing need for the Bottom 40% (B40) and Middle 40% (M40). Approximately 80% of the households fall into this group. There are many interpretations for affordable housing (Gyourko and Tracy 1999, Jewkes and Delgadillo 2010). But the two most cited are briefly illustrated next. According to the "30/40" rule, housing is affordable if it does not cost more than 30% of the household disposable income. This principle evolved from the United States National Housing Act of 1937 (Schwartz and Wilson 2008). For rental purposes, the 30% includes rental cost and, utilities bills. But for mortgage, the amount includes the actual payment, tax, insurance, utilities and maintenance costs. Spending more than 30% implies "cost-burdened". In Malaysia, the monthly expenditure for the rental for the low and middle income earners is 35% (Department of Statistics 2011). The 'median multiplier' method is also common. Demographia compares the median house price to median household income (Demographia 2015). In this method housing affordability is classified into 5 categories. The computation involves the accumulation of the median income for 3 years. In 2014 the median income in Malaysia was RM4, 585 (Department of Statistics 2014). But, considering the annual growth rate of 11.7% experienced from 2012 to 2014, the median income in 2016, would increase to some RM 5720. Therefore, to contextualise, the median multiplier principle with the statistics, the total median income for three years would be about RM 205,945. Therefore, based on this figure, a question that might follow is what type of a house can this amount procured in the housing market? The answer is probably none considering the associated legal charges, interests and fees. However, the prices of properties have increased by 94% for condominium and by 80% for low cost house (Table 1). Prices have increased twice during the period, but without a corresponding increase in household incomes. Medium income only increased by 11.7% from 2012 to 2014. But, it is to those in low and medium income group that the 30% ratio is and income multiplier is an indicator of true affordability problems. Therefore, the question of the affordable housing price determinants requires demonstration.

Table 1. Median sales price of residential properties for selected cities (RM/unit).

Location	**Condominium/Apartment**		**Low-Cost House**	
	2003	**2013**	**2003**	**2013**
Johor	180,000	253,000	50,000	60,000
Penang (PP)	216,250	472,000	83,000	138,800
Selangor	140,000	260,000	74,500	112,500
WP Kuala Lumpur	230,000	500,000	120,000	278,000

(NAPIC 2014)

3 DETERMINANTS OF AFFORDABLE HOUSING PRICE

Like any other good or service, house has a price tag. The price denotes its economic value. But the price of a house depends on many factors (Abelson *et al,* 2005, IMF 2010, Abelson et al. 2005, Balázs and Dubravko 2007, Abelson, et al. 2005, Andrews 2010, Galati et al. 2011, Borowiecki 2009, Pillaiyan 2015, Xu and Tang 2014, Zou and Chau 2015, Zhang, *et al.* 2012,. Aziz 2011, Glindro et al. 2011, Panagiotidis and Printzis 2015). These factors are called determinants of housing price. Each of the determinants has a differentiated impact on the price of the house in given time. For instance, the impact of the determinants could be positive or negative and high or low. Collectively, the academic literature suggests that a number of determinants for housing price determinants are enormous. On one part, the many determinants are also a source of the complexities in the accurate modelling of the housing costs. However, the review of the literature suggests that the main focus of the previous studies on housing markets was on the advanced economies. Also, most of the research focused on both the demand and supply sides. The focus of the previous studies is concerned with the macroeconomic determinants, with the exception of a few (i.e., Peek and Wilcox 1991, Zietz *et al.* 2007) that focus on the building characteristics or configuration. However, studies on the housing price determination in Malaysia are scanty. The examples of the studies that focused on Malaysian housing price determinants include: Aziz's (2011), Ong's, (2013), Osmadi *et al.*'s (2015) and Pillaiyan's (2015) but yet, are mainly concerned with the macroeconomics determinants. Importantly, these studies are not concerned with the affordable housing. But the determinants are useful references for defining the determinants of affordable housing. Hence some were adapted to suit the affordable housing context. The salient aspect of this study is to model the price function for affordable housing. In other word, it seeks to provide a statement of the relationship between the price for affordable housing and the factors that influence its price.

However, price for affordable housing is the price the owners paid for the finished houses. This price excludes payment for utility bills and other expenditure incident to the operation of the houses. The determinations of the prices are attributable to both monetary and nonmonetary dimensions. Also the determinants could be internal or external dimensions. The internal dimensions are those that are specific to the buildings while the external dimension are factors that have an impact on the price of the house, but not explicit to the building or their owners/ users directly. With this background in mind, it is very unlikely that the determinants of the housing development in general will be useful for. For example, affordable housing development, is commonly funded partly by the federal government, and though, requires a larger state and local commitments. Household income, housing price and locations are some of the major determinants that increase affordability gaps. Affordable housing owners or users have constrained in term of a portion of their disposable incomes they use to finance their homes.

4 RESEARCH DESIGN

The primary data was based on survey questionnaire. The survey was based on the convenience sampling method. Convenience sampling is appropriate when respondents are not easily accessible and the population size is not available or accurate. The survey was administered through hand delivery. Respondents were asked various questions, but with regards to housing price determinants, respondents were asked based on their current experience of the housing market, to place a tick on the extent to consider the factors will determine housing price. The different variables included in the survey were adopted from what was outlined in the previous section and the authors' experiences. The survey was administered on homeowners/users of a low cost housing estate consisting of 500 units in Pulau Pinang. Households use the housing price to measure performance of their house or its affordability. The levels of their importance are measured on a Likert scale of 1 to 5. Where 1 represents not important and 5 denotes extremely important, 3 denotes important and 2 and 4 fall in between. The surveyed was administered to existing householders using hand delivery in a housing estate. The housing estate is located in Penang. Each of the low cost housing consists of 2 bedrooms and 1 bathroom with a total built up area of 550 square feet. About 300 units were surveyed. All the questions on the surveyed were positively worded. The ranking of these factors was determined by mode and Average Relative Index (ARI). The index is based on the cumulative weighting of the initial frequency score of each of the roles.

5 ANALYSING THE RESULTS

Altogether 71 valid survey forms were returned and analysed for this study. Most (63.2%) of the respondents were parents. Three percent have no 'formal' education. The percentage of those with a Diploma or Bachelor degree is 36.6. 85.9% of the surveyed respondents have been occupying the houses for more than five years. 76% of the houses are occupied by more than two occupants and about 19.7% have more than 5 occupants. None of the occupants' family income is less than RM4, 000. More than 80% earned between RM6, 000 to RM9, 999. 67.6% owned their units. The monthly utility bills for some 80% of the respondents exceed RM100 but many pay more than RM300 monthly. The results to the questions about housing determinants costs were broadly consistent with our expectations. The determinants have high validity and reliability scores. The KMO of 0.718 indicates that the factors are a very good measure of housing cost and the results is equally significant. Furthermore, t-test was performed to determine inclusion of each of determinant in the survey. For each factor, the null hypothesis was the factor was unimportant (H_0: $U=U_0$) and the research hypothesis was the factors important ($H_{r:}$

$U>U_0$). U_0 is the population mean. A 3.5 importance level was set. In other words, a determinant must be significant to justify its inclusion. The results of the t-test were displayed in Table 2. As may be seen all of the determinants are significant. The findings on the determinants are very interesting, albeit not surprising. Location factors and housing size are the main dominants. The energy cost and commuting to place of works and children's schools are possible indications of the respondents' decisions. In fact, access to means of transport is also crucial. Despite the relative family income of the respondents and considering that most of units housed more 3 occupants can be alluded to the fact that the proximity or access to places of works and children's schools are critical issues due to their economic implications.

Table 2. Distribution of One-Sample Test and Descriptive Statistics.

Factor	Test Value = 3.5						Std. Error Mean	Weightage mean
	t	df	Sig. (2-tailed)	Mean Difference	95% Confidence Interval of the Difference			
					Lower	Upper		
Strategic (e.g: School, Hospitalities, Place of Job, Market & etc.)	6.131	70	0.000	0.5423	0.3659	0.7186	0.088	0.792
Size of the House	7.508	70	0.000	0.6127	0.4499	0.7754	0.082	0.778
Location – Urban/Rural	7.407	70	0.000	0.7535	0.5506	0.9564	0.102	0.749
Leasehold / Freehold House	8.359	70	0.000	0.7676	0.5844	0.9508	0.092	0.747
Recreational facilities	14.50	70	0.000	1.2747	1.0994	1.4499	0.088	0.645
Layout of the House	14.14	70	0.000	1.2747	1.0948	1.4545	0.090	0.645
Transportation	11.04	70	0.000	1.3028	1.0674	1.5382	0.118	0.639
Land Acquisition	13.71	70	0.000	1.5423	1.3179	1.7666	0.113	0.592
Levy	22.83	70	.0000	1.7254	1.5746	1.8761	0.0756	0.555
Permit Fees	22.05	70	0.000	1.7394	1.5821	1.8968	0.079	0.552
Stamp Duty	24.56	70	0.000	1.7394	1.5982	1.8807	0.071	0.552
Government Policy	25.72	70	0.000	1.8944	1.7475	2.0413	.0737	0.521
Workmanship (eg: Finishing)	24.47	70	0.000	1.9507	1.7917	2.1097	0.080	0.510

Cronbach's Alpha reliability = 0.743; Validity = 0.641- 925Kaiser-Meyer-Olkin Measure of Sampling Adequacy = 0.718, Bartlett's Test of Sphericity = 491.707; Significance =0.000,

6 CONCLUSIONS AND FURTHER RESEARCH

The housing cost is a major concern for households and developers due its economic impact. This study has found that 13 cost determinants are significance household decision making in affordable housing delivery. Government needs to reduce her charges/fees on the developers to increase home ownership. Implicit in the findings is that the households make choices to reduce their financial burden. The more than 50% of loan rejection rate by the banks is critical. Future work on this topic might increase the sample size and various housing units with multiple compositions should be included to provide broader findings.

Acknowledgments

This research is part of a larger research under the FRGS grant: FRGS/1/2015/TK06/UTAR/02/2A (i) Vote No. V21000. Therefore the supports of FRGS are highly acknowledged.

References

Abelson, P., Joyeux, R. Milunovich, G. and Chung, D., Explaining House Prices in Australia: 1970–2003. *Economic Record*, Vol. 81, pp. S96–S103, 2005.

Aziz, A.A. F., Causality of Residential Properties Price Movements in Malaysia. MPRA Paper No. 47682, 2011.

Andrews, D., Real house prices in OECD Countries: the role of demand shocks and structural and policy factors economics department working. Paper No. 831, 2010.

Balázs, E. and Dubravko, M., Determinants of house prices in central and eastern Europe, BIS Working Papers No. 236, Basel, 2007.

Ball, M., Meen, G. and Nygaard, C., Housing supply price elasticities revisited: evidence from international, national, local and company data. Journal of Housing Economics, 19, 255- 268, 2010.

Borowiecki, K. J. The determinants of house prices and construction: An empirical investigation of the Swiss housing economy, International Real Estate Review. 12 (3), 193– 220, 2009.

Demographia, (11th Annual International Housing Affordability Survey, 2015) 2015 Ratings for Metropolitan Market. Available at: www.demographia.com/dhi.pdf. [accessed on 05 March 2015]

Department of Statistics Malaysia, Report on household expenditure survey, 2011

Department of Statistics Malaysia, Report of household income and basic amenities survey 2014

Galati, G., Teppa, F., & Alessie, R. Macro and Micro Drivers of House Price Dynamics: An Application to Dutch Data. De Nederlandsche Bank Working Paper, No.288, March, 2011.

Glindro, E.T., Subhanij, T., Szeto, J., Zhu, H., Determinants of house prices in nine Asia–Pacific economies. International Journal *of Central Banking*, 7, pp.163–204, 2011.

Gokm, Y. I and Keceli, A determinants of house prices in turkey: comparative analysis of development regions. *The 2015 WEI International Academic Conference Proceedings*, Harvard, USA , The West East Institute p110- 122, 2015

Government of Malaysia (2010), Tenth Malaysian Plan 20011-2015. Putrajaya: Economic Planning Unit Prime Minister's Department.

Gyourko, J., Tracy, J., A look at real housing prices and incomes: Some implications for housing affordability and quality, *Federal Reserve Bank of New York Economic Policy Review* 5, 63–77, 1999.

IMF (International Monetary Fund, 2008) Interest rate elasticity of residential housing prices. IMF Working paper, WP?08)/247.

Jewkes, D. M. and Delgadillo, M. L., Weaknesses of Housing Affordability Indices Used by Practitioners. *Journal of Financial Counseling and Planning*, Vol.21, Iss. (1)43-52, 2010.

NAPIC (NAPIC, 2014). The Malaysian House Price Index by House Type.

Ong, S. T. Factors Affecting the Price of Housing in Malaysia .Journal of Emerging Issues in Economics, Finance and Banking .Vol.1 No.5, pp.414- 429, 2013.

Osmadi , A., Kamal, M. E., Hassan, H. and Abdul Fattah1, H., Exploring the Elements of Housing Price in Malaysia, *Asian Social Science*, (11):24,26-38, 2015.

Panagiotidis, T. and Printzis, P., On the macroeconomic determinants of the housing market in Greece: A VECM approach. Paper No.88 Hellenic Observatory Papers on Greece and Southeast Europe, 2015.

Peek, J. and Wilcox, J. A., The Measurement and Determinants of Single-Family House Prices. *American Real Estate and Urban Economics Association Journal,* 19, (3): pp.353-82, 1991.

Pillaiyan, S. Macroeconomic Drivers of House Prices in Malaysia. *Canadian Social Science*, 11 (9), 119-130, 2015.

Schwartz, M. and Wilson, E. (2008) Who Can Afford to Live in a Home? A Look at Data from the 2006 American Community Survey, US Census Bureau, Washington, DC.

Yong, T.C (The Star, 25 Wednesday May 2016) Our household debt stands at RM1bil.

Xu, L., & Tang, B., On the determinants of UK house prices. International Journal *of Economics and Research*, 5(2), pp.57-64, 2014.

Zhang, Y, Hua, X, and Zhao, L. (2012). Exploring determinants of housing prices: A case study of Chinese experience in 1999–2010, Economic Modelling, 29(6), pp.2349-2361.

Zietz, J., Zietz, E., Sirmans, G., Determinants of house prices: a quantile regression approach. *Journal of Real Estate Finance and Economics*, 37, pp.317–333, 2008.

Zou, L. G. and Chau, W. K., Determinants and Sustainability of House Prices: The Case of Shanghai, China. *Sustainability*, 7, pp.4524-4548, 2015.

Integrated Solutions for Infrastructure Development
Edited by Lau, H. H., Tang, F. E., Ng, C. K., and Singh, A.
Copyright © 2016 ISEC Press

CAUSES OF VARIATION ORDERS FOR MEGA CONSTRUCTION PROJECTS

SAMER EZELDIN and AMR KHALED EL-SADEK

Dept of Construction Engineering, The American University in Cairo, Egypt

Variations or change orders are inevitable in any construction project. They are defined as any change that happens in the scope of the project. Change is an additional scope of work, omission, or even alteration. This paper presents the main causes of variation orders in MENA region represented by Egypt. The research investigates the causes, as identified by clients, consultants, and contractors. It classifies them into four main categories: owner related, contractor related, consultant related, and other causes. The top 10 most important causes are ranked among the opinion of each party. The study reveals that the top 3 causes are: 1) The client instruction of additional works; 2) The contractor using ambiguous areas in the contract to request variations, and 3) continuous changes in the project schedule. A comparison about causes of variations was performed between Egypt, Palestine, Malaysia, and the United Kingdom. Furthermore, a model is proposed to provide the user with a scale that predicts the effect of the events triggering variations on both project cost and time.

Keywords: Change, Construction, Civil, Middle East, Variations.

1 INTRODUCTION

A variation order is any modification in the original scope of the contract. Changes may be issued by a contractor's claim, a designer's modification of drawings, or the owner's change in his requirements. A variation order could be as small as changing a window type to as large is omitting the construction of a building in a complex of buildings. The variation order is formally issued to announce the modification of the contract between the contractor and the owner. It recognizes that there is a change in the assigned work without having to resort to a new contract (Hester 1991).

In any construction project, time and money are usually of an extreme essence. Variations are one of the processes that prolong the end time of the project and extrapolate the value of it as well. Many authors such as: Keanne *et al.* (2010), Memon *et al.* (2014) have highlighted the most recurrent problems that are generated due to rise of variations.

Limited studies were carried to understand the causes of variations in the Middle East and more specifically in Egypt. Hence, there is a gap of understanding the events that lead to such variations.

The aim of this study is to identify and rank the most common types of events that lead to variation orders for projects designed and constructed in Egypt. A guidance model, using the collected data, is developed to predict the severity of the combined effect of causes of variations. The model is further verified and validated with a case study.

The followed research methodology in this study starts with a literature review of events that gives rise to variations, their effects, and their rectification methods. A comprehensive survey is conducted among experts to identify the most common types of variations and their ranking in the Egyptian construction industry. Finally, the proposed model is developed, verified, and validated.

In this paper, events that lead to variations are categorized into four categories, namely; client related, contractor related, consultant related, and others.

2 DATA GATHERING AND CALCULATION OF RELATIVE IMPORTANCE FACTOR

The purpose of the gathered data is to identify and rank the top ten most frequent events that give rise to variations. The main components of the survey are:

1 A top ten list for most common events that give rise to variations according to the client, the contractor, and the consultant

2 Probability of the effect of the event that gives rise to variation on project cost

3 Probability of the effect of the event that gives rise to variation on project time

Given the complexity of the survey, and the importance of getting reliable data, only professionals with more than 15 years of experience in the market were included. 25 professionals were selected; all of them with previous experiences in projects with a value exceeding 100 million USD. The participants representing clients were nine and participants representing contractors and consultants were eight each. The reason why there are equal numbers of surveys for each party is to be able to create a reasonable comparison of results between them.

2.1 Calculation of the Average Values Using Relative Importance Factor

In order to obtain the average values for the causes of variation orders, the relative importance factor method (RII) was proposed. The RII was adopted in many previous studies to rank the events that lead to variations: Memon *et al.* (2014), Alnuaimi *et al.* (2010), and Muhwezi *et al.* (2014). The equation of the relative importance factor used was:

$$Score = \frac{\Sigma W}{H \times N} \times 100 \quad (1)$$

The RII is basically an average of the values obtained ranging from 1 (being lowest) to 5 (being highest), where:

W: summation of all the values given from the experts

H: Highest rating score, which is 5 in this case

N: The total number of respondents

Equation 2 presents the calculation method for confidence intervals to test the upper and lower boundaries of data calculated to rank the causes where:

$$\text{Confidence Interval} = \overline{X} \pm 1.96 \frac{\sigma}{\sqrt{n}} \quad (2)$$

$\overline{X}$ = Average value of the cause; σ = Standard deviation; n = Sample size; 1.96 = constant coefficient (For 95% confidence level)

3 RESEARCH RESULTS

The analyzed results based on the collected data are presented in Table 1.

Table 1. Results of top 10 overall important factors that lead to variations.

Top 10 overall important factors	Score	Confidence Int.
Client instructs additional works	84.44	77.63-91.17
The contractor uses the grey areas in the contract to request variations	78.89	72.36-86.04
Continuous change in project schedule	77.22	68.95-85.45
Conflicts between contract documents	76.94	70.67-82.93
Lack of coordination among project parties	76.94	70.67-82.93
Client's brief before the design stage are unclear or not well defined	75.28	69.68-80.72
Contractor's financial difficulties	75.28	69.23-81.17
Client fails to make decisions at th6e right time	72.50	65.33-80.27
Lack of contractor's involvement in design	72.22	64.01-80.79
Modification of scope	71.67	64.49-79.51

Table 1 presents the overall assessment of events. The overall assessment is the average of the contractor, client, and consultant's data. Topping the list is *client's instruction of additional works* which has always monopolized the top two spots among the results. It can be confidently drawn that this is the main and the most frequent cause of variations. Its average value among all the parties scored a value of 84.44. *The contractor using the grey areas in the contract to request variations* took number two in the list because it was given a high score value of 78.89 from the client and from the consultant. *Continuous changes in project schedule* scored 77.22 while *conflicts between contract documents and lack of coordination* came in fourth and scored 76.94.

3.1 Comparison between the Top Ten Important Factors for Each Party

The analyzed results based on the collected data for each party are presented in Table 2.

Table 2. Results of top 10 overall important factors for every party.

#	Top 10 important factors for client	Top 10 important factors for contractor	Top 10 important factors for consultant
1	The contractor uses the grey areas in the contract to request variations	Client instructs additional works	Client instructs additional works
2	Client instructs additional works	Client's brief before the design stage are unclear or not well defined	Continuous change in project schedule
3	Client fails to make decisions at the right time	Conflicts between contract documents	The contractor uses the grey areas in the contract to request variations
4	Modification of scope	Lack of coordination among project parties	Contractor's financial difficulties

#	Top 10 important factors for client	Top 10 important factors for contractor	Top 10 important factors for consultant
5	Short period for design stage	Poor construction management by contractor	Change in specification by owner/client
6	Lack of contractor's involvement in design	Non availability of professional engineers to maintain the quality of consultancy services	Lack of contractor's involvement in design
7	Continuous change in project schedule	Poor workmanship	Conflicts between contract documents
8	Client's brief before the design stage are unclear or not well defined	Contractor's financial difficulties	Lack of coordination among project parties
9	Contractor's financial difficulties	Continuous change in project schedule	Low consultancy fees leading to hiring less experienced designers
10	Conflicts between contract documents	Change in specification by owner/client	Client fails to make decisions at the right time

If a conclusion can be drawn here, it would be that each party is biased towards its own benefit. An example can be stated by observing that both the client and the consultant think that "*the contractor uses the grey areas to request variations*" is one of the top reasons why variations are generated. Another example can be noticed is that both the contractor and the consultant think that one of the main reasons variations occur is because the tendency of the client to "*change project specifications*".

3.2 Comparison of the Overall Results with Other Studies

The results in table 3 are showing a comparison of the top five causes of variations in four countries.

Table 3. Comparison between the results in four different countries.

Rank	Gaza	Egypt (current study)	Malaysia (Memon, Abdul Rahman, & Abul Hasan, 2014)	UK (Keanne, Sertyesilisik, & Ross, 2010)
1	Lack of construction materials & equipment	Client instructs additional works	Unavailability of equipment	Errors and omissions in design
2	spare parts due to closure and siege	The contractor uses the grey areas in the contract to request variations	Poor workmanship	Little involvement in design from contractor
3	Change in design by consultant	Continuous change in project schedule	Design complexity	Inadequate project objectives
4	Lack of consultant's knowledge of available materials and equipment	Conflicts between contract documents	Change in schedule	Poor design
5	Errors and omission in design	Lack of coordination among project parties	Impediment to prompt decision making process	Conflicts between contract documents

It can be observed that there is no solid overall similarity between these top five causes for the different countries. This would support a conclusion that events giving rise to variations are very sensitive to regional conditions and would differ from country to another.

4 MODEL DEVELOPMENT AND VALIDATION

A model was developed using the data collected through the research to allow the user to select a number of events which he/she thinks might happen to the project. The selected events are compared against the database collected and the user gets two results, one for the cost and the other for the time. The results displayed estimate the effect of the events that give rise to variations on project cost and project time. The results depend on the severity of the events on cost and time and ranges from low effect to moderate, up to important and critical. The user gets the display shown in table 4.

The scale calculates the severity of events on project cost and time, if the effect is less than 10% from the maximum calculated score; the result is "low". If the effect is less than 30%, the result is displayed is "moderate", if it is less than 50%, the result is "important". Finally, if the effect is more than 50% of the calculated severity score, then the result displayed is "critical".

Table 4. User display.

Effect on Project Cost	MODERATE
Effect on Project Time	IMPORTANT

4.1 Validation Using Case Study

The model was used against an existing retail shopping center that was finished in Cairo in 2013. The effect of events on cost and time were previously known since it was an already finished project. When the data was entered by an independent user, the model showed comparable results.

5 CONCLUSION

Variation orders are a very important factor in construction projects. They affect the project cost, time, quality, productivity and can have other adverse effects on organizations. Through the course of this study, it was found that the top event that leads to variation orders in Egypt is: client's instruction of additional works. It was also found that the top ranked events leading to variation can vary from one country to another. Finally, a simple easy-to-use model is proposed to evaluate the impact of the variation events on the time and cost of projects.

References

Alnuaimi, A. S., Taha, R. A., Al Mohsin, M., & Al Harthi, A. S., (May 2010), Causes, Effects, Benefits, and Remedies of Change Orders on Public Construction Projects in Oman, *Journal of Construction Engineering and Management* , 615-622.

Hester, T., Kuprenas, A., & Chang., (1991), *Construction Changes and Change Orders: Their Magnitude & Impact,* Construction Industry Institute.

Keanne, p., Sertyesilisik, B., & Ross, A. D., (2010), Variations and Change Orders on Construction Projects, *Journal of Legal Affairs and Dispute Resolution in Engineering and Construction*, 89-96.

Memon, A. H., Abdul Rahman, I., & Abul Hasan, M. F., (2014), Significant Causes and Effects of Variation Orders in Construction Projects, *Research Journal of Applied Sciences, Engineering and Technology*, 4494-4502.

Muhwezi, L., Acai, J., & Otim, G., (2014), An Assessment of the Factors Causing Delays on Building Construction Projects in Uganda, *International Journal of Construction Engineering and Management*, 13-23.

Integrated Solutions for Infrastructure Development
Edited by Lau, H. H., Tang, F. E., Ng, C. K., and Singh, A.
Copyright © 2016 ISEC Press

SAFETY AND HEALTH SITE INSPECTIONS FOR ON-FIELD RISK ANALYSIS AND TRAINING

MARCO L. TRANI[1], LUCA BERETTA[2], GIADA NICOLINI[3], and MANUELE CASSANO[1]

[1] *Politecnico di Milano, A.B.C. Dept, Milan, Italy*
[2] *CPT Milano, Assago, Italy*
[3] *Research Consultant, Lesmo (MB), Italy*

The field of construction is always affected by a large number of accidents at work that have many different causes and responsible. Therefore, it is of utmost importance to focus on all these issues, in order to reduce all risk factors that can undermine individuals' safety on building sites. The objective of the research is then the development of a method for quick on site analysis of all critical issues that can create accidents and identification of the related causes in order to directly provide a correct and focused training identified as the best method to act on the causes to reduce accidents. The research was carried on during construction of the Universal Exhibition of Milan – Expo 2015 – that counted almost 70 contemporary construction sites. To reach the goals further research steps has been followed and in particular: (i) inspections on building sites through all the Expo area; (ii) analysis of the main identified problems; (iii) development of a methodology to quickly identify the cause of problems; (iv) validation of the method through back office analysis of site documents; (v) correct on-site training according to found problem. During the whole construction site, the improvements in criticalities solving have been visible thanks to the focused training. The developed method, carried on in a high-risk environment, is applicable in any other building sites and environment as independent from the boundary conditions of the place.

Keywords: Construction critical issues, Site inspections, Training.

1 INTRODUCTION

Nowadays the construction appears to be in Italy one of the fields with the largest number of accidents at work (INAIL 2015). Our experience told us that training courses are the main vehicle to develop a proper safety skill, but only if deal directly with potentially dangerous situations. Today the two main standards for health and safety training in Italy are: the Legislative Decree no. 81/2008 and the State-Regions Agreement of 2011. In particular, the latter defines duration, contents and methods of training based on the risk considered for the specific activity: low, medium, high. However, we think that it is also essential to calibrate topics and duration of training courses in order to be perfectly aligned with the critical issues present in the construction site and that workers have to face every day. Demirkesen and Arditi (2015) think that effectiveness of safety training practices is very important and analyzed if this consciousness is present in professionals' perceptions in order to understand how training can be improved. Choudhry (2014) worked in particular on critical behavior of workers in order to improve safety performances. Behavior-based Safety (BBS) was considered also by Zhang and

Fang (2013) that try to introduce this practice into management routine. Finally Yang *et al.* (2012) recognized training as one of the requirements for accident identification and implemented a real-time identification system. It is visible then how on site direct supervision and training can improve site safety by acting directly on accident causes determined on field. The presented research aims in particular to categorize the main accident causes in order to find the best solution in terms of workers training. Then site inspections permitted to develop a method for a direct evaluation of the cause and the responsible of the problem. Finally tests and office analysis on the documentation permitted to validate the developed method in order to be used consistently in future construction site. The output of the research is an "on the job" training focused on the causes of the faced problems. These research project arise from a partnership between bilateral bodies (CPT Milano) and INAIL (The National Insurance Institute Against Accidents at Work) of the Lombardy region and has been developed in teamwork with Politecnico di Milano.

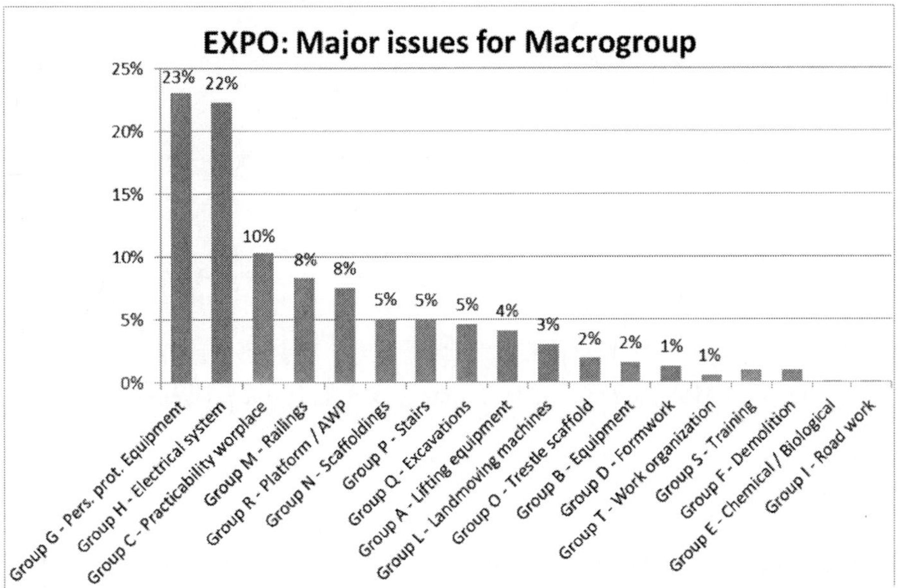

Figure 1. Major issues for Macro group checking during inspections.

2 MONITORING ANALYSIS

The experimental analyzes have been conducted by the task group of Politecnico di Milano on the basis of data obtained during the inspections carried out in the sites in Milan - EXPO 2015 by the CPT technicians. The main work of the CPT was to do inspections on sites aimed to a continuous assistance on safety matters. This help was offered to construction companies and to workers all over the Expo site, and was directed to the identification of any critical situation. Statistics on visits carried out by CPT engineers and trainers are the following:

- Number of sites monitored: 69
- Number of companies involved: 799
- Number of visits carried out: 683
- Number of employees trained: 1314
- Number of total of problems found: 4829

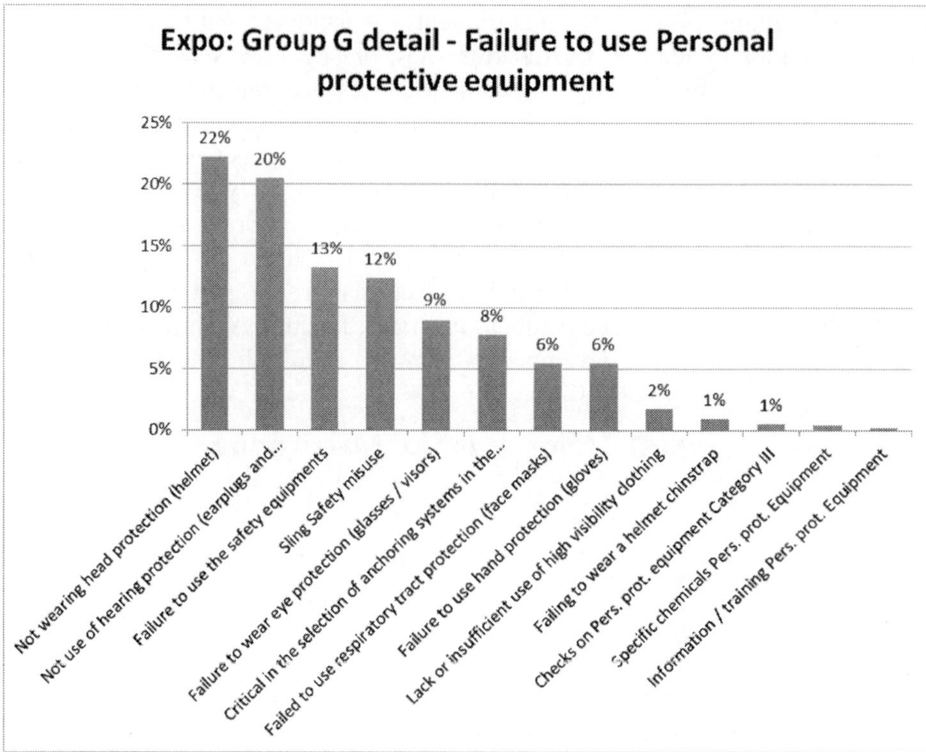

Figure 2. Macro-group details relating to personal protective equipment.

During inspections, the technicians analyzed and controlled all the works in progress, checking the proper performances compared to existing regulatory dictate, thus offering advices to firms' technicians or training support directly to workers. Once completed the monitoring activities, we proceed with the preparation of two documents for the check of construction activities, the first to be compiled by the technician and the second, based on the first one, to be compiled by the trainer. These verification cards were drawn up for each individual work in action (for example: mounting scaffolding, excavations, etc.), for each site and for each specific company. The technical report, developed from extensive experience of established systems of CPT, takes into account all the main activities that are potentially carried out in the construction site and shows, for each of them, all potential sources of risk. This creates the basic matrix consisting of two levels. The first is related to a specific subject area while the second is related to a specific critical issue. Table 1 shows the 18 captions that includes about 180 criticalities.

Table 1 . List of identified subject areas.

"A" - Lifting equipment	"G" – Pers. prot. equipment	"O" – Trestle scaffold
"B" – Equipment	"H" – Electrical system	"P" – Ladder
"C" – Practicability workplace	"I" – Road works	"Q" – Excavations
"D" – Formwork	"L" – Land moving machines	"R" – Platform / AWP
"E" – Chemical / Biological	"M" – Railings	"S" – Training
"F" – Demolition Total / Partial	"N" – Scaffolding	"T" – Organization of work

From these data it was possible to perform an analysis which allowed to have a global and detailed view of the whole criticalities. It has been conducted a first level of analysis on major nonconformity found for broad groups (see graphic in Figure 1 for the major issues found). It was then possible to deepen the analysis of non-compliances detected down to a second level of detail. Graphic in figure 2 shows the details of the personal protective equipment macro-group.

3 EVALUATION METHOD

Starting from the outcome of technical assistance on site, we tried to find unique ways to bring works non-conformities at three different conditions specified in these areas:

- Behavioral area, where the problem is directly due to the worker's working method;
- Organization corporate area, where the problem is due to deficient equipment provision;
- Area related to the organization of the workplace, where the problem is due to a wrong control of individual processes and work contexts in which they occurred.

A specific method has been set up for on-field evaluation of the non-conformities and the related areas. It is mainly based on the analysis of the time required to resolve problems encountered. In the pipeline it was assumed the amenability of the same problem to the three areas described above. Three resolution times were considered and at the three levels were associated different assignments, related to the causes of the critical issue:

- ST = Short time: 5/10 min = Behavioral problems (L), whose almost immediate solution can be fielded directly by the worker with tools that they already have
- MT = Medium time: medium time due to the duration of the monitoring carried out = Labor Organization (O), the resolution of which potentially requires the intervention of multiple resources, human and technical, which can therefore be put in place by those who have functional power directly on site
- LT = Long time: Attributable to cases in which the resolution is greater than the monitoring activity = Organization (M), the solution of which necessarily requires the intervention of company top management.

The method, used to be able to support the above-mentioned hypothesis, takes place starting from the monitoring carried out in the construction site followed by back office analysis. The starting point for this investigation is developed considering primarily the inspections carried out by the technicians and CPT trainers in the various construction sites with the detection of any critical issues for each machining process. After reporting the criticality to the worker and/or to the person in charge all timing resolution of every problem were monitored. The analysis consists to check how quickly the critical issues was resolved, compared to the same period of monitoring, distinguishing them in the three times above. This method is established with the aim of being able to carry out the feedback on the cause of the criticality directly on site, quickly, without losing days behind the cards.

4 VALIDATION

In order to validate this method, we proceed with a deeper back office analysis. In fact we must be sure that the time consumed for a criticality resolution can be associated to one of the above mentioned areas in order to be certain where intervene.

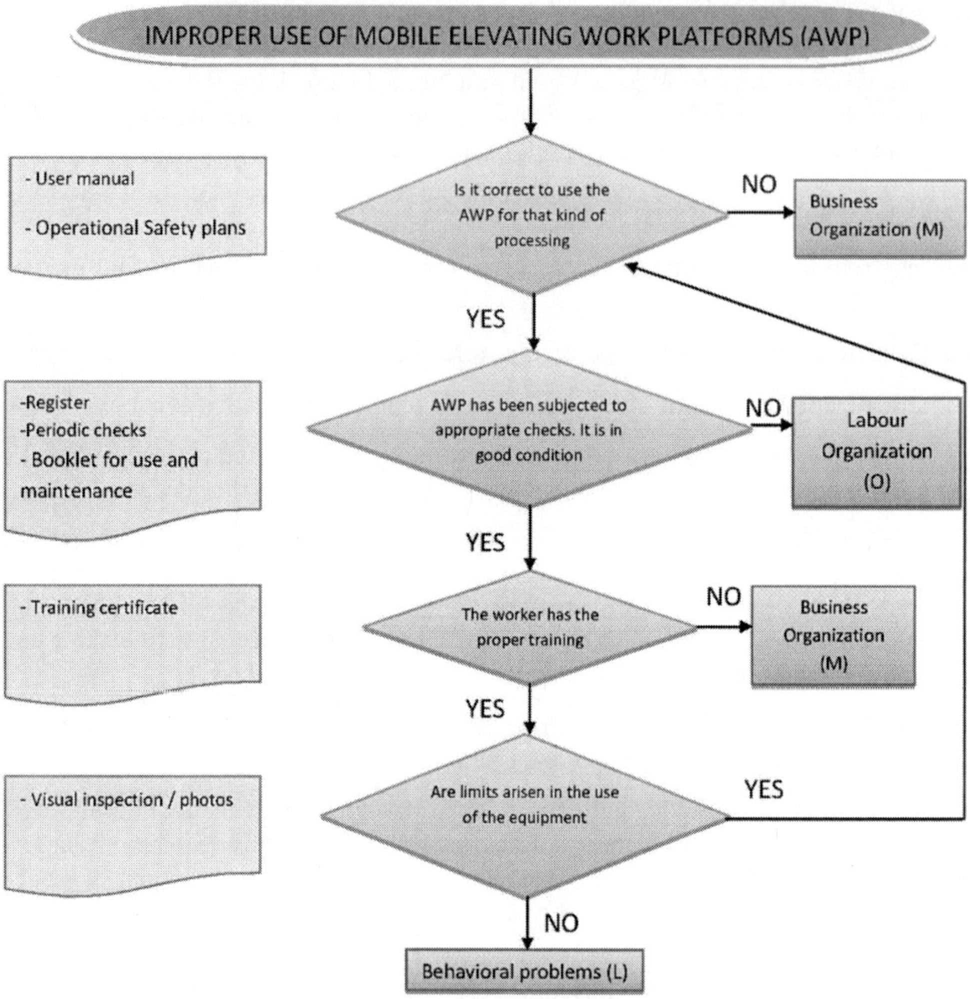

Figure 3. Example of flow chart: improper use of AWP.

This analysis involves, in the first instance, the use and study of documents provided by the companies (Operational Safety plans, Safety procedures, Training certificates, etc.) and secondly the use of flow charts made for this purpose for each single Macro-group. Each Macro-group has its own flow chart to figure out the cause of the criticality. The flow charts concern each examined item during the technical assistance activities. For each operation described in the diagrams, we have identified different outputs, so that we could arrive to identify the subjects that made the non-compliance. Figure 3 shows one of the twelve flow charts created for this purpose.

Applying this methodology, over a sample of issues found in the construction sites, we obtained an estimated percentage useful to determine the capabilities of the instrument adopted. Figure 4 shows the summary graphs of such analysis. The 96% of the solved situations in the short term relates with the behavioral sphere of the individual worker. It remains a 4% due to the necessary intervention of a subject with organizational power than the other assigned work. As in the previous case, you may notice that there is no automatic allocation of a responsibility given based on the time taken to solve a problem. Each situation should be carefully considered in order to understand from where derives the error returned. In the case shown the mismatch is even more striking: 30% of the situations resolved in a high amount of time, by default attributed to the apex of the company, is actually due to an intervention required to responsible business.

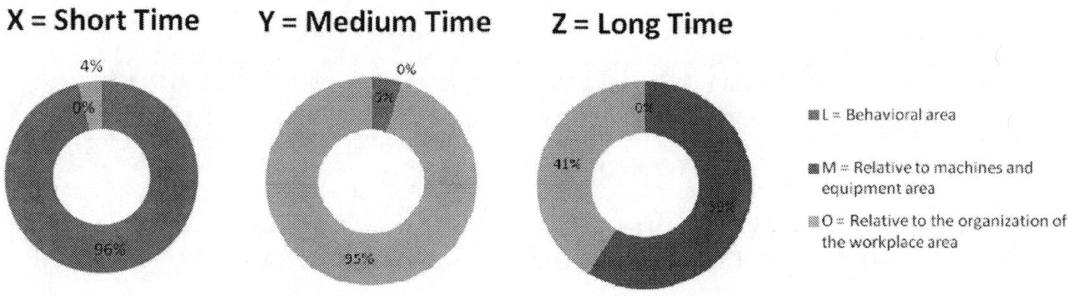

Figure 4. Percentage about situation solved in Short / Medium / Long time.

It's possible to underline that the adopted instrument responded positively to the objective.

5 CONCLUSIONS

In conclusion, this research has allowed us to create a model to neatly categorize, and then rework in statistical form, the data collected during the monitoring conducted in the Expo sites, aimed at business support and training for workers. This model, referred to the problems encountered, allowed to obtain evidence of risk situations identified above and, to bring out the causes that generated them. The knowledge of the causes allowed to direct interlocutors to the best methods of troubleshooting, also suggesting actions to be taken to prevent their recurrence in time. It has thus developed a first proposal for the basic training course for workers in the field of health and safety, able to optimize the timing and subject matter, based on the obtained statistical data and the comparison with the teachers of the training courses themselves.

Furthermore, this research has allowed us to confirm that the training provided in the 2011 State-Regions must be developed according to the task, and integrated with the introduction of greater specificity in terms of risk classes and type of work. This model, made from data obtained in a high risk environment, can be used in any other type of construction site and can be implemented in national and international level. Also well suitable for application in any type of environment as independent of the boundary conditions of the place, ensuring that it can be seen as an expendable standard course for all companies in the construction sector.

References

Choudhry, R. M., Behavior-based safety on construction sites: A case study, *Accident Analysis & Prevention*, Volume 70, 14-23, 2014.

Demirkesen, S., Arditi, D., Construction safety personnel's perceptions of safety training practices, *International Journal of Project Management*, 33(5), 1160-1169, 2015.

Italian Law 81/2008 on workers' health and safety and its following modifications (2008).

Italian State-Regions Agreement about safety training (2011).

INAIL (The National Insurance Institute Against Accidents at Work). Database of accidents at work (2015).

Yang, H., Chew, D.A.S., Wu, W., Zhou, Z., Li, Q., Design and implementation of an identification system in construction site safety for proactive accident prevention, *Accident Analysis & Prevention*, Volume 48, 193-203, 2012.

Zhang, M., Fang, D., A continuous Behavior-Based Safety strategy for persistent safety improvement in construction industry, *Automation in Construction*, Volume 34, 101-107, 2013.

Integrated Solutions for Infrastructure Development
Edited by Lau, H. H., Tang, F. E., Ng, C. K., and Singh, A.
Copyright © 2016 ISEC Press

CHALLENGES IN THE IMPLEMENTATION OF QUALITY CONTROL TECHNIQUES DURING SITE INSPECTION

NUR NABIHAH ABD RAZAK[1], INTAN ROHANI ENDUT[1], SITI AKMAR ABU SAMAH[2], AHMAD RUSLAN MOHD RIDZUAN[1], and SABIHAH SAAIDIN[1]

[1]*Faculty of Civil Engineering, Universiti Teknologi MARA, Shah Alam, Selangor*
[2]*Akademi of Language Studies, Universiti Teknologi MARA, Shah Alam, Selangor*

Quality has been the main issue in construction projects. In handling it, the quality control techniques during construction process are required to thoroughly carry out in order to avoid any mistakes which lead to bigger problem such as poor quality of project outcomes. The implementation of quality control techniques during site inspection is able to discard all possible negative effects towards construction projects. There are three quality control techniques that have been used during site inspection known as Project Quality Plan (PQP), Inspection Test Plan (ITP) and Checklist. In order to implement these techniques, there are some matters that need to be identified and overcome. Therefore, this study has been carried out to identify the challenges in implementation of quality control techniques during site inspection. To achieve the objective, the survey questionnaire was distributed among G5, G6 and G7 contractors' in Malaysia. A total of 82 useable questionnaires were received and analyzed using factor analysis. The challenges were clumped and categorized into respective groups. The seven groups were renamed, namely: (1) attitude, (2) documents, (3) operations, (4) management, (5) workers, (6) information and (7) methods. This study helps to improve the quality control techniques especially during site inspection among Malaysian contractors.

Keywords: Challenges, Implementation, Quality control, Site inspection, Techniques.

1 INTRODUCTION

Construction Industry Development Board Malaysia (CIDB) has established the Construction Industry Standard known as CIS 7:2006 for Malaysia's construction projects. The standard focus on quality of structural works comprises the site inspection for formwork and the laboratory test on concrete and materials required for construction (CIDB 2006). Inspection is the process of observation at construction site that outlines divergence of components based on the material specifications, tests and techniques implemented (Building Inspection Guideline 2012). Tamaki (2014) points out that activity of assuring construction works complies with standard is by doing physical scrutiny called inspection. Hence, the identified quality control techniques are needed to be implemented during site inspection. In order to do so, the challenges in its implementation need to be clarified and overcome.

2 QUALITY CONTROL TECHNIQUES

In this study, the techniques of quality control used during site inspection are Project Quality Plan (PQP), Inspection Test Plan (ITP) and Checklist. PQP is a planned set of construction activities that helps in controlling the project quality. Prior to the commencement of construction works, PQP must be developed and required an approval from the Engineer (Noel 2015). There is emphasis on the necessity of clear statements in the PQP to assure the information is deliverable for everyone as well as smooth progression of the project. PQP is one of major documents for construction projects to evaluate the project quality.

Meanwhile, Wauchope (2014) indicates that Inspection Test Plan (ITP) highlights the critical control points for each construction phase such as scheduled inspection or verification activity by securing the progression of works. According to Mande (2013), ITP must consist of detailed document for each construction phase that has been verified, details on required equipment for the test and finally concluded the results from each conducted test.

The purpose of the checklist is for identification, revision and correction found during inspection which does not follow the standards and requirement. The essential matters in the checklist are all information of project which details are compiled and upgraded via the checklist followed by construction processes (Ruin 2009).

3 CHALLENGES

The quality control techniques namely PQP, ITP and Checklist are identified and the challenges implementing these quality control techniques in Malaysian construction projects are investigated.

Documents are records required at construction site since it contains various construction activities. The main purpose of developing the documents is an act of communication between project team. These documents require contractors obligation so as to fulfill all the standards stated in the contracts. Thus, all crucial documents must be presented during inspection at construction site. However, if the documents are not properly managed and organized it could lead to other problems. As been clarified by Low and Ling-Pan (2004), the restriction and inefficiency in updating these documents can inflect the effectiveness of quality control techniques. The exact information must be conveyed and reported so as to assure that there will be no same mistake made as well as to avoid carry out jobs redundancy. As been agreed, there was a case involved in Indonesian construction project which the main causes lead to that problem, that is, absence in quality information. In line with Chin-Keng and Abdul-Rahman (2011), the stumbling block in implementation of this system is synchronized with the involved documents for construction projects.

An approach from top management can be seen from its support, commitment, and suitable quality policy that need to be accepted by employee (Sambasivan and Yun 2007). Most of barriers to implement the new quality control techniques or system come from lack of support from top management (Shehu and Akintoye 2010). Insufficient support from top management has been a known issue since it has been agreed and declared by most researchers such as Escanciano *et al.* (2001), Wiele *et al.* (2005), Hesham and Magd (2007), as well as Al-Najjar and Jawad (2011). The commitment and dedication from top management can influence the continuous improvement if it has been shown on the quality aspects.

According to Low and Ling-Pan (2004); as well as Chin-Keng and Abdul-Rahman (2011), the support from top management is crucial factor in execution quality control system. Furthermore, top management has an obligation in establishing a rigid quality policy as well as continuous assessment on the effectiveness of quality control system.

The scope involved with workers such as the accessibility of training, relationship between team members, the awareness of quality control systems, the willingness of receiving new system among team members and the human resources given (Wiele *et al.* 2005). Poor accessibility of training was a hindrance in ensuring better implementation of quality control techniques (Ibrahim 2005). Team member is a group of people connected in finishing the construction projects and all the decisions made consist of consents in each team member. As been declared by Balzarova *et al.* (2002), one huge barrier or hindrance towards a successful quality control technique is lack of team work. Meanwhile, awareness is something that required companies to ensure its employees to feel that each team member is part of quality organization (Crosby 1996). This indicates that without team members' consciousness towards the importance of quality control system, it can cause negligence and unwillingness to follow the provided quality policy.

To establish, introduce and implement these quality control techniques do consume time and cost. It is supported by the companies that require time and cost to productively implement these techniques thoroughly, since the processes vary and are large to cover and at the same time to assure there is a specified improvement to achieve (Dory and Schier 2002). Meanwhile, it has been found that the most difficult elements to comply in better implementation of quality control techniques are cost and time (Stevenson and Barnes 2001, Dickenson *et al.* 2004).

4 METHODOLOGY

The survey questionnaire was distributed by hand or email and a total of 82 useable questionnaires were collected. The targeted respondents were Project Manager, QA and QC Manager, Site Engineer, Site Supervisor and Construction Manager from G5, G6 and G7 contractors within the area in Selangor, Malaysia. They were registered under Construction Industry Development Board Malaysia (CIDB). Meanwhile, the questionnaire forms were produced by using the Likert Scale range from one until five developed by Rensis Likert in 1932. Factor analysis was used for data analysis using SPSS to identify, manage and reduce wide items to the smallest item from the questionnaire (Chua 2009).

5 RESULT AND DISCUSSION

Kaiser-Meyer-Olkin Measure of Sampling Adequacy (KMO) test was carried out to check the data suitability for Factor Analysis. According to Field (2009), the value of KMO varies according to a range such as value above 0.9 being excellent, values between 0.9 and 0.8 are great, values between 0.8 and 0.7 are good and values between 0.7 and 0.5 are mediocre. Meanwhile, for Bartlett's Test of Sphericity the value must be less than 0.05. Table 1 indicates that the values of both KMO and Barlett's Test of Sphericity are satisfactory for factor analysis.

Table 1. KMO and Barlett's Test of Sphericity.

Kaiser-Meyer-Olkin Measure of Sampling Adequacy		.768
Bartlett's Test of Sphericity	Approx. Chi Square	1012.032
	df.	.276
	Sig.	.000

The result from factor analysis has been separated according to the respective groups and there were seven groups. Table 2 presents the rotated component matrix which contained seven components.

Table 2. Loading for rotated factor matrix.

Barriers	Component	Loading
Lack of trust-based relationship between project team.	1	0.808
Lack of supervision.	Attitude	0.753
Unwillingness of project team to accept and adapt the quality control system.		0.680
Lack of training and motivation.		0.569
Allocation of financial and human resources by top management is taken lightly.		0.547
Lack of awareness in benefits of implementing quality control techniques consistently.		0.503
Negligence on Technical Report in order to pursue construction works.	2	0.751
An unorganized Resources Plan.	Documents	0.713
The provided Quality Manual is difficult to comply and understand.		0.548
Top management rarely conducts management reviews on project quality.		0.534
Time consuming.	3	0.841
High operating cost.	Operations	0.780
Lack of knowledge in managing certain job.		0.520
Inadequate attention on setting quality policy by top management.	4	0.752
Inadequate top management support.	Management	0.606
Unclear description for each worker's roles and responsibilities as has been stated in Manpower Record.	5 Workers	0.709
Do not implement on a full scale.		0.658
The Progress Report is not up-to-date	6	0.858
Lack of information regarding defects and corrective action taken in Progress Report	Informative	0.654
Contractors do not comply with Method Statement	7	0.759
The unclear and unspecified information of Method Statement	Methods	0.748

- **Component 1: Attitude** – This principal consists of seven components. The first component consists of six variables that contribute to difficulty in implementing the quality control system that lacks trust-based relationship between project team, lack of supervision, unwillingness of project team to accept and adapt the quality control system, lack of training and motivation, allocation of financial and human resources by top management is taken lightly and lack of awareness in benefits of implementing quality control techniques consistently. The main point is lacking in human behavior itself to comprehend and understand the whole purpose of quality control techniques. It reflects the nature and culture of working in construction projects which must not be taken lightly.
- **Component 2: Documents** – The second factor consists of four components that need to be filed for construction projects. These four variables are negligence on Technical Report in order to pursue construction works, an unorganized Resources Plan, the provided Quality Manual is difficult to comply and understand, and top management rarely conducts management reviews on project quality. Although some may conclude that these files are not as important towards effective quality control system, due to its small purpose in construction projects may contribute to huge unseen problem if neglected.
- **Component 3: Operations** – The third factor consists of three components, namely time consuming, high operating cost, and lack of knowledge in managing certain job. It is due to the focal point that operates these quality control system that many have agreed on the

overshadowing in terms of time and cost which is unbearable if not followed. In addition, because of less knowledge in handling a specific job, this may be a hindrance towards effective implementation of these quality control systems.

- **Component 4: Management** – The fourth factor consists of two components, namely inadequate attention on setting quality policy by top management and inadequate top management support. These two components reflect the sole purpose of top management establishing in construction projects so as to assure that quality methods or procedures are followed. At the same time, developed or created a department helps in enhancing the quality control for construction projects.
- **Component 5: Workers** – This fifth factor consists of two components, namely unclear description for each worker's roles and responsibilities, as has been stated in Manpower Record and have not been implemented on a full scale. Those two factors briefly said that the responsibilities of workers involved in construction projects and due to vague picture of tasks, the workers may be in non-systematical orders. Moreover, the quality control technique is lack of encouragement since it has been taken lightly and lack of thorough implementation in large scale that emphasizes on the quality control.
- **Component 6: Informative** – The sixth factor consists of two components, namely Progress Report that is not up-to-date and lacks of information regarding defects and corrective action taken in the Progress Report. The Progress Report has been agreed by many experienced personnel in construction projects that give most impact in conducting the quality control because it is a consistent and frequent updated file for construction process. Furthermore, it contains the identified defects or any error that could bring mishap during construction. Briefly said, it gives important information and it is so informative for construction projects. If the information in the report is disorganized, it becomes the problem for better implementation of quality control techniques.
- **Component 7: Methods** – The seventh factor consists of two components, namely do not comply with Method Statement and the unclear and unspecified information of Method Statement. Method Statement comprises all procedures for construction projects and it gives a clear definition for each procedure. Due to lack in explanation in Method Statement, this can lead to misunderstanding of construction projects objective. Moreover, the descriptions in Method Statement give a lot of similarities in method for construction projects.

6 CONCLUSION

Most of the results have shown that the challenges comes from human factors, such as lack of supervision, lack of strong relationship among team members, unwillingness to accept new quality policy and lack of awareness in benefits to implement these techniques. Lack of financial supports from top management and lack of training are also part of the challenges to implement the quality control techniques. The result indicates that in order for these techniques to be implemented, there has to be commitment and dedication from all personnel to assure its implementation. In addition, disorganized and outdated information in documents implicated better implementation of these techniques. Even though the techniques used are advanced techniques, without an experience personnel to supervise it and proper documents, the system will not give any help in controlling the quality of construction projects.

References

Al-Najjar, S. M.; and Jawad. M. K. (2011). ISO 9001 Implementation Barriers and Misconception: An Empirical Study. International Journal of Business Administration, E-ISSN 1923-4015.

Balzarova, M.; Bamber, C. J.; McCambridge, S.; and Sharp. J. M. (2002). The Factors Affecting Successful Implementation of Process-Based Management: A UK Housing Associate Experience. 2nd International Conference on System Thinking, University of Salford, UK.

“Building Inspection Guideline,”.(2012).Rwanda Housing Authority.Republic of Rwanda.

Chua, Y. P. (2009). Partial Least Square, Factor Analysis and SEM Test. McGraw Hill,Malaysia.

Chin-Keng, T.; and Abdul-Rahman, H. (2011). Study of Quality Management in Construction Projects. Chinese Business Review, vol. 10(7), 542-552.

CIDB, Construction Industry Development Board. (2014). Retrieved from http://www.cidb.gov.my/ on April 22, 2015.

Crosby, P. B. (1996). Quality is free. Penguins Book Ltd.,London.

Dickenson, R., Campbell, D. and Azarov, V. (2004) ‘Quality management implementation in Russia: strategies for change’, International Journal of Quality & Reliability Management, Vol. 17, No. 1, pp.66–82.

Dory, J.; and Schier, L. (2002). Perspectives on the American quality movement. Business Process Management Journal, Vol. 8 Iss: 2, pp.117 – 139.

Escanciano, C., Fernandez, E. & Vazquez, C. (2001). Influence of ISO 9000 certification on the progress of Spanish industry towards TQM. International Journal of Quality and Reliability Management, (18)5,481-494. doi:10.1108/02656710110392629.

Hesham, A., and Magd, E. (2007). ISO 9001: 2000 Certification Experiences in Egyptian Manufacturing Sector:Perceptions and Perspectives. International Journal of Quality & Reliability Management, (25)2, 173-200. doi:10.1108/02656710810846934.

Ibrahim, M. S. (2005). The Barriers Affecting the Implementation of Quality Management System-ISO 9000 in Libyan Manufacturing Public Sector Organisations. Management Research Institute School of Management, Faculty of Business and Informatics, University of Salford, UK.

Low S. and Ling-Pan H., (2004), "Critical Linkage Factors between Management and Supervisors Staff for ISO 9001: 2000 Quality Management Systems in Construction". 9th International Conference on ISO 9000 & TQM.

Mande, S. (2013). What is the Inspection and Test Plan?.Retrieved from http://ngsuyasa.wordpress.com on October 27, 2015,

Noel, M. (2015). Project Quality Plan, ITP and Checklist Documents Construction. Retrieved from http://www.qualityengineersguide.com/ on October 12, 2015.

Ruin, E. J. (2009). The Art and Soul of Project Management. Selangor: Leeds Publication.

Sambasivan, M. and Yun, F. N. (2007). Evaluation of critical success factors of implementation of ISO 14001 using analytical hierarchy process (AHP): A case study from Malaysia. J. Cleaner Prod, 16,1424-33.

Shehu, Z. and Akintoye, A. (2010). Major Challenges to the successful implementation and practice of programme management in the construction environment: A critical analysis. International Journal of Programme Management, 28, 26-39. doi:10.1016/j.ijproman.2009.02.004.

Stevenson, T & Barnes, F.C. (2001). Fourteen years of ISO 9000: Impact, criticisms. Costs and benefits. Business Horizon, May-June, 45-51.

Tamaki, M. (2014). How Building is Inspected. Auckland Council.Vol.3.

Wauchope, C. (2014). Quality System. Retrieved from http://www.qualitysystem.com on November 20, 2015.

Wiele, T., Warden, J., Williams, R. & Dale, B. (2005). Perceptions about the ISO 9000 (2000) quality system standard revision and its value: the Dutch experience. International Journal of Quality and Reliability Management, (22)2,101-119. doi:10.1108/02656710510577189.

Integrated Solutions for Infrastructure Development
Edited by Lau, H. H., Tang, F. E., Ng, C. K., and Singh, A.
Copyright © 2016 ISEC Press

NEAR ZERO ENERGY HOUSE (NZEH) DESIGN VARIABLES FOR HOUSING DEVELOPMENT

YUSUF LATIEF, MOHAMMED ALI BERAWI, LENI SUPRIADI, JADE PETROCEANY, AYU HERZANITA and ARIO BINTANG KOESALAMWARDI

Civil Engineering Dept, Universitas Indonesia, Depok, Indonesia

Housing development, as part of the economic development must be supported by energy availability in order to obtain a sustainable growth. One of the approaches to support the renewable energy promotions is designing and building energy efficient housing. However, optimal design of energy efficient buildings is facing two conflicting requirements, namely cost effective consideration and minimum environmental impact. The high costs from energy efficient building such as the Near Zero Energy House (nZEH) is due to high price of materials and equipment used, such as solar panel, insulation and other supporting materials. Indonesia is situatued at the equator and received sunlight throughout the year. Nonetheless, this potential has not been fully discovered due to the high cost of the solar generated energy technology for housing. Moreover, this technology is not integrated with the main electricity network. Thus, the objective of this study is to identify the design variables for nZEH that suit the tropical climate condition in Indonesia. Experiments and Case Study are used for the study, and the validated design variables for nZEH, which includes building orientation, PV, fenestration, and passive design, will be the basis for optimum nZEH design.

Keywords: Near Zero Energy House, Energy, Tropical climate, Indonesia.

1 INTRODUCTION

Economic development of a developing country such as Indonesia depends on energy availability in a sustainable manner that requires attention to environmental aspects. In a smaller scope as housing, energy performance of a dwelling depends on the key element of carbon dioxide emissions reduction and energy savings. In the matter of improving the energy performance of a house, nZEH is proven as a cost-effective solution to overcome climate change and increase energy security. Moreover, it can support energy conservation which could address issues due to climate change such as the increase of global temperature and the reduction of fossil fuel energy (Rattanongphisat and Rordprapat 2014). In adoption of nZEH concept, using the energy collected from the sun, enable the building to generate its own saved energy which minimizes the utilization of energy from sources outside the building.

Located in the equator and illuminated by the sun throughout the year makes Indonesia an ideal location for development of renewable energy i.e., solar energy and its implementation and application in nZEH technology. However, this potential has not been fully explored due to the high cost of solar power generation systems technology (Zhou *et al.* 2001, Evola *et al.* 2014). Accordingly, the use of solar panels on the building technology in Indonesia, is still very rare. The practical uses of solar panel technology are limited to a street lamp, or a building located in remote places far from the PLN[1] grid.

This problem also poses a contradiction since design of energy efficient buildings typically uses expensive materials and technology, which directly affects the overall cost of construction (Milajic *et al.* 2013, Panão *et al.* 2013). It requires compromise and optimization of the use of "green" technology and construction cost, so that energy-efficient building technologies affordable to society. Several studies have conducted to examine the design optimization of nZEH in sub-tropical climate to four seasons climate environment. The purpose of this study is to identify design variables for nZEH to be optimized; particularly in the tropical climate therefore it uses design variables corresponding to the condition of Indonesia, which has a tropical climate.

2 LITERATURE REVIEW

2.1 Definition nZEH

Basically, the concept of Zero Energy Building (ZEB) is a building that able to meet its energy needs from cheap, clean, easy to obtain, and renewable source. A clearer definition states a ZEB concept generates renewable energy on site as much as or even exceed its annual energy consumption. A zero energy building can be defined in several different ways, usually depending on the project objectives. Thus explains more building owners think about energy spending these days. Torcellini *et al.* (2006) introduces four different types of commonly used definitions, namely: net zero site energy, net zero source of energy, net zero energy cost, and net zero energy emissions. Meanwhile Haslam and Farrell (2014) highlighted an energy-plus buildings and buildings that consume slightly more energy than they produce are called near-zero energy buildings (nZEB).

2.2 nZEH Practice in the World

One of the NZEH designs is Ecoterra nZEH in Canada which equipped by Building-Integrated Photovoltaics (BIPV). In addition, Pearl River Tower in China implies wind turbines, radiant slabs, microturbines, ground heat pump, ventilated facades, BIPV, and natural lighting control system in its building concept. Inherit the number of floors and spacious available, efforts are made to bring the concept of Zero Energy Building (ZEB) by reducing the energy account for the use of building lighting and air conditioning system. In order to reduce the energy consumption, the building was designed by combining site orientation, shape of the building, natural lighting and automated building control systems. Moreover, in Germany there are also Canadian Solar and Green City Energy that gather the installation of solar power plants with a capacity of 1 MWp on 16 rooftops, with a total area of 17,000 square meters in the city Haertensdorf, Germany.

Various design variables for nZEH used in previous studies are as follows (Boeck *et al.* 2013): (1) Renewable Energy Technologies, namely the use of PV (a) the ratio of the panel area - roof area and (b) Directions azimuth (orientation) of PV panels; (2) The shape of the building (a) building orientation, (b) the slope of the roof; (3) building envelope that affect insulation; (4) fenestration (windows and doors) with the components of the window glazing

---

[1] Indonesia State-Owned Electricity Company

type, window-to-wall ratio, type window sills, window direction, internal and external shading; (5) Heating-Ventilating-Air Conditioning (HVAC) which affected Natural Ventilation and mechanical Cooling / heating; (6) Lighting.

Boeck *et al.* (2013) also conducted a study to muster 65 scientific journals that discuss about energy efficiency in housing and apartment buildings. The results of this study are used as a reference in the preliminary variables comply to this research whereas resulted in 5 groups of variables other than the use of PV panels: building shape, building envelope, fenestration, HVAC, and lighting. Additional variables, i.e., PV panel, obtained from Bucking *et al.* (2013) who studied nZEH that uses PV panels.

3 METHODOLOGY

The research strategy used are experiment and case study - due to the absence of nZEH in Indonesia. Experiments conducted assessment of 400 design variable combinations calculated manually with the assistance of EnergyPlus™ software. EnergyPlus™ is a whole building energy simulation program that engineers, architects, and researchers use to model both energy consumption—for heating, cooling, ventilation, lighting and plug and process loads—and water use in buildings.

Variable designs which can be attributed to energy efficiency (adapted from Boeck *et al.* (2013)) are utilized in nZEH simulation, with parameters of tropical climate. The experiments conducted at the beginning of the study was the development of a house building model. Indicators of a schematic model of energy in a house is checked by EnergyPlus™ simulation using two sample models of houses – one with PV house and the other, non-PV house.

A number of 400 scenarios design variables that affect the energy buildings along with varieties of building azimuth, window-wall ratio (WWR) and glazing type were used in the experiment. The combination of design variables and its variations produce: data distribution and nZEH zone, conventional and PV house, optimal cost; the combination of design variables on the energy consumption; and the combination of design variables on cost of electricity. Therefore, the design variables that can be used and optimized, among others: (1) Renewable Energy Technologies, in the study, Photovoltaic (PV); (2) Building through the sub-variable of building orientation; (3) Fenestration (windows and doors) with subvariable window glazing type and window-to-wall ratio. These design variables may be combined with other design variables and varied into several sub-variables then used as input data in building a nZEH house model, where the software help simulate energy consumption of the building site as well as resources and NPV calculations of life cycle costs.

In addition to the experiment, a case study was conducted using Listijono's designated house in Pluit Residential House as a secondary data, an ASEAN-winning Energy Efficiency and Conservation (EE & C) Best Practice Competition in Buildings in ASEAN Energy Awards - 2014.

Pluit Residential House occupies a site area of about 250 square meters and was completed in March 2013 whereby equipped with three storeys and one flat roof for Solar cell, one cold water tank, one hot water tank, one Virtual Routing and Forwarding (VRF) Air Conditioning unit with hot water generation, three parabolas, one Pyramid meditation, two outlets voids, one sky lighting, Green roof and one Gazebo for leisure area. The total gross floor area of the building are 781 square meters designed by Aria Architect and John Budi H Listijono as its M&E Engineer (Listijono 2014).

4 RESULT AND DISCUSSION

4.1 Optimum Design Cost of nZEH

From the experiment, it was found that for optimum nZEH design costs, combination of azimuth is known to maximize the generation of electricity from the PV panel. Azimuth to intensify generation of electricity from the PV panel ranging from -10 azimuth to 10 from the North face orientation. Then, this azimuth is combined with other design variables to minimize the consumption of energy, i.e., electrical energy from air conditioning.

Variables that could minimize the use of air conditioning are WWR and glazing. Determination of the ratio of WWR to occupy the right selection type of glazing, enable minimum amount of heat energy derived from solar radiation, without lessen the natural lighting during the day (daylighting). For tropical areas such as Indonesia, WWR ranged from 15 to 20 percent believe to be enough for optimum natural lighting during the day without letting excessive intrusion the heat energy from solar radiation entering the house. Besides WWR, the choice of glazing may affect the amount of heat intensity of solar radiation to the house. Hence, to minimize the use of air conditioning for nZEH in the tropical area, a wise choice of glazing on the windows would limit the intensity of solar radiation.

4.2 nZEH Design Variable (for Tropical Climate)

Development of the building design with Near Zero Energy House (nZEH) in Indonesia is influenced by several variables. Based on the experiment, some variables which affect nZEH optimum design in Indonesia are building orientation direction, Panel Photovoltaic (PV), and fenestration. Geographically, Indonesia lies in 6°North Latitude – 11°South Lalitude and 95°Eastern Longitude – 141°Eastern Longitude, consequently this brought tropical climate to the nation. A prominent feature in the tropical climate area are high average daily temperatures and higher air humidity and relatively slow air flow for achieving thermal comfort. The use of energy in Indonesia focused on air cooling and dehumidification (reduction of air humidity) that commonly pursued using passive design to obtain a comfortable temperature inside the building (Wimmer 2013).

Moreover, the case study of Pluit Residential House exposes the passive design concept. Passive design concept acknowledge climate in Jakarta Area and natural ventilation with description of building is facing south (facing the road) where all the doors and windows are located as the openings of natural ventilation and designed from the early stage of using sky lighting to create natural ventilation to induce air speed of >0.5m/s in each floor system and day lighting system for non-air and air conditioned areas. Thus possible for having a house with natural ventilation of the temperature ranging from 24 °C-32.2 °C where relative humidity house ranging from 30% - 95%, to maintain the thermal comfort condition provided by air speed maintained at 0.4- 1.0m/s.

The air speed induced by this natural ventilation of sky lighting is very powerful that the curtains were blown with air speed of more than 1.0m/s even if the doors are fully opened. While the sky lighting produce day lighting in none air conditioned area, for 1st, 2nd and 3th floor respectively, it is also function to suck the hot air. With dimension of 1st and 2 nd floor void area are 4m x 5m, and the 3th floor void area is 1.5m x 5m help prevent the solar radiation from entering to the lower floor.

Other highlight to the design is the utilization of LED light that covers 98% of light. Using particular design make a big saving for having day lighting generated from the sky lighting, and two others void in this building whereby the light load in this building is only 3.4

Watt per square meter. As part of green technology, the solar PV panels contribute 25.7% of the total electric energy from the grid, therefore it is very important to monitor and maintain the performance of this solar PV Panel. Consequently, WIFI Plug installed in those 3 solar inverters to monitor and record the total electricity generated in these solar PV to keep the electric energy generated above 8 kWH/day with the average of 12 kWH/day (Listijono 2014).

Thus, from the experiments and case study, there are four design variables for nZEH that significantly suit the tropical climate and support energy efficiency, which are: building orientation (building azimuth), panel PV, fenestration dan passive design. These findings are also relevant with the studies of Wimmer (2013) and Torcellini *et al.* (2006). Firstly, to its tropical climate with high average temperatures caused by vertical positioning of the sun, Indonesia bound to have set up on building orientation toward the north 0° to get comfortable indoor temperatures and reduce expenditure due to electricity costs for air cooling system and lighting. Secondly, for tropical countries, such as Indonesia where the sun shines all year round, a renewable energy source derived from sunlight is seen as a potential energy. Solar radiation can be used to generate electricity (Photovoltaic) for building utilities, and heating the water (Photovoltaic Thermal). Thirdly, fenestration, materials and installation also contribute to the energy channeled through the windows, doors, or skylights, together with the airflow in the window components to the means of increased energy efficiency. Last but not least, passive design is apt to be applied in tropical countries since the passive cooling system obtain a comfortable temperature (thermal comfort) in the building (Wimmer 2013, Torcellini *et al.* 2006).

5 CONCLUSION

As a result of this research by looking at experiment of modeling 400 scenarios and case study of Pluit Residential Housing that agreed upon expert validation through literature, nZEH design variables that can be used in the tropical climate are: building orientation, PV, fenestration, and passive design. Whereas the variables comply proposition of the north 0° facing of building orientation; maintain PV Solar performances; fenetsration through WWR ranged from 15 to 20 percent; performed passive design that affirm the temperature ranging from 24 °C-32.2 °C, humidity house ranging from 30% - 95%, to maintain the thermal comfort condition provided that the air speed is maintained at 0.4- 1.0m/s with its natural ventilation. These variables can further be used to optimize the design NZEH in Indonesia in order to increase performance and achieve energy efficiency with optimum life cycle cost.

Acknowledgments

This paper is part of Research Grant from Ministry of Reseach, Technology and Higher Education Republic of Indonesia (Kemenristek) 2016.

References

Boeck, L. D., Audenaert, A. & Mesmaeker, L. D., Improving The Energy Performance Of Residential Buildings: A Literature Review, HUBRUSSEL RESEARCH PAPER, 2013.

Bucking, S. *et al.*, Design Optimization Methodology for a Near Net Zero Energy Demonstration Home, Quebec: s.n, 2013.

Evola, G, Margani, G, and Marletta, L, Cost-effective design solutions for low-rise residential Net ZEBs in Mediterranean climate, Energy and Buildings 68 (2014) 7–18.

Haslam, Michael P.G. and Farrell, Anthony, Natural Ventilation Strategies in Nearzero-Energy Building: A design primer for students and professionals, Across: Architectural Research through to Practice: 48 International Conference of the Architectural Science Association, pp. 619–630, 2014.

Listijono, J.B.H., Pluit Residential House, ASEAN Energy Efficiency and Conservation (EE&C) Best Practice Competition in Buildings, ASEAN Energy Awards - Vientiane, 2014.
Milajic, A., Milosevic, P. & Belkovic, D., Optimal Green Building Design Using Multi-Objective Genetic Algorithm. Reporting For Sustainability, pp. 119-123, 2013.
Panão, M.J.N, Rebelo, M.P., and Camelo, S.M.L., How low should be the energy required by a nearly zero-energy building? The load/generation energy balance of Mediterranean housing, Energy and Buildings 61 (2013) 161–171.
Rattanongphisat, Waraporn and Rordprapat, Wathanyoo, Strategy for energy efficient buildings in tropical climate, Energy Procedia, Elsevier, 52, 10 – 17, 2014.
Torcellini, P., Pless, S. & Deru, M., Zero Energy Buildings: A Critical Look at The Definition, Pacific Grove, California, National Renewable Energy Laboratory, 2006.
Wimmer, Robert, Adapting Zero Carbon House for Tropical Climate – Passive Cooling Design in the Philipines, SB13 Sustainable Buildings, Infrastructure and Communities in Emerging Economies, Manila, 1 – 8, 2013.
Zhou, Aiming, Alleng, G., and Byrne, John, Using Economic Incentives to Accelerate Development of Green Technologies, Green Building Workshop Sponsored by the Pacific Energy Center (China) and Environmental Market Solutions, Inc. (U.S.), Shanghai, China, July 9-10, 2001.

Integrated Solutions for Infrastructure Development
Edited by Lau, H. H., Tang, F. E., Ng, C. K., and Singh, A.
Copyright © 2016 ISEC Press

AN ENTROPY-BASED STUDY ON THE EFFECTIVENESS OF PREFABRICATION IN ACHIEVING SUSTAINABLE CONSTRUCTION

PETER S. P. WONG, LUKE EDMONSON, and MICHAEL PHILLIP KANELLOPOULOS

School of Property, Construction and Project Management, RMIT University, Melbourne, Australia

Prefabrication has been widely adopted in the construction projects in recent years. It has also been advocated as a greener and more sustainable approach of project delivery. However, with a wealth of evidence supporting other reasons as the drivers of using prefabrication, it is questionable whether the real goal of prefabrication is to achieve sustainable construction. This study aims to investigate the roles of prefabrication in fostering sustainable construction. The effectiveness of prefabrication on fostering sustainable construction was evaluated on the five aspects enlisted by the UK's Green Construction Board: Waste, Water, Carbon, Materials, and Biodiversity. An industry survey was conducted in Melbourne, Australia for data collection. 200 questionnaires were sent via email or post to a variety of fields within the construction industry. Entropy ranking analysis was adopted to analyze the effectiveness of using prefabrication in construction. The results suggest that prefabrication is effective in reducing construction waste and sourcing energy efficient materials. However, the utilization of prefabrication was found unsuccessful in reducing carbon emissions during the construction process. It is suggested that proactive actions should be taken at the design stage to unleash the potential of prefabrication in construction.

Keywords: Prefabrication, Sustainable construction, Effectiveness, Entropy analysis.

1 INTRODUCTION

Prefabrication can be defined as the completion of substantial parts of construction works prior to their installation on-site (Blismas *et al.* 2006). It was mainly applied in producing precast concrete components like facades, staircases, partition walls and slabs (Gibb 1999). This situation did not change until recent years when the advancement of building information modeling, 3-dimensional printing and volumetric preassembly technologies unleashed the potential of a more extensive use of prefabrication in construction projects (Pan *et al.* 2012). The advancement of prefabricated technology greatly reduces on-site assembly time as compared to conventional construction methods. Fully manufactured building facilities including; Bathroom pods, kitchen pods and modular units are now possible to be manufactured off-site (Jaillon and Poon 2014). By adopting more radical methods, completely finished modular buildings can be factory made and, once complete, transported to site for installation on a developed substructure (Lu 2007). The aforementioned methods reduce the requirements for skilled on-site labor associated with in-situ construction.

Other benefits of greater reliance on prefabrication have also been extensively reported (Blismas *et al.* 2006). These include optimizing integration between planning and design, improving end product quality control and improving site safety (Nadim and Goulding 2010). Interestingly, in recent years, researchers also advocated prefabrication as the effective means of achieving sustainable construction (Osmani *et al.* 2006). Aye *et al.* (2012) conducted an embodied carbon estimate and justified that prefabrication as a better alternative in regard to embodied energy savings. Based on a questionnaire survey conducted in the United Kingdom, Osmani *et al.* (2006) revealed that prefabrication fosters sustainable construction by reducing construction wastes and encouraging proactive planning for greener designs.

While some researchers advocated the use of prefabrication as a more sustainable option, ironically other literatures indicate that perhaps getting rid of laborers and reducing construction time are the true motives for pushing prefabrication forward (Blismas *et al.* 2006). The mix findings reported above make it interesting to investigate the effectiveness of prefabrication in achieving sustainable construction. This study helps provide insight into choosing appropriate strategy to achieve sustainable construction. This helps address how prefabrication can be truly embraced as one the effective strategies to achieve such goal. The study reported in this paper aims to investigate the effectiveness of prefabrication on fostering sustainable construction. The proceeding sections of this paper are organized as follows. Firstly, the measures of evaluating the effectiveness of prefabrication in sustainable construction are identified. Secondly, the research methodologies of this study are presented. Finally, the findings and its implications are discussed.

Table 1. Evaluators of effectiveness of prefabrication in achieving sustainable construction (adopted from Green Construction Board 2015).

Aspects	Respective Operational Statements:- *Implementing prefabrication in your project is effective to achieve the following goals:*	
Waste	WS1-	Waste minimization during the design and construction phase of the project
	WS2-	Enable greener designs to be established that create less waste
	WS3-	Limit the amount of packaging used to protect material
	WS4-	Allow for the recycling of previously used materials to thus avoid landfill
Water	WT1-	Allow for water targets to be set to minimize water usage both during the construction and operation of the project
	WT2-	Reduce water consumption through the life of the project
	WT3-	Allow for water usage to be tracked during the construction phase
	WT4-	Prevent water pollution at during construction and operational life
Carbon	CB1-	Allow for carbon targets to be set to monitor effective performance
	CB2-	Reduce carbon emissions in the design process by taking into account the whole life of the project
	CB3-	Allow for carbon emissions to be cut during the construction and manufacturing stage
	CB4-	Enable carbon saving by giving occupants the ability to cut their carbon usage.
Materials	MT1-	Allow for new and more environmentally friendly materials to enter the market
	MT2-	Enable material to go further through more effective designs and resource efficiency
	MT3-	Have a low environmental impact whilst improving the performance of the project
	MT4-	Allow for the responsible sourcing of materials
Biodiversity	BV1-	Provide better results when paired with sites of low ecological and agricultural value
	BV2-	Enable the assessment and monitoring of natural habitat
	BV3-	Incorporate features such as green roofs and walls that are able to protect the environment
	BV4-	Benefit from consultations of specialists who can develop long term management plans that meet the needs for people and wildlife

2 IDENTIFYING EVALUATORS OF SUSTAINABLE CONSTRUCTION

The concept of sustainable construction has been described in different contexts (Kilbert 1994; Ding 2008; Miller *et al.* 2015). Kilbert (1994) described it as 'the creation and responsible maintenance of a healthy built environment, based on ecological principles, and by means of an efficient use of resources'. Miller *et al.* (2015) stated sustainable construction is generally about 'delivering development by any definition that appropriately and equally benefits economic and social considerations, while concurrently minimizing related environmental impacts'. Ding (2008) argued that the interpretation of sustainable construction varies depending on the parties involved. For instance, a building owner may view this from an economical viewpoint whereas an occupant may be more interested in the in-door quality and safety impacts of buildings. The above review indicates that the concepts of sustainable development and sustainable construction are sometimes mixed up (Green Construction Board 2015). Sustainable development, while being a globally used expression of idea, combines a range of environmental issues with socio-economic issues (Kilbert 1994). In contrast, sustainable construction focuses on how sustainability can be achieved throughout the building life cycle (Green Construction Board 2015). In this aspect, the United Kingdom's Green Construction Board (2015) conducted a comprehensive review and developed measures that evaluate the effectiveness of sustainable construction in five aspects: Waste; Water, Carbon, Materials and Biodiversity. This study adopts the work of Green Construction Board (2015) and proposes that the effectiveness of prefabrication in achieving sustainable construction can be evaluated under these five aspects. The related operational statements are outlined in Table 1.

3 THE RESEARCH METHODOLOGIES

Data was collected by way of questionnaire. The questionnaire designed for this study consists of two parts. Part 1 includes questions designed to solicit the respondents' demographic information. Part 2 of the questionnaire consists of the 20 operational statements as shown in Table 1 for evaluating the effectiveness of prefabrication in achieving sustainable construction. Respondents were asked to assess the degree of agreement of the statements by using a five-point Likert scale (from 1:"Strongly Disagree" to 5: "Strongly Agree"). Respondents targeted for this study are directors, project managers and professional grade staff from consultants, main contractors, sub-contractors and suppliers which are listed in the latest edition of builders' directories and the official web pages of the professional institutes. On random basis, 200 questionnaires were sent via e-mail to the respondents who are working in the greater Melbourne region in Australia. To assist evaluation, mean scores of the five aspects suggested by the Green Board Construction (2015) are compared: waste, water, carbon, materials, biodiversity. The aspects scores are computed using the Eq. (1):

$$F_i = \frac{\sum_{j=1}^{n} A_{ij}}{n} \tag{1}$$

Where F_i is the Aspect score, A_{ij} is the mean score of the jth operational statement of Aspect i. For example, the Aspect score of "Waste" is computed as follows:
F_{waste} = [3.02 (WS1: Waste minimisation during the design and construction phase of the project) + 2.98 (WS2: Enable greener designs to be established that create less waste) + 3.41 (WS3: Limit the amount of packaging used to protect material) + 3.03 (WS4: Allow for the recycling of previously used materials to thus avoid landfill)] / 4 = 3.11.

Entropy ranking analysis is adopted to present in five major aspects the effectiveness of prefabrication in achieving sustainable construction. Entropy ranking analysis is a quantitative measure that compares the importance of the attributes in terms of their respective normalized weightings (Tang *et al.* 2012). Eq. (2) presents how priority ratings of the attributes can computed.

$$P_x = \frac{E_x}{e^{H_x}} = \frac{\sum_{i=l}^{n} S_i \cdot T_{mi}}{e\sum_{i=l}^{n} T_{mi} log_5(T_{mi})} \quad (2)$$

Where: P_x = Priority Rating of the operational statement (OS); E_x = Expected value of the OS; H_x = Entropy of the OS; i = A constant from 1 to n; S_i = scales of degree of significance; T_{mi} = probability of a scale; n = number of scales.

4 RESULTS AND DISCUSSIONS

A total of 200 questionnaires were sent to the identified respondents by e-mails. A total of 56 usable responses were used in the. The valid response rate, therefore, is 27%. Similar research study performed by Tang *et al.* (2012) had a sample size of 35. The sample size of this study is considered comparable to similar studies.

Among the five aspects suggested by the Green Construction Board (2015), 'Waste' received the highest mean at 3.30 out of 5. Second place was Materials (Mean = 3.02, S.D. = 0.92). The results indicate that the respondents generally agree that prefabrication is effective to achieve sustainable construction in terms of using more environmental friendly materials and waste reduction. However, mean scores of Water (Mean = 2.59, S.D. = 0.94), Carbon (Mean = 2.26, S.D. = 0.99) and Biodiversity (Mean = 2.12, S.D. = 0.85) are all lower than 3 on a 5-point Likert scale, this indicates that prefabrication may not be effective to achieve sustainable construction in these aspects. The results of Entropy ranking analysis further verify such findings (Table 2).

Those attributes related to Biodiversity (BV1 to BV4), Carbon (CB1 to CB4) and Water (WT1 to WT4) are ranked among the lowest. The results indicate that prefabrication is generally ineffective in reducing carbon emissions, water consumptions and maintaining the biodiversity of the environment. However, it is worth noting that previous studies that attempted to describe prefabrication as sustainable option usually focused on carbon reduction and water saving during the construction process (Jaillon and Poon 2014). Such assumptions failed to fully appreciate indirect embodied carbon emissions covering manufacture and transporting prefabrication units. Similar results were obtained when attributes of Biodiversity (BV1 to BV4) were tested, their almost the lowest normalized weightings and the low mean score indicate that prefabrication cannot effectively mitigate the ecological impact the construction project brings onto the environment. Indeed, a number of studies have already pointed out that despite construction practitioners might aware the negative impacts brought by the construction developments to the environment, they do not necessarily need to respond to the call from the governments and the societies to go green (Wong and Zapantis 2013).

The comparative high normalized weightings of WT1 to WT4 (ranged from 0.059 to 0.068 as shown in Table 2) indicate the ability to minimize waste as a key reason that cause prefabrication to be effective in sustainable construction. However, this may be due to the fact that the construction practitioners were concerned with the cost of construction waste disposal (Wong *et al.* 2015). However, on a positive note, attributes related to 'Materials' are seen to be the second most significant category behind 'Waste' through the entropy normalized weightings that are ranged from 0.055 to 0.059 (Table 2). The results perhaps indicate that the respondents see prefabrication as a tool to introduce new environmental friendly materials to enter the market that foster sustainable construction.

Table 2. Entropy rankings on effectiveness of prefabrication to achieve sustainable construction.

	Expected Value E_x	Priority Rating P_x	Normalized Weighting	Entropy Ranking
WS1-	3.59	3.49	0.068	1
WS2-	3.22	3.13	0.061	2
WS3-	3.11	3.03	0.059	4
WS4-	3.26	3.17	0.061	2
WT1-	2.50	2.43	0.047	12
WT2-	2.59	2.52	0.049	10
WT3-	2.57	2.50	0.048	11
WT4-	2.69	2.61	0.050	9
CB1-	2.46	2.39	0.046	13
CB2-	2.30	2.23	0.043	15
CB3-	2.33	2.27	0.044	14
CB4-	1.96	1.92	0.037	20
MT1-	3.15	3.05	0.059	4
MT2-	3.09	3.0	0.058	6
MT3-	2.91	2.82	0.055	7
MT4-	2.94	2.86	0.055	7
BV1-	2.02	1.96	0.038	19
BV2-	2.11	2.05	0.040	18
BV3-	2.15	2.08	0.040	18
BV4-	2.24	2.18	0.042	17

5 THE CONCLUDING REMARKS

Prefabrication has become a more popular construction alternative in recent years. However, prefabrication may not be adopted deliberately for sustainable construction purpose. Some studies pointed out that prefabrication can be an effective way to foster sustainable construction. The results of this study indicate that such perceptions may be valid in terms of reducing construction wastes and sourcing more environmental friendly materials to building prefabrication components. However, prefabrication might not be conducive to sustainable construction in terms of reducing carbon, water consumption as well as maintaining the biodiversity of the surrounding environment. With due caveats on the limitations of the sample size as well as the geographical constraints, this finding may be though provoking for those who believe prefabrication is a panacea for attain sustainable construction. The true motives of using prefabrication can be unrelated to sustainable construction. The findings of this study help the policy makers, practitioners and researchers to focus on the true motives that can foster sustainable construction more effectively.

References

Aye, L., Ngo, T., Crawford, R. H., Gammampila, R., and Mendis, P., "Life cycle greenhouse gas emissions and enegy analysis of prefabricated reusable building modules. " *Energy and Buildings,* 47, 159-168, 2012.

Blismas, N., Pasquire, C., and Gibb, A. "Benefit evaluation for off-site production in construction." *Construction Management and Economics*, 24(2), 121-130, 2006.

Ding, G. K. C., " Sustainable construction—The role of environmental assessment tools." *Journal of Environmental Management,* 86(3), 451–464, 2008.

Gibb, A. G. F., *Off-site Fabrication: Prefabrication, Preassembly and Modularisation*, Whittles Publishing, Caithness, Scotland, 1999.

Green Construction Board, 2015, "20 steps to greening our industry" The Green Construction Board, United Kingdom. Retrieved from http://www.greenconstructionboard.org/images/gcbtoptips.pdf.

Jaillon, L., and Poon, C. S., "Life cycle design and prefabrication in buildings: A review and case studies in Hong Kong." *Automation in Construction*, 39, 195-202, 2014.

Lu, N., 2007, "Investigation of designers and general contractors perceptions of off site construction techniques in the United States construction industry." PhD Dissertation, Graduate School of Clemson University. Retrieved from: http://tigerprints.clemson.edu/cgi/viewcontent.cgi?article=1081&context=all_dissertations.

Miller, D., Doh J-H., and Mulvey, M., "Concrete slab comparison and embodied energy optimisation for alternate design and construction techniques." *Construction and Building Materials,* 80, 329-338, 2015.

Nadim, W. and Goulding, J. S., "Offsite production in the UK: the way forward? A UK construction industry perspective. ." *Construction Innovation*, 10(2), 181-202, 2010.

Osmani, M., Glass, J., and Price. A., Architect and Contractor Attitudes to Waste Minimisation, Waste and Resource Management, 2(1), 65-72, 2006.

Pan, W., Gibb, A. G. F., and Dainty, A. R. J., "Strategies for integrating the use of off-site production technologies in house building." *Journal of Construction Engineering and Management*, 138(11), 1331-1340, 2012.

Tang, L. C. M., Atkinson, B. and Zou, R. R., "An entropybased SWOT evaluation process of critical success factors for international market entry: a case study of a medium-sized consulting company." *Construction Management and Economics*, 30(10), 821-834, 2012.

Wong, P. S. P., Lindsay, A., Crameri, L., and Holdsworth, S., "Can energy efficiency rating and carbon accounting foster greener building design decision? An empirical study." *Building and Environment*, 87, 225-264, 2015.

Wong, P. S. P. and Zapantis, J., "Driving adoption of carbon reduction strategies in the construction projects – the moderating role of organizational culture." *Building and Environment*, 66, 120-130. ISSN: 0360-1323, 2013.

Integrated Solutions for Infrastructure Development
Edited by Lau, H. H., Tang, F. E., Ng, C. K., and Singh, A.
Copyright © 2016 ISEC Press

THE NEW TOLL ROAD FINANCING MODEL DETERMINATION

AYOMI RARASATI[1], LUKAS SIHOMBING[1], ANDREAS WIBOWO[2], and YUSUF LATIEF[1]

[1]*Dept of Civil Engineering, Universitas Indonesia, Depok, Indonesia*
[2]*School of Graduate Program-Civil Engineering, Parahyangan Catholic University, Bandung, Indonesia*

There are three main problems in Indonesia toll road financing model. First, it is the lack of state budget and delayed funding distribution for land acquisition. The second problem is the lack of financial ability of the winning company which leads to financial closing failure. The third problem is related to the land acquisition process that a special purpose company has to obtain for 100% before it can propose a loan from a bank. Therefore, it is essential to develop a new financial modeling to be implemented in Indonesian toll road development. Seven financial models were identified as having been implemented successfully and effectively. These models are earmarked tax revenue, deep discount bonds, take out financing, tax increment financing, land leases, deferred debt, and private donation. However, these models have never been utilized for toll road projects in Indonesia. Therefore, this research aims to explore those seven financial models and determine the most suitable one to be implemented in Indonesia. Literature review and in-depth interview with the key stakeholders were conducted to collect the data. The results subsequently show that there are four models that have never been implemented in Indonesia. However, the combination of deep discount bond and land leases is the most suitable financing model for Indonesian toll road industry.

Keywords: Financial modeling, Deep discount bonds, Land leases, Earmarked tax revenue, Take out financing, Tax increment financing, Deferred debt, Private donation.

1 INTRODUCTION

In 2005, the Government of Indonesia, through the Toll Road Authority, published Indonesian toll road development planning for 2005-2025. It was planned that the total length of Indonesian toll road would be 6,115 km with approximately 72.3 billion USD total costs. However, according to the Indonesian Toll Road Authority in 2015, there were only 949 km of toll road operated and the increased of toll road development since 1978 was only 8.79%.

Presently, the financing model used in Indonesia is based on Law No. 2 Year 2012 on Land Acquisition for public interest development, Government Regulation No. 15 Year 2005 on Toll Road, and Government Regulation No. 43 Year 2013 on the second change of Government Regulation No. 15 Year 2005. The financing is regulated based on the appointment of a state owned enterprise to use the budgeting from central and local government for land acquisition purposes.

In order to identify the problems in the existing financial model, interviews were conducted with resource persons representing both the government and the state owned enterprises. There are three main problems in Indonesia toll road financing model. Firstly, it is the lack of state budget and delayed funding distribution for land acquisition. The second problem is the lack of financial ability of the winning company to provide equity which leads to its financial closing failure. The government requires such company to provide a three-year financial statement, information on its involvement in other projects and its financing track record. These requisitions can make it more complicated for a company to invest in toll road. The third problem is related to the land acquisition process that a special purpose company has to obtain for 100% before it can propose a loan from a bank. In addition, the bank imposes high interest rates, short tenor and high draw down requirements.

To optimize the existing financial models, they have to be developed in a way that can accelerate Indonesian toll road development. In order to fill the funding gap, there are previous relevant studies in toll road financing models, namely earmarked tax revenue system (Favero 2006, Nambu 2006, Blackwell *et al.* 2006, Doll and van Essen 2008, Jackson 2008, Zhao and Chengxin 2011, Kunz and O'Leary 2012, Nguyen-Hoang 2014); deep discount bonds (Kistner 1991, Dailami and Hauswald 2003, Tourrucôo 2004, Lakhmani and Sikroria 2012, Ovsiannykova 2014), take out financing (Srivastava 2011, IIFCL 2014); tax increment financing (Farris and Horbas 2009, Xueming 2012), land leases (Xueming 2012, Weidong and Xiaolong 2012), deferred debt (Xueming 2012), and private donation (Xueming 2012, Tecklin and Sepulveda 2014). This research aims to explore those seven financial models and determine the most suitable model to be implemented in Indonesia.

2 LITERATURE REVIEW

Earmarked tax revenue is a financing obtained from pre-construction and construction phases by utilizing the tax revenue from vehicle and fuel taxes. The tax revenue is used to finance land acquisition, cost of finance or legal fees during the pre-construction phase (Favero 2006, Nambu 2006, Blackwell *et al.* 2006, Doll and van Essen 2008, Jackson 2008, Zhao and Chengxin 2011, Kunz and O'Leary 2012, Nguyen-Hoang 2014). Several countries have successfully implemented earmarked tax revenue model to finance their infrastructure development: United Kingdom (Wilkinson 1994), France (Favero 2006), Japan (Nambu 2006), Switzerland and German (Doll and van Essen 2008), USA (Kunz and O'Leary 2012, Jackson 2008, Ulbrich 2011, Nesbit and Kreft 2009), and China (Zhao and Chengxin 2011).

Deep discount bond is obtained from financial markets in the form of zero coupon bonds (Kistner 1991, Dailami and Hauswald 2003, Tourrucôo 2004, Lakhmani and Sikroria 2012, Ovsiannykova 2014). The use of deep discount bonds was the most innovative financing model in 1980s in USA (Varma and Chamber 1990) and it was also successfully implemented in India (Lakhmani and Sikroria 2012).

Take out financing is a financing that is taken over by a financial institution of a lender due to the long construction period (up to 10-15 years) while the lender is only able to finance up to 5 years (Srivastava 2011, IIFCL 2016).

Tax increment financing is obtained from the incremental value of the building near the toll road before the toll road is constructed and when the toll road is operated. The value is calculated from the land and property tax value and the taxable value of the building (Farris and Horbas 2009, Xueming 2012). Tax increment financing was first successfully used in California and is currently adopted in 40 states in USA (Johnson and Scott 2004, Farris and Horbas 2009).

Land leas financing model is obtained from the rental use of toll roads such as the use of toll road's land and property during the operation phase (Xueming 2012, Weidong and Xiaolong

2012). The model was successfully implemented in Beijing, China when the land leases increased by approximately 39.7% from 2003 to 2010 (Weidong and Xiaolong 2012).

Deferred debt is obtained by suspending the debt payment and the payment is paid when there is positive earning during operational phase (Xueming 2012). Private donation can be implemented when donation is granted from industrial or residential area because they gain access to the toll road (Xueming 2012, Tecklin and Sepulveda 2014).

3 METHODOLOGY

A qualitative method is used in this research to find the most suitable financial model while in-depth interviews were conducted in data collection. Additionally, this research also collect secondary data from associated documents. The aim of the interviews was to explore the contents of the study based on the historical perspective of the respondents, to explore in detail based on the reflection of the respondents' experience, and to get an understanding based on the experiences of the respondents (Seidman 2006).

Interviews were conducted to build an in-depth understanding on toll road financial models that will be used for the toll road construction in Indonesia as well as to determine the perception of the respondents based on their knowledge and experience. There are two different approaches to analyze qualitative data, namely inductive and deductive approaches. The deductive approach is conducted by coding and emphasizing on themes derived from the literature, while the inductive approach is based on the theory and more focused on what is actually found in the study (Thomas 2006).

4 RESULTS

Resource persons were experts on infrastructure financing, especially in toll road financing. They were first contacted via e-mail and instant messages (short message service) to confirm their willingness to participate in this research and their availability to be interviewed. When agreed, the interview was conducted at the location specified by the interviewee. All interviews were conducted during the period between April 29, 2015 and May 15, 2015.

Overall, there are five expert practitioners who are willing to participate in this study as respondents. There is a lack of experts in the field of road infrastructure funding, especially in the area of toll road, thus respondents were selected based on their experience and track record. In accordance to a statement letter signed by the respondents, as well as for maintaining the respondents' confidentiality, the participated respondent was given the code name R. The details of these respondents are shown in Table 1.

Table 1. Respondents' background.

No	Code	Experience (years)	Education	Sector
1	R1	15	PhD	Infrastructure financing
2	R2	20	Master	Infrastructure financing
3	R3	10	Master	Toll road investment
4	R4	8	Master	Infrastructure financing
5	R5	20	PhD Candidate	Toll road investment

Interviews with a qualitative research approach can be reported in a structure where the questions are followed by comments from the respondent, and then put in a description followed by the quote sources, or to quote and interpret the results of interviews with warm words including the use of tables. In this study, the analysis of the interviews will be shown as long as it is considered important to select interviewees' quotes. In order to have a focus interview, any proposed financial models were discussed based on the respondents' expertise as shown in Table 2.

Table 2. Interview topic for each respondent.

Financial Model	R1	R2	R3	R4	R5
Earmarked Tax Revenue	☑	☑	☐	☑	☐
Deep Discount Bonds	☑	☑	☐	☐	☐
Take Out Financing	☑	☑	☑	☐	☐
Tax Increment Financing	☑	☑	☐	☑	☐
Land Leases	☑	☐	☑	☑	☑
Deferred Debt	☐	☐	☑	☐	☐
Private Donation	☐	☐	☑	☐	☐

The results analysis of the interviews from each model was transformed into a matrix as shown in Table 3.

Table 3. Toll road financial model in Indonesia.

	Has been implemented in Indonesia	Never been implemented in Indonesia	Existing regulations	Institutional support	Budgeting / revenue management
Earmarked Tax Revenue	☐	☑	None for toll road	None for toll road	Mixed in government budgeting
Deep Discount Bonds	☐	☑	☑	☑	☑
Take Out Financing	☑	☐	☑	☑	☑
Tax Increment Financing	☐	☑	None for toll road	None for toll road	Mixed in government budgeting
Land Leases	☐	☑	☑	None for toll road	Mixed in government budgeting
Deferred Debt	☑	☐	☑	☑	☑
Private Donation	☑	☐	☑	☑	☑

Based on Table 3, of the seven funding models proposed, three models (take out financing, deferred debt and private donation) have been applied in Indonesia while four models (earmarked tax revenue, deep discount bonds, tax increment financing, and land leases) have never been applied in Indonesia. In terms of regulations, five models (deep discount bonds, take out financing, land leases, deferred debt, and private donations) have existing regulations, while two models (earmarked tax revenue and tax increment financing) have existing regulations that exclude the toll road sector. With regards to institutional support, four models (deep discount bonds, take out financing, deferred debt, and private donations) are supported, while three models (earmarked tax revenue, tax incremental financing, and land leases) have no special institution supporting the toll road sector. Furthermore, in terms of budgeting/revenue management, four models (deep discount bonds, take out financing, deferred debt, and private donations) can be managed privately, while three models (earmarked tax revenue, tax increment financing, and land leases) are managed together within the government budgeting.

From the above analysis, four financial models (earmarked tax revenue, deep discount bonds, tax increment financing, and land leases) are potential and have never been applied for toll roads in Indonesia. In addition, two financial models (earmarked tax revenue and tax increment financing) have no special regulations for the toll road and three financial models (earmarked tax revenue, tax increment financing, and land leases) have no special institutional support for toll roads.

5 CONCLUSION

Based on the results, to fill a funding gap for the toll roads development in Indonesia, there are four possible financial models that have not been applied in Indonesia. However, the most probable models are deep discount bonds and land leases. Both funding models were chosen because regulations exist to support the model, namely Law No. 8 Year 1995 on Capital Markets and Law No. 25 Year 2007 on Investment. Additionally, there is a financial service agency available to support these financial models. For a developing areas, land prices are still relatively low, thus the land acquisition for the toll road can be wider. Therefore, there is a potential to design a combination model between deep discount bonds and land leases for toll road financial model. This deep discount bonds can be applied as project bonds, where the bonds are issued by the Project Company to bond investors through Lead Arranger and Credit Rating Agency, and repayment will be paid during the operation phase when they receive revenue from toll road users. Meanwhile, land leases can be applied during the operation phase, where the land authority receives money from tenants. The money collected by this land authority can be used for land acquisition funding to other lands for the toll road.

Acknowledgments

The authors would like to express gratitude to Universitas Indonesia for granting a support through PITTA scheme in order to assist the researchers in completing this research.

References

Blackwell, C., Crotts, J. C., Litvin, S. W., and Styles, A. K., Local Government Compliance with Earmarked Tax Regulation, *Public Finance Review*, 34 (2), 212-228, March, 2006.

Dailami, M., and Hauswald, R., *The Emerging Project Bond Market: Covenant Provisions and Credit Spreads*, Policy Research Working Paper, Vol. 3095, World Bank, University of California, 2003.

Doll, C., and van Essen, H. *Road Infrastructure Cost and Revenue In Europe*, Report to Internalisation Measures and Policies for all external cost of Transport (IMPACT), 2008.

Farris, S., and Horbas, J., Creation vs. Capture: Evaluating the True Costs of Tax Increment Financing, *Journal of Property Tax Assessment & Administration*, 6 (4), 23, 2009.

Favero, G., *Fiscal Decentralization and Urban Public Transport*, MIT Thesis, 2006.

India Infrastructure Finance Company Limited (IIFCL), *Takeout Finance Scheme for Financing Viable Infrastructure Projects*, retrieved from http://www.iifcl.org/Content/mtfc.aspx on December 24, 2014.

Jackson, J. J., *The Political Economics of Earmarked Taxation*, Washington University PhD Thesis, 2008.

Johnson, T. G., and Scott, J. K., A Comprehensive Approach to The Assessment of Tax Increment Financing (TIF) Projects, *Journal of Public Budgeting, Accounting & Financial Management*; 16 (3), 395, Fall, 2004.

Kistner, W. G., Uncertain Times Point to Zero Coupon Bonds, *Healthcare Financial Management*, 45 (4) April, 1991.

Kunz, K. and O'Leary, S., The Importance of Federal Earmarks to State Coffers: An Examination of Distribution Trends over the Decade, *Journal of Public Budgeting, Accounting & Financial Management*, 24 (4), 578-608, 2012.

Lakhmani P., and Sikroria, R., Infrastructure Financing Instruments with a Special Emphasis on Highways and Road, *International Journal of Management Research and Review*, 2 (9), 1668-1677, 2012.

Nambu, T., *History of Road Development, Finance and Investment in Japan*, 2006.

Nguyen-Hoang, P., Volatile Earmarked Revenues and State Highway Expenditures in The United States, *Transportation*, 42, 237-256, 2014.

Nesbit, T. M., and Kreft, S. F., Federal Grants, Earmarked Revenue, and Budget Crowd-Out: State Highway Funding, *Public Budgeting & Finance*, Summer, 2009.

Ovsiannykova, I. Project Bonds in Financing Public-Private Partnerships in Ukraine, Vìsnik, KiïVs'Kogo Nacìonal'Nogo Unìversitetu Ìmenì Tarasa Ševčenka, *Ekonomìka*, 11 (164), 45-52, 2014.

Seidman, I., *Interviewing as Qualitative Research: A Guide for Researchers in Education and The Social Sciences*, Third Edition, Teachers College Press, Teachers College, Columbia University, New York and London, 2006.

Srivastava, V. Project Appraisal in Infrastructure Financing by Indian Banks, *The IUP Journal of Infrastructure*, 9 (1), 2011.

Tecklin, D. R., and Sepulveda, C., The Diverse Properties of Private Land Conservation in Chile: Growth and Barriers to Private Protected Areas in a Market-friendly Context, *Conservation and Society*, 12 (2), 203-217, 2014.

Thomas, D. R., A General Inductive Approach for Analyzing Qualitative Evaluation Data, *American Journal of Evaluation*, 27 (2), 237-246, 2006.

Tourrucôo, F., *Perturbation Methods in Mathematical Finance: Zero-Coupon Bonds and Equivalent Volatilities*, PhD Thesis, University of Delaware, 2004.

Ulbrich, H. H., *Public Finance in Theory and Practice*, Second Edition, Routledge Taylor & Francis Group, London and New York, 2011.

Varma, R. S., *Capital Structure Changes and Shareholder Wealth: The Case of Deep Discount Debt*, PhD Thesis, The Pennsylvania State University, 1987.

Weidong, Q., and Xiaolong, L., Assessing the Performance of Chinese Land Leas Auctions: Evidence from Beijing, *The Journal of Real Estate Research*, 34 (3), 291-310, July-September, 2012.

Wilkinson, M., Paying for Public Spending: Is There a Role for Earmarked Taxes?, *Fiscal Studies*, 14 (4), November, 1994..

Xueming, C., Managing Transportation Financing in an Innovative Way, *Management Research and Practice*, 4 (3), 5-17, 2012.

Zhao, Z. J., and Chengxin, C., Funding China's Urban Infrastructure: Revenue Structure and Financing Approach, *Public Finance and Management*, 11 (3), 284-305, 2011.

Integrated Solutions for Infrastructure Development
Edited by Lau, H. H., Tang, F. E., Ng, C. K., and Singh, A.
Copyright © 2016 ISEC Press

MULTIPLE CRITERIA EVALUATION OF ASSEMBLING BUILDINGS FROM STEEL FRAME STRUCTURES

RUTA MINIOTAITE

Dept of Civil Engineering Technologies, Kaunas University of Technology, Kaunas, Lithuania

Steel frame structures are often used in the construction of public and industrial buildings. They are used for: all types of slope roofs; walls of newly-built public and industrial buildings; load bearing structures; roofs of renovated buildings. The process of assembling buildings from steel frame structures should be analysed as an integrated process influenced by such factors as construction materials and machinery used, the qualification level of construction workers, complexity of work, available finance. It is necessary to find a rational technological design solution for assembling buildings from steel frame structures by conducting a multiple criteria analysis. The analysis provides a possibility to evaluate the engineering considerations and find unequivocal solutions. The rational alternative of a complex process of assembling buildings from steel frame structures was found through multiple criteria analysis and multiple criteria evaluation. In multiple criteria evaluation of technological solutions for assembling buildings from steel frame structures by pairwise comparison method the criteria by significance are distributed as follows: durability is the most important criterion in the evaluation of alternatives; the price (EUR/unit of measurement) of a part of assembly process; construction workers' qualification level (category); mechanization level of a part of assembling process (%), and complexity of assembling work (in points) are less important criteria.

Keywords: Steel frame structure, Technological solution, Assembling work, Network model, Multiple criteria evaluation.

1 INTRODUCTION

Steel frame structures are often used in the construction of public and industrial buildings. Buildings from insulated metal structures have the following advantages: excellent architectural look that meets the strictest requirements for modern buildings; simple and fast mounting enabling to erect a big building in a very short time calculated in months or even weeks; good thermal-technical characteristics as modern metal structures enable to avoid thermal bridging at connections and the thickness of heat insulating layer is selected according to applicable standards; good operation characteristics as reliable elements ensure the required tightness of the building, metal elements are protected from corrosion by more than one layer of coating.

Modern steel frames are made of the following components: bearing metal elements of the structural framework (partitions and load bearing elements); roof, roof-wall and wall cladding; partition insulation packages (thermal insulation, noise insulation, wind and vapour barriers).

Elements of partitions structures and external wall insulation systems are presented in Figures 1, 2.

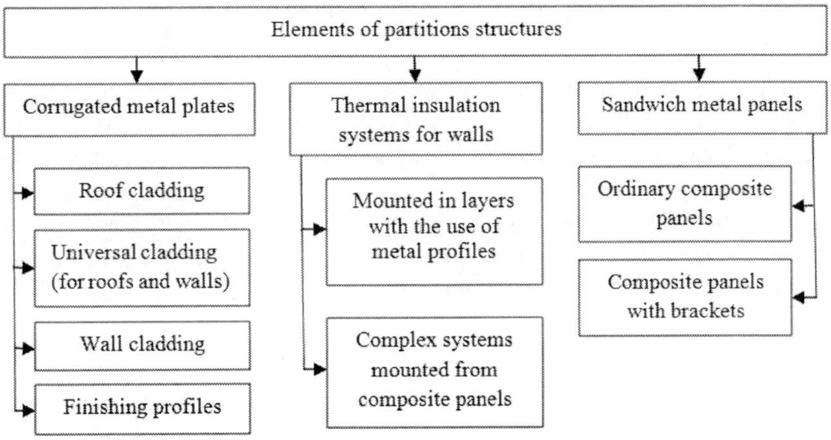

Figure 1. Elements of partitions structures.

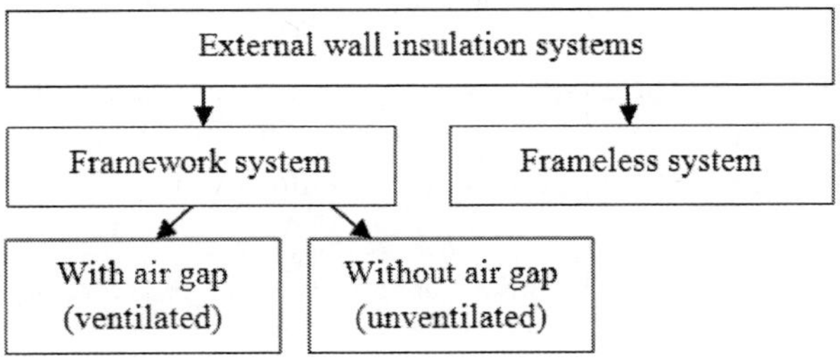

Figure 2. External wall insulation systems.

Methods of insulating walls of existing industrial buildings:

- Thermal insulating layer is fixed directly to the façade that is afterwards finished by reinforced (with mesh or fibre) plastering. It is a frameless insulation system.
- Wood or metal studs are fixed to the wall, thermal insulation is placed into the spaces between studs and various finishing panels are fixed to the studwork. It is a framework system (ventilated). It is a framework system (ventilated).
- Finished elements made of joined thermal insulation and finishing layers (composite panels) are fixed to the wall.

2 DESIGNING THE NETWORK MODEL FOR ALTERNATIVE MOUNTING SOLUTIONS IN STEEL FRAME BUILDING

2.1 Making Combinations of Complex Processes

The main stages for designing network models for steel frame building are as follows: buildings:

- making combinations of complex processes used in steel frame building;
- finding possible alternatives of partial processes in steel frame building;

- finding technological links between the alternatives of partial processes in steel frame building;
- drawing networks for steel frame building technology.

In terms of system approach, steel frame building is a complex process made of various partial (work) processes that can be completed by different working methods, which are determined by work object characteristics, construction materials, work tools, equipment, number of workers and their qualification. Each of the above factors is described by certain technical and economic indicators.

The complex process of steel frame building can be divided into the following partial technological processes: *F* – steel frame mounting, *B* – mounting of bearing structures (beams and trusses); *I* – installing connections (Figure 3).

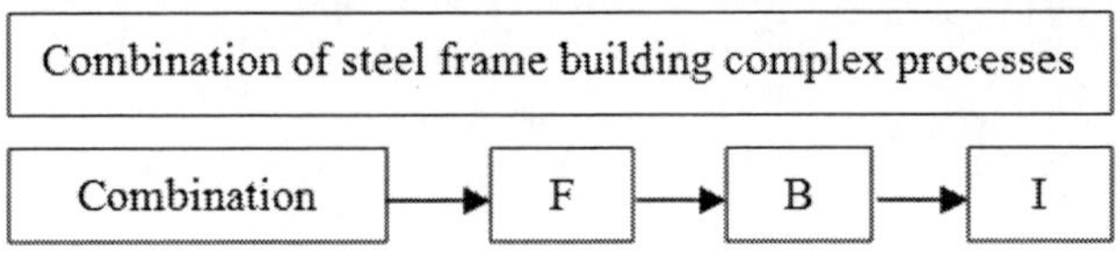

Figure 3. Diagram of steel frame building complex process combination.

The complex process of roof erection can be divided into the following partial technological processes: *P* – mounting of purlins, *IL* – inner layer mounting, *S* – sound insulation mounting, *T* – thermal insulation mounting, *W* – wind insulation mounting, *FL* – finishing layer (Figure 4).

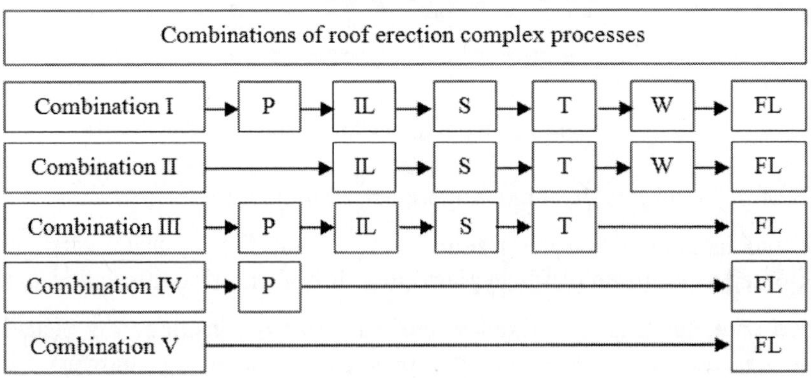

Figure 4. Diagram of roof erection complex process combination.

2.2 Alternatives of Partial Processes in Steel Frame Building

The analysis of steel frame building reveals many technological systems of this complex process. Many alternative solutions can be found by changing work methods. Different work methods result from the change of building structures, work tolls and mechanisms used.

Available alternatives of work processes used for the building a metal framework are presented in the Table 1 below.

Table 1. Alternatives of work processes used for the building a metal framework.

Partial process alternative code	Title of partial process in steel frame building: partial process alternatives (short description)
	Mounting of columns:
F1	manually, bolted connections
F2	manually, welding
F3	mechanically, crane lifting, bolted connections
F4	mechanically, crane lifting, welding
F5	mechanically, hoist lifting, bolted connections
F6	mechanically, hoist lifting, welding
	Mounting of bearing structures:
B1	mounting beams manually, bolted connections
B2	mounting beams manually, welding
B3	mounting beams, crane lifting, bolted connections
B4	mounting beams, crane lifting, welding
B5	mounting beams, hoist lifting, bolted connections
B6	mounting beams, hoist lifting, welding
B7	mounting trusses, hoist lifting, assembling on the ground
B8	mounting trusses at designated height, hoist lifting
B9	mounting trusses, assembling on the ground, crane lifting
B10	mounting trusses at designated height, crane lifting
	Making connections:
I1	making connections manually, welding
I2	making connections manually, bolting
I3	making connections, hoist lifting, welding
I4	making connections, hoist lifting, bolting
I5	making connections, crane lifting, welding
I6	making connections, crane lifting, bolting

2.3 Technological Links between Partial Processes in Steel Frame Building and Network Model Design

Network modelling of construction processes significantly improves operations management, work culture and efficiency, shortens the commissioning term, and reduces construction costs.

The method is beneficial only if the following conditions are met: well-organised data collection, transfer and processing; expedient system of decision making and task delegation to operators supported by computerised network and specialists highly competent in network planning and control.

The graphical representation of the variety of works in a construction process with marked technological and organizational links is called a network model (Janusaitis 1998). The network model with computed space, time and technological parameters is called a network. Alternatives of separate (partial) technological processes and dependency relationship between them must be set in the design of network technological model. Technological links are made for three building alternatives: steel frame (Figure 5) and roof (Figure 6). The quantitative and qualitative characteristics are shown in Table 2.

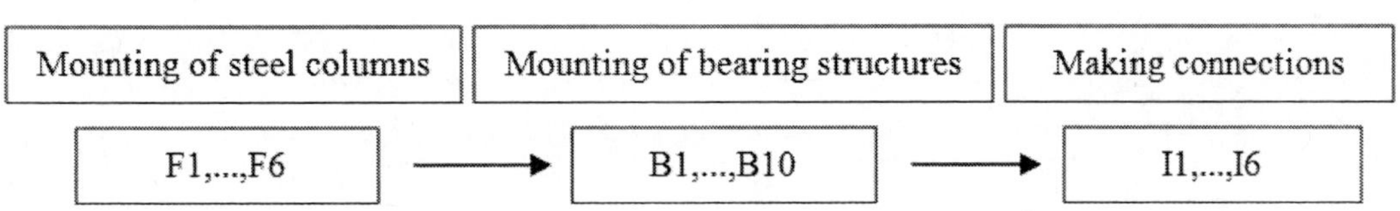

Figure 5. Technological links between partial processes in steel frame building.

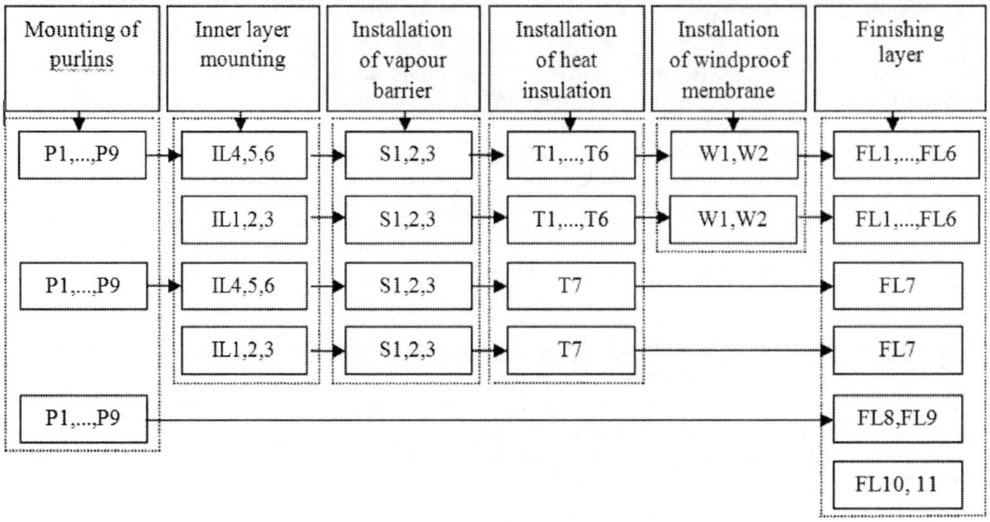

Figure 6. Technological links between partial processes in roof building.

Table 2. Quantitative and qualitative characteristics.

Criteria	Unit of measurement	Definition
The price of a part of assembly process	(EUR/unit of measurement)	Price (in EUR) per conventional unit of measure of the analysed partial process alternative
The qualification level of construction workers	category	Evaluation criterion indicating the workers' ability to do the work of the relevant complexity
Mechanization level of a part of assembling process	%	
Durability	in year	Life cycle of steel frames
Complexity of assembling work	in points	Complexity of work evaluation criterion

3 MULTI-CRITERIA EVALUATION OF ALTERNATIVE SOLUTIONS IN STEEL FRAME BUILDING

The rational alternative for steel frame building is found from the diagram presented in Fig. 7:

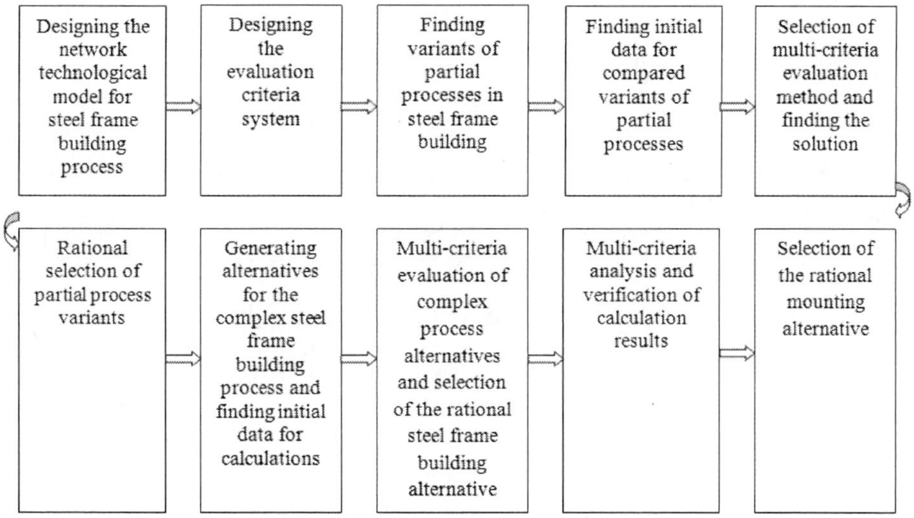

Figure 7. Algorithm for the selection of the rational alternative for the complex process.

The method of proximity to an ideal point can be applied to finding the most effective engineering solution alternative (Miniotaite 2016).

4 CONCLUSIONS

Taking into consideration the factors that affect the rationality of steel frame building process solutions it is feasible to do the technological modelling of multi-criteria evaluation of alternative buildings.

A criteria system for the evaluation of alternative partial processes must be designed in order to find the rational steel frame building alternative. Criteria values and importance may be subsequently adjusted taking account of priorities and the current situation.

In practice it is possible to find the most rational technological alternatives for metal framework, roof and wall structures separately with the help of network technological model and multi-criteria analysis of steel frame building solutions.

In multiple criteria evaluation of technological solutions for assembling buildings from steel frame structures by pairwise comparison method the criteria by significance are distributed as follows: durability is the most important criterion in the evaluation of alternatives; the price (EUR/unit of measurement) of a part of assembly process; construction workers' qualification level (category); mechanization level of a part of assembling process (%), and complexity of assembling work (in points) are less important criteria.

References

Janusaitis R., The evaluation of wall heating alternative solutions, applying the method of technological net model of the "component cutting" method, *Civil engineering*, IV, 2, Technika, Vilnius,161-170, 1998.

Miniotaite, R., Multi-criteria decision analysis of up-to-date construction technology, in *Interaction between Theory and Practice in Civil Engineering and Construction,* Komurlu, R., Gurgun, A. P., Singh, A., and Yazdani, S (eds.), ISEC Press, 2016.

Integrated Solutions for Infrastructure Development
Edited by Lau, H. H., Tang, F. E., Ng, C. K., and Singh, A.
Copyright © 2016 ISEC Press

DEVELOPMENT OF EMBEDDED MINI-SENSOR FOR DIAGNOSIS OF CORROSION OF STEEL REINFORCEMENT IN CONCRETE NONDESTRUCTIVELY

MOHAMED A. ISMAIL[1], HAN-SEUNG LEE[2], and MOHD WARID HUSSIN[3]

[1]*Dept of Civil and Construction Engineering, Curtin University Sarawak, Sarawak, Malaysia*
[2]*Dept of Architectural Engineering, Hanyang University, Erica Campus, Ansan, Republic of South Korea*
[3] *Construction Research Center (CRC), Institute for smart infrastructure and innovative construction, UTM, Skudai, Malaysia*

Corrosion of steel reinforcement embedded in concrete is one of the main causes of degradation of reinforced concrete structures. Degradation occurs in reinforced concrete structures from corrosion caused by the Chloride ingress into concrete. That degradation has a severe impact on the structure in terms of maintenance and rehabilitation costs. Therefore, early detection of reinforcement corrosion is important for efficient maintenance, repair and planning. Meanwhile, the evaluation of the corrosion of reinforcement by non-destructive measurements have been used a lot. In particular CM-II (corrosion meter) is used to measure the polarization resistance, but has some disadvantages. Embedded mini-sensor has been developed in order to overcome these disadvantages. In this study, measurement of corrosion by using the mini-sensor is compared with the measured results by CM-II to verify the validity of the newly developed mini senor. Results show that there are agreement in trends of the parameters measured and as such the developed mini sensor has a promising start to be used.

Keywords: Structure, Polarization, NDT, Detection, Resistivity.

1 INTRODUCTION

Occurrence of corrosion in reinforced concrete structures begins at the inner surface of reinforced concrete. Therefore, when there is crack or rust on the concrete surface it means that the corrosion of rebar is considerably advanced. Corrosion of the embedded rebar in concrete has become a major cause of early deterioration and collapse of reinforced concrete structures (Seok 2008). Deterioration caused by corrosion that occurs in reinforced concrete structures causes enormous costs in terms of maintenance and rehabilitation of structures in addition to the time required that is also difficult to determine. Therefore, prediction and early detection of corrosion of reinforcement in concrete is critical for establishing effective repair plan (Song *et al.* 2009).

CM-II (corrosion meter) which can measure the polarization resistance is being utilized a lot. The principle of CM- II meter is to obtain the polarization resistance to diagnose the degree of corrosion from the AC impedance of the reinforcement in the concrete by calculating a corrosion rate (Mecater 2004). However, due to the fact that CM-II meter measures from concrete surface, there are some drawbacks that generate a number of errors in the measurements by the different environments of the concrete and that decrease the accuracy of the corrosion evaluation.

In order to overcome these problems, a proposed mini sensor attached to reinforcing bar in concrete is used. Although mini-sensor can be used to directly measure corrosion development of the steel surface in concrete; it is still in early stage of development and as such more research is needed.

2 EXISTING LITERATURE

2.1 Polarization Resistance Method

Polarization resistance method is practical because it can be applied in estimating the corrosion status in both laboratory measurement and on site. Table 1 shows the relationship between the polarization resistance and corrosion rate. High rate of corrosion can be evaluated when polarization resistance is low. Table 2 shows the criteria for evaluating the corrosion of the reinforcement according to the polarization resistance. It can be determined by measuring the polarization resistance, corrosion current density and the corrosion depth. On the other hand, the main electrical field is being utilized for the measurement of corrosion rate and chemical polarization method with a long measurement time DC polarization resistance and ac resistance (AC impedance).

Table 1. Relationship between polarization resistance and corrosion rate (BRE, 2001).

polarization resistance Rp (kΩcm²)	corrosion rate
more than 250	passive state
250~25	low/middle
25~2.5	high
2.5~0.25	very high

Table 2. Corrosion of rebar by polarization (BRE, 2001).

current density of corrosion i_{corr},(μA/cm²)	corrosion depth i_{corr},(μA/cm²)	corrosion rate
Up to 0.2	Up to 2	passive/low
0.2 ~ 0.5	2 ~ 6	low/middle
0.5 ~ 1.0	6 ~ 12	middle/high
> 1.0	> 12	high

2.2 AC Impedance Method (A /C impedance)

AC impedance method is to measure the resistance when applying 10 ~ 20 mv (Alternative Current) on a specimen. The frequency of the alternating current is between 100 kHz ~ 1m Hz. Figure 1 shows a schematic illustration of the measurement method using CM- II. In the case of AC impedance measurements include measuring time is relatively long, and frequency response analysis device or a transfer function analyzer is required additionally, by the practicability requires less technical knowledge to interpret the test results.

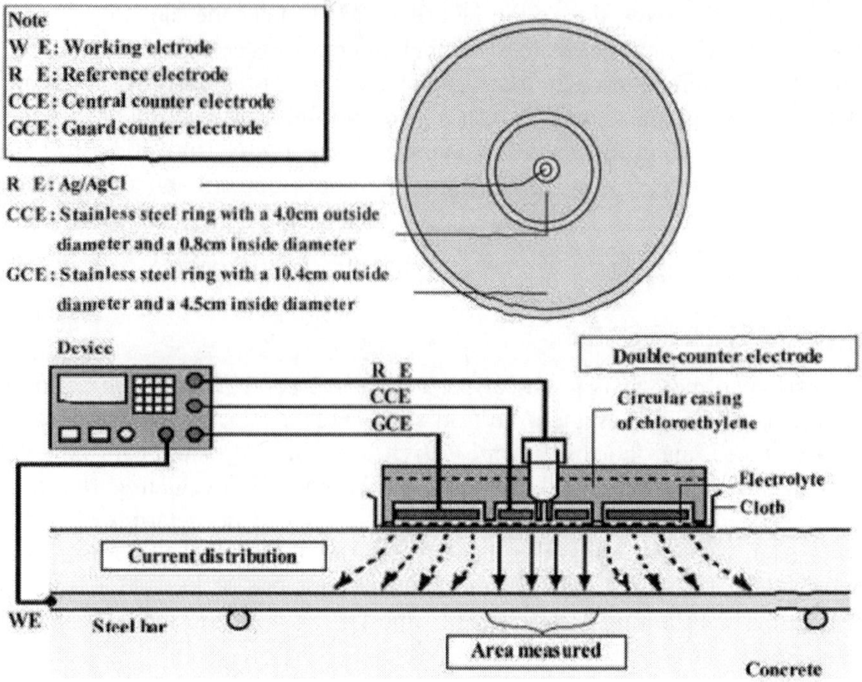

Figure 1. Measurement using CM-II.

3 OVERVIEW AND EXPERIMENTAL METHODS

3.1 Experimental Program

Mortar was used to fill in between mini sensor and rebar. Mortar made with cement and sand ratio of 1:2 and, W / C was set to 60%. The size of a specimen was 200 x 200 x 100 mm. from the specimen surface, reinforcement installed at depth of 20 mm. 13 mm diameter rebar was used. Settlement was reinforced by a cable tie to attach the sensor is integral with the reinforced mortar. Mini Sensors reinforced mounting process is shown in Figure 2. Figure 3 illustrates a measuring method and a shape. CM-II is measured at the concrete surface, and is embedded in a mini-sensor measures the specimens.

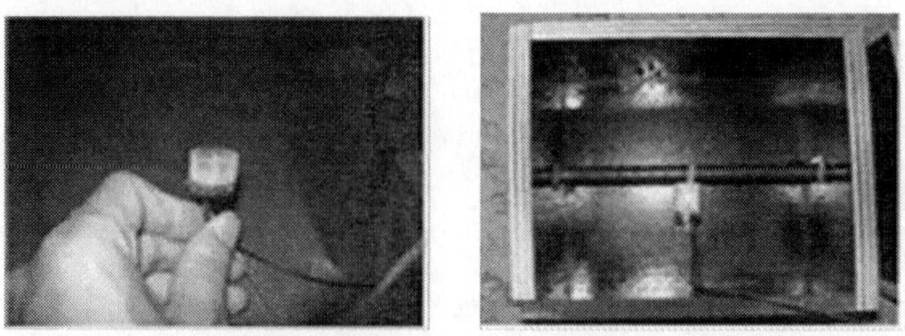

(a) Keep 5 mm depth (b) Adhere to rebar

Figure 2. Processing mini-sensor adhere to rebar.

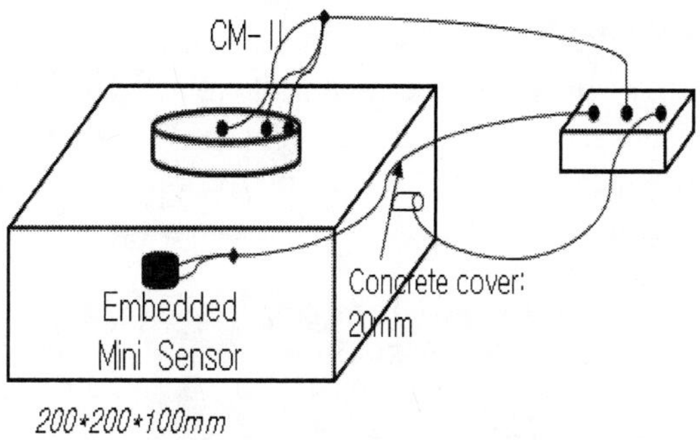

Figure 3. Measurement method and figuration.

Details of concrete mix proportions are shown in the Table 3.

Table 3. Concrete mix proportion.

W/C (%)	Unit weight (kg/m³)			
	Water	Cement	Sand	NaCl
60	180	300	600	18

A humidity chamber was used to apply accelerated corrosion cycle. One cycle consists of RH 35% - 60 ℃ for 24 hours, and caught up after 24 hours at RH 95% -60 ℃ then measurement was done in the following 24 hours. Schematic diagram of accelerated corrosion cycle is shown in Figure 4.

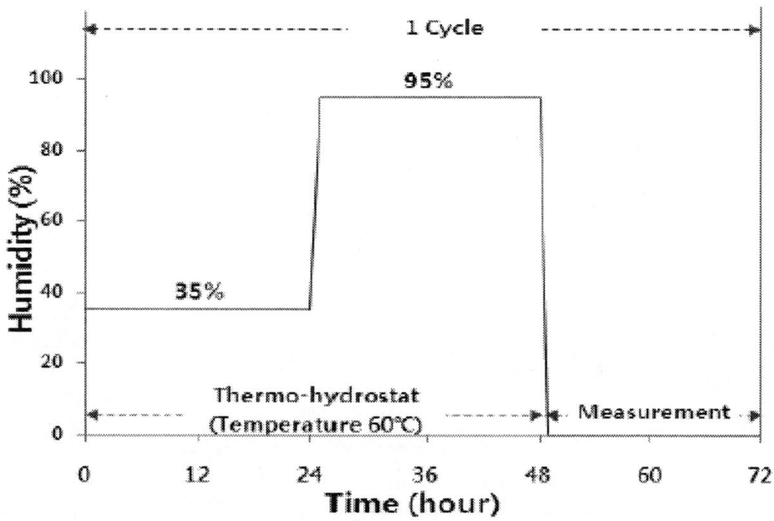

Figure 4. Condition of accelerated corrosion cycle.

3.2 Measuring Parameters

After promoting corrosion, CM-II (corrosion meter) measured the polarization resistance by an AC impedance method and also mini-sensor was used to measure the same parameter.

4 EXPERIMENTAL RESULTS

Figure 5 shows the corrosion of the reinforcement at the end of the experiments. While corrosion of the reinforcement is noticed, corrosion did not occur in the electrodes of the mini sensor. Mini sensor through which it was found that there is no problem caused by the corrosion of the electrode material for measuring the corrosion of rebar in a mortar.

Figure 5. The status of rebar and mini-sensor at the end of experiment.

4.1 Polarization Resistance Measurements

Figure 6 shows the actual measurement data values of polarization resistance by CM II. The frequency range of the measured data from the body between 10 Hz and 0.03162 Hz. The horizontal axis represents the real part (real) and the vertical axis is the imaginary part data appeared as (imaginary).

Each point on the plot was calculated for each data of the polarization resistance at the same frequency, the dotted part in the actually measured value and the data of each frequency, to shape the diameter (Nyquist Plot) of a semi-circular polarization resistance to the diameter of the semicircular polarization resistance Rp that was calculated.

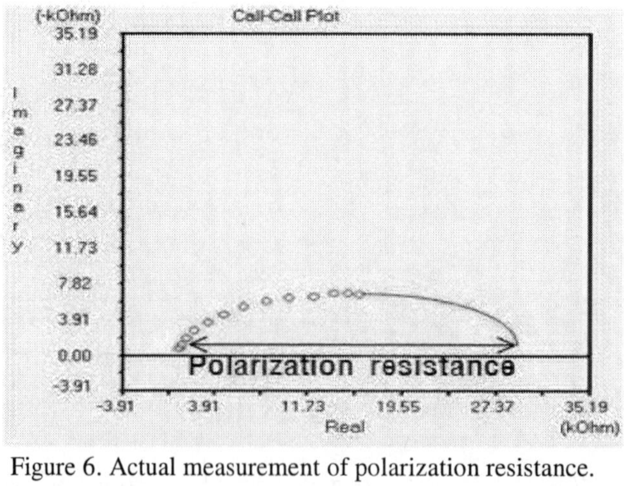

Figure 6. Actual measurement of polarization resistance.

Figure 7 shows the relationship between the polarization resistance obtained by the CM-II and that obtained by mini-sensor. The correlation coefficient R^2 is 0.992.

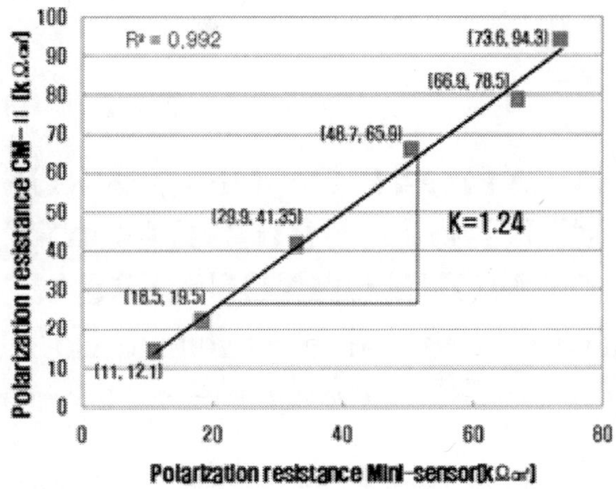

Figure 7. Polarization resistance correlation between CM-II and mini-Sensor.

5 CONCLUSIONS

This study was carried out by non-destructive corrosion promotion tests to verify the validity of measuring polarization resistance by CM-Ⅱ in comparison with the developed mini sensor. Results showed that polarization resistance can be measured by both technique and similar trend was observed.

References

Building Research Establishment (BRE), Corrosion of Steel in Concrete; Investigation and Assessment, 2001.

Mecater, W. John, "Review; Sensor System for Use in Reinforced Concrete Structure", Construction and Building Materials, 2004, pp.351-358.

Seok, S. Yeong, "measured the corrosion rate of rebar in reinforced concrete structures-Experimental study on the use of direct current electrical resistance method" Architectural Journal of the Society, 24 No. 6 (No. XIV 236), 2008. 6, pp.93-100.

Song, Ha-Won; Lee, Chang-Hong; Lee, Kewn-Chu; Ann, Ki-Yong, "A Study on Chloride Threshold Level of Blended Cement Mortar Using Polarization Resistance Method", Journal of the Korean Concrete Institute, Vol.21, No 3, pp.245-253, June, 2009.

Integrated Solutions for Infrastructure Development
Edited by Lau, H. H., Tang, F. E., Ng, C. K., and Singh, A.
Copyright © 2016 ISEC Press

AN INVESTIGATION OF CORROSION ENVIRONMENT OF A BRIDGE CONSTRUCTED IN MOUNTAINOUS AREA

KAZUTO MARUYAMA and TOSHIHIKO ASO

Dept of Civil and Environmental Engineering, Yamaguchi University, Ube, Japan

Corrosion occurred on steel bridges which were constructed in mountainous area. However, few studies have investigated the corrosive environment of steel bridge in mountainous area. In this research, Temperature, humidity, wind conditions and amount of airborne salt have been observed at bridge site. In order to comprehensively assess corrosive condition, exposure test was also performed. Observation and exposure test were conducted at the steel plate girder bridge with 60m length and three main girders. During observation period, anti-freezing agent (Na-Cl) is sprayed in winter. From the observation, time of wet on the bridge is longer than bridges those were constructed at coastal area. It was clearly observed that flying behavior of anti-freezing agent around superstructures is influenced by the prevailing wind. Furthermore, distribution of amount of adhesion salt at the surface of steel in the bridge section was also clarified. Rust thickness of exposure specimen was revealed that the influence of anti-freezing agent, and at lower flanges becomes larger than that of web. These results can be applied to corrosive environmental simulation of steel bridges.

Keywords: Corrosion, Weathering steel, Bridge, Exposure test, Anti-freezing agent.

1 INTRODUCTION

In Japan, from 1978 to 2014, weathering steel bridges has increased to about 20% of the total steel bridges (JBA 2015). If it is generating protective rust on the design plan in weathering steel bridges, it is possible to reduce the corrosion rate. In other words, if it is possible to reduce the corrosion rate by a protective rust, eliminates the need for corrosion protection by paint, it can be the reduction of life-cycle-cost (Yamaguchi *et al.* 2006). However, in some bridges that were constructed at coastal area, protective rust did not formed due to large amount of airborne salt and high humidity. Therefore, in the construction of the weathering steel bridge, it is important to determine the propriety of the corrosive environment in the installation point. Conventionally, it has been paying attention to airborne salt as one of determine the propriety, thus there are lot of researches about corrosive environment of steel bridges constructed in coastal area. But corrosion due to anti-freezing agent also occurred on steel bridges which constructed in mountainous area (Iwasaki *et al.* 2012, NILIM *et al.* 2014), and few studies have investigated the corrosive environment of steel bridge in mountainous area. This paper aims to clarify the corrosive environment of the mountainous area bridge. Temperature, humidity, wind conditions and amount of airborne salt have been observed at bridge site.

In order to comprehensively assess total corrosive condition, exposure test was also performed.

2 SITE OBSERVATION

On-site observation was conducted at Chigashira Bridge. Figure 1 shows location of the bridge and other steel bridges which constructed in coastal area compare with the bridge. Figure 2 and Figure 3 displays outline of the bridge and Table 1 indicates overview of the bridge. The bridge is located in Ato town, Yamaguchi city, Japan. The bridge length is 60 m, and it is a plate girder bridge with 3 main girders. In Fig 2, main girders of the bridge assumed G1, G2, and G3 from the south side, and strain girders of it assumed S1, S2 from the south side.

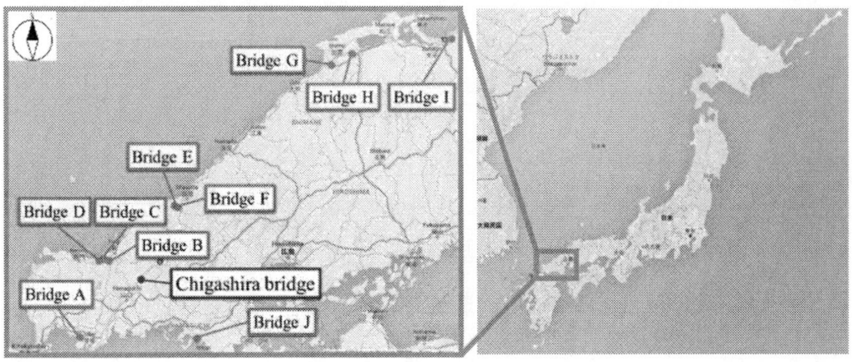

Figure 1. Location of bridges.

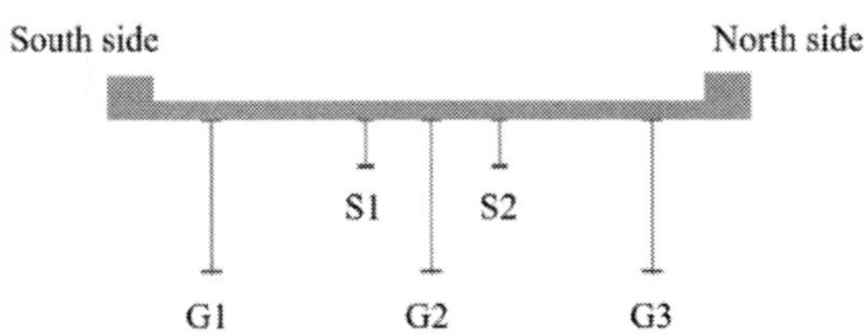

Figure 2. Cross section of Chigashira bridge.

Figure 3. Chigashira bridge.

Table 1. Overview of Chigashira bridge.

Location	Root 9, Ato town
Bridge length (m)	60
Offshore distance(km)	20
Direction of the bridge axis	Northwest-Southeast
Type	Plate girder bridge with 3 main girders

On the occasion of elucidation of the corrosion environment at the bridge, airborne salinity was measured by dry gauze method which installed in nine places. Adhering salinity was measured by steel sheet which installed in fifteen places. Wind direction and wind speed were measured by anemometer which installed in south end of the inspection passage. Temperature and humidity were measured by logger which installed in five places. Exposure test was performed by exposed specimens (10mm×10mm) which installed in the place same as a steel sheet (six pieces). Rust thickness was measured by film thickness meter. Average thickness of each specimen were computed by average of rust thickness of five points on the specimens. An observation at the bridge was started from September 29, 2015, and it is still ongoing.

3 RESULT OF OBSERVATION

3.1 Time of Wet and Wind Condition

In Chigashira bridge, ratio of time of wet was calculated from temperature and humidity, and Fig.4 shows compared with ratio of time of wet of steel bridges constructed in coastal place (past of the same period). The ratio of time of wet was found by dividing total of measurement time from total of time of wet. The time of wet defined as the temperature at 0°C or more and a humidity of 80% or more of time by ISO.

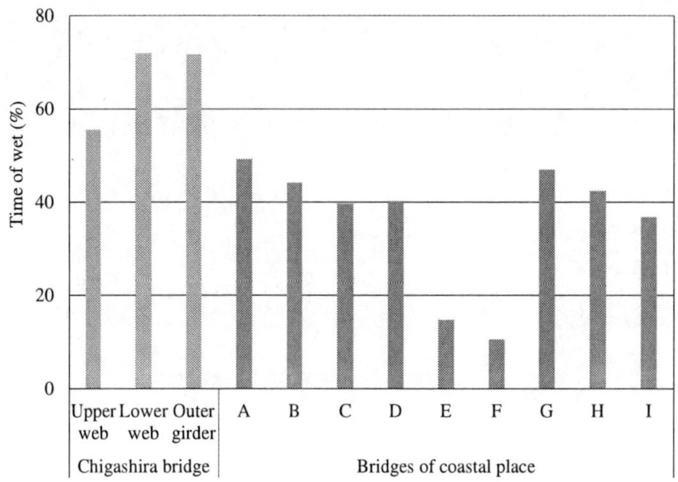

Figure 4. Comparison of time of wet.

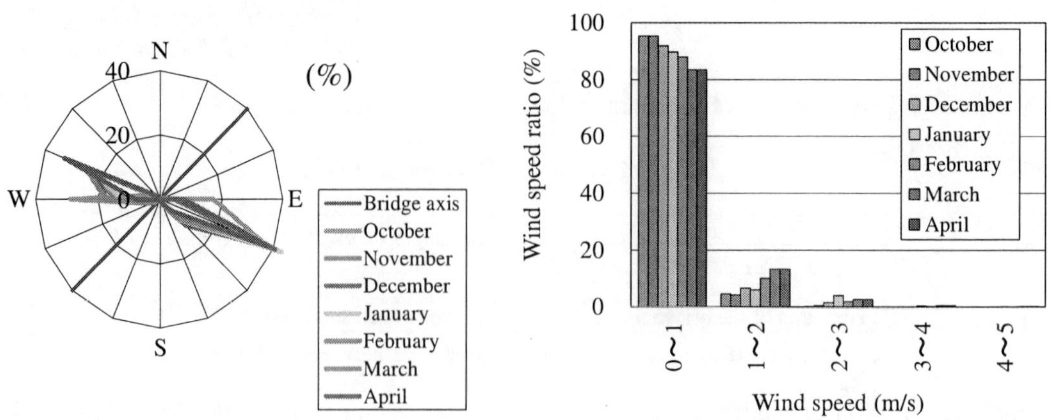

(a) Wind rose (b) Wind speed ratio

Figure 5. Wind condition of Chigashira bridge area in different month.

From Fig.4, ratio of time of wet of all parts in Chigashira bridge were bigger than that of steel bridges constructed in coastal place. Figure 5 (a) shows wind rose, and Figure 5 (b) shows wind speed ratio. From Fig.5 (a), prevailing wind was from east southeast every month. It is bridge axial orthogonal direction. From Fig.5 (b), wind speed has become stronger every month, but wind ratio of 0~1 m/s indicates a more than 80% every month. From, wind conditions of around Chigashira bridge, it can be seen that gentle wind is mainly blowing bridge axial orthogonal direction.

3.2 Airborne Salinity

Figure 6 shows the installation position of dry gauze and the status of inflow wind. Fig.7 shows airborne salinity that measured from October, 2015 to March, 2016, and Fig.8 shows transitions of anti-freezing agent and airborne salinity of No.5. From Fig.7, airborne salinity of under the girder (No.1, 5 and 9) and part of inner girder (No2 and 6) increased from December, 2015 to January, 2015. This type of phenomenon is due to the anti-freezing agent. Because from Fig.8, airborne salinity has also increased with the increasing of anti-freezing agent. As a result, anti-freezing agent scattered to the girder. In addition, amount of airborne salt of No.2 and 6, the prevailing wind is affecting. Because prevailing wind blows from the G1 side, and flows into the girder as shown in Fig.6.

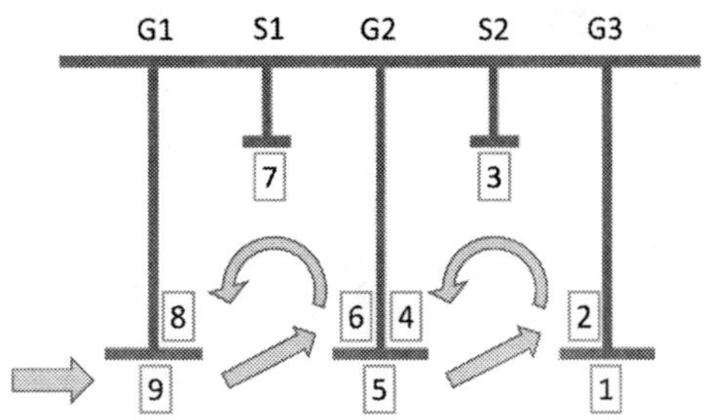

Figure 6. Installation position of dry gauze and the status of inflow wind.

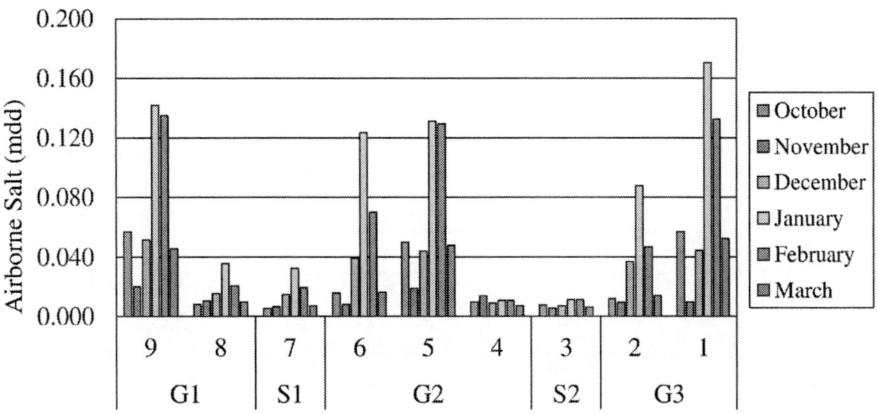

Figure 7. Airborne salinity in different month.

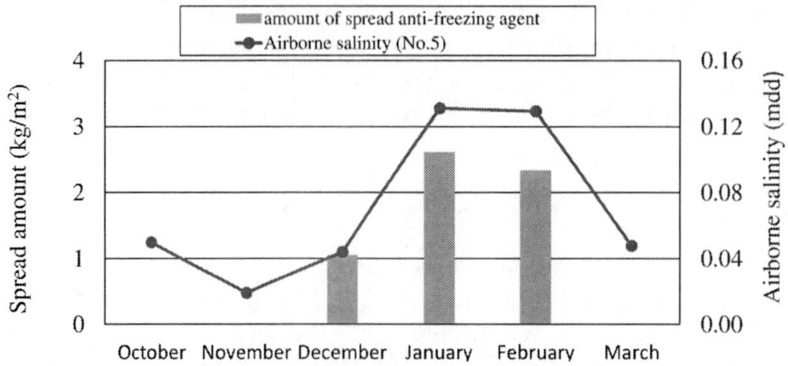

Figure 8. Transitions of anti-freezing agent and airborne salinity (No.5).

3.3 Exposure Test

Figure 9 shows installation position of exposed specimens. Figure 10 shows rust thickness of exposed specimens that measured from October, 2015 to April, 2016. From Fig.10, focusing on exposed specimens of inner girder (No.4, 5, 6, 7, 9, 10, 11 and 12).

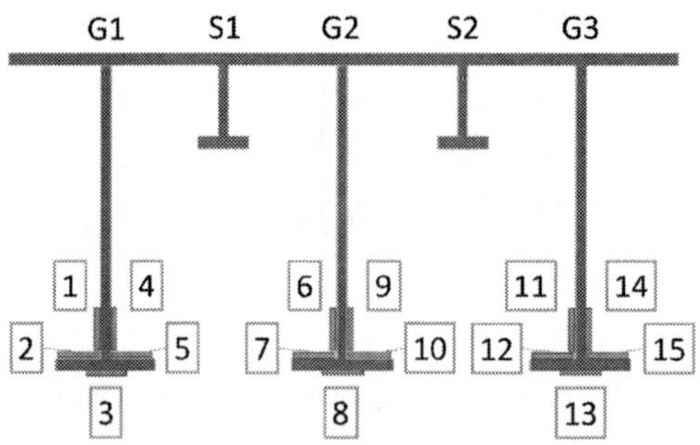

Figure 9. Installation position of exposed specimens.

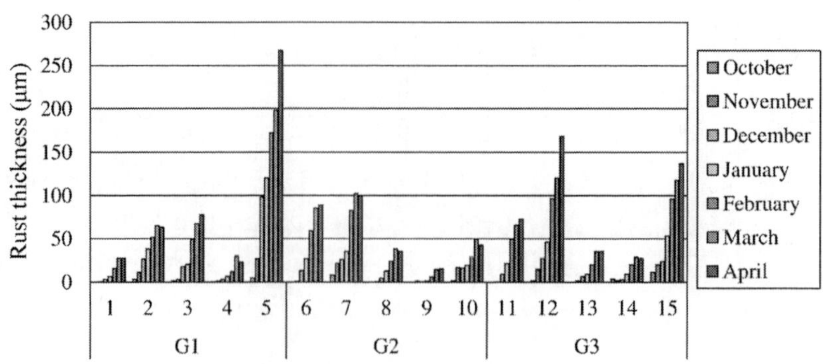

Figure 10. Rust thickness on exposed specimen.

In inner girder of G1G2, rust thickness of No.5 is shown maximum because rust liquid had been deposited in November. But in web, rust thickness of No.6 of susceptible to prevailing wind is more growth than rust thickness of No.4. And in inner girder of G2G3, rust thickness of 11 and 12 of susceptible to prevailing wind were more growth than rust thickness of No.9 and 10. From these, rust thickness of inner girder is affected by the anti-freezing agent that scattered by prevailing wind. Figure 11 shows the time course of rust thickness at Chigashira bridge and steel bridges constructed in coastal place in the L.flg lower surface. From Fig.11, rust thickness of Chigashira bridge was bigger than that of Bridge A to Bridge F. This is assumed to be related to the fact that there are more time of wet at Chigashira bridge compared with that of steel bridges constructed in coastal area, and anti-freezing agent scattered to the girder in Chigashira bridge.

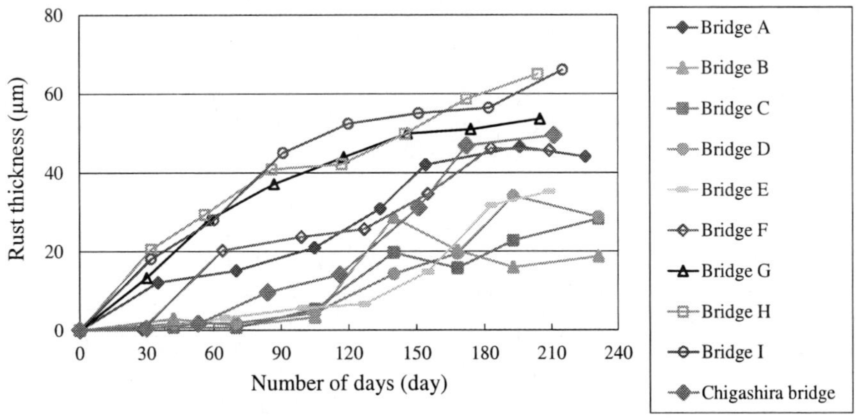

Figure 11. Comparison of rust thickness.

4 CONCLUSION

The corrosive environment investigated in Chigashira bridge, which is located in mountainous area, and ratio of time of wet and rust thickness, were compared with steel bridges constructed in coastal areas. From the fact that airborne salinity has also increased with the increasing of anti-freezing agent, anti-freezing agent was scattered to the girder. In addition, scattered of anti-freezing agent by prevailing wind is the main, prevailing wind affected airborne salinity of especially under the girder and part of inner girder. For rust thickness, prevailing wind affected rust thickness of especially part of inner girder. In comparison with steel bridges constructed in coastal place (Bridge A to Bridge I), rust thickness of Chigashira bridge was bigger than that of Bridge A to Bridge F. This is assumed to be related to the fact that the length of the time of wet and anti-freezing agent is affected to rust thickness. These results can be applied to corrosive environmental simulation of steel bridges.

References

Iwasaki E., Nagato T., Yuasa A., and Nishi T, *Scattering of Deicing Salt and Corrosion of Steel Bridges*, Journal of structural engineering, Vol 58A, Mar 2012. (in Japanese)

Japan Bridge Association., *Weathering Steel Bridge Actual Book.*, Japan Bridge Association.,No.21, Nov, 2015. (in Japanese)

National Institute for Land and Infrastructure Management., Japan Iron and Steel Federation., Japan Bridge Association., *Research on Evaluation Technique of Applicable Condition for Weathering Steel Bridge (II) Study about Effect of the Scattered Deicer on the Bridge*, Technical note, No.778, Jan., 2014. (in Japanese)

Yamaguchi E., Nakamura S., Hirokado K., Morita C., Sonoda Y., Aso T., Watanabe H., Yamaguchi K., and Iwatsubo K, *Performance of Weathering Steel in Bridges in Kyushu-Yamaguchi Region*, Journal of JSCE A, Vol.62 No.2, 243-254, Apr 2006. (in Japanese)

Integrated Solutions for Infrastructure Development
Edited by Lau, H. H., Tang, F. E., Ng, C. K., and Singh, A.
Copyright © 2016 ISEC Press

DIAGNOSIS OF BRIDGE BEARINGS BASED ON FIELD MEASUREMENTS

NAHO SHIBASAKI[1], MARIKO IKEDA[2] and MASAHIRO SAKANO[3]

[1] *Kawakin Core-Tech CO., Ltd., Osaka, Japan*
[2] *Kajima Corporation, Tokyo, Japan*
[3] *Kansai University, Osaka, Japan*

Bridge bearings are the structural members which are installed at the connection between superstructure and substructure of a bridge, thus it is crucial to maintain bearing functions as structural members. If bearing functions are declined, it is possible for the structure of the whole bridge to be changed and have an effect on functions of superstructure and substructure. Thus, maintaining functions of bearings is not negligible. Generally, "Visual inspection" is indicated as a standard method. However, it is difficult to evaluate the deterioration degree of bearing functions with "Visual inspection" only. Therefore, we have studied to establish a qualitative evaluation method regarding basic functions required for bridge bearings as structural members based on measuring displacement and strain. In this study, measuring displacement and strain in the bridge in service was carried out, and evaluating the deterioration degree of bearings with the evaluation method proposed was attempted.

Keywords: Bearing, Highway bridge, Displacement, Strain, Fatigue life.

1 INTRODUCTION

Bridge bearings are the structural members which are installed at the connection between superstructure and substructure of a bridge. Their functions are expected to transfer load which acts on superstructure to substructure and accommodate expansion and rotation of superstructure, due to live load and temperature variations, in order to absorb relative displacement of superstructure and substructure. Bridge design is based on whole analysis models which the bearing functions are reflected. If bearing functions are deteriorated, the structure system of the whole bridge can be possibly changed, and it may affect on functions of superstructure and substructure. Thus, it is crucial to maintain bearing functions as structural members. According to the reference (Ministry of Land, Infrastructure, Transport and Tourism 2014) for a bearing inspection, "Visual inspection" is shown as a standard method to inspect bearings, as well as "Measuring displacement" is shown as an example which can be taken if necessary. However, in reality, it is difficult to evaluate the deterioration degree of bearing functions with "Visual inspection" only. Moreover, any specific evaluation method for "Measuring displacement" has not been established. In this case, with consideration of these points above, we have studied to establish qualitative evaluation methods regarding required basic functions (load supporting, horizontal moving and rotating function) for bridge bearings as structural members based on measuring displacement and strain.

In this study, measuring displacement and strain in the bridge in service was carried out to attempt evaluating the deterioration degree of bearings with the evaluation method we proposed.

2 METHOD

The evaluation method for the basic functions that are required to bearings and the measurement method are shown as below.

2.1 Evaluation Method

The evaluation method was set for basic functions required for bearings as structural members for load supporting, horizontal moving and rotating function respectively. The measurement items were vertical and horizontal displacement of bearings, and deterioration degree of bearings was evaluated with sequential transition of displacement measured. Table 1 shows details of the evaluation method and measurement items. In addition, in order to verify effects of deterioration of the bearing functions on superstructure, strain of the girder lower flange was also measured.

Table 1. Evaluation method and measurement item for bridge function.

Basic functions required for bearings	Evaluation method	Measurement items
Load supporting function	Checking if vertical supporting function is deteriorated that the whole bearing is moved up and down	Vertical displacement in the front and back in the longitudinal direction
Rotating function	Checking if the bearing can accommodate rotation due to girder deformation due to live load	Vertical displacement in the front and back in the longitudinal direction
Horizontal moving function	Checking if the bearing can accommodate horizontal movement due to girder deformation due to live load and temperature changes	Horizontal relative displacement of superstructure and substructure in the longitudinal direction
Effect on girder	Checking if large stress fluctuation occur at lower flange due to bearing function deterioration	Stress fluctuation at lower flange bottom of the girder located 20mm from weld toes of the sole plate

2.2 Measurement Method

Figure 1 shows the measurement position of displacement and strain. Displacement transducers were used to measure vertical and horizontal displacement, and stress was measured with a strain gauge, which is installed just before measuring, at the bottom of the lower flange. Measurements were conducted in two conditions, short-term and long-term. Short-term measurement focuses on live load, whereas long-term measurement focuses on temperature variations. Table 2 shows the detailed measurement conditions respectively.

Table 2. Evaluation method and measurement items for bridge functions.

Measurement type	Action	Period	Frequency
Short-term	Live load	20 minutes	200 Hz
Long-term	Temperature Change	1 day	0.5 Hz

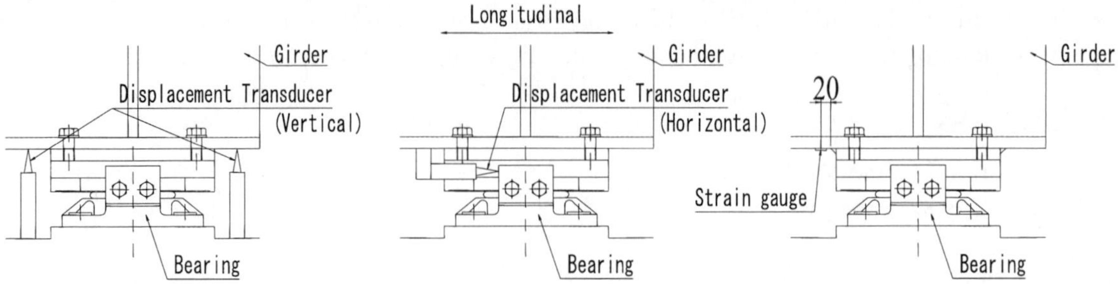

Figure 1. Measurement positions.

2.3 Measurement Object

Table 3 shows the measurement objects, the four bridges have bearings of different types. All of the bearings whose corrosions and cracks were not recognized were evaluated as sound in a visual inspection.

Table 3. Evaluation method and measurement items for bridge functions.

Bridge	Bearing Type	Span length (m)	Horizontal alignment(m)	Completion year	ADT (Vehicles/day)
A	Spherical bearing	30.0	R = 35	1986	21,179
B	Rubber bearing	202.65	R = ∞	2006	85,684

3 RESULTS AND CONSIDERATIONS

For respective object bridges, the bearing functions were evaluated by the measurement results verified.

3.1 Spherical Bearing

Figure 2 shows the measurement results of Bridge A with Spherical bearing. Graphs of (a) to (c) are extracted for 15 seconds from the short-term measurement results, while the graphs of (d) to (f) are for the long-term measurement results. Since vertical displacement was occurred in the vertical opposite direction in the front and back of the bearing, it was inferred that the bearing accommodated rotation, due to live load and temperature variations in short-term and long-term measurement. Meanwhile horizontal displacement was occurred in neither short-term nor long-term measurement, which indicates that the bearing did not move horizontally. It was because curvature radius of the horizontal alignment is small, thus, the moving direction of superstructure is inconsistent with the horizontal movement of the bearing because superstructure is just moved in the longitudinal direction. It is considered that the fixing structure in the transverse direction restrains deformation of superstructure. Moreover, in both short-term and long-term measurement, horizontal displacement was not occurred when vertical displacement was occurred, and stress fluctuation was occurred in the girder lower flange, thus it can be inferred that the horizontal moving function has some issues.

3.2 Rubber Bearing

Figure 3 shows the measurement results of Bridge B with Rubber bearing. In the short-measurement, from the fact that vertical displacement was occurred downward in the front and back of bearing, it can be inferred that bearing was compressed, due to live load. Due to that magnitude of vertical displacement in the front and back of bearing was different; it can be also presumed that the bearing accommodated rotation by live load. In the short-measurement, the fact, which vertical displacement in the front and back of the bearing was occurred upward with the temperature rise and downward with temperature fall, indicates that the bearing was expanded and contracted with the temperature variations. However, stress fluctuation was recognized when rotation and horizontal movement was occurred in both of the short-term and long-term measurements, thus it is confirmed that the bearing does not completely accommodate deformation of superstructure.

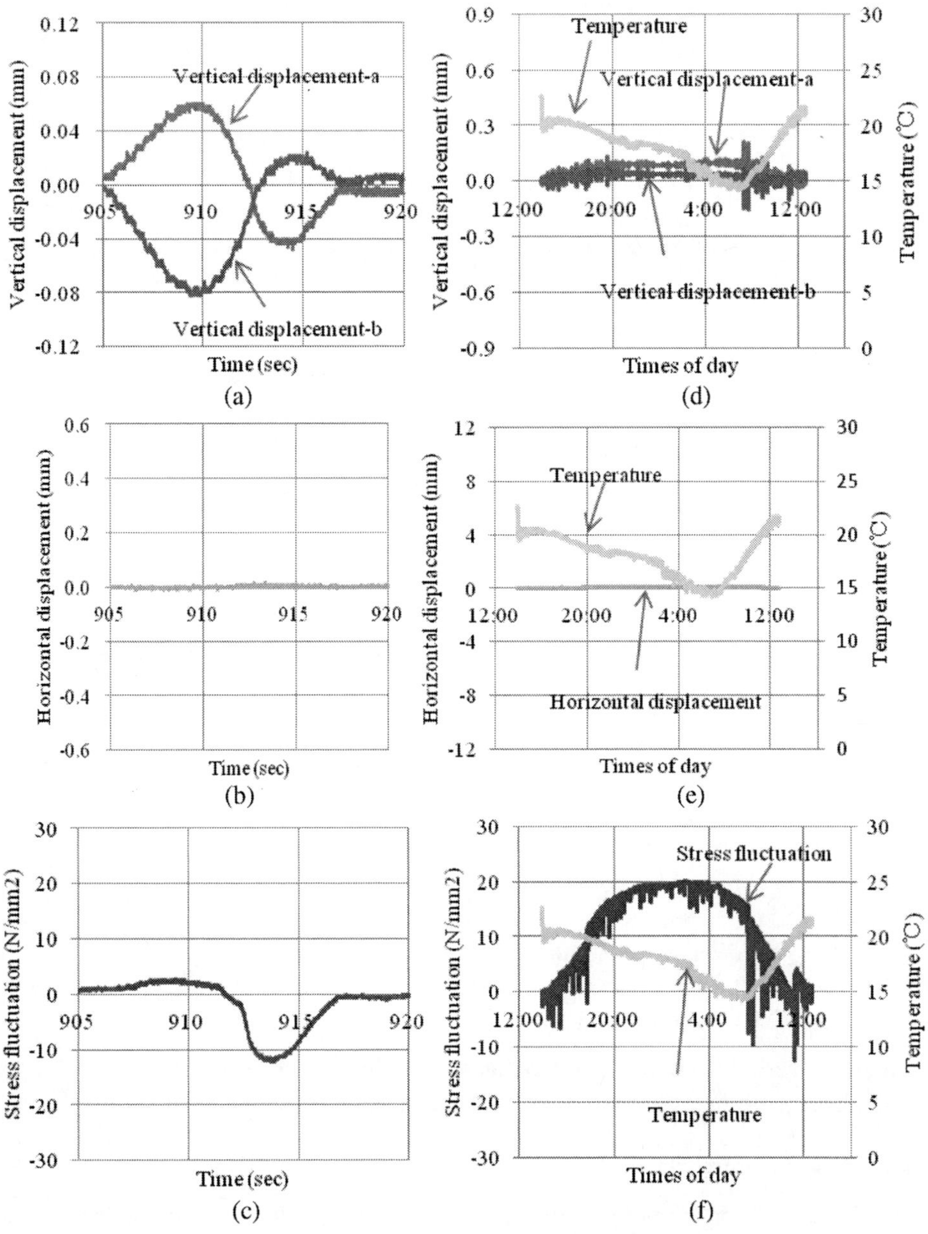

Figure 2. Results of Bridge A with Spherical bearing.

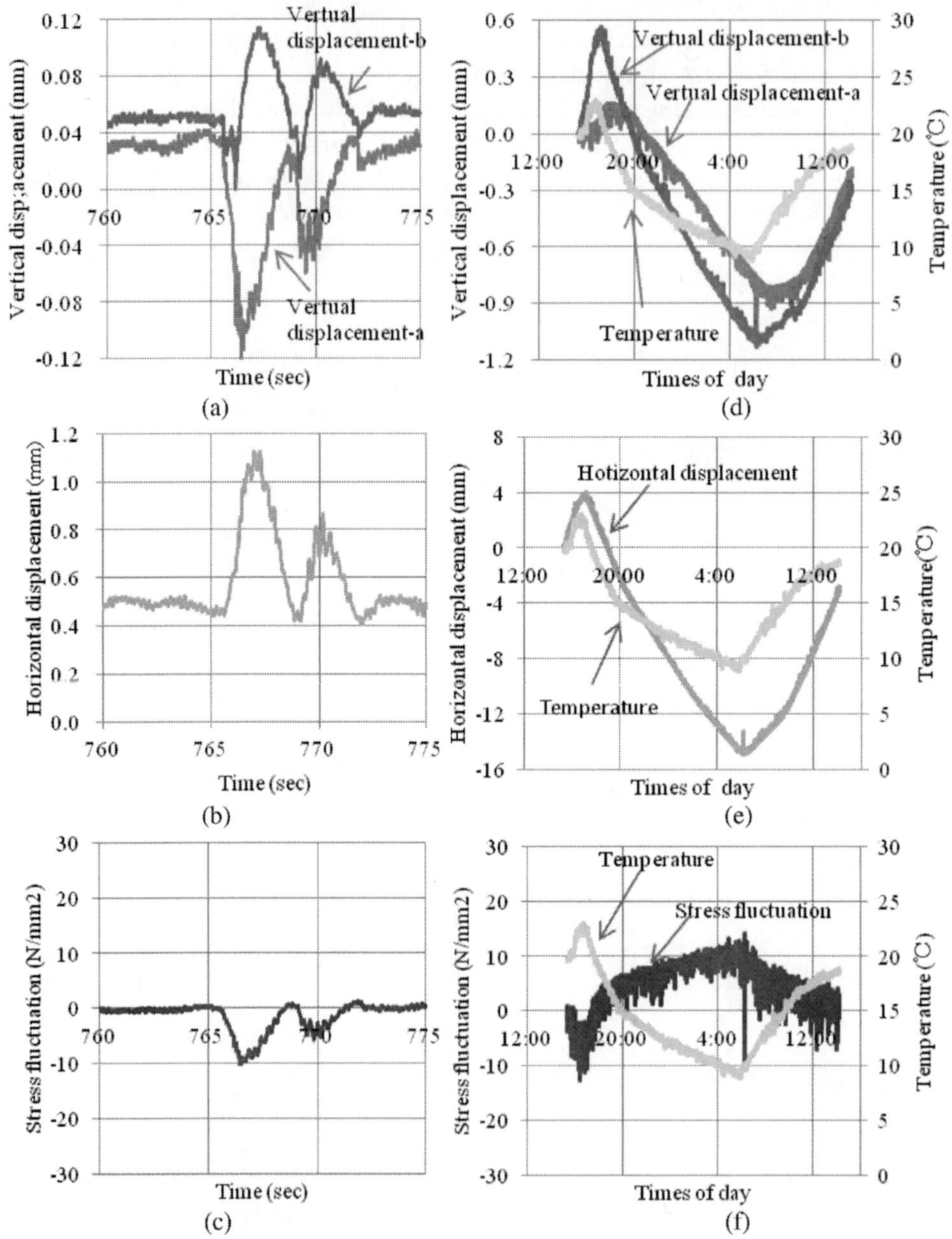

Figure 3. Results of Bridge B with Rubber bearing.

4 RESULTS AND CONSIDERATIONS

We attempted to verify the correlation between the functional evaluation and the fatigue life which is calculated with measurement results of the stress fluctuation.

4.1 Calculating Fatigue Life

The fatigue curve used for the fatigue life calculation was the one, which was the lowest category H' in the reference (Japan road association 2002), and the endurance limit was used as 7 MPa.

Moreover, for the calculation of frequency distribution of the stress range, the rain-flow method, which is the most common, was used. In addition, the correction coefficient by the structure details was not taken into account.

4.2 Correlation between the Functional Evaluation and the Fatigue Life

Regarding the fatigue life of Rubber bearing which rotates and moves horizontally, is relatively long, meanwhile, the fatigue life of Spherical bearing which rotates and does not move horizontally at all, is relatively short. The fatigue life of Rubber bearing which has rotating and moving functions, is the shorter than Spherical bearing. It can be considered that what constant stiffness of Rubber bearing occurs resistance force against rotation and movement of superstructure is one of the causes. Moreover, it can be inferred as one of the factors that the traffic volume of large vehicles in the object bridges is heavy. In case of Rubber bearing, even if bearings have rotation and moving functions in certain levels, the fatigue life can become relatively short dependent on traffic volume of large vehicles.

Table 4. Fatigue life.

Bridge	Bearing Type	ADTT (Vehicles/day)	Fatigue life (years)	
			Short-term	Long-term
A	Spherical bearing	4,294	233	1691
B	Rubber bearing	7,800	71	864

5 CONCLUSIONS

As a result of the measurements of vertical and horizontal displacements and the strain of the girder lower flange with functional evaluation methods, the following knowledge was gained.

(1) As measuring vertical and horizontal displacement for one day, it becomes possible to evaluate rotation and horizontal moving functions to accommodate deformation of superstructure due to temperature variations.

(2) As measuring vertical and horizontal displacements for 20 minutes, it becomes possible to evaluate rotation and horizontal moving functions to accommodate deformation of superstructure due to live load.

(3) Even if rotation and movement is occurred in a bridge which may have heavy traffic with large vehicles and whose fatigue damages are concerned, necessity of measurements of stress fluctuation in the girder lower flange as well as vertical and horizontal displacements is still considered.

From above, we consider that effectiveness of the bearing function evaluation method proposed in this study is validated. Hereafter, we would like to establish more precise and practical evaluation methods with increasing a number of measurement times.

This research was financially supported by the Kansai University Grant-in-Aid for progress of research in graduate course, 2015.

References

Japan Road Association, Fatigue Design Recommendations for Steel Highway Bridges, 2002.
Ministry of Land, Infrastructure, Transport and Tourism, Bridge Inspection Manual, 2014.

Integrated Solutions for Infrastructure Development
Edited by Lau, H. H., Tang, F. E., Ng, C. K., and Singh, A.
Copyright © 2016 ISEC Press

REMAINING LIFE AND CORROSION LIMIT OF CORRODED TRUSS BRIDGE FOCUSING ON THE ULTIMATE STRENGTH

HIROAKI FUKUDA[1], NAOYUKI ASAO[2] and KATASHI FUJII[1]

[1]*Hiroshima University, Hiroshima, Japan*
[2]*Pacific Consultants Co., Ltd, Tokyo, Japan*

This study is targeted for the estimation method for the remaining life of a corroded truss bridge based on the ultimate strength through one of case studies. The remaining life can be estimated by the corrosion limit which is defined as a corrosion state based on the assumption that the bridge collapse occurs when reaching the load required of the bridge performance. The deterioration curve of the corroded truss bridge is proposed by conducting nonlinear finite element analyses, in which the overall truss bridge is modeled as a rigid frame which consists of beam elements. According to the analytical results in this study, the overall bridge model at a non-corroded state has the safety factor as about 5 times of the design live load. In general, almost actual bridges have the concealed safety factors which were not considered in the design. It can be said that this is the reason why the bridges don't collapse even if the severe corrosion occurs in members. However, corrosion progressions cause the strength reductions of bridges, then the bridges finally collapse when they reach the corrosion limit. Therefore, this study shows the fact that it is very important to evaluate the current safety of the actual bridges quantitatively by estimating the corrosion limit and remaining life.

Keywords: Ultimate strength, Remaining strength, Nonlinear FEM analysis, Deterioration curve, Overall structure, Maintenance.

1 INTRODUCTION

Recently in Japan, there are a lot of steel bridges with conspicuous corrosion due to the aging of themselves. 20 years later from now, over 70% of approximately 700,000 existing road bridges with at least 2 m in length will reach the age of more than 50 years. As a result, deteriorated bridges are thought to increase even more, at the same time, in light of recent socioeconomic conditions, they would be needed to maintain more rationally and efficiently to ensure the sustainable use and the safety. Thus, it is required for civil engineers to evaluate the safety of corroded bridges based on the load bearing capacity. However, in fact, the safety of bridges has usually been evaluated by visual inspections without judgment based on a quantitative way, such as the remaining strength. It is very difficult for engineers to evaluate the corroded bridges numerically according to the corrosion progression because they often encounter the following questions; 1) How to evaluate the remaining load bearing capacity of corroded bridges, 2) How to judge the extent of corrosion damages can be allowed, and 3) How to estimate the remaining life of existing bridges. As an available method for resolving the questions above in terms of the maintenance field, it is considered that the safety of structures can be evaluated based on the strength of overall bridges. This is why the bridges actually exist as total systems that consist of various structural members, although the bridge design usually is based on the design for each member.

In this study, the remaining strength and corrosion limit of corroded truss bridges are analyzed and evaluated based on the ultimate strength and behavior of the 3D model, and the remaining life is estimated by deciding the required performance. Finite element analyses are performed considering geometrical and material nonlinearities, then the rationality and applicability of the overall 3D model is shown by comparing the results of both models.

2 METHOD FOR ESTIMATING THE REMAINING LOAD BEARING CAPACITY, CORROSION LIMIT, AND REMAINING LIFE

First, in order to estimate the load bearing capacity of a truss bridge at the present situation, an analytical model which can express current corrosion conditions is made up. In this study, the remaining load bearing capacity as the ultimate strength can be obtained by FEM analyses under the conditions of incremental loads by adding the live loads to the dead load. Corrosion damages are given to the model by thickness reductions of steel plates which compose members. Then, it is assumed that the corrosion damages increase proportionally according to the elapsed time. From the analytical results accompanied with the corrosion damages, the relationships between thickness reductions and remaining load bearing capacities can be obtained. The corrosion limit would be given as the minimum values of remaining steel plate thicknesses when meeting the required performance, then the required strength of the truss bridge is assumed that it can bear the load as three times of the live load used for the original design. Finally, it is estimated the elapsed years to reach the corrosion limit by using the relationships between the exposure time and the thickness reductions of steel plates. The remaining life is defined as the difference between the elapsed years to reach the corrosion limit and the elapsed years at the time.

The corrosion state is shown by the thickness reduction ratio R as shown in Eq. (1).

$$R = \frac{\text{Reduction of cross sectional area}}{\text{Original cross sectional area}} \times 100\ [\%] \qquad (1)$$

In other words, '$R=R_n\%$' indicates the status that the thicknesses of every steel plates composed each member section are reduced by Rn% respectively. 'R=100%' means a non-existence of elements, that is to say, a member fracture. In addition, it is used a parameter called the reduction magnification ratio β as one corrosion index of the truss bridge because the thickness reduction ratios are different for each member. 'β=1' indicates the current corrosion state, the thickness reduction ratios R of each member increase with proportional factors of the reduction magnification ratios β.

3 CONDITIONS OF ANALYSIS

Skeleton structures and respective member names of the bridge applied in this study are shown in Figure 1. This shows a steel Warren truss bridge that has been designed according to the Japanese specifications for highway bridges.

In order to obtain the corrosion limit, the ultimate strength of the overall bridge is calculated by the large deformation elastic-plastic finite element analysis performed with ABAQUS, a general-purpose FEM analysis program by overall (3D) bridge model Figure 2. The bridge model is made by reference to the literature (Tetsuya Nonaka et al, 2010), where the steel and deck parts are modeled using beam elements and laminated shell elements consisting of concrete layers and rebar layers respectively. The material constitutive law of the steel is shown in Figure 3 based on Eq. (2) as the formula of Ludwik.

$$\sigma = \begin{cases} 20000\,\varepsilon & (\varepsilon \geq \varepsilon_y) \\ 235 + 500\,\varepsilon^{0.5} & (\varepsilon < \varepsilon_y) \end{cases} \tag{2}$$

Herein, ε_y indicates a yield strain of the steel. Figure 4 shows the constitutive laws related to the RC slab that take into account the rebar tension stiffening effect and the adhesion of the rebar and concrete (Japan Society of Civil Engineers, 2013). Slab anchors of connections between the concrete slab and the steel girder are expressed as non-linear springs. The relationship between horizontal slip forces and relative slip displacements is shown in Figure 5. For boundary conditions, one end has a pinned bearing, the other end has a roller bearing.

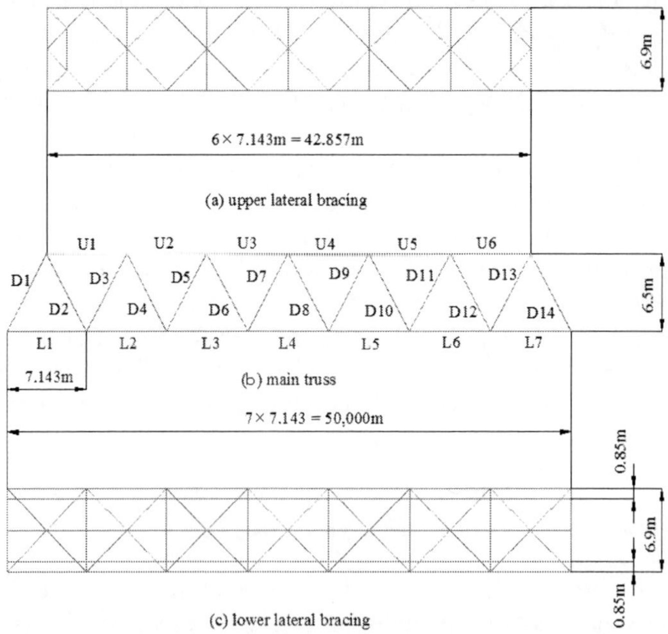

Figure 1. Skeleton and respective member names of bridge.

Figure 2. Overall (3D) bridge model.

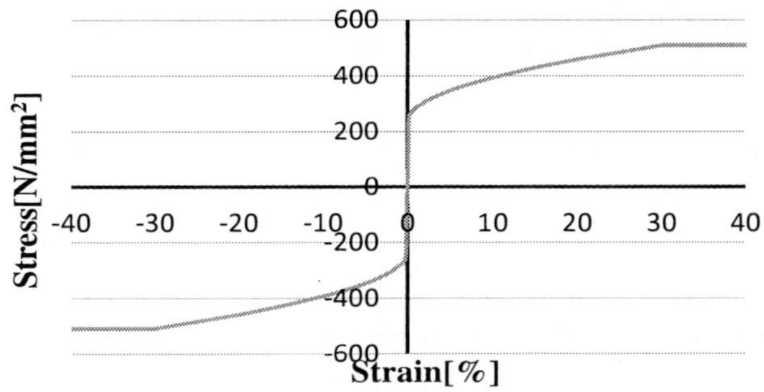

Modulus of elasticity [N/mm^2]	200000
Yield stress [N/mm2]	235
Poisson's ratio	0.3
Unit weight [kN/m^3]	77

Figure 3. Material constitutive law of steel.

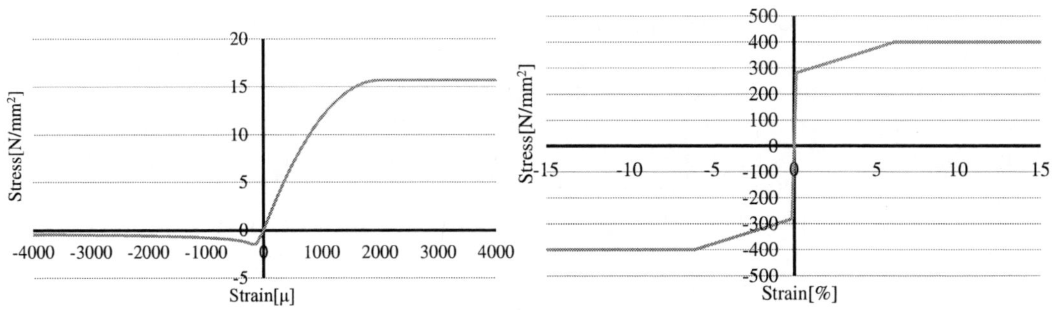

(a) Concrete layer (b) Rebar layer

Figure 4. Material constitutive law of RC slab.

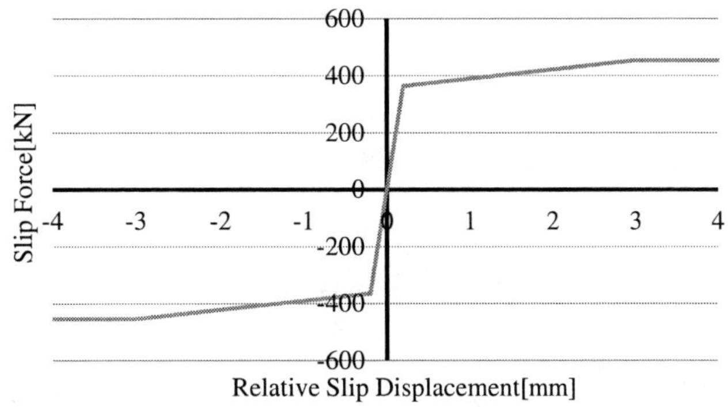

Figure 5. Relationship of horizontal slip force and relative slip displacement of slab anchor.

First, the dead load L_d is loaded, then the design live loads L_l increase gradually to the extent of the maximum load by using load increment parameters α as shown in Eq. (3), where α is called the live load magnification and shows a safety factor against the design live load. In this study, the arrangement of the live load is set as the situation that the member force of L3 is maximized.

$$L_{all} = L_d + \alpha L_l \tag{3}$$

To estimate the ultimate strength, corrosion limit and remaining life, the following assumptive situations are applied to this bridge model; for the thickness reduction ratios R, the

D4, L2, L3 members are given as 10%, and the D5 member is given as 20%, under the condition that 15 years have passed since the beginning of bridge use.

4 ESTIMATIONS OF THE REMAINING LOAD BEARING CAPACITY, CORROSION LIMIT, AND REMAINING LIFE

In order to evaluate the corrosion limit, the live load magnifications α at the ultimate strengths of the 3D bridge model are estimated by analyzing the total 10 cases that the thickness reduction rates R are gradually increased by the reduction magnifications β as shown in Table 1. The relationships of the reduction magnifications and the ultimate strengths are shown in Figure 6. According to the figure, when α is equal to 3 that was defined as the corrosion limit of the bridge, β is almost equal to 6.7. That is to say, it can be said that the bridge may lose a structural functional capacity when β reaches almost 6.7 (R of the D4, L2, L3 members is 67%, R of the D5 member is 100%). Then the relationships between the exposure time and the thickness reductions of steel plates with consideration for the anti-corrosion effect of coating film and its degradation can be given as Figure 7 based on the corrosion surface simulation model (Katashi Fujii et al, 2004). According to the figure, the elapsed time to reach the corrosion limit as the reduction magnification "β=6.7" is around 31 years. Thus, the remaining service life is almost 16 years because nearly 15 years have already passed.

Furthermore, Figure 8 shows the bridge can secure the safety factor as about four times of the design live load even though the D5 member completely fractures with the reduction magnification "β=5". However, it is to be noted as fact that the strength itself has been decreasing as compared to the sound condition.

Table 1. Corrosion state of each analysis model.

Corroded Member	Reduction Magnification β									
	1	2	3	4	5	6	7	8	9	10
	Thickness Reduction Rate R [%]									
L2, L3, D4	10	20	30	40	50	60	70	80	90	100
D5	20	40	60	80	100	100	100	100	100	100

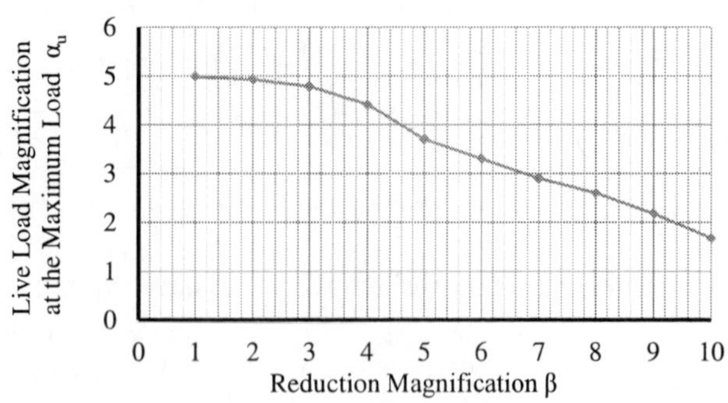

Figure 6. Relationships between reduction magnifications β and ultimate strengths.

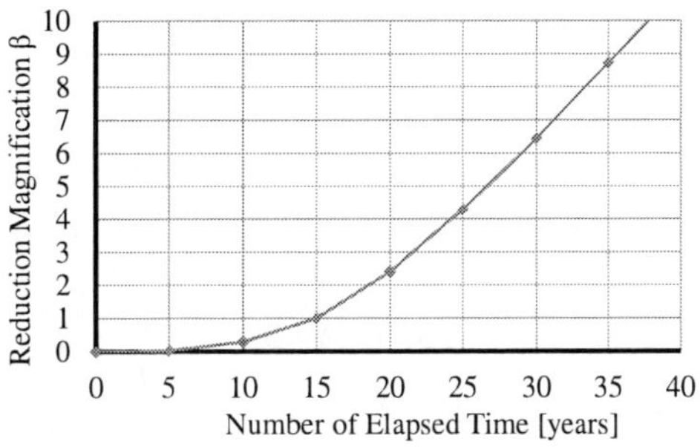

Figure 7. Relationship of elapse time and reduction magnifications β.

5 CONCLUSIONS

(1) By reference to the literature 1), the steel Warren truss bridge is modeled that beam elements and laminated shell elements are applied to the steel and deck parts respectively. The ultimate strengths of the overall bridge under several corroded conditions are calculated by using the composite nonlinear analytical method with consideration for the material and geometric nonlinearity.

(2) The corrosion limit of the corroded truss bridge would be given as the minimum values of remaining steel plate thicknesses when meeting the required performance, then the required strength of the truss bridge is defined that it can bear the load as three times of the live load used for the original design. It is estimated the elapsed years to reach the corrosion limit by using the relationships between the exposure time and the thickness reductions of steel plates. The remaining life is defined as the difference between the elapsed years to reach the corrosion limit and the elapsed years at the time.

(3) According to the analytical results in this study, the overall bridge model at a non-corroded state has the safety factor as about 5 times of the design live load. Then the bridge can secure the safety factor as about four times of the design live load even though the D5 member completely fractures with the reduction magnification "β=5". This fact indicates that actual bridges have the concealed safety factors which were not considered in the design. It can be said that this is the reason why the bridges don't collapse even if the severe corrosion occurs in members. Therefore, it is very important for a near future maintenance theory to evaluate the current safety of the actual bridges quantitatively by estimating the corrosion limit and remaining life.

References

Tetsuya Nonaka, Tsutomu Usami, Maki Iwamura, Atsushi Hirozumi, and Hiroichi Yoshino, A *Proposal for a Redundancy Analysis of Steel Bridges including Progressive Member Failures*, Japanese Journal of Structural Engineering, Vol.56A, 779-791, March, 2010.

Japan Society of Civil Engineers, *STANDARD SPECIFICATIONS FOR CONCRETE STRUCTURES-2012*, Design, Maruzen Co., Ltd., Japan, March, 2013.

Katashi Fuji, Tatsumasa Kaita, Hideharu Nakamura, and Ichiro Ario, *A model generating surface irregularities with consideration of corrosion progress in aging*, Japanese Journal of Structural Engineering, Vol.50A, 657-665, March, 2004.

Integrated Solutions for Infrastructure Development
Edited by Lau, H. H., Tang, F. E., Ng, C. K., and Singh, A.
Copyright © 2016 ISEC Press

THE BENEFIT OF THE NEW SUPER U-BEAM USED FOR SECTION 4 OF A WEST COAST EXPRESSWAY

ONG JIUNDAR and LOKMAN HASSAN

Jacobs Engineering Group Malaysia Sdn Bhd, Kuala Lumpur, Malaysia

The West Coast Expressway is approximately 315.8 km in length and is separated into 15 sections of varying lengths. The concessionaire is West Coast Expressway Sdn. Bhd. Jacobs was involved in Section 4, totaling 5.59 km in length and including 7 directional ramps. These structures feature a new 'Super U-Beam' section whereas standard pre-tensioned precast beams such as the 'inverted T', 'I', 'M' and 'U' beams are typical options for span lengths ranging from 25 m to 40 m. The 'Super' U-beams are larger than standard U-beams and feature wide flanges. This arrangement results in a significant reduction in the number of beams necessary as well as a reduction in all the major material quantities. This paper will further discuss the technical details and advantages of the new 'Super U-Beam'.

Keywords: Pre-tensioned, Precast, Saving, Efficient, Quantity, U-beam.

1 INTRODUCTION

In 2012, Jacobs was appointed by West Coast Expressway Sdn. Bhd. to provide bridge and structural consultancy services and associated technical support during the construction stage for the Sg Rasau-NNKSB Viaduct in Section 4 of the West Coast Expressway (WCE). Section 4 – which runs from the Padang Jawa KTMB rail line to the New North Klang Straits Bypass Interchange – is approximately 5.59 km in length and features elevated structures over the majority of its length. In addition, there are 7nos of ramps which comprise elevated sections between RE Wall embankment approaches. The scale and characteristics of this project were such that the application of a new structural form could be made cost effective, the remit being, viaduct decks that would need predominantly off-site casting, could be as aesthetic as standard U-beam bridges while being cost competitive with post-tensioned T-beam bridges. This led to the development of the precast 'Super' U-beam section.

In order to standardize the design, multi-span viaducts with simply supported girders were proposed with spans ranging from 20 to 40 m, with most spans in the 28-32 m range. For spans up to 32 m, precast pre-tensioned Super U-beams with an in-situ concrete slab are used for the bridge decks. For longer spans, precast post-tensioned Super U-beams are used. The client's preference for off-site casting and lack of space along the alignment led to the preference for pre-tensioned precast beams rather than post-tensioned beams cast on or near the site.

A similar type of precast beam known as the "Super T" or "Teeroff" Girder is in widespread use in Australia. This new beam shape featuring a U with an extended top flange was introduced in the late 1980's and early 1990's. (Gray, Gaby, Grown, Kirkcaldie, Cato and Sweetman 2003) Super T girder's standard shapes have since been developed with depths ranging from 1.0 to 1.80 meter, for spans ranging from 18 to 35 meter. The Super T girder now dominates the precast beam market and is used nationwide in Australia for most bridge spans in the 22 to 35 meter range.

Based on our extensive experience in designing bridges using Super T girders in Australia we have further developed the Super 'T' Girder concept into the Super 'U' beam to be used as a first in Malaysia in Section 4 of the West Coast Expressway.

2 ADVANTAGES OF THE SUPER U-BEAM

2.1 Efficient Structural Form

The Super U-beam is markedly different from the standard U-beam. It has an outstand wing on each side of the webs, providing permanent participating formwork for the casting of the in-situ concrete deck slab. This wide top flange is part of the precast section and is typically 100mm thick on average. Between two adjacent girders, a gap not more than 1m is left between the edges of the top flanges. Permanent non-participating formwork, in the form of cement fiberboard, is provided in the gap between the edges of the top flanges as formwork to the in-situ slab casting. This lightweight permanent formwork allows hassle-free concrete casting without the need for erecting and removing formwork.

A wider top flange coupled with greater beam depth increases the lever arm between the precast section's centroid and the prestressing strands' centroid as well as creating a significantly larger second moment of area. This coupled with increased space in the bottom flange allows a greater amount of pre-tensioning to be applied in its most efficient location during precasting and leads to the possibility of using more widely spaced beams. This does not necessarily require thicker webs as the standard U-beam webs are sized for the practicalities of concreting rather than for strength. The Super U-beam, like the standard U-beam, has a closed box section when the slab is cast, which provides high torsional rigidity and shear resistance. With wider beam spacing, its web capacity in resisting shear and torsion can be fully utilized.

In terms of aesthetics, the Super U-beam provides an attractive box shape that can be used in a variety of situations and is comparable in this respect to the standard "U" beams. The wider beam spacing also facilitates bridge maintenance.

The overall system results in reduced concrete volume reduced prestressing and similar reinforcement quantities when compared with a standard U-beam deck.

2.2 Reduction in Beam Quantity

More widely spaced girders results in significant savings in the overall beam quantity, as elaborated in Table 1. Even though each beam will be more expensive due to greater concrete, prestress and reinforcement quantity, this is more than offset by the saving in construction time at casting yard, storage space, transportation and handling of girders.

With the reduction in beam quantity, the effort required for beam erection, storage and transportation reduces. Even though the Super U-beam requires heavier cranage due to its heavier weight, the deck construction cycle at site and the off-site precast beam production is significantly reduced as the quantity of beams to be lifted is reduced. The reduction of construction time also means reduced exposure to potentially hazardous site activities thereby minimizing construction risk.

3 BEAM PROPERTIES COMPARISON

The goal of the Super U-beam was specifically to improve the efficiency of the existing standard precast U-beam commonly used in Malaysia. The typical characteristics of the two types of beam are compared for a typical span length of 28 m. For this purpose, pre-tensioned standard U10 beam is selected as representative of the typical size chosen for this span length.

3.1 Beam Arrangement and Section Properties

First of all, comparison is made on the quantity of beams required for a range of deck widths, which is shown in Table 1. In general, the typical U-beam spacing is approximately 1.5 - 2.5 m while the optimum spacing for Super U-beams is approximately 3.5-4.5 m. From Table 1, it is evident that the number of Super U-beam required would be significantly less than for the standard U-beam. The optimum beam spacing mentioned dictates the quantity of beams required.

Table 1. Comparison of Beam Quantity Required for Super U-beam and "U10" U-beam.

Deck Width (mm)	Nos "U10" U-beam Required	Nos Super U-beam Required
7900	3	2
9700	4	2
11700	5	3

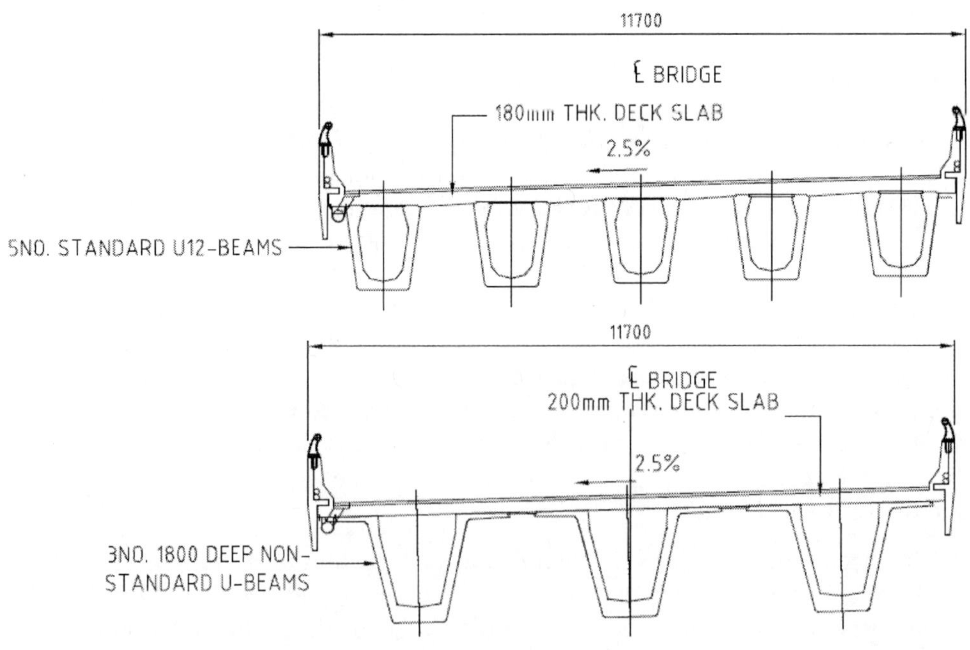

Figure 1. Typical cross sections of precast Super U-beam and "U10" U-beam.

Figure 1 shows the typical cross sections for an 11.7 m wide (twin traffic lanes + hard shoulder) Super U-beam deck which consists of three girders, and the equivalent "U10" U-beam deck which consists of five girders. Figure 2 and 3 show the concrete outline of the Super U-beam and "U10" standard U-beam respectively. The comparison of the detailed beam section properties for a specific deck width of 11.7 m is shown in Table 2.

From Table 2, it can be seen that the Super U-beam is about 48% heavier than a "U10" U-beam. The increased weight of Super U-beam is however, compensated by the reduced number of beams, especially for the wider decks where more than three U-beams are required. For instance, an 11.7 m wide deck requires three Super U-beams with total weight of 7.2 ton per meter length, while five "U10" U-beams with total weight of 8.2 ton per meter length are required.

The 48% increase in concrete volume is accompanied by a 48% increase in cross-sectional area and a 200% increase in section modulus. This means that for the same stress limit Super U-beams can sustain significantly higher bending moments.

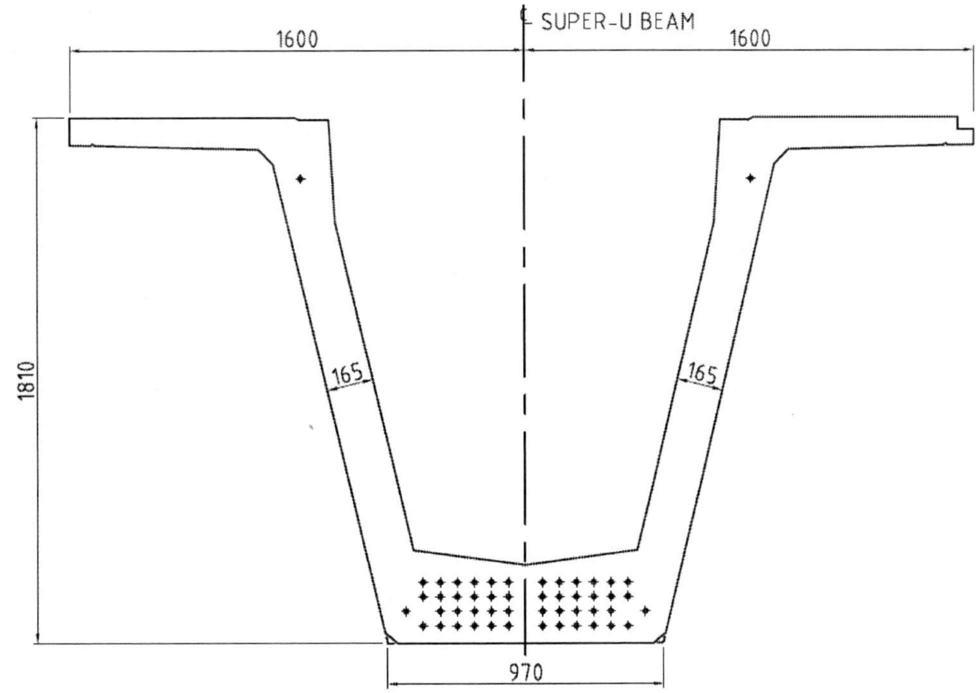

Figure 2. Concrete outline of precast Super U-beam.

Table 2. Comparison of Section Properties of Super U-beam and "U10" U-beam.

Section Properties	**Super U-beam**		**"U10" U-beam**	
	Precast	**Composite**	**Precast**	**Composite**
Beam depth (mm)	1810	2010	1400	1600
Web thickness (mm)	165		165	
Thickness of top flange (mm)	110 – 95 (tapered)		N/A	
Thickness of bottom flange (mm)	270		150	
Cross section area (mm^2)	984955.81	1762494	664755	1132341
Approximate beam weight (Ton)	69	123	47	79
Second moment of area, Ixx (mm^4)	4.02E+11	7.76E+11	1.41E+11	3.35E+11
Depth of centroid from soffit, Yb (mm)	888.64	1265.74	640	976.42
Section modulus about X-axis, Zxt (mm^3)	4.37E+8	1.04E+9	1.86E+8	3.43E+8
Section modulus about X-axis, Zxb (mm^3)	4.53E+8	6.13E+8	2.20E+8	5.37E+8

The top flange of Super U-beam consists of a pair of 110mm thick cantilever slabs at each side of the webs which then taper to 95 mm at the tip. The top flange is combined with a 200 mm thick in-situ RC slab topping to form the composite section. The resulting composite deck slab varies in thickness from 310 mm to 295 mm at the precast flange locations and is 200 mm thick elsewhere (over 1m between beam flanges and in between the webs of each Super 'U' Beam).

The thicker bottom flange coupled with a higher section modulus allow more prestressing strands to be provided without causing excessive tension in the beam's top fiber. This is beneficial during construction stage, where superimposed dead and live loads are not in place.

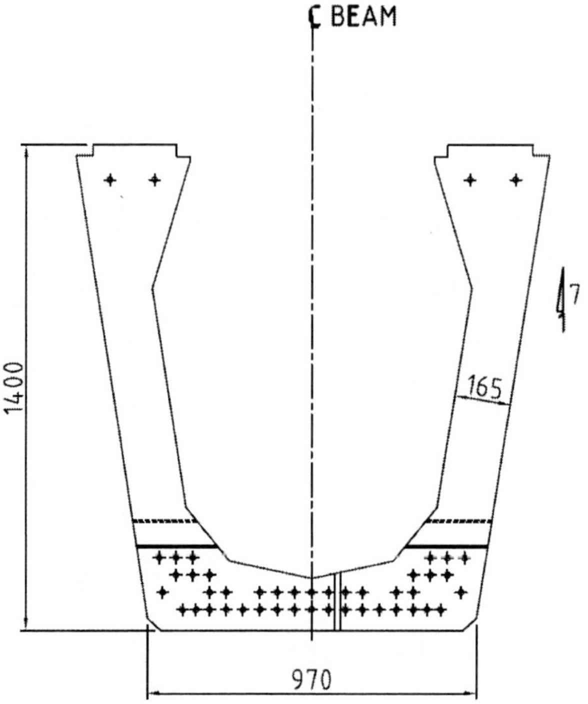

Figure 3. Concrete outline of "U10" standard U-beam.

4 BEAM STRESS COMPARISON

Consider an 11.7 m wide (10.8 m carriageway width), 28 m span simply supported deck for the comparison of SLS beam stress. For an 11.7 m wide deck, three number of Super U-beams at 3.82 m spacing or five number of "U10" U-beam at 2.34 m are required. Assume the deck is subject to gravity loads such as beam concrete self weight, deck surfacing and HA loading. The distributed load applied and the corresponding mid-span bending moment is shown in Table 3.

Table 3. Vertical loads applied on Super U-beam and "U10" U-beam.

Load	Super U-beam	"U10" U-beam
Beam selfweight (kN/m)	25*0.985 = 24.6	25*0.665 = 16.6
Deck in-situ concrete selfweight (kN/m)	25*0.2*3.82 = 19.1	25*0.2*2.34 = 11.7
Surfacing (kN/m)	2.2*3.82 = 8.4	2.2*2.34 = 5.1
HA (kN/m)	9.49*3.82 = 36.3	9.49*2.34 = 22.2
Total SLS load (kN/m)	97.3	61.1
SLS moment at midspan (kNm)	9539	5984

In this case study, the precast beams are designed as class 1 under load combination 1. Concrete grade 60 (strength at transfer of 50 N/mm^2) losses are assumed to total 10 % at transfer and 25% for long term. Based on optimized design, a total of 44nos and 39nos 15.7 mm diameter low relaxation bottom strands, with an initial force of 209 kN, are provided for the Super U-beam and "U10" beam respectively. The beam stress comparisons are summarized in Table 4.

For same deck area of 28 m span and 11.7 m width, the total prestressing strand required for Super U-beam and "U10" U-beam is 4.4 tons and 6.4 tons; the total concrete volume required for the precast Super U-beam and "U10" U-beam is 82.3 m^3 and 93.1 m^3. This leads to saving in 31% of steel strand and 12% of concrete. The saving becomes even more prominent with the increase in the deck width. The advantage of Super U-beam is clearly manifested through the increase in cross section area, second moment of inertia and concrete lever arm.

Table 4. SLS1 beam stress for Super U-beam and "U10" U-beam (Positive for compressive stress).

Load	Midspan Beam Stress at transfer	Midspan Beam Stress at SLS1
Super U-beam – Top fiber	0.62 N/mm^2	5.11 N/mm^2
Super U-beam – Bottom fiber	17.08 N/mm^2	0.5 N/mm^2
"U10" U-beam – Top fiber	0.89 N/mm^2	6.07 N/mm^2
"U10" U-beam – Bottom fiber	20.69 N/mm^2	0.39 N/mm^2

5 DECK CONTINUITY

For multispan bridges, the designer has the choice between isolated individual spans or a succession of continuous structures. In this project, reinforced concrete link slabs have been used to create continuity of the running surface between simply supported spans. Expansion joints are then placed 4 to 6 spans apart creating continuous "modules" between joints up to 160m long. The deck slab between the ends of the precast Super U beams transmits longitudinal loads through axial stiffness and local vertical loads in bending. The flexibility of the link slab is such that no longitudinal bending moment is generated from loads applied in the spans; however, the link slab must be designed to accommodate the imposed rotations occurring as a result of the deflections of the adjacent beams as well as the local bending moments from wheel loads (Benaim 2008).

6 CONCLUSION

This paper has described the characteristics of an innovative new Super U-beam used in the design of viaducts in Section 4 of the West coast Expressway Project. Comparisons have been drawn between this new beam and the commonly used standard precast U-beams and it can be concluded that significant material savings are achievable while maintaining the aesthetic advantages of the U-beam.

Acknowledgments

We would like to thank the concessionaire West Coast Expressway Sdn. Bhd, Malaysian Highway Authority (LLM), and Jacobs Engineering Group Malaysia Sdn Bhd, for supporting the publication of this paper. We would like to express our heartiest gratitude to Mr. Akram Malik, Jacobs ex-Technical Director for his review and contribution to the development of the Super U-beam concept.

References

Gray, A., Gaby, P., Grown, G., Kirkcaldie, D., Cato, R., Sweetman, P. *New Standard Precast Concrete Bridge Beams Stage 1 – Research & Identify Proposed Standard Beam Shapes and Spans*, Transfund New Zealand, New Zealand, 2003.

Benaim, R., *The Design of Prestressed Concrete Bridges Concepts and Principles*, Taylor & Francis, USA, 2008.

Integrated Solutions for Infrastructure Development
Edited by Lau, H. H., Tang, F. E., Ng, C. K., and Singh, A.
Copyright © 2016 ISEC Press

SCISSORING ORIGAMI INSPIRED DEPLOYABLE BRIDGE FOR A DISASTER

KOTARO ADACHI[1], ICHIRO ARIO[1], YUKI CHIKAHIRO[2], and SHUICHI ONO[3]

[1] *Graduate School of Engineering, Hiroshima University, Higashi-Hiroshima, Japan*
[2] *Dept of Water Environment and Civil Engineering, Shinshu University, Nagano, Japan*
[3]*Japan Construction Method and Machinery Research Institute, Shizuoka, Japan*

Many natural disasters cause not only critical situations for facilities and resident's life, but also significant damage to economy. It is obvious that quick rescue action must be undertaken and that there are many problems due to the occurrence of secondary disasters at rescue work-site. For example, many bridges were damaged by a huge earthquake in 2011 in Japan's Tohoku areas, called the Great East Japan Earthquake. We need to develop a new rescue structure to survive these disasters. We have to consider how to rebuild damaged infrastructures and how to build a new type of rescue system. Therefore, we suggest a new type of emergency bridge, Mobile bridge(MB). In this paper, we discuss the scissors type of bridge in order to evaluate its numerical approach including a reinforced strut and characteristics. Moreover, we analyze the design method of calculation model based on theoretical equilibrium theory based on origami-folding engineering.

Keywords: Origami, Mobile bridge, Emergency bridge, Scissors structure, Rescue system, Strut reinforce.

1 INTRODUCTION

In recent years, natural disasters such as earthquakes, floods and tsunamis have caused widespread social damages. For example, many bridges were damaged by a huge earthquake in 2011 in Japan's Tohoku areas, called The Great East Japan Earthquake. We need to develop a new rescue structure to survive these disasters. We have to consider how to rebuild damaged infrastructures and how to build a new type of rescue system (Ario *et al.* 2011). Therefore, we suggest a new type of emergency bridge, Mobile Bridge(MB), with a scissors structure as shown in Figure 1 (Ario *et al.* 2011). This experimental MB can expand and fold main structural frame, and its characteristic provides rapid construction on site (Ario 2006). However, the MB is a flexible structure because of consisting many number of hinge connection. That is, the MB has an engineering issue such as wind vibration, earthquake shaking. Therefore we consider the strut reinforcement to raise more performance of the MB. In this paper, we suggest strut reinforcement for the MB after the expanding, and try to build a calculation technique. From the result of numerical simulation, we inspect mechanical property of MB by the reinforcement.

Figure 1. Mobilebridge ver 2.0 at CMI.

2 MECHANICS OF SCISSORS STRUCTURE

In this section, it is reviewed mechanics of scissors structure. FBD of a unit scissors structure is shown in Figure 2 (Chikahiro *et al.* 2011). When the length of the members is L_0 and the expanding angle of inclination is θ, the sectional length λ and height $2h$ are $L_0\sin\theta=\lambda$ and $L_0\cos\theta=2h$. So, the construction and storage of such a structure can be shown by the angle θ.

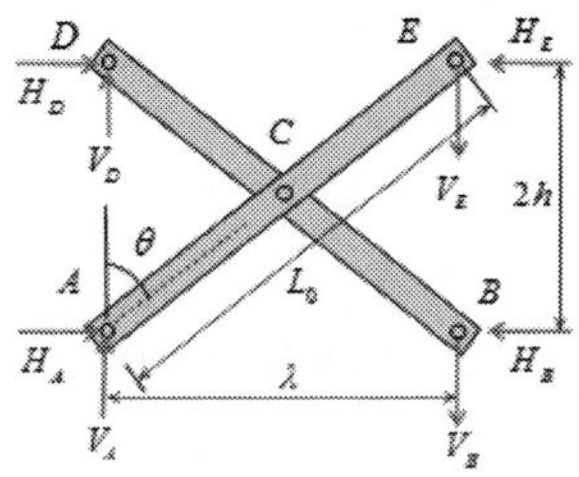

Figure 2. FBD of scissors structure.

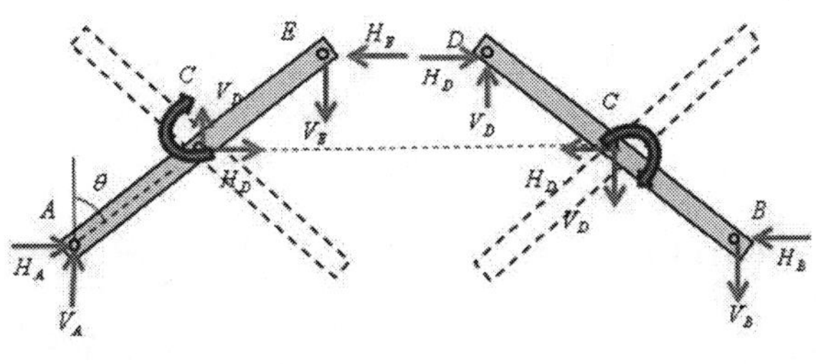

Figure 3. Continuity conditions of each member for AE and BD.

This unit scissors structure can be designed by using the equilibrium equations. The equilibrium equations concerning each external force V_A, H_A and V_E, H_E are given as two expressions, and the equilibrium equation of the moment concerning each nodal point from A to D is set up with four expressions, as shown in the following:

$$\sum H = H_A - H_B + H_D - H_E = 0 \tag{1}$$

$$\sum V = V_A - V_B + V_D - V_E = 0 \tag{2}$$

$$\sum M_{atA} = (H_D - H_E)2h + (V_B + V_E)\lambda = 0 \tag{3}$$

$$\Sigma M_{atB} = (H_D - H_E)2h + (V_A + V_D)\lambda = 0 \tag{4}$$

$$\Sigma M_{atD} = (-H_A + H_B)2h + (V_B + V_E)\lambda = 0 \tag{5}$$

$$\Sigma M_{atE} = (-H_A + H_B)2h + (V_A + V_D)\lambda = 0 \tag{6}$$

Looking at the members AE and BD that intersect as shown in Figure.3, it is apparent that the equilibrium equations of a couple of moments occur at Point C.

$$\textit{Member} \text{ of } AE: M_C = -H_A h + V_A(\lambda/2) = H_E h - V_E(\lambda/2) \tag{7}$$

$$\therefore \left(-H_A - H_E\right)h + (V_A + V_E)(\lambda/2) = 0$$

$$\textit{Member} \text{ of } DB: M_C = H_D h + V_D(\lambda/2) = -H_B h - V_B(\lambda/2) \tag{8}$$

$$\therefore \left(H_B + H_D\right)h + (V_B + V_D)(\lambda/2) = 0$$

It can be present following matrix by arranging the eight calculated equilibrium equations.

$$\begin{bmatrix} 1 & 0 & -1 & 0 & 1 & 0 & -1 & 0 \\ 0 & 1 & 0 & -1 & 0 & 1 & 0 & -1 \\ 0 & 0 & 0 & \lambda & \lambda & 0 & -2h & \lambda \\ 0 & \lambda & 0 & 0 & 2h & \lambda & -2h & 0 \\ -2h & 0 & 2h & \lambda & 0 & 0 & 0 & \lambda \\ -2h & \lambda & 2h & 0 & 0 & \lambda & 0 & 0 \\ -2h & \lambda & 0 & 0 & 0 & 0 & -2h & \lambda \\ 0 & 0 & 2h & \lambda & 2h & \lambda & 0 & 0 \end{bmatrix} \begin{Bmatrix} H_A \\ V_A \\ H_B \\ V_B \\ H_D \\ V_D \\ H_E \\ V_E \end{Bmatrix} = \begin{Bmatrix} 0 \\ 0 \\ 0 \\ 0 \\ 0 \\ 0 \\ 0 \\ 0 \end{Bmatrix} \tag{9}$$

However, the pivot is set up from the condition of continuity of the member. $2h = L_0\cos\theta$, and $\lambda = L_0\sin\theta$. An unknown reaction force can be solved by thinking about the loading condition and the boundary condition for the equations of equilibrium.

3 MECHANICS OF A TWO-UNIT SCISSORS STRUCTURE

A two-unit scissor structure under the cantilever condition, which includes pinned support at points B^L_1 and A^L_1 and a load P at point A^R_2, as shown in Figure 4 is considered as an example.

In this problem, it is possible to treat the external forces as the left and rights sides of internal forces operating on the hinges at points $B_{1,2}$ and $A_{1,2}$, Hence, these relationships are expressed as,

$$\begin{Bmatrix} (B_{1.2})_x \\ (B_{1.2})_y \\ (A_{1.2})_x \\ (A_{1.2})_y \end{Bmatrix} = \begin{Bmatrix} (B_1^R)_x \\ (B_1^R)_y \\ (A_1^R)_x \\ (A_1^R)_y \end{Bmatrix} + \begin{Bmatrix} (B_2^L)_x \\ (B_2^L)_y \\ (A_2^L)_x \\ (A_2^L)_y \end{Bmatrix} \tag{10}$$

$$\{(B_{1.2}), (A_{1.2})\}^T = \{(B_1^R), (A_1^R)\}^T + \{(B_2^L), (A_2^L)\}^T \tag{11}$$

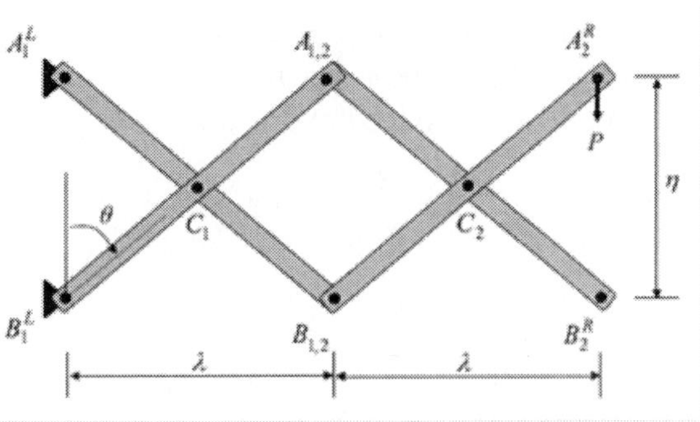

Figure 4. Two-unit cantilever model.

The matrix form for each scissor unit in the two-unit scissor problem can be also obtained by following a procedure similar to the one described for the unit scissor model. The equilibrium equation for the first unit can be expressed as

$$\mathrm{L}\{(B_1^L),(A_1^L)\}^T = -\mathrm{R}\{(B_1^R),(A_1^R)\}^T - \{(C_1,0)\}^T \tag{12}$$

Similarly, the equilibrium equation for the second unit can be expressed as

$$\mathrm{L}\{(A_2^L),(A_2^L)\}^T = -\mathrm{R}\{(A_2^R),(A_2^R)\}^T - \{(C_2,0)\}^T \tag{13}$$

Substituting Eq. (12) and Eq. (13) into Eq. (11) and rearranging them, the matrix form for the two-unit scissor problem under the cantilever condition is obtained:

$$\{(B_1),(A_1)\}^T = -(L^{-1}\mathrm{R})^2\{(B_2),(A_2)\}^T - L^{-1}\mathrm{R}\{(B_{1,2}),(A_{1,2})\}^T - L^{-1}\mathrm{R}L^{-1}\{(C_2,0)\}^T - L^{-1}\{(C_2,0)\}^T \tag{14}$$

By substituting the initial condition of $(A^R{}_2)_y = P$ and the other nodal forces $= 0$ into Eq. (14), the unknown reaction forces for the two-unit scissor structure can be expressed as $(A^L{}_1)_x = -(B^L{}_1)_x = -2P\tan\theta$ and $(A^L{}_1)_y = P$, $(B^L{}_1)_y = 0$. When the theoretical results for the one-unit and two-unit scissor structures are compared, we can see that the vertical reaction forces are the same, but the horizontal reaction forces are doubled. It is clear that the horizontal reaction forces may become very large if the number of scissor units is increase.

4 MECHANICS OF A N-UNIT SCISSORS STRUCTURE

The reaction and section forces for a scissor structure with n units can be calculated in a manner similar to that for the two-unit scissor problem. If the relationship between the sectional stresses and the angle θ that acts on each member are considered, we can see that the bending stress that depends on the length of each member is also the predominant sectional stress. From these concepts, it is possible to design a scissor structure simply by considering only the predominant sectional stress, that is, the bending stress.

5 MECHANICS OF A REINFORCED SCISSORS STRUCTURE

5.1 Effect of a Vertical Reinforcement

In this section, the effect of scissors structure with the reinforcement is examined. Although the reinforced scissors structure is statically indeterminate problem, the section forces of scissors structure with the reinforcement also can be solved once statically indeterminate force is solved. Figure 5 (a) shows the scissors structure with vertical member. A unit scissor structure under the cantilever condition, which includes pinned support at left side hinges and a load P at upper-right hinge is considered. The section forces of the vertical member have statically indeterminate force X. Statically indeterminate force X is introduced by the unit load method.

$$X = -\frac{\sum\int\frac{M_0M_1}{EI}dx+\sum\frac{N_0N_1}{EA}\frac{L}{2}}{\sum\int\frac{M_1M_1}{EI}dx+\sum\frac{N_1N_1}{EA}\frac{L}{2}+\sum\frac{N_1N_1}{EA}\frac{L}{2}} \tag{15}$$

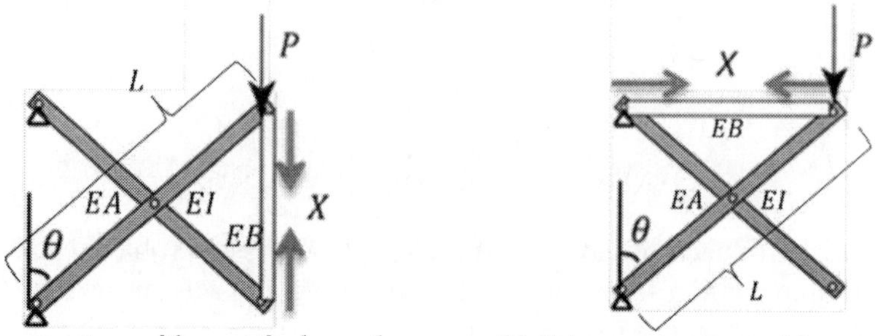

(a) Scissors structure with a vertical member (b) Scissors structure with a upper member

Figure 5. A unit scissors structure with a reinforcing member.

Here, P = load, E = Young's modulus, A = cross-sectional area of scissors structure, I = second moment of area of scissors structure, L = length, θ = expanding angle, B = cross-sectional area of reinforcement structure, N_0 = axis force of statically indeterminate, M_0 = bending moment of statically indeterminate, N_1 = axis force of first system, M_1 = bending stress of first system. Statically indeterminate force X of vertical member is

$$X = -\frac{P}{2} + \frac{6Cos\theta}{12Cos\theta+\alpha(12+\beta L^2+(2-\beta L^2)Cos\theta)} \tag{16}$$

where $\beta=A/I$, and we define α parameter of stiffness. $\alpha=EB/EA$. From Eq. (16), a statically indeterminate force X depends on not only the cross-sectional area of scissors structure but also the area of reinforcement structure. Final section forces are obtained by statically indeterminate force X and found by using principle of superposition (17) and (18).

$$N = N_0+N_1X \tag{17}$$

$$M = M_1+M_1X \tag{18}$$

And, there is a change in comparison with the no reinforced pattern. Bending stress has large change. All bending moment has one member which works load before reinforcement. However, two members have same value of bending moment separately after reinforcement.

5.2 Effect of Upper Reinforcement

An upper reinforced structure is shown in Figure 6(b). Boundary condition and load condition are the same in the previous one. Statically indeterminate force X is introduced by the unit load method.

$$X = -\frac{\alpha P(-12\gamma+L^2+24\gamma csc^2\frac{\theta}{2})sin\frac{\theta}{2}}{24\gamma cos\theta+\alpha L^2\left(1+cos\frac{\theta}{2}\right)+3\alpha\gamma(7+cos\frac{\theta}{4})csc^2\frac{\theta}{2}} \tag{19}$$

where $\gamma = I/A$. Using two equations (17) and (18), and final section forces are solved. By adding the upper reinforcement, the section forces are decreased. The decrease ratio of the section forces by adding the upper reinforcement is larger than vertical one. Bending moment of work one member is almost disappear and the other member has no bending moment. It leads to support point through a reinforcing structure. First one member is worked all bending moment, parameter α is increased, and suddenly bending stress is decreased and becomes horizontally asymptotic.

6 CONCLUSION

The points which became clear from this research are followed as:

- We introduce the equation from equilibrium of force and moment.
- We can expand equilibrium equations as statically indeterminate problem, and lead statically indeterminate force and sectional forces based on two analytical examples.
- The section forces in a scissor structure can be decreased by the strut reinforcement. Especially, the upper reinforcement is more effective than the vertical reinforcement.
- Reinforcing effect does not so increase even if the stiffness ratio between scissor and strut member is change.

References

Ario, I., Structure with the expanding and folding equipment Patent No.2006-037668, Japan, 2006.

Ario, I., Nakazawa, M., Tanaka, Y., and Ono, S., Development of a prototype deployable bridge based on origami skill, AUTOMATION IN CONSTRUCTION, 32, 104-111, 2011.

Chikahiro, Y., Ario, I., and Nakazawa, M., Theory and Design Study of a Full-Scale Scissors-Type Bridge, and M. Nakazawa, Journal of Bridge Engineering (ASCE), 2016.

Integrated Solutions for Infrastructure Development
Edited by Lau, H. H., Tang, F. E., Ng, C. K., and Singh, A.
Copyright © 2016 ISEC Press

THE INFLUENCE OF TEMPERATURE AND HUMIDITY ON RUST GROWTH ON WEATHERING STEEL

RYOYA NAGATA[1], SHUSAKU TACHIBANA[2] and TOSHIHIKO ASO[1]

[1]*Dept of Civil and Environmental Engineering, Yamaguchi University, Ube, Japan*
[2]*Nippon Steel and Sumikin Engineering, Osaka, Japan*

In this study, to develop a growth prediction method of rust on weathering steel, the rust progress characteristics under constant temperature and humidity have been conducted. In order to achieve a long life of weathering steel bridges, determination of several problems are required. Especially, rust evaluation technology, rust prediction method and repair technology of corroded weathering steel are significant. Different concentration (0.3%, 1.0%, 3.0%) of salt water sprayed on steel test pieces and,- were placed in a constant temperature and humidity test vessel, to observe the progress of rust in in-plane direction. Four patterns of exposure environment were tested. Exposure environment were combination of temperature (40°C, 20°C) and relative humidity (50%, 95%). To calculate the increase of rust area, specimen surface was confirmed. Higher salt water concentration (3%) brought rapid rust development. Furthermore, the increase of rust area was highly dependent on temperature than humidity. From time observations, it becomes clear that increase of rust area can be approximated to a logarithmic curve.

Keywords: Weathering steel, Rust, Salinity, Temperature, Humidity.

1 INTRODUCTION

In recent years, cost in construction has been based on life-cycle-cost. Life-cycle-cost consists of not only initial construction cost but also maintenance cost (Kihira 2007). Generally, a large part of maintenance cost of steel structures is paint. Paint has been used for corrosion control of steel structures.

Weathering steel is a steel which can reduce the corrosion rate by generating a dense protective rust on the steel surface and exert anti-corrosion performance. Thereby, weathering steel could be provided corrosion protection performance without painted layer. Due to these advantages in maintenance cost, number of weathering steel bridges has been increased in Japan (Kage *et al.* 2007). However, in some cases, very severe surface corrosion which be able to reduce load bearing capacity has been observed (JSSC 2006, JSSC 2009).

In order to achieve a long life of weathering steel bridges, rust rating technique, method of rust predictive simulation and repair technique of corroded weathering steel member must be established.

To establish a growth prediction method of rust, relationship between rust growth speed and the corrosive environment has to be clarified. In this study, exposure tests of weathering steel were performed under constant temperature and humidity environment. During exposure test, area of rust was observed by image analysis to estimate spread speed of rust on specimen surface. From the results of test, horizontal spread of rust could be approximate as logarithmic curve.

2 MATERIALS AND METHODS

2.1 Exposure Test

Exposure tests were started from October 20, 2015. A growth of rust is horizontal spread on steel surface and three dimensional progress by occurrence of steel wear and corrosion product. In exposure test, weathering steel of 70×70×6(mm) was used as specimen. Salt water of 0.02 ml was dropped on the specimen. Salt water was made from sodium chloride and deionized water. One specimen was supplied only one kind of salt water. Then moisture on the specimen surface was dried. The specimen was installed in a small environment tester. The concentration of salt water was set to 0.3%, 1% and 3%. In this study, the specimen which is dropped salt water was exposed. Furthermore, to clarify the effect of salt to steel corrosion in this tests, non-supply specimens ware also set. Table 1 gives conditions of the test. Temperature and humidity were decided due to the performance of small-environmental-tester and observations of bridge sites. Formed rust on each specimen were photographed every day. The image was trimmed to 280 × 280 pixels around center of a salt water drop part. The area of this image was 1238.278mm^2. Shot images are divided into three components, red, green and blue(RGB). Then using analysis software, the image was taken out only green component from the RGB components. The area ratio of rust part was calculated by extracted image. The area of rust part was the ratio of rust and trimmed image area.

Table 1. Test conditions.

	Temperature-Humidity	Number of specimen	Salt water concentration	Period of exposure
Case1	40°C-95%	4	Not dropped	30
Case2	40°C -50%	4	0.3%	29
Case3	20°C -50%	4	1%	18
Case4	20°C -95%	4	3%	20

3 TEST RESULTS

3.1 Appearance Evaluation

Figure 1 shows the appearance of the specimen which is exposed of day1 and day 30 in temperature 40°C and relative humidity 95% (3% salt water). The color of the salt water drop part changed to black during 30 days. Therefore, it was possible to confirm clearly that the rust grew. Except for the salt water drop part, the occurrence of uniform corrosion was recognized. However, this uniform corrosion was not verification target. Figure 2 indicates the banalization of green component of Figure 1(b). This process is effective to enhance the rust region in images. Figure 3 shows the image of particle analysis. The area ratio of all images was calculated from this step.

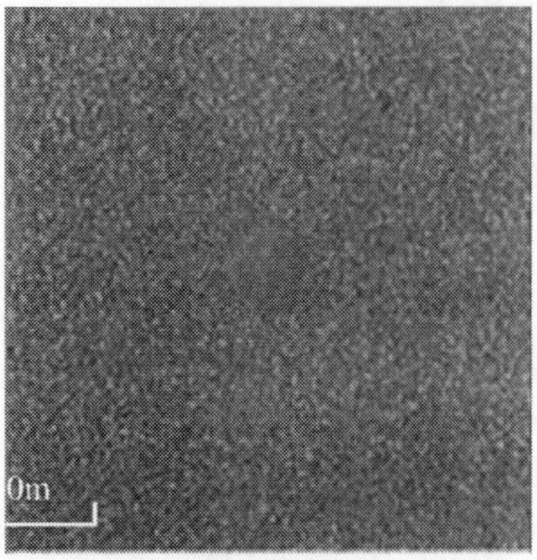

(a) Day1

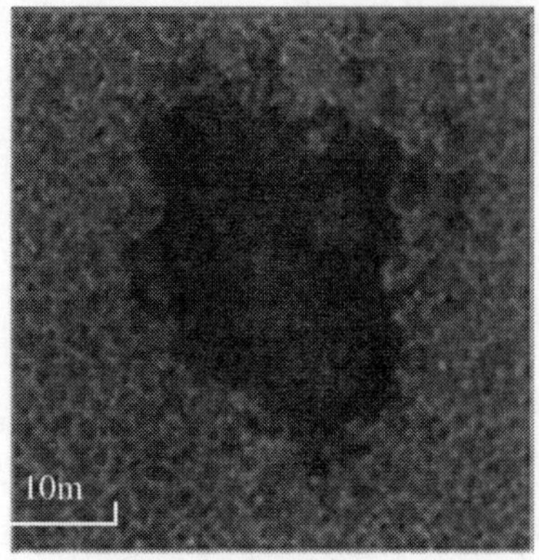

(b) Day30

Figure 1. Variation of area ratio at day1 and day30 (3% salt water)

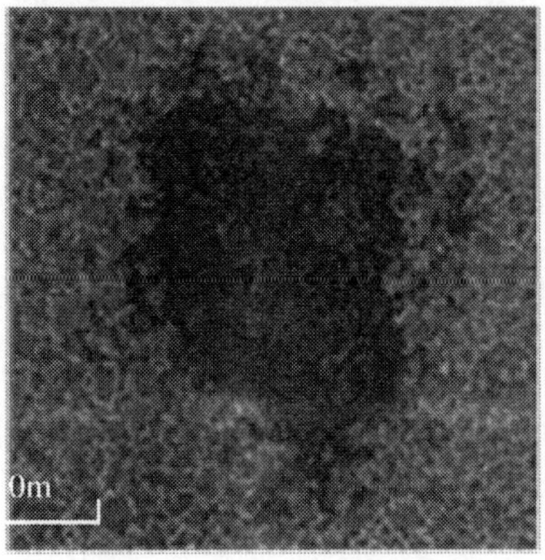

Figure 2. Image of green component

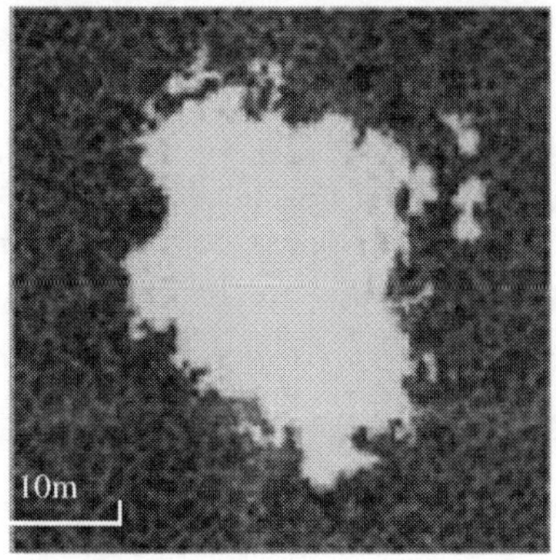

Figure 3. Particle analysis image

3.2 Evaluation of Area Ratio of Rust Part

Influence of corrosion environment is evaluated from the ratio of rust-region. Figure 4 (a) and Figure 4 (b) illustrates the variation of the area ratio with duration in Case1 and Case4, Case2 and Case3, respectively. In Case1 and Case 4, relative humidity is set for 95% in both case but temperature is different (40°C in Case1, 20°C in Case4). In Case2 and Case 3, relative humidity is 50%. In all cases, the area ratio of the specimen which was dropped 3% salt water was big. Therefore, the influence of salt on the rust growth could be confirmed. However, increasing tendency of the area ratio was different in Case1 and Case4. The area ratio which is dropped 3% salt water in Case1 was 25% in day 20. In contrast, the area ratio in Case4 was 3%. It was found that tendency did not depend on the salinity. Furthermore, it was observed that the influence of temperature to growth of rust is big from Fig.4 (a) and Fig.4 (b). This tendency could be confirmed in Case2 and Case3 that are low relative humidity. From the comparison with Case1 and Case 2, in the same temperature and same salt water concentration, high relative humidity provides large area ratio. However, this tendency could not have observed in low temperature (Case3 and Case4). The logarithmic approximation formula is shown to Fig.4 (a) and Fig.4 (b) dashed line. But the approximation was impossible from data of Case3 and Case4. The calculated approximate value from these approximate formula was close to the actual measured value. In all cases, the coefficient of determination was more than 0.75.

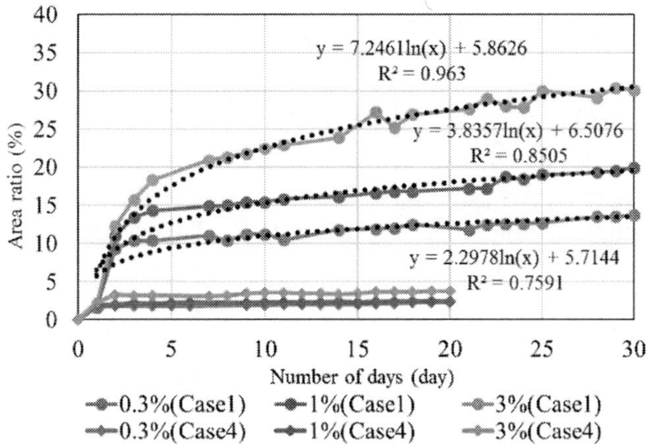

(a) Case1 and Case4

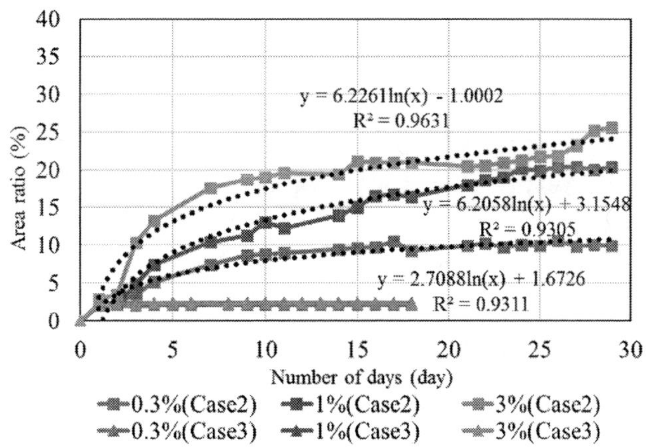

(b) Case2 and Case3

Figure 4. Variation per day of area ratio with duration.

4 GROWTH PREDICTION METHOD OF RUST

This paper also aims to establish a growth prediction method of rust to quantitatively evaluate a corrosion process of weathering steel. From the results of exposure tests it became clear that rust growth is due to temperature, humidity, salinity. The relationships between horizontal growth of rust and exposure duration can be assumed by using logarithmic function. The rust growth prediction would be expressed by equation (1)

$$Y = A\ln(X) + B \tag{1}$$

Where, Y: rust thickness, X: exposure days, A and B: Constants.

In Case1 and Case2, rust growth could be confirmed clearly. The relationship between concentration of salt water and coefficient or intercept of the logarithmic formula was verified.

Figure 5 shows the relationship between coefficient of the logarithmic formula and concentration of salt water in Case1 and Case2. Also Figure 6 shows the relationship between the intercept of the logarithmic formula and concentration of salt water. In Case1, linear relationship was recognized between the coefficient and concentration of salt water. In addition, three values of intercept are almost the same in the case of concentration. And on the other hand, the tendency was not recognized in Case2. In Case2, the value of 1% salt water was protruded. The reason is that the rust growth of the specimen which was dropped 1% salt water was big in 10 days later in Case2. In Case1, the rust growth prediction equation was calculated. It became clear that A is $1.805C$ (C: concentration of salt water) and B is 6.027. The rust growth prediction equation was expressed by equation (2) in Case1.

$$Y=1.805C\ln(X)+6.027 \tag{2}$$

Where, Y: area ratio of rust, X: exposure days, C: concentration of salt water.

Figure 7 shows measured value and calculated value from equation (2). Equation (2) shows good agreement to measured value. Therefore, equation (2) is applicable as rust growth prediction equation in Case1. Accuracy of approximation could be improved by increasing the experimental results.

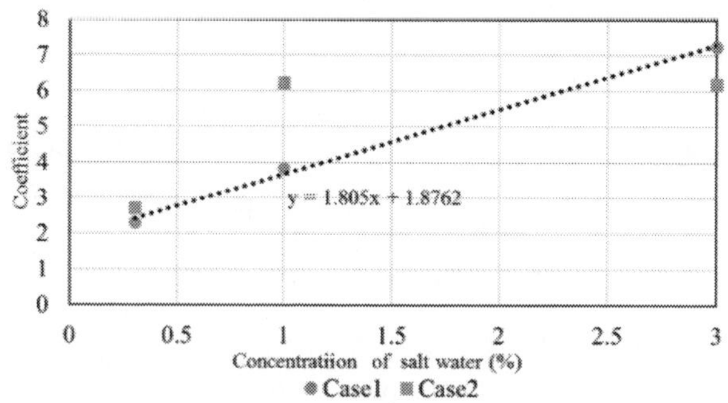

Figure 5. Coefficient and concentration of salt water.

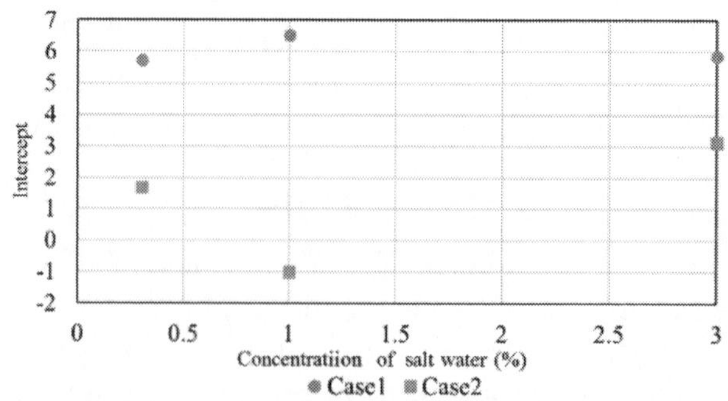

Figure 6. Intercept and concentration of salt water.

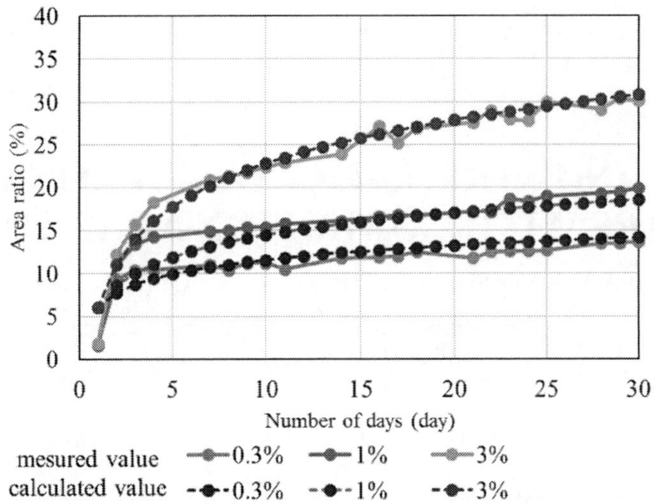

Figure 7. Measured value and calculated value.

5 CONCLUSION

Rust growth on weathering steel under several corrosion environment was evaluated by ratio of rust region. Results of this research could be pointed as below.

1. Rust growth is depending on temperature and salt condition strongly. An effect of relative humidity was small in this test.

2. Ratio of rust region could be approximated by logarithmic formula. Accuracy of approximation could be improved by increasing the experimental results.

3. Coefficient of approximated curves are related to concentration of salt water.

To apply the result of this research to actual bridges, corrosion environment (temperature, relative humidity and amount of airborne salt) of these bridges are required. A database of corrosion environment should be constructed for planning and design of weathering steel bridges.

References

Japan Society of Steel Construction, Potentiality of Unpainted Weathering Steel Bridges, Technical Report No.73, Japan Society of Steel Construction, Tokyo, 2006. (in Japanese)

Japan Society of Steel Construction, Proactive Maintenance of Weathering Steel Bridges, Technical Report No.86, Japan Society of Steel Construction, Tokyo, 2009. (in Japanese)

Kage I., Kyono K., and Matuda Y., Prediction for Corrosion Loss of Weathering Steels, JFE Technical Report, No.18, 62-66, Nov 2007. (in Japanese)

Kihira H., Systematic approaches toward minimum maintenance risk management methods for weathering steel infrastructures, Corrosion Science 49, 112-119, 2007

Integrated Solutions for Infrastructure Development
Edited by Lau, H. H., Tang, F. E., Ng, C. K., and Singh, A.
Copyright © 2016 ISEC Press

REMAINING STRENGTH EVALUATION OF PLATE GIRDERS WITH CORROSION NEAR SUPPORTS

NAOYUKI ASAO[1], KATASHI FUJII[2], HIROAKI FUKUDA[2], and RYUICHI YOKOTA[2]

[1]*Pacific Consultants Co., LTD., Tokyo, Japan*
[2]*Hiroshima University, Hiroshima, Japan*

In Japan, a lot of steel railway bridges were constructed during the period of rapid economic growth in the 1960s. Now, since over 50 years have passed, the aging of these structures has become conspicuous and a major social problem. In particular, for steel plate girders with open deck system, deterioration and damage due to corrosion is becoming obvious. Moreover, the remaining strength evaluation method of corroded structures is not always sufficient. Therefore the purpose of this study is to propose an analytically-based evaluation method for the maintenance of steel railway bridges. The following are noted: 1) Regarding the shear strength of plate girders with corrosion near supports, the remaining strength depends on the corroded surface condition of each component such as a web or a vertical stiffener. And, according to the form and degree of corrosion, the fracture mode also changes from the buckling of web to the buckling of cross-shape column consisting of the web and vertical stiffener. The result shows that the corroded web has more effect on the shear strength than the vertical stiffener. 2) Regarding the peeling between the web and bottom flange due to corrosion near supports, the remaining strength is deteriorated more rapidly than above case 1, when the peeling range lengthens more than the sole plate area.

Keywords: Shear strength, Local buckling, Web, Vertical stiffener, Peeling.

1 INTRODUCTION

As of 2012, there were approximately 102,300 railway bridges of at least 1 m in length in Japan. Figure 1 shows the stock pyramid which consists of the number of these railway bridges separated by age groups. Of them, steel plate girders are one of the oldest bridges, their corrosion damage due to age deterioration is becoming more obvious. Partial repairs and reinforcement as well as total replacement of girders have taken place in recent years, but it cannot be said with certainty that these measures were selected after a proper, quantitative evaluation of the remaining strength of the girders. In light of recent socioeconomic conditions caused by budgetary and labor shortages, there is an even more pressing need to establish streamlined, economically viable methods for maintaining these bridges into the future.

One characteristic of plate girder corrosion is the occurrence of corrosion near supports. The complicated structure of members near supports at girder edges prevents the passage of wind and often limits exposure to sunlight, leading to the pooling of water atop abutments, thus facilitating the progress of corrosion than in other parts. In general, there are many observed cases of corrosion of webs and vertical stiffeners near supports. There are also some cases of peeling at joints between webs and lower flanges due to local corrosion (Figure 2).

This study focuses on steel plate girders, presents a full understanding of the conditions of corrosion and damage observed relatively near the supports, and analytically examines remaining shear capacity and declining strength near steel bridge supports based on specific corrosion models. The study also explains and evaluates how the remaining shear capacity of steel plate girders and forms of collapse change in response to the location and progression of corrosion.

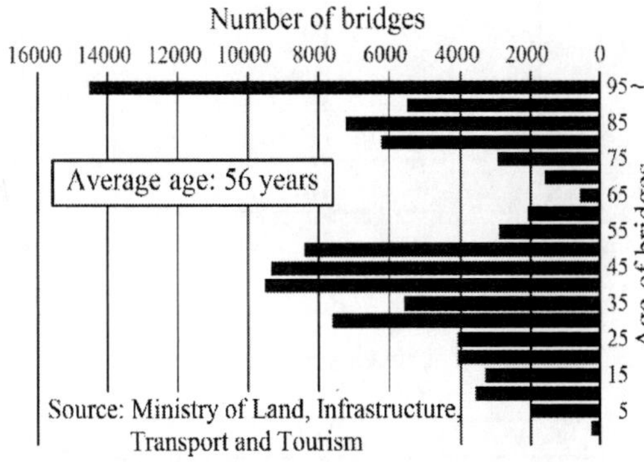

Figure 1. Stock pyramid of existing railway bridges in Japan (As of 2012).

Figure 2. Peeling conditions between webs and lower flanges near supports.

2 CONDITIONS OF ANALYSIS

2.1 Analytical Models

This analysis is conducted using elastic-plastic finite deformation analysis from the commonly used analysis program NASTRAN. SS400 steel per Japanese Industrial Standards (JIS) is used as the material for the analytical models, and material properties have a yield stress σ_y of 235 MPa (tensile strength of 400-510 MPa), an elastic modulus E of 200 GPa, and a Poisson ratio v of 0.3. The constitutive law of materials is applied based on the von Mises yield criterion.

Figure 3 shows the dimension of main plate girders, the target of the analysis. The purpose of this study is to provide evaluations focused on shear capacity, therefore the analytical models are configured such that they account for the symmetry of main girders by applying loads near supports in order to lessen the effects of bending to the extent possible while working with the analytical models. In addition, rigid elements are applied to the supporting sole plate. Shell elements with 4 nodal points are applied to these analytical models, with each element divided into a size of approximately 10 mm (Figure 4). In addition, Figure 5 shows boundary conditions. These analytical models take into account initial imperfections in the out-of-plane direction on the web between the vertical stiffeners. Modeling procedures like these are listed in a reference[1].

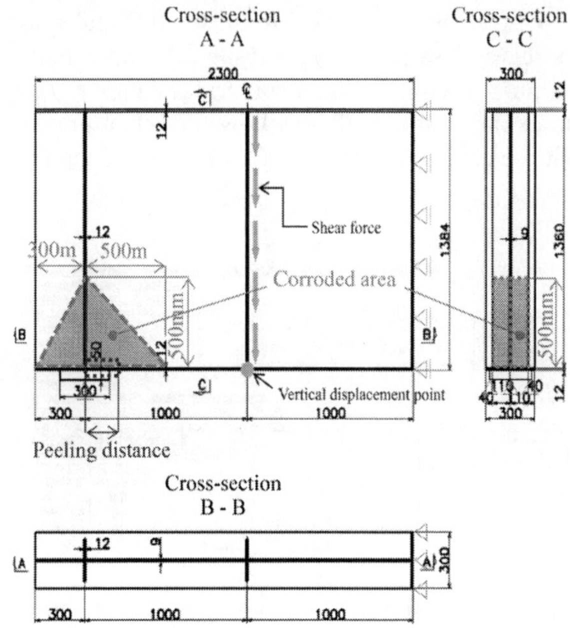

Figure 3. Dimensions of main plate girders targeted for the analysis.

Figure 4. Analytical model.

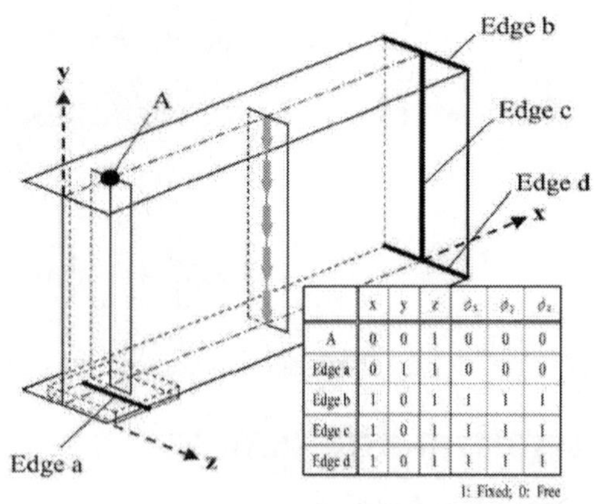

	x	y	z	ϕ_x	ϕ_y	ϕ_z
A	0	0	1	0	0	0
Edge a	0	1	1	0	0	0
Edge b	1	0	1	1	1	1
Edge c	1	0	1	1	1	1
Edge d	1	0	1	1	1	1

Figure 5. Boundary conditions.

2.2 Corrosion Patterns

The results of investigation shown in Table 1 have been reported as the main forms of corrosion of plate girders near supports[1]. With reference to the results, the analyses are conducted on the assumption that uniform mean thickness reductions occur within the typical areas of local corrosion of webs and vertical stiffeners as shown in Figure 3. Case studies where remaining shear load bearing capacity is calculated with respect to corrosion location and remaining thickness as shown in Table 2 are analyzed. Furthermore, the same analyses are conducted on cases of peeling of the joints between webs and lower flanges directly above supports (Figure 3).

---

1 Tamakoshi, T., Keita Nakasu, K., Ishio, M., Tatsuya Takeda, T., and Suizu, N., Research on local corrosion of Highway steel bridges, TECHNICAL NOTE of National Institute for Land and Infrastructure Management, No.294, January 2006.

Table 1. Characteristics of corrosion at girder ends.[1]

	Main corrosion patterns and number of cases			
Corrosion configurations	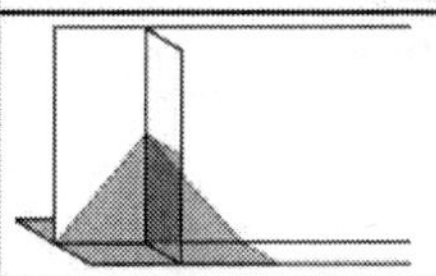	Number of cases 3 / 20 examples (15 %)	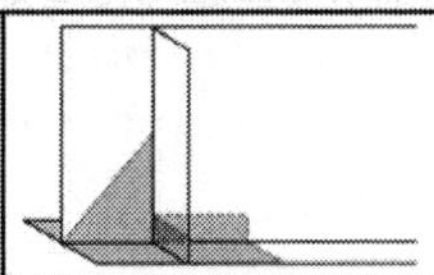	Number of cases 4 / 20 examples (20 %)
	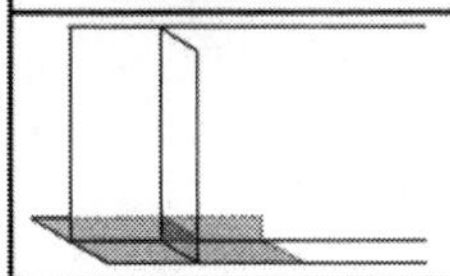	Number of cases 6 / 20 examples (30 %)	Others	Number of cases 7 / 20 examples (35 %)

Table 2. Analytical cases (Non-corroded or corroded, peeling).

I. Non-corroded (Original thickness)

Analytical case	**(1)**
Web (mm)	**9**
Vertical stiffener (mm)	**12**

II. Web (Corroded)

Analytical case	**(2)**	**(3)**	**(4)**
Web (mm)	**7** (-2)	**5** (-4)	**3** (-6)

III. Vertical stiffener (Corroded)

Analytical case	**(5)**	**(6)**	**(7)**
Vertical stiffener (mm)	**10** (-2)	**8** (-4)	**6** (-6)

IV. Web and vertical stiffener (Simultaneously corroded)

Analytical case	**(8)**	**(9)**	**(10)**
Web (mm)	**7** (-2)	**5** (-4)	**3** (-6)
Vertical stiffener (mm)	**10** (-2)	**8** (-4)	**6** (-6)

V. Non-corroded (Original thickness) / Peeling between web and lower flange

Analytical case	**(11)**	**(12)**	**(13)**
Web (mm)	**9**		
Vertical stiffener (mm)	**12**		
Peeling distance (mm)	**75**	**150**	**200**

VI. Web and vertical stiffener (Simultaneously corroded) / Peeling between web and lower flange

Analytical case	**(14)**	**(15)**	**(16)**	**(17)**	**(18)**	**(19)**	**(20)**	**(21)**	**(22)**
Web (mm)	**7** (-2)			**5** (-4)			**3** (-6)		
Vertical stiffener (mm)	**10** (-2)			**8** (-4)			**6** (-6)		
Peeling distance (mm)	**75**	**150**	**200**	**75**	**150**	**200**	**75**	**150**	**200**

* Above ( ) values show thickness reduction

3 RESULTS OF ANALYSIS

3.1 Corrosion Models

Figure 6 shows the relationship between thickness reduction and remaining strength for models in which only one of webs or vertical stiffeners near supports corrodes, and for models in which they both corrode simultaneously. The figure shows that web corrosion has the largest effect on shear capacity at girder edges. This rapid reduction of strength is caused by the occurrence of local buckling of webs before the shear force is transferred to the diagonal tension field of the webs (Figure 7(a)). The figure also shows that, for the models in which only vertical stiffeners corroded, shear strength reduced only around 10% even when thickness was reduced 50%; the effect is not very significant. On the other hand, shear strength reduced much more dramatically for the models in which webs and vertical stiffeners corroded simultaneously than for the models in which only webs corroded. This likely has a greater effect on shear capacity than cases where only vertical stiffeners corroded because shear force is intensively redistributed to vertical stiffeners when the corroded parts of webs suffer local buckling.

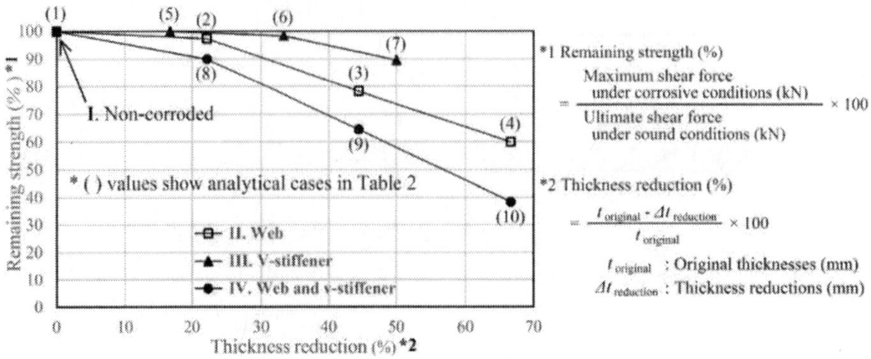

Figure 6. Shear strength reduction due to corrosion.

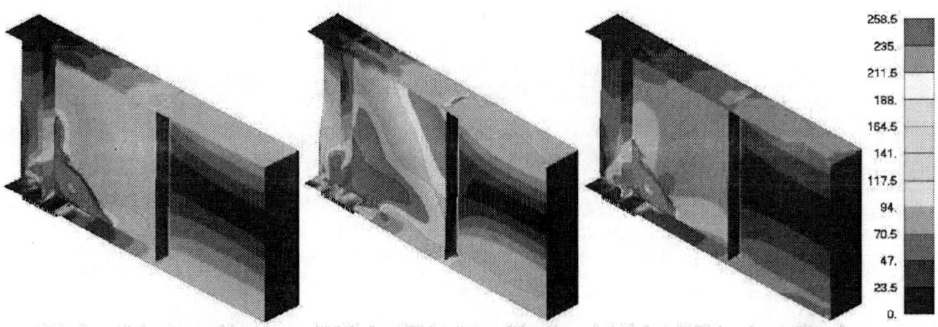

(a) Point-(4) in above Fig. 6 (b) Point-(7) in above Fig. 6 (c) Point-(10) in above Fig. 6

Figure 7. Stress distribution and deformation diagram at maximum shear force (Corrosion).

3.2 Peeling Models

Figure 8 shows the relationship between the distance of peeling from support centers and remaining strength for models in which peeling occurs at joints between webs and lower flanges directly above supports, and for models in which they both corrode simultaneously. In the non-corroded models, shear capacity decreases suddenly when the peeling distance exceeds 150 mm. In addition, deformation stress diagrams of those cases show high concentrations of stress toward the peeling areas (Figure 9(b), (c)). Sole plates exist up to 150 mm from support centers, and if peeling occurs within that area, the webs effectively work as stress transferring mechanism between vertical stiffeners and lower flanges. However, if that function fails once and peeling exceeds that area, stress becomes concentrated on the lower flange and sole plate boundaries, which likely renders lower flanges incapable of withstanding shear deformation. Furthermore, the models show that peelings with local corrosion of webs and vertical stiffeners simultaneously have smaller effects on the strength reduction according to peeling extension as the thickness reduction due to corrosion increases. This is likely because extreme local corrosion near supports causes local buckling at areas with reduced thickness earlier than in peeling areas.

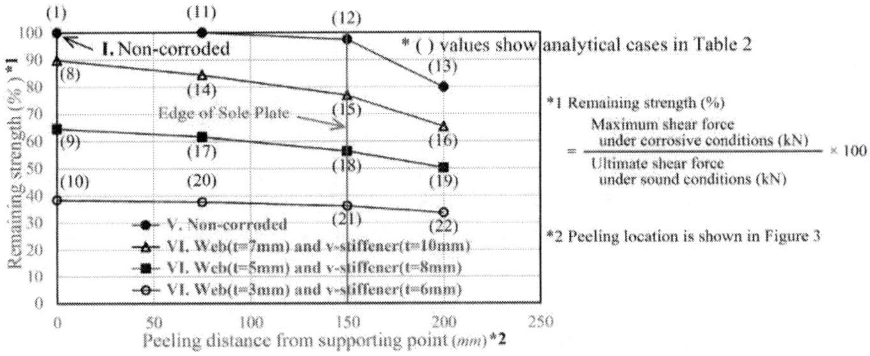

Figure 8. Shear strength reduction due to peeling (Incl. corrosion).

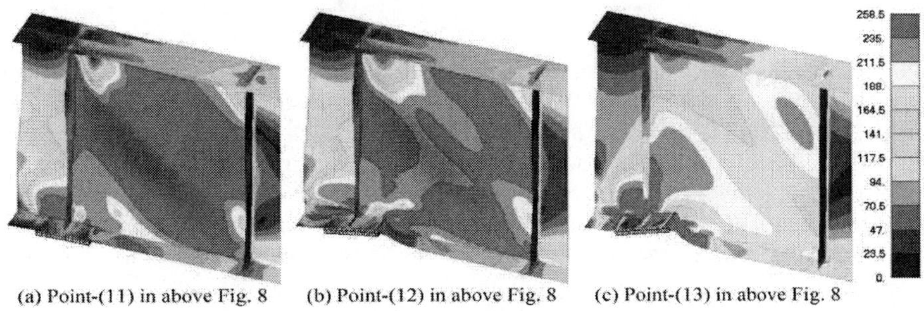

Figure 9. Stress distribution and deformation diagram at maximum shear force (Peeling).

4 CONCLUSIONS

1) Thickness reduction of only webs due to local corrosion near supports has more effect on shear capacity than that of only vertical stiffeners.

2) Collapse due to corrosion takes the form of local buckling of webs or vertical stiffeners, and occurs when the cross-shaped columns comprised of webs and vertical stiffeners become unable to withstand the support reaction. Thus, the strength reduction becomes even more pronounced when both webs and vertical stiffeners corrode simultaneously as compared to when only vertical stiffeners corrode.

3) When peeling occurs and extends past the range of sole plates, stress concentrates on the lower flanges above the sole plate edges, causing sudden shear deformation and strength reduction. And peelings with extreme local corrosion have smaller effects on the strength reduction according to peeling extension because local buckling at areas with reduced thickness occurs earlier than in peeling areas.

References

Tamakoshi, T., Keita Nakasu, K., Ishio, M., Tatsuya Takeda, T., and Suizu, N., *Research on local corrosion of Highway steel bridges*, TECHNICAL NOTE of National Institute for Land and Infrastructure Management, No.294, January 2006.

Integrated Solutions for Infrastructure Development
Edited by Lau, H. H., Tang, F. E., Ng, C. K., and Singh, A.
Copyright © 2016 ISEC Press

SEISMIC PERFORMANCE OF AN OLD JAPANESE-STYLE TWO-STORY WOODEN HOUSE UNDER A STRONG EARTHQUAKE GROUND MOTION

TOMIYA TAKATANI[1] and HAYATO NISHIKAWA[2]

[1]*Dept of Civil Engineering & Architecture, National Institute of Technology, Maizuru College, Maizuru, Kyoto Prefecture, Japan*
[2] *Education and Research Supporting Center, National Institute of Technology, Maizuru College, Maizuru, Kyoto Prefecture, Japan*

In Japan, there is a serious and urgent issue on seismic retrofit for a lot of old Japanese-style two-story wooden houses built by a Japanese traditional framed-construction method. In order to investigate the seismic performance of an old Japanese-style two-story wooden house, 3-D non-linear collapsing process analysis of this wooden house was conducted against a strong earthquake ground motion with the Japan Meteorological Agency seismic intensity of "6 upper" level. The effect of post fixing condition under the floor of wooden house on the seismic response of an old Japanese-style two-story wooden house was numerically investigated in this paper. As a result, it was found that seismic collapsing behavior of the wooden house strongly depends on the post fixing condition under its first floor.

Keywords: Collapsing process analysis, Japanese traditional framed-construction method, Post fixing condition.

1 INTRODUCTION

In Japan, there is a serious and urgent issue on seismic retrofit for a lot of old Japanese-style two-story wooden houses, which were built by a Japanese traditional framed-construction method more than 80 years ago. It is very important for structural engineers to accurately evaluate a seismic response of the Japanese-style two-story wooden house in the design process of seismic retrofit, because the seismic response of two-story wooden house can play a key role to propose an effective countermeasure in the structural design process of seismic retrofit for the wooden house. In order to investigate the seismic performance of an old Japanese-style two-story wooden house, 3-D non-linear collapsing process analysis (Nakagawa and Ohta 2010) of timber structure was conducted against a strong earthquake ground motion with the Japan Meteorological Agency seismic intensity of "6 upper" level. A non-linear behavior of timber elements in the wooden structure during a strong earthquake ground motion can be simulated by this 3-D non-linear collapsing process analysis based on the Distinct Element Method (Cundall and Strack 1979). In general, the posts under the floor of an old Japanese-style wooden structure are erected on their foundation stones and are not fixed. Consequently, this system is likely to be a seismic base isolation. The effect of the post fixing condition under the floor of wooden structure on the seismic response of an old Japanese-style two-story wooden house was numerically investigated in this paper.

2 OUTLINE OF SEISMIC COLLAPSING PROCESS ANALYSIS

In this paper, a structural analysis software of "Wallstat" (Nakagawa and Ohta, 2010) is employed in order to investigate seismic response behavior and collapsing process of two-story wooden house during a strong earthquake ground motion. This software has an original analysis technique using the basic theory of the Distinct Element Method (Cundall and Strack, 1979), and can be taken into consideration the extremely non-linear properties of timber members breaking or being dispersed. In the collapsing process analytical calculation, two-story wooden house can be modeled by lumped mass and the weight of each floor in two-story wooden house model can be obtained from each structural element as illustrated in Figure 1. Timber characteristics of the compression and tensile elasto-plastic springs consist of an elastic part and slip-type part.

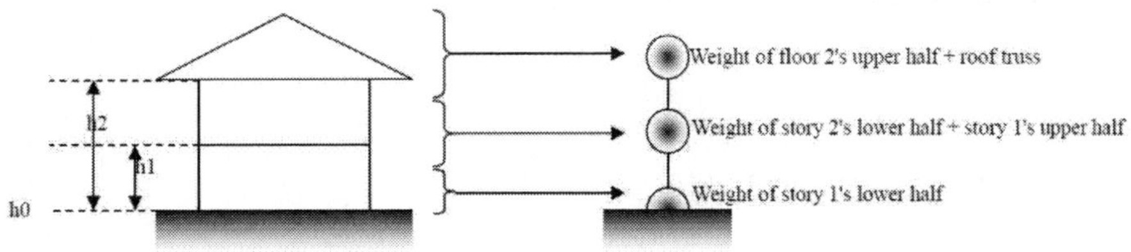

Figure 1. Weight of floor in the analytical model of wooden house.

2.1 Modeling of Structure Element (Nakagawa and Ohta, 2010)

2.1.1 Modeling of Frame

Timber frame is modelled by two elasto-plastic rotational springs (plastic hinge) and an elastic beam component as shown in the left-hand side of Figure 2. The spring can be defined by a relationship between bending moment M and angle of rotation θ with the skeleton curve indicated in the right-hand side of Figure 2.

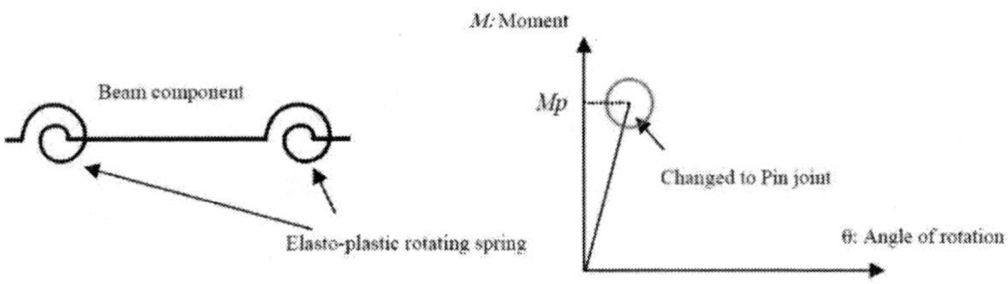

Figure 2. Schematic diagram and skeleton curve of frame spring.

2.1.2 Modeling of Joint Spring

Joint spring can be modelled by both an elasto-plastic spring and a rotational spring as shown in Figure 3(a). Timber characteristic of the compression and tensile elasto-plastic spring consist of an elastic part and slip-type part indicated in Figure 3(b), and also timber characteristics of the rotational spring is assumed to be a slip-type relationship between ending moment M and angle of rotation θ shown in Figure 3(c).

2.1.3 Modeling of Shear Wall and Bracing

Vertical shear wall indicated in Figure 4 can be modelled by the replacement of truss component with a load-displacement non-linear relationship shown in Figure 6. Also, bracing shear wall illustrated in Figure 5 can be modelled by the replacement of compression and tensile truss components defined by a set of bi-linear and slip skeleton curve shown in Figure 6, too.

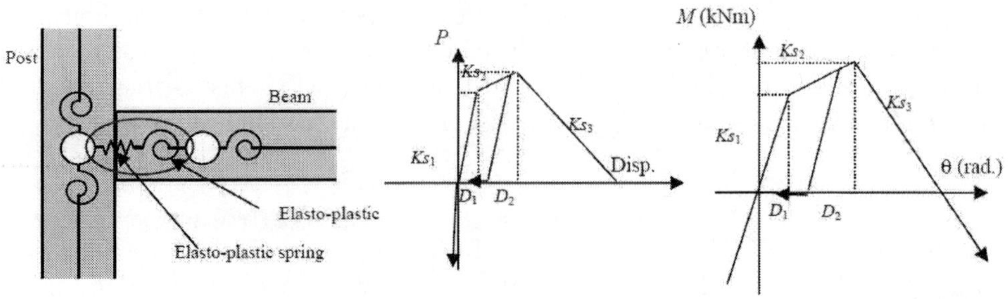

(a) Schematic diagram of joint element (b) Characteristic of elastoplastic spring (c) Characteristic of rotational spring

Figure 3. Outline of joint modelling.

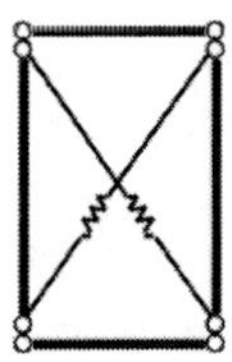

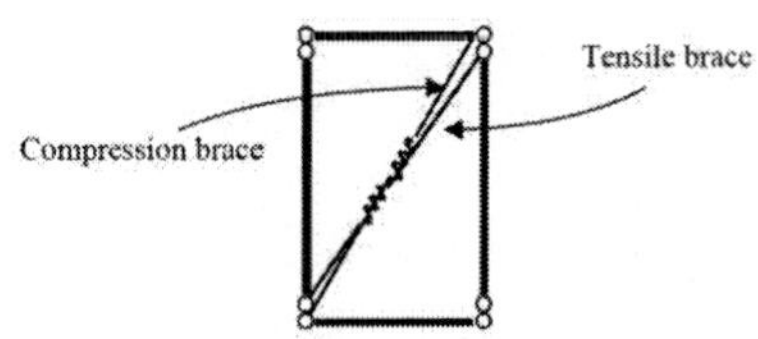

Figure 4. Shear wall spring. Figure 5. Outline of bracing shear wall.

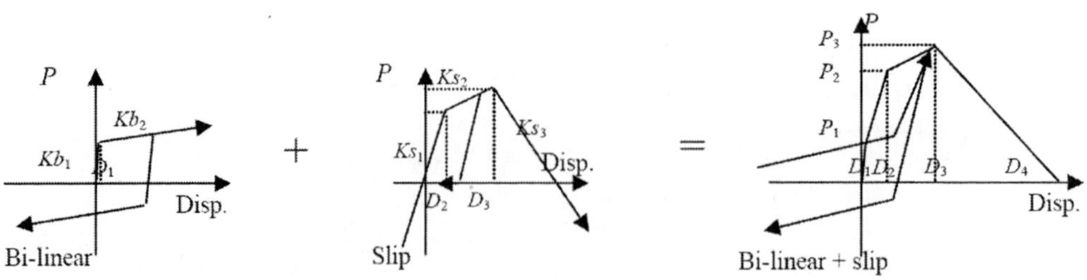

Figure 6. Hysteretic characteristics of shear wall and bracing.

3 SEISMIC RESPONSE RESULTS

3.1 Seismic Collapsing Frame Model and Input Earthquake Ground Motion

Figure 7 shows two wooden frame models without and with wall for seismic collapsing analysis against a strong earthquake ground motion, which consist of wooden columns and beams and also are made according the framing plan of two-story wooden house.

Table 1 shows three earthquake ground motion measured in the past earthquakes, and the JMA seismic intensity, peak acceleration, peak velocity, peak frequency and duration in each earthquake motion are indicated in this table. In this paper, three earthquake ground motion with the JMA seismic intensity of "6 upper" level are employed in seismic collapsing analysis of two-story wooden house. Figures 8 – 10 illustrate the displacement waves and the Fourier spectra of their acceleration waves for JMA Kobe wave, JR Takatori one and Kariwa one, respectively.

3.2 Seismic Collapsing Results

In this section, the effect of post fixed condition on the seismic performance during a strong earthquake ground motion is numerically investigated.

Figure 11(a) and 11(b) show final states after seismic collapsing simulation against JMA Kobe wave under both post fixed condition and unfixed one, respectively. In this paper, a dynamic friction coefficient, $\mu_d = 0.3$, between the base of wooden house and the ground is employed for post unfixed condition. Because JMA Kobe wave has a peak velocity of 91 kine, the roof truss on the second floor of two-story wooden house collapses under post fixed condition shown in Figure 11(a). While, there are few damages in two-story wooden house collapsing behavior under post unfixed condition shown in Figure 11(b). This is because this post unfixed condition is likely to simulate a seismic base isolation.

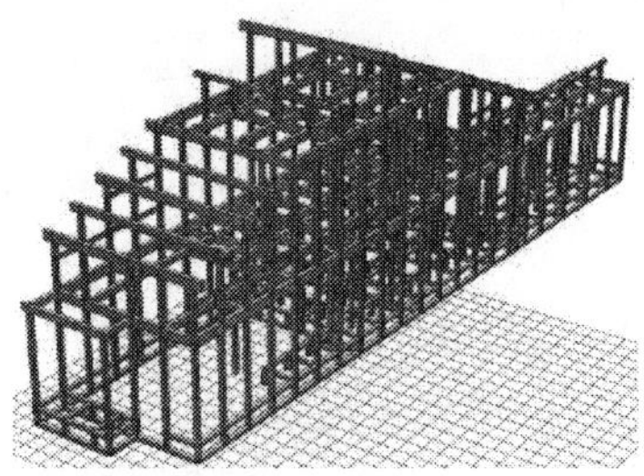

(a) Without wall

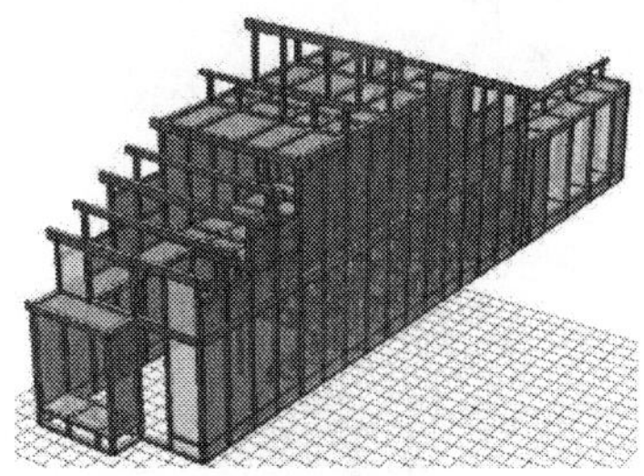

(b) With wall

Figure 7. Frame model of two-story wooden house for seismic collapsing analysis.

Table 1. Earthquake ground motion records.

Earthquake Motion Name	I_{JMA}	Peak Acceleration (cm/s^2)	Peak Velocity (cm/s)	Peak Frequency (Hz)	Duration (s)
JMA Kobe (1995)	6.4	818	91	1.43	30
JR Takatori (1995)	6.4	657	126	0.81	30
Kariwa (2004)	6.0	465	122	0.38	45

Final states after seismic collapsing simulation against JR Takatori wave under both post fixed condition and unfixed one are indicated in Figure 12(a) and 12(b), respectively. The second floor of two-story wooden house collapses under post fixed condition shown in Figure 12(a). While, there are few damages in the two-story wooden house collapsing behavior under post unfixed condition shown in Figure 12(b).

Figure 13(a) and 13(b) illustrate final states after seismic collapsing simulation against Kariwa wave under both post fixed condition and unfixed one, respectively. Two-story wooden house completely collapses under post fixed condition shown in Figure 13(a). While, there are few damages in the two-story wooden house collapsing behavior under post unfixed condition shown in Figure 13(b), and this final state is the same as Figures 11(b) and 12(b). The effect of the post unfixed condition on the seismic performance is analytically confirmed through some simulations.

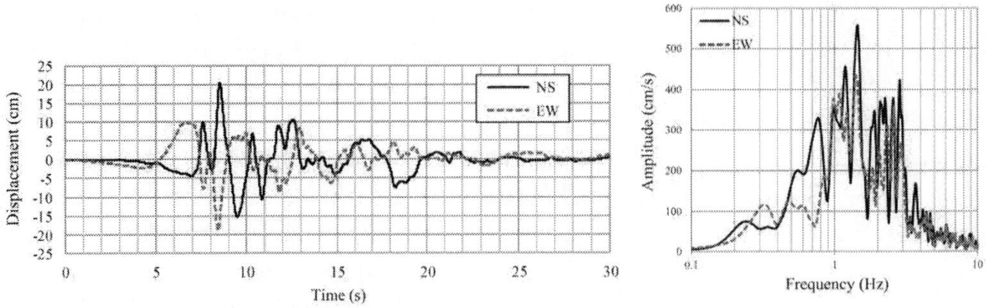

Figure 8. Earthquake motion and its Fourier acceleration spectrum (JMA Kobe).

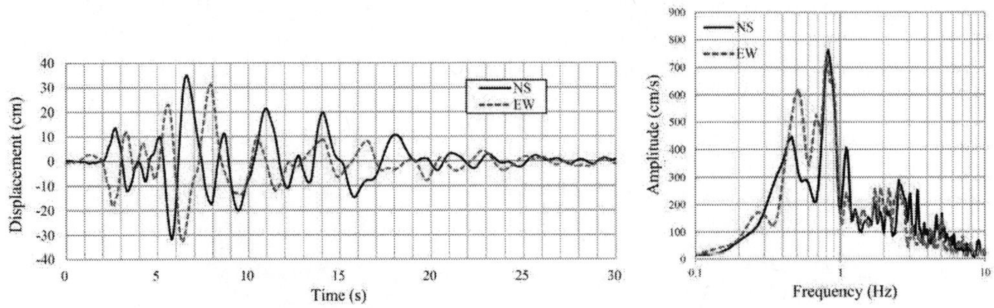

Figure 9. Earthquake motion and its Fourier acceleration spectrum (JR Takataori).

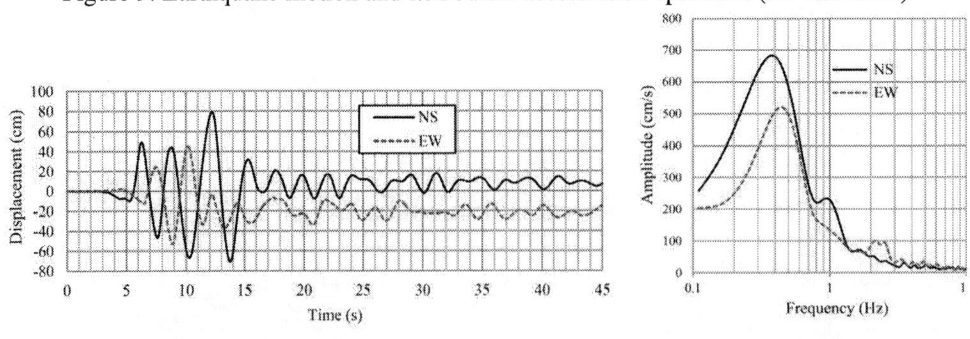

Figure 10. Earthquake motion and its Fourier acceleration spectrum (Kariwa).

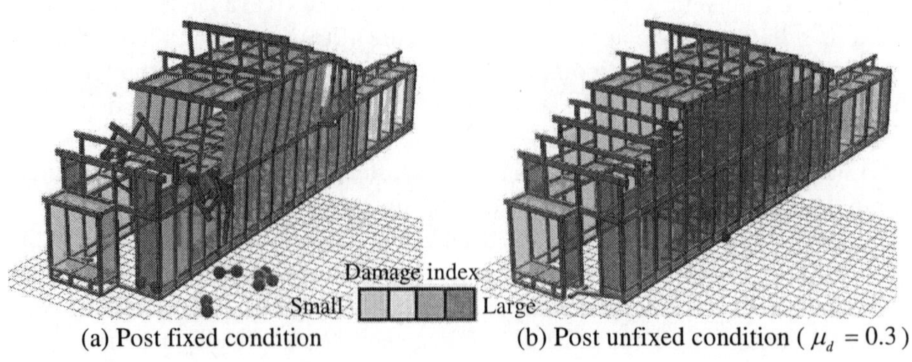

(a) Post fixed condition (b) Post unfixed condition ($\mu_d = 0.3$)

Figure 11. Seismic collapsing result after earthquake motion (JMA Kobe).

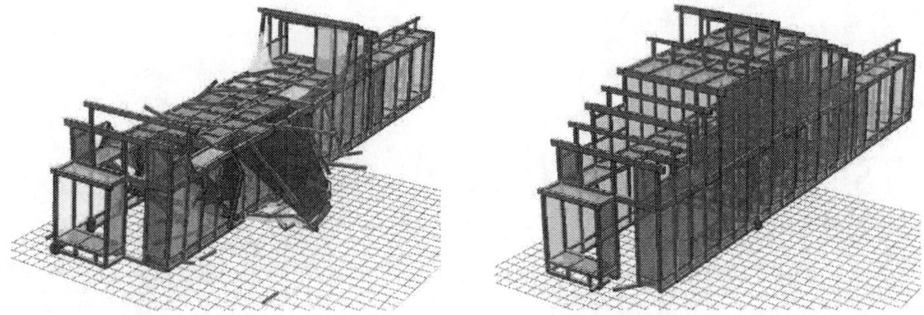

(a) Post fixed condition (b) Post unfixed condition ($\mu_d = 0.3$)

Figure 12. Seismic collapsing result after earthquake motion (JR Takatori).

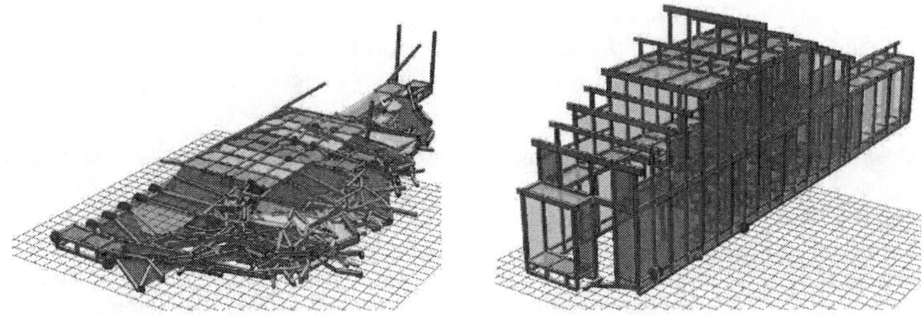

(a) Post fixed condition (b) Post unfixed condition ($\mu_d = 0.3$)

Figure 13. Seismic collapsing result after earthquake motion (Kariwa).

4 CONCLUTIONS

Investigation of the seismic performance of Japanese-style, two-story wooden houses was done using 3-D non-linear collapsing process analysis with simulated seisic activity. The effect of post fixing condition under the floors of wooden houses was also numerically investigated.

A summary of the results obtained by this study is as follows:

(1) Seismic behavior of old Japanese-style, two-story wooden houses can be numerically simulated by 3-D collapsing process analysis.

(2) The application of 3-D collapsing analysis to this style of wooden houses may play a key role in the seismic retrofit design process for these two-story wooden houses.

(3) Two-story wooden structure may be collapsed by a strong earthquake ground motion with the JMA seismic intensity of "6 upper" level.

(4) Seismic collapsing behavior strongly depends on the post fixing condition under the floor of an old Japanese-style two-story wooden house.

References

Cundall, P. A. and Strack, O. D. L., A Discrete Numerical Model for Granular Assemblies, *Géotechnique*, 29(1), 47-65, 1979.

Nakagawa, T. and Ohta. M., Collapsing Process Simulations of Timber Structures under Dynamic Loading III: Numerical Simulations of the Real Size Wooden Houses, *Journal of Wood Science,* 56(4), 284-292, 2010.

Integrated Solutions for Infrastructure Development
Edited by Lau, H. H., Tang, F. E., Ng, C. K., and Singh, A.
Copyright © 2016 ISEC Press

SEISMIC DAMAGE PREDICTION OF OLD JAPANESE-STYLE WOODEN HOUSE USING PREDOMINANT PERIOD DISTRIBUTION OF GROUND SURFACE LAYER

TOMIYA TAKATANI[1] and HAYATO NISHIKAWA[2]

[1]*Dept of Civil Engineering & Architecture, National Institute of Technology, Maizuru College, Maizuru, Kyoto Prefecture, Japan*
[2]*Education and Research Supporting Center, National Institute of Technology, Maizuru College, Maizuru, Kyoto Prefecture, Japan*

In general, the evaluation of a site amplification effect is very important in earthquake engineering when a seismic damage to wooden house with a low seismic performance against a strong earthquake will be predicted by an accurate estimation of the seismic intensity at surface ground. In this paper, both horizontal and vertical microtremors at 51 measuring sites in the west district in Maizuru city were measured by servo type accelerometers, and also the predominant periods at 727 sites in the same area were numerically estimated from the predominant periods measured at 51 sites using the Inverse Distance Weighting method. Moreover, a seismic damage prediction of wooden house against a strong earthquake ground motion was conducted by a relationship between a seismic damage function and a maximum drift angle of wooden house.

Keywords: Microtremor observation, Vulnerability function, Maximum drift angle of wooden house, Inverse distance weighting method.

1 INTRODUCTION

The evaluation of a site amplification effect, that is, a predominant period of surface ground layer plays a very important key role in the earthquake engineering when a seismic damage distribution of wooden house with low seismic performance will be predicted by an accurate estimation of the seismic intensity. In general, the site amplification effect has been analytically evaluated by the multiple reflection theory using a surface ground layer model with the soil characteristics at each observation site. However, it is so difficult to uniformly and accurately evaluate the site amplification effect for a wider area, because there is a limited and available information data and this evaluation procedure needs a great amount of work. Therefore, some site amplification effects can be evaluated from a relationship between the geological features/topography obtained from much simpler information and ground amplification characteristics. S-wave amplification spectrum at the observation site without any ground information was evaluated based on the microtremor measurements (Nishikawa and Takatani 2014).

In this paper, both horizontal and vertical microtremors at 51 sites in the west district in Maizuru city were measured by servo type accelerometers, and also the predominant periods at 727 sites in the same area were numerically evaluated from the predominant periods measured at

51 sites. Moreover, seismic damage prediction of wooden house against a strong earthquake ground motion with the Japan Meteorological Agency seismic intensity of "5 lower to 6 upper" level was conducted by the relationship between the vulnerability function and the maximum drift angle of wooden house.

2 OUTLINE OF MICROTREMOR MEASUREMENT

2.1 Microtremor Measurement System

Figure 1 shows a microtremor measurement system, which consists of a preamplifier, a data logger, three servo-type accelerometers, a rechargeable portable battery, and a PC. Sampling frequency of a microtremor measurement is 160Hz, and its measurement time is 51.2s per one set (8,192 data number). Table 1 shows an outline of the microtremor measurement instruments shown in Photo 1 which were used in the microtremor H/V spectral ratio measurement at 51 measuring sites in the east district in Maizuru city.

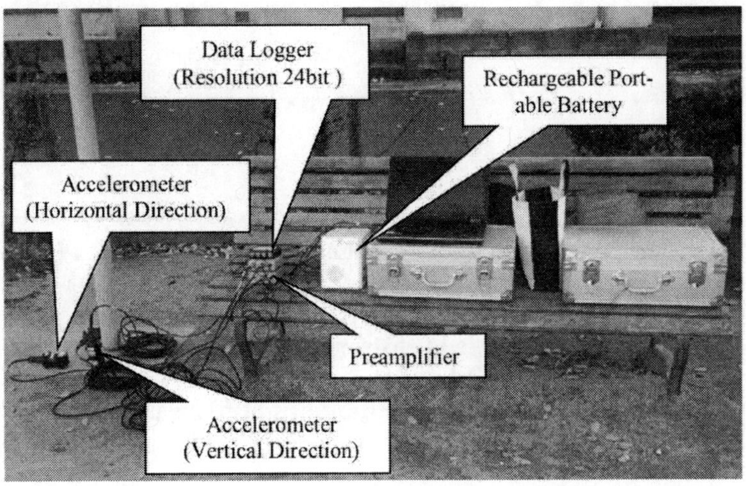

Figure 1. Microtremor measurement system.

Table 1. Outline of instruments used in microtremor measurement.

Instrument Name	Outline
Real Time Vibration Analysis Device (DSA-PHOTON)	Frequency Range : Maximum 21,000Hz A/D Transformation : 24-bit resolution D/A Transformation : 24-bit resolution
Real Time Vibration Wave Controlling System (DSA-RTPro)	Vibration Output Function, FFT Analysis Function, Long Term Vibration Recording Function, Measurement Data Editing Function
Servo-type Accelerometer (V405-BR)	Measurement Range : $\pm$ 30m/s^2, Resolution 1×10^{-6} m/s^2
Preamplifier (PA-9102)	Frequency Range : 0.3 – 45 Hz

2.2 Microtremor H/V Spectral Ratio

In this paper, Fourier spectrum of a microtremor acceleration is numerically obtained from the microtremor acceleration data of 10s section selected from microtremor measurement data.

Microtremor H/V spectral ratio can be obtained from both horizontal and vertical components of Fourier spectrum of microtremor acceleration, and then Fourier spectrum of microtremor can be smoothed by Parzen window with 0.4Hz band width. The microtremor H/V spectral ratio used in this paper is given by the following equation.

$$\frac{H}{V} = \sqrt{\left(\frac{H}{V}\right)^2_{NS-UD} + \left(\frac{H}{V}\right)^2_{EW-UD}} \tag{1}$$

where, H/V is an average spectral ratio, $(H/V)_{NS-UD}$ and $(H/V)_{EW-UD}$ are NS and EW components of spectral ratio, respectively.

3 ESTIMATION OF H/V SPECTRAL RATIO BY THE INVERSE DISTANCE WEIGHTING METHOD

In this paper, a microtremor H/V spectral ratio at unmeasured site can be numerically estimated by the following Inverse Distance Weighting method (Shepard, 1968).

$$H/V = \sum_{i=1}^{N} w_i (H/V)_i \ , \ w_i = \frac{1/r_i^2}{\sum_{i=1}^{N} 1/r_i^2} \tag{2}$$

where, $(H/V)_i$ is a microtremor H/V spectral ratio at i-th site, is a weight at i-th measuring site, r_i is a distance between i-th measuring site and unmeasured one, and N is a total number of measured sites.

Figure 2 shows the microtremor H/V spectral ratios at 51 measured sites, and Figure 3 indicates a microtremor H/V spectral ratio distribution map estimated by the Inverse Distance Weighting method described above. Microtremor H/V spectral ratios at 727 estimating sites were numerically obtained from the H/V spectral ratios at 51measured sites shown in Figure 3. It should be noted that H/V spectral ratio at the estimating site can be obtained by Equation (2)

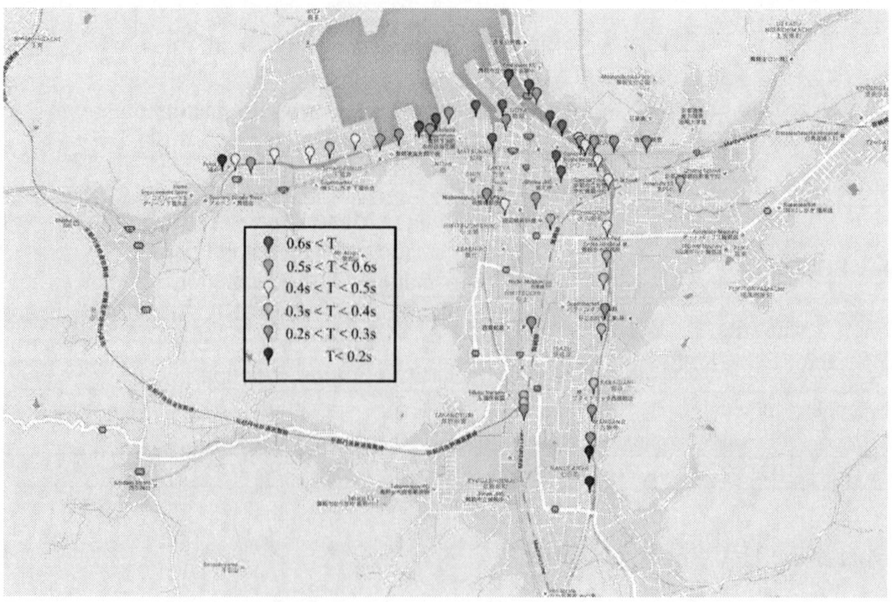

Figure 2. Microtremor measuring points in the west district in Maizuru city.

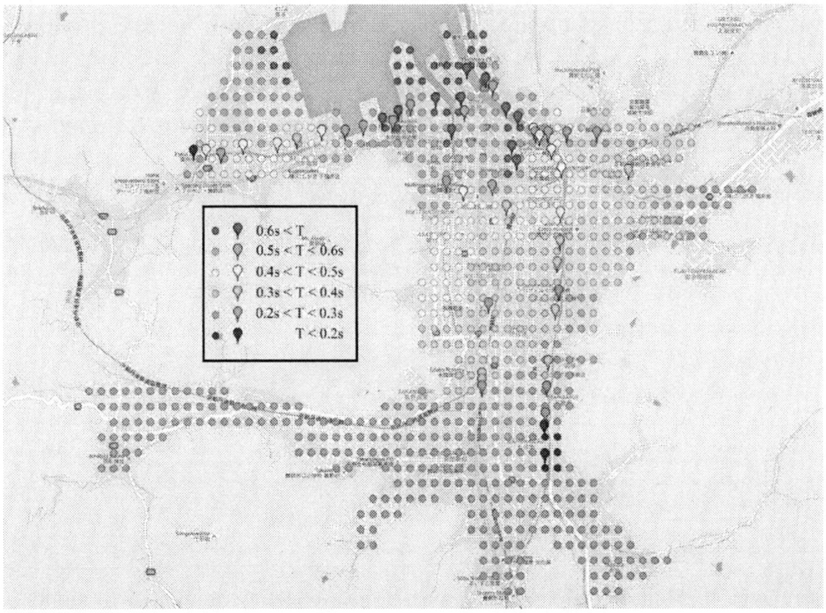

Figure 3. Predominant period interpolation result by microtremor measuring points.

under the limited microtremor observations of 51 sites. It is found from Figure 3 that the microtremor H/V spectral ratio in the seaside area in the east district in Maizuru city trends to have a long predominant period, where an extensive damage to the old wooden structure with a low seismic performance may occur during a strong earthquake ground motion.

4 SEISMIC DAMAGE ESTIMATION OF WOODEN HOUSE WITH LOW SEISMIC PERFORMANCE

4.1 Seismic Damage Function

In general, there is a certain significant relationship between a predominant period of the surface ground layer and a seismic damage to wooden house in past earth-quakes in Japan. In addition, a base shear coefficient C_y of wooden structure may be greatly related with the seismic damage against a strong earthquake ground motion. Based on the seismic damage function for a base shear coefficient C_y, the seismic damage estimation of wooden house with a low seismic performance is conducted using a predominant period of surface ground evaluated in the previous section.

Figure 4 indicates a relationship between a base shear coefficient C_y and a drift angle R. Several signs of M_e, H_e, Q, R, R_y, and $M_e g C_y$ in Figure 4 mean an effective mass and an effective height of structure, a horizontal load, a drift angle, a maximum drift angle, and a yielding load, respectively. It is found from Figure 4 that a seismic performance of structure is higher with increase of the base shear coefficient C_y. Seismic damage function is evaluated for three base shear coefficients $C_y = 0.2$, 0.3, and 0.4 in this paper.

The maximum drift angle of a two-story wooden structure is analytically calculated using a simulated earthquake ground motion wave due to a ground boring data in this paper. Based on a calculation result for each predominant period of surface ground, a total collapse ratio of wooden structure in this paper is defined as a percentage that the maximum drift angle is over 1/30.

A seismic damage function for each base shear coefficient C_y of wooden structure can be obtained from a curve fitting technique, which a cumulative distribution function given by a logarithmic normal distribution is applied to the relationship between the predominant period of surface ground and the total collapse ratio of wooden structure. Seismic damage to wooden house in the east district in Maizuru city can be numerically evaluated using the seismic damage function of wooden structure obtained from the procedure previously mentioned.

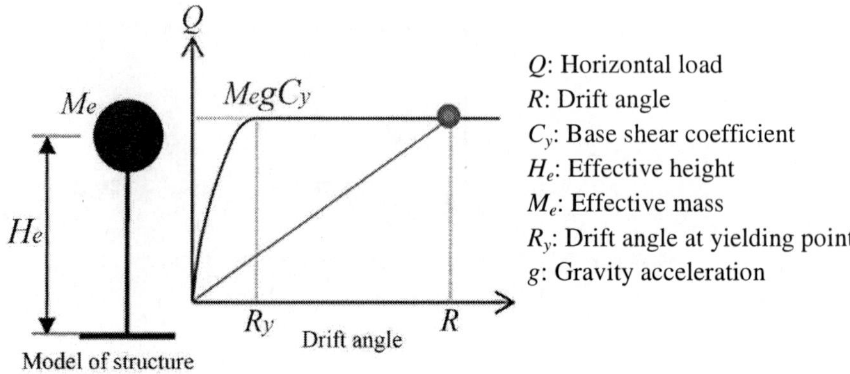

Figure 4. Relationship between a base shear coefficient and a drift angle.

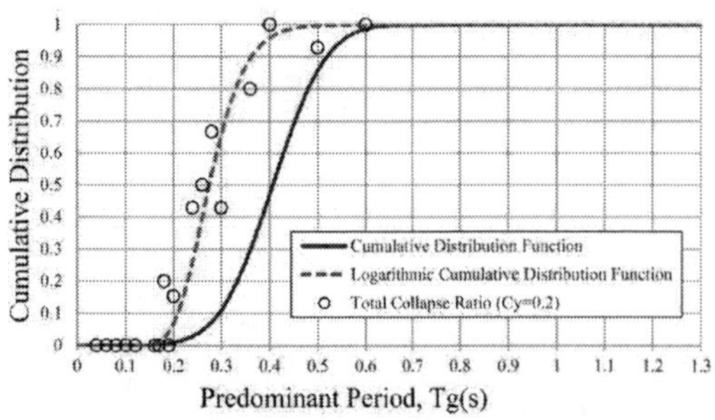

(a) Collapsing ratio vs. predominant period

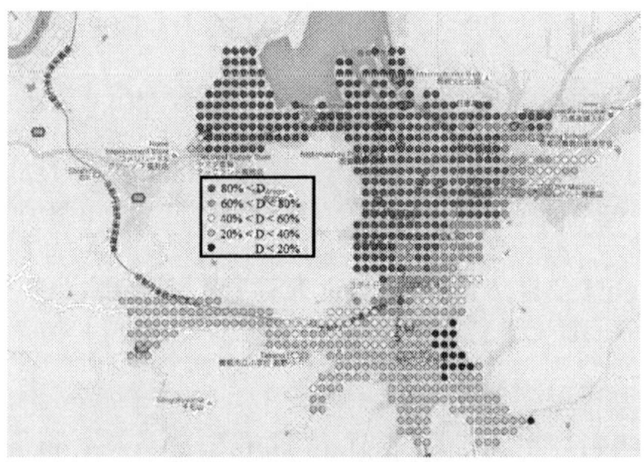

(b) Interpolation result

Figure 5. Relationship between a base shear coefficient C_y=0.2 and a drift angle.

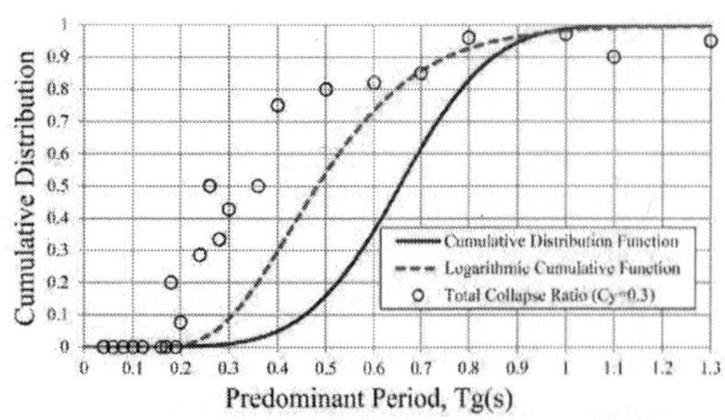

(a) Collapsing ratio vs. predominant period

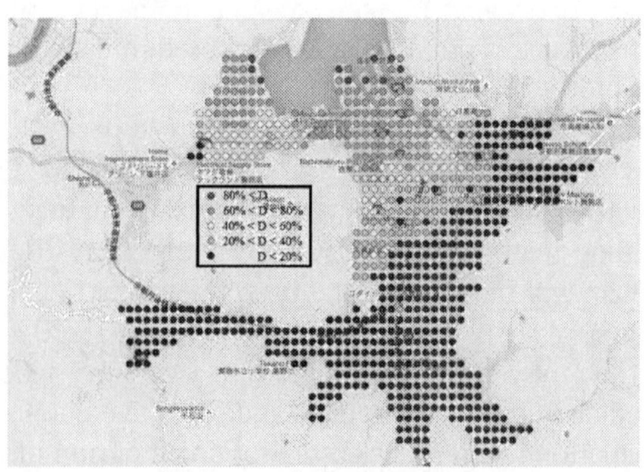

(b) Interpolation result

Figure 6. Relationship between a base shear coefficient C_y=0.3 and a drift angle.

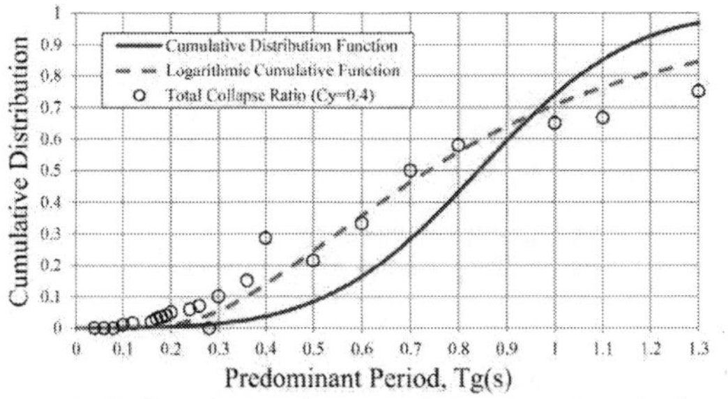

(a) Collapsing ratio vs. predominant period

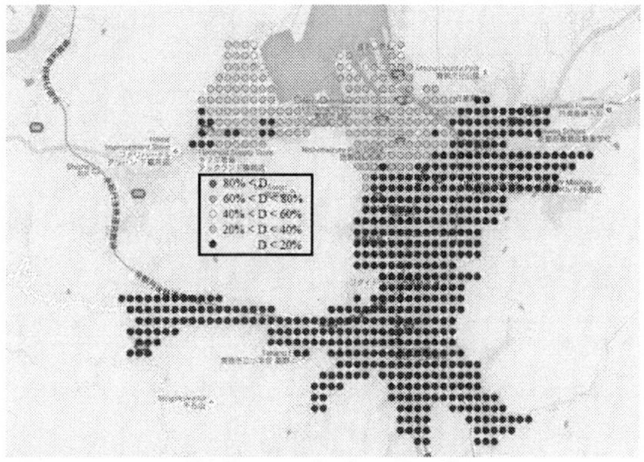

(b) Interpolation result

Figure 7. Relationship between a base shear coefficient C_y=0.4 and a drift angle.

Figures 5, 6 and 7 show seismic damage functions for three base shear coefficients C_y =0.2, 0.3, and 0.4, respectively. Using these seismic damage functions, seismic damage for each base shear coefficient C_y of wooden house can be numerically evaluated against an assumed strong earthquake ground motion.

5 CONCLUSIONS

The evaluation of a site effect is very important in the earthquake engineering when a seismic damage distribution of wooden house will be predicted by an accurate estimation of seismic intensity. In this paper, both horizontal and vertical microtremors at 51 sites in the west district in Maizuru city were measured by servo type accelerometers, and also the predominant periods at 727 sites in the same area were numerically evaluated from the predominant periods measured at 51 sites. Moreover, seismic damage prediction of wooden house against a strong earthquake ground motion with the Japan Meteorological Agency seismic intensity of "5 lower to 6 upper" level was conducted by the relationship between the vulnerability function and the maximum drift angle of wooden house.

The summary obtained in this paper is as follows:

(1) A predominant period at the site without any ground information can be easily evaluated from microtremor *H/V* spectral ratio obtained from microtremor measurement.

(2) Using the Inverse Distance Weighting method, a distribution map of predominant period of ground can be numerically estimated under the limited microtremor observation values.

(3) Vulnerability function with a base shear coefficient can be obtained from the maximum drift angle of wooden house, which is evaluated by the predominant period of ground. Also, collapse rate of wooden house for each base shear coefficient can be predicted by the vulnerability function.

References

Nishikawa, H. and Takatani, T., Evaluation of S-wave Amplification Spectrum using Microtremors in Maizuru City, *Proceedings of the 2nd Australasia and South East Asia Conference in Structural Engineering and Construction* (ASEA-SEC2), 109-114, Bangkok, Thailand, 2014.

Shepard, D., A Two-dimensional Interpolation Function for Irregularly-spaced Data, *Proceedings of the 1968 23rd ACM National Conference*, 517-524, 1968.

Integrated Solutions for Infrastructure Development
Edited by Lau, H. H., Tang, F. E., Ng, C. K., and Singh, A.
Copyright © 2016 ISEC Press

INFLUENCE OF BASEBOARD HEIGHT ON RESISTANCE TO WIND FORCE OF SCAFFOLDS AT WINDWARD SIDE OF BUILDINGS

HIROKI TAKAHASHI[1], MIZUKI AOKI[2], KATSUTOSHI OHDO[3], and KAZUO OHGAKI[2]

[1]*Construction Safety Research Group/National Institute of Occupational Safety and Health, Kiyose, Japan*

[2]*Dept of Building Technologist/Institute of Technologists, Gyoda, Japan*

[3]*Industrial Accident Investigation Center/National Institute of Occupational Safety and Health, Kiyose, Japan*

The Japanese Industrial Safety and Health Law was revised in March 2009 to introduce new measures by which to prevent accidental falls in the construction industry. As part of this revision, regulations on the installation of guard rails, toe boards, mesh sheets, and other components in appropriate positions on scaffolds were established. When scaffolds are installed in construction sites, their resistance against wind force needs to be calculated. Japanese design guidelines stipulate a specific scaffold resistance against wind force, but such regulations are applicable to conventional scaffolds. The problem with outdated regulations is that scaffolds are used during building construction without practitioners knowing whether the existing guidelines are suitable for new-style scaffolds. Accordingly, this study was conducted a wind tunnel test to examine the wind force exerted on building scaffolds, with the parameters being baseboard height and distance between scaffolds and a building. The relationship between the wind force coefficient of the scaffolds and baseboard height was proportional only on the scaffolds. As the distance between the scaffolds and the building lengthened, however, the relationship between the parameters reflected a steeper curve as baseboard height increased. Whenever the scaffolds were set near the building, negative pressure acted on the scaffolds as a consequence of the downwind structure. This study was examined the correction factor of the wind force coefficient of the scaffolds.

Keywords: Fall accident, Projected area, Wind force coefficient, Wind tunnel test.

1 INTRODUCTION

The Japanese Industrial Safety and Health Law was revised in March 2009 to introduce new preventive measures for accidental falls in the construction industry (Japan Construction Occupational Safety and Health Association 2009). Part of the revision was the establishment of regulations regarding the installation of guard rails, toe boards, mesh sheets, and other components in appropriate positions on scaffolds. Additionally, regulations on the installation of leading handrails mandate that handrails be erected before construction work commences to protect workers against falls (Ministry of Health, Labor, and Welfare 2003). To satisfy this requirement, practitioners use a special structure called a handrail frame. Figure 1 shows examples of modern scaffold designs.

There are some papers on countermeasures for scaffold resistance against wind force (Yoshida *et al.* 1980, Ministry of Labor the Research Institute of Industrial Safety 1981, Chino 1998). When scaffolds are installed at construction sites, their resistance against wind force needs to be calculated (Scaffolding and Construction Equipment Association of Japan 2004). Japanese design guidelines require a specific scaffold resistance against wind force, but these papers and regulations are applicable to old-style scaffolds. A number of risks are presented by the existing guidelines. First, new-style scaffolds are used in construction sites without practitioners knowing whether the design guidelines are appropriate for modern building components. Second, scaffolds are set near buildings, but workers are unaware that the distance between scaffolds and buildings affects the resistance of the former against wind force. Finally, conventional designs feature the use of baseboards as scaffold components.

With consideration for the above-mentioned issues, this study was carried out a wind tunnel test to examine the wind force exerted on scaffolds erected near a building. The parameters used in the test were baseboard height and the distance between the scaffolds and the building.

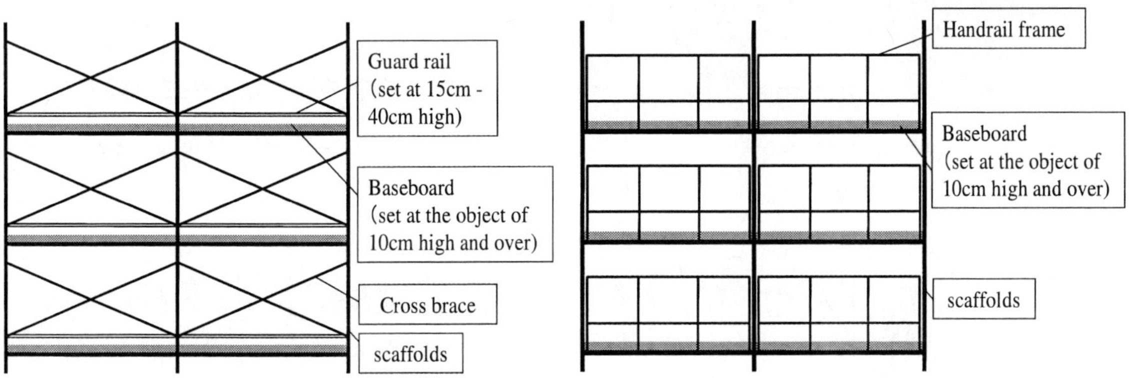

(a) Cross brace, guardrail, and toe board.

(b) Handrail frame and toe board.

Figure 1. Examples of modern scaffolds.

2 OVERVIEW OF WIND TUNNEL TEST

2.1 Wind Tunnel Device and Model

The wind tunnel device is 74,900 mm long, and the device interior is 2,300 mm wide and 2,000 mm high. The test setup is shown in Figure 2. A six-component force balance was used to measure wind force. The model was placed on the balance, which is shown in Figure 3.

The model features scaffolds that are 1/10 the size of buildings and used at general construction sites. Specifically, the scaffolds are three stories high and one span wide. The scaffolds and the structure used for the wind tunnel test are illustrated in Figure 4. The vertical frame direction of the scaffolds is denoted by X, the cross-brace direction of the scaffolds is represented by Y, and the height of the scaffolds is Z. A baseboard was situated on one side of the Y–Z face of the scaffolds. The scaffold model was placed in the wind tunnel, and the structure was positioned near the scaffold model. The distances between the scaffolds and the structure (D) are 30 and 200 mm. For a clear comparison, only the results obtained on the scaffolds were examined in the test.

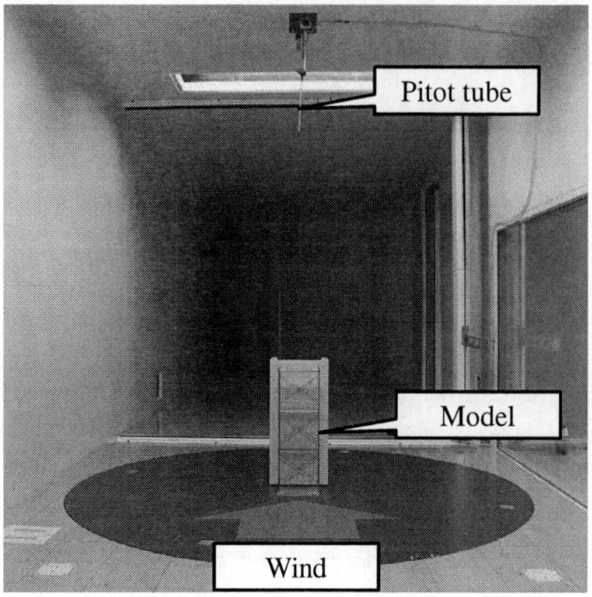

Figure 2. Wind tunnel test.

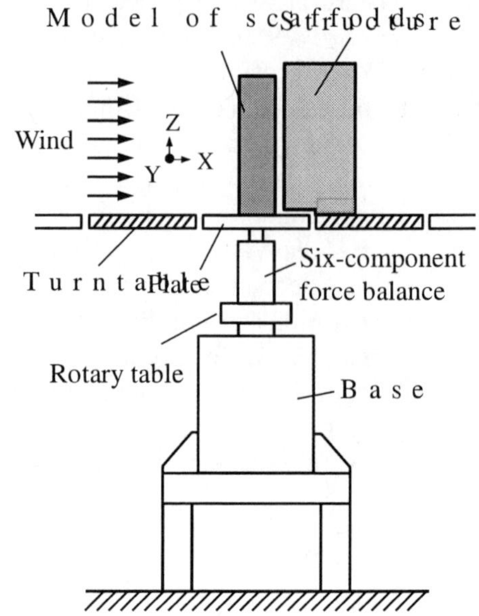

Figure 3. Six-component force balance.

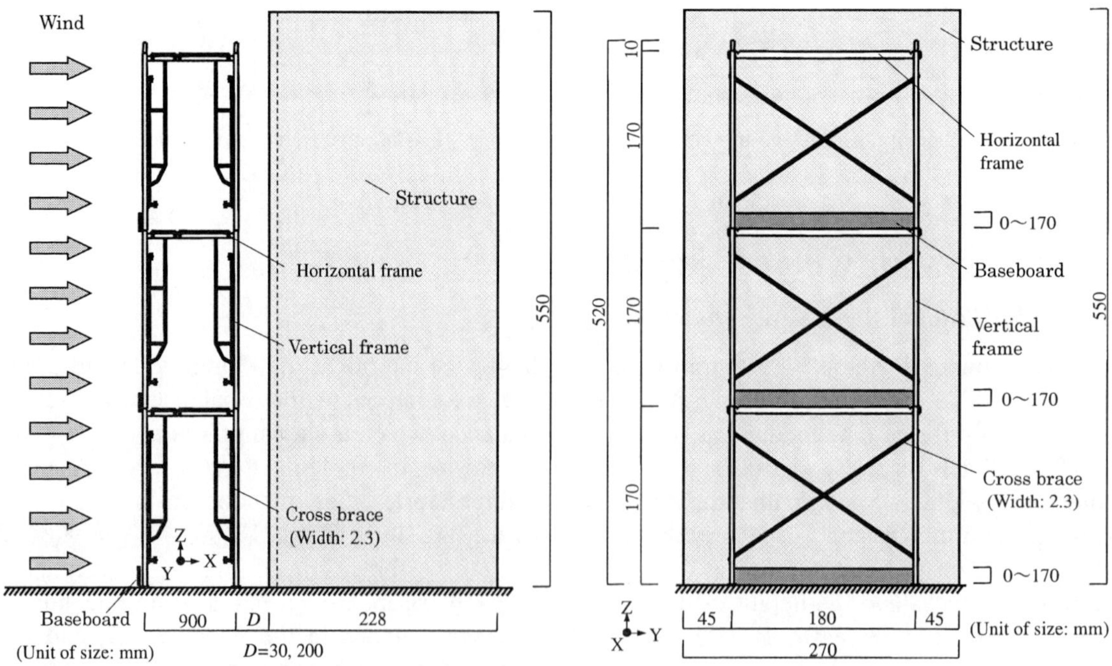

(a) X–Z face

(b) Y–Z face

Figure 4. Scaffolds and structure used for the wind tunnel test.

2.2 Projected Area and Reynolds Number

The ratios of the baseboard height and the projected area spanned by the scaffolds are expressed as dimensionless coefficients, thus:

$$\tau = h_b / h_V \tag{1}$$

$$\eta = A_b / A_0 \tag{2}$$

where τ is the ratio of the baseboard height, h_b is the baseboard height, h_V represents the height of the vertical frame, η denotes the ratio of the projected area of the scaffolds, A_b is the projected area of the scaffolds with a baseboard, and A_0 is the projected area covered by the scaffolds with a 170-mm baseboard. Figure 5 shows the relationship between η and τ; that is, η increased as τ increased.

The pitot tube was positioned 550 mm from the ceiling of the wind tunnel, after which wind speed was measured. The wind speed was set at a uniform flow of 10 m/s. Characteristic length D was positioned 5 mm along the diameter of a leg member. The Reynolds number, Re, was approximately 3.5×10^3, determined as follows:

$$\mathrm{Re} = \frac{UD}{v} \tag{3}$$

where U is the wind speed (in m/s), D represents the characteristic length (in mm), and v denotes the coefficient of kinematic viscosity [$v = \mu/\rho$, where μ is the coefficient of viscosity ($\mu = 1.82 \times 105$(N s/m^2), and ρ is the air density].

The scaffolds are made of ring-shaped steel pipes, whose lengths are proportional to their diameters. These features necessitate the calculation of the wind force coefficient of two-dimensional cylinders (Scaffolding and Construction Equipment Association of Japan 2004). The wind force coefficient of each cylinder changes in accordance with the Reynolds number.

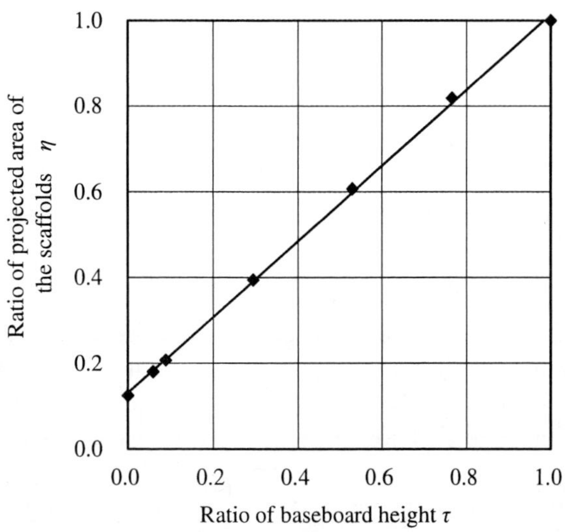

Figure 5. Relationship between the ratio of the projected area of the scaffolds and the ratio of the baseboard height.

3 RESULTS AND DISCUSSION

Figure 6 presents the results of the wind tunnel test. In the figure, the vertical axis pertains to the wind force coefficient of the scaffolds (C), and the horizontal axis refers to the ratio of the baseboard height (τ). The wind force coefficient of the scaffolds in the X direction was calculated as follows:

$$C = \frac{F}{q_F A} \tag{4}$$

where F is the force exerted on the scaffolds, q_F denotes the reference speed pressure ($=1/2\rho V_H^2$, where ρ denotes the air density, and V_H is the wind speed), and A represents the reference area (projected area by the Y–Z direction of the scaffolds).

Only on the scaffolds was the relationship between C and τ proportional. On the scaffolds by the building, however, the distance D between the scaffolds and the building was longer, thereby producing a stronger (steeper curve) relationship between C and τ as τ increased. Negative pressure acted on the scaffolds as a consequence of the downwind structure.

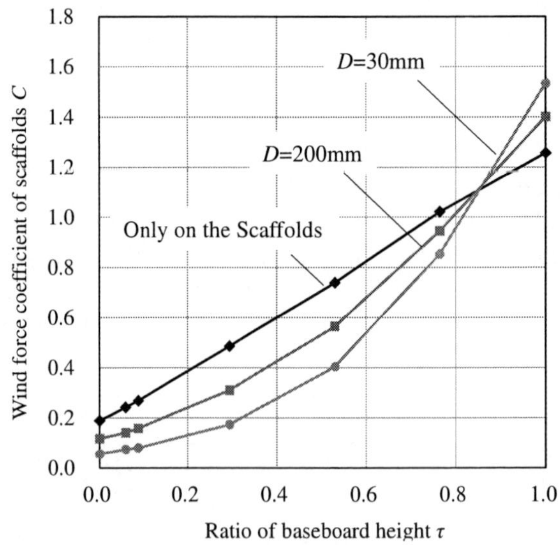

Figure 6. Relationship between the wind force coefficient and the ratio of the baseboard height.

4 EXAMMINATION OF CORRECTION FACTOR

This study was examined the correction factor of the wind force coefficient of the scaffolds with baseboards at the windward side of the building. From Figure 6, the calculation used was (C for distance D = 30 mm) - (C for only on the scaffolds) = (Correction factor of the wind force coefficient of scaffolds for results of test S') at the windward side of the building). Figure 7 illustrates the relationship between S' and τ. The points in the figure refer to the test results, and the curved line is the approximate value derived using the least squares method. The approximate equation was as follows:

$$S = 1.76\tau_0^2 - 1.41\tau_0 - 0.1 \tag{5}$$

Equation (5) can be reflected in one of the correction factors of C at the windward side of the building.

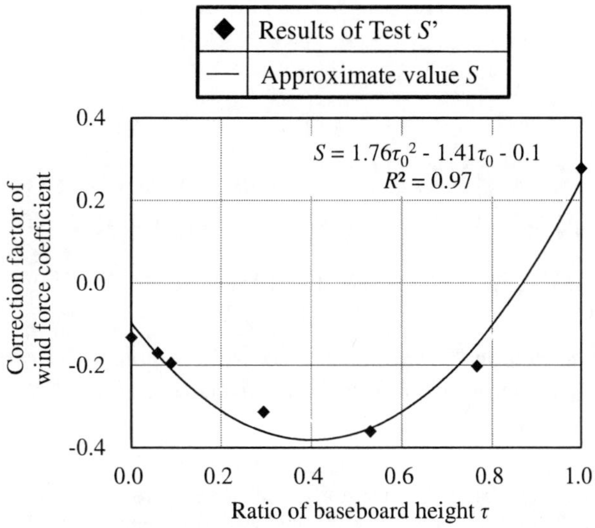

Figure 7. Relationship between the correction factor of the wind force coefficient of the scaffolds and the ratio of the baseboard height.

5 CONCLUSION

The wind force acting on scaffolds equipped with baseboards was investigated by wind tunnel testing. On the scaffolds at the windward side of the tested building, the plot of the relationship between the wind force coefficient and the baseboard height reflected a steep curve as the baseboard height increased. The wind force coefficient differed only in accordance with modifications to the scaffolds, that is, when the scaffolds were set near the downwind structure.

Figure 7 shows the correction factor of the wind force coefficient of the scaffolds with baseboards at the windward side of the building.

Acknowledgments

This work was supported by the JSPS Grant-in-Aid for Scientific Research C (grant number 15K01238). We are grateful for the financial aid provided.

References

Chino, N., A Study on Basic Design Wind Speeds for Scaffold, *Summaries of technical papers of Annual Meeting Architectural Institute of Japan. B-1*(in Japanese), Architectural Institute of Japan, 223-224, 1998.

Japan Construction Occupational Safety and Health Association, *Yearbook of Safety and Health in Construction 2009* (in Japanese), Tokyo, 75–110, 2009.

Ministry of Health, Labor, and Welfare, *Guidelines on the Handrail Preceding Method* (in Japanese), Tokyo, 2003.

Ministry of Labor The Research Institute of Industrial Safety, Special Research on Countermeasures for Safety of Scaffolds etc. against Strong Wind, *Special Research Report* (in Japanese), 1981.

Scaffolding and Construction Equipment Association of Japan, *Recommendations for Safety Technology of Scaffolds to Wind Loads* (in Japanese), Tokyo, 2004.

Yoshida, M., Sanada, H., Hongo, T. and Nakamura, O., Experimental Study of Wind Forces on the Scaffold with the Protective Sheets or Soundproof Panes, *Annual report of Kajima Institute of Construction Technology* (in Japanese), Kajima Corporation, 129-136, 1980.

Integrated Solutions for Infrastructure Development
Edited by Lau, H. H., Tang, F. E., Ng, C. K., and Singh, A.
Copyright © 2016 ISEC Press

VIBRATION PROPERTIES OF TRADITIONAL MASONRY WALLS OF CORTES CHURCH

NORIKO TAKIYAMA[1], AKARI YAMAGUCHI[1], KOHEI HARA[1], and VERDEJO JUAN RAMON JIMENEZ[2]

[1]*Division of Architecture and Urban Studies, Tokyo Metropolitan University, Tokyo, Japan*
[2]*Graduate School of Environmental Planning, University of Shiga Prefecture, Shiga, Japan*

The Santo Niño Church in Bohol Island was partially damaged in the Bohol Earthquake in October 2013. In this study, we investigated the structural property of the Santo Niño Church to understand the vibration characteristics of Philippine masonry constructions by checking the main damage and conducting measurement surveys and microtremor measurement on the structure and ground. The major findings of this study can be summarized as follows: (a) Collapsed walls, peeling wall stone, corner deformation, and collapsed ceiling were the main damages that were observed. The collapsed wall was measured, and the make-up and material thickness of the wall was verified. (b) By performing microtremor measurements on the walls, the vibration property of the church, such as the natural frequency and vibration mode of the longest wall A was clarified. The 1st natural frequency of the longest wall A is the lowest.

Keywords: The 2013 Bohol Earthquake, Masonry structure, Microtremor measurement, Natural frequency, H/V spectra.

1 INTRODUCTION

The Earthquake that hit Bohol Island, Philippines in October 2013, killed 222 people; more than 3 million people all over Philippines suffered its effects (Ministry 2016). According to a previous disaster investigation, 23 churches in Bohol Island with masonry structures were completely or partially damaged (Japan 2014) owing to their poor seismic performance.

The walls of a traditional masonry church in Bohol is constructed using the fixed rammed-earth method; mud which containing stones, shells, gravel, and lime paste is piled up into layer form and is tamped dawn (Hanazato 2000, Rammed 2016).

These churches are used as places of assembly; thus, they need to have sufficient seismic performance to minimize damages caused by an earthquake. However, these churches were constructed without considering. Moreover, some structures were damaged by the earthquake, while some structures were not.

In this study, we clarify the vibration characteristics of Philippine masonry constructions by conducting a measurement survey and microtremor measurements on the structure and ground. The Santo Niño Church in Bohol Island was selected as the target structure; the official name of the church is St. Niño Parish Church in Cortes.

2 DAMAGE OF SANTO NIÑO CHURCH

The Santo Niño church was constructed on a hill by Augustinians Recollect in 1880 (Japan 2014). Its location is shown in Figure 1. The outside appearance of the church before earthquake and its drawing are shown in Figure 2. Photographs showing the main damages suffered by the church are shown in Figure 3.

2.1 Santo Niño Church

The church has a cruciform plan and an octagonal bell-tower. The construction of the bell-tower was finished in the 20th century. The roof frame consists of a wooden structure. The wooden floor is arranged only on the east side, and the other area consists of a large space with a high ceiling.

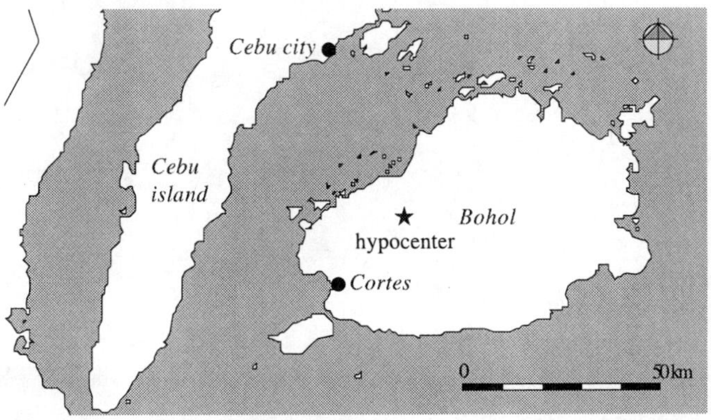

Figure 1. Location of Santo Niño Church in Cortes.

(a) Southeastern part of church before Bohol earthquake

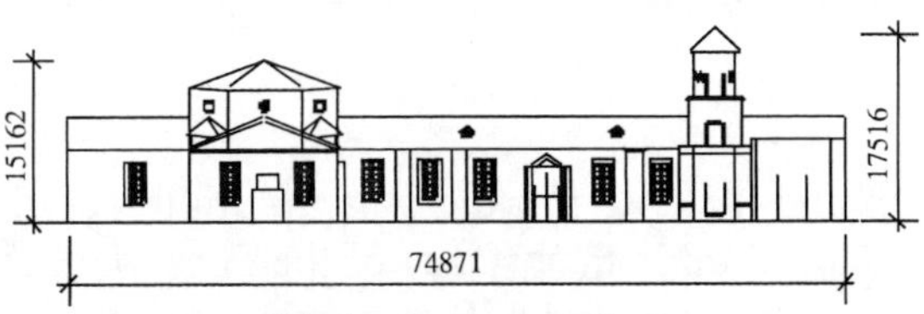

(b) Southern elevation

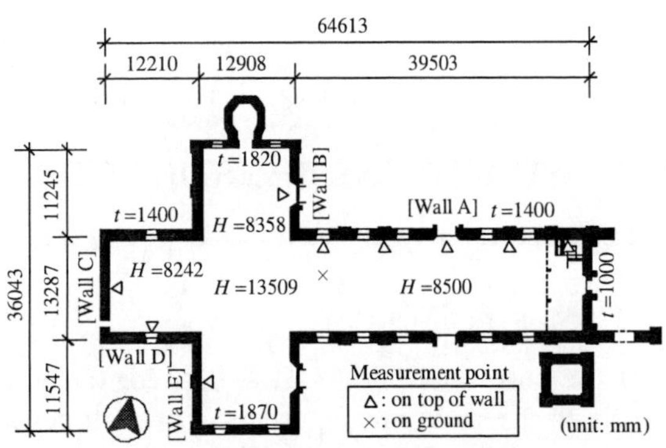

(c) Plan and measurement point

Figure 2. Drawing of Santo Niño Church in Cortes and microtremor measurement points.

2.2 Main Damage

Collapsed walls, peeling wall stone, corner deformation, and collapsed ceiling were the main damages observed. The points of main damage are illustrated by dashes line and arrow head in Figure 3 (a). Then, (b) to (f) say the photographing point of Figure 3 (b) to (f).

The eastern wall of the second floor collapsed, as shown in Figure 3 (b); significant damage was observed around the opening and top of the southern gable wall, as shown in Figure 3 (c) and (d). The northern gable wall collapsed completely, as shown in Figure 3 (e). The shear deformation of the eastern wall of the ground floor is shown in Figure 3 (f). Such types of damage were observed throughout the church.

Figure 3. Damage due to Bohol earthquake; photographed on August 1, 2014.

3 STRUCTURAL INVESTIGATION

The structural investigation was conducted in September 19, 2015.

3.1 Masonry Wall

As the entire northern wall collapsed, the make-up and material thickness of the wall could be verified, as shown in Figure 4. Coral stones are arranged on both surfaces of the thick wall; the radius and thickness of each stone was approximately 300 mm and 150 mm, respectively. The space between the coral stones is filled up with mud mortar, which had a thickness of approximately 1000 mm.

We measured the dimensions of the church, and the length, thickness, and height of each wall, as shown in Figure 2 (c) and Table 1. Then, estimated values of the nearest wall, as shown in '( )', were used as thickness value for walls whose thickness could not be measured.

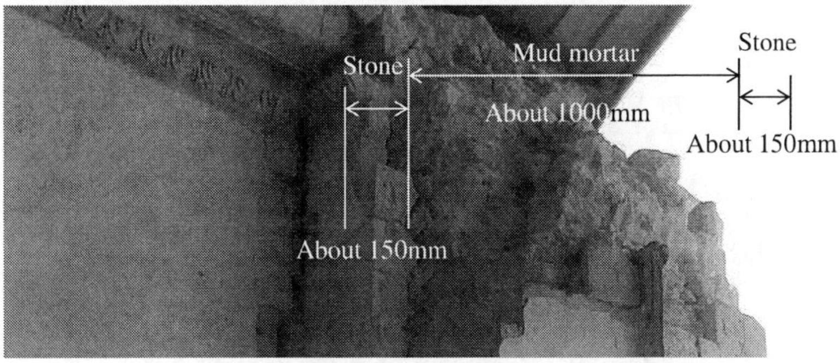

Figure 4. Construction of wall.

Table 1. Size and natural frequency of wall.

Wall	Wall size [mm]			1st natural frequency [Hz]	
	Length	Height	Thickness	Out-of-plane	In-plane
A	39,503		1,400	3.3	12.6
B	11,245		(1,820)	4.6	12.1
C	13,287	7,300	(1,400)	4.5	12.6
D	12,210		(1,400)	5.3	11.9
E	11,547		1,870	5.2	12.1

3.2 H/V Spectra on Ground

We conducted microtremor measurements of the ground to clarify surface conditions. The measurement was performed on the free ground at x point, as shown in Figure 2 (c). The frequency of the ground ranged between 7–9 Hz, as shown in Figure 5.

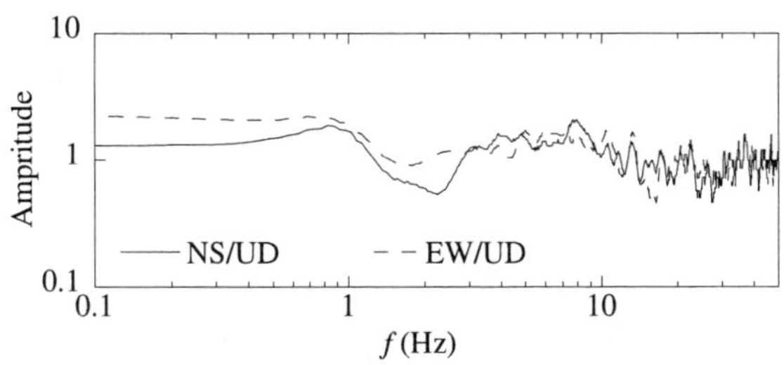

Figure 5. H/V spectrum on the ground point.

3.3 Natural Frequency and Vibration Mode of Masonry Wall

We conducted microtremor measurements on the walls to clarify the vibration property of the church. The measurement was performed at several points on the top of walls, A–E, as shown in Figure 6. In Figure 2 (c), the measurement points are marked by triangles. The simultaneous measurement was conducted at multiple points by installing one velocity sensor on the ground and some sensors on the wall. Five sensors were installed at wall A to understand the vibration mode. Then, the Fourier spectral ratio is defined as the Fourier spectra on the wall divided by on the ground.

Fourier spectral ratios in the out-of-plane and in-plane directions, obtained using the three sensors installed in wall A, are illustrated in Figure 7. The peaks are clear in both directions; the 1st, 2nd, and 3rd natural frequencies in the out-of-plane direction are 3.3 Hz, 4.5 Hz, and 5.6 Hz, respectively. In addition, the 1st and 2nd natural frequencies in the in-plane direction are 12.6 Hz and 15.5 Hz, respectively. The 1st natural frequencies of walls A to E in both the directions are indicated in Table 1.

Figure 6. Measurement point.

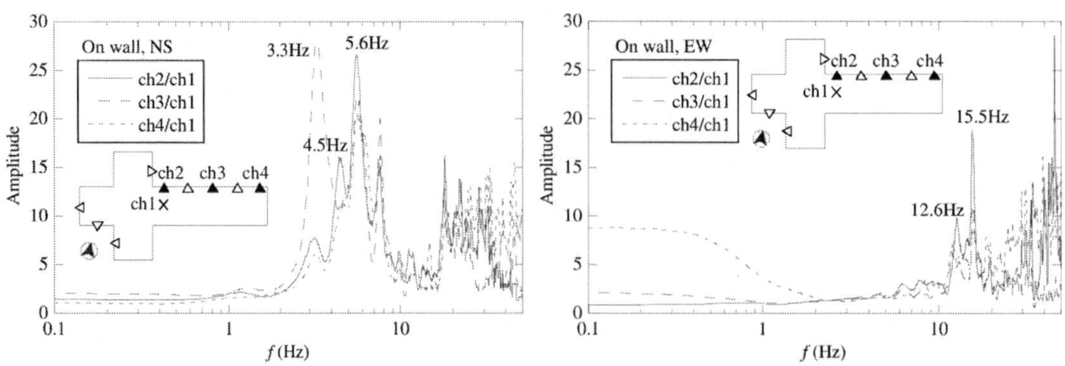

(a) For walls in the NS direction (b) For walls in the EW direction

Figure 7. Fourier spectral ratio.

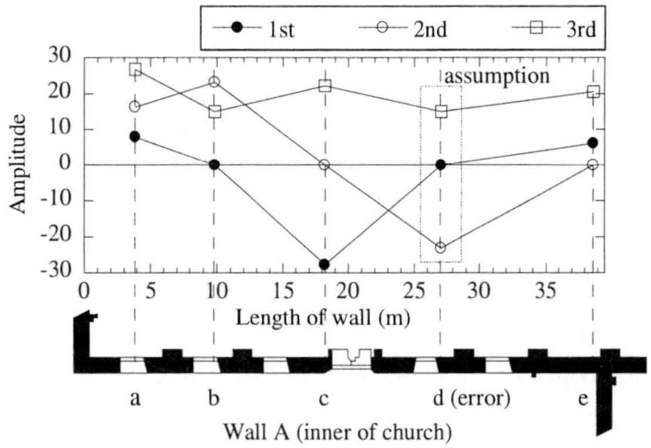

Figure 8. Vibration mode of Wall A.

The vibration mode of wall A is shown in Figure 8. The five sensors are named a to e, and the 1st to 3rd modes are illustrated. As measurement could not be performed using sensor d, the value of sensor b is used, based on a past study (Takiyama 2012).

4 CONCLUSIONS

In this study, we clarify the vibration characteristics of Philippine masonry constructions by conducting a measurement survey and microtremor measurements on the structure and ground. The Santo Niño Church in Bohol Island was used as the target structure, which was partially damaged by The October 2013 Earthquake. The major findings of this study can be summarized as follows:

(a) Collapsed walls, peeling wall stone, corner deformation, and collapsed ceiling were the main damages observed. The collapsed wall was measured, and the make-up and material thickness of the wall was verified.

(b) By conducting microtremor measurements on the walls, the vibration property of the church, such as the natural frequency and vibration mode of the longest wall A, was clarified. The 1st natural frequency of the longest wall A was the lowest.

Acknowledgments

We thank Ikumi OSAWA and Fumiaki OKAMURA, the students of Tokyo Metropolitan University.

References

Hanazato, T., Monitoring to the Paoay Church, the Philippines, ICOMOS, 2000.

Japan Consortium for International Cooperation in Cultural Heritage, *Survey Report on the Protection of Cultural Heritage in Republic of the Philippines*, Japan, 2014.

Ministry of Land, Infrastructure, National Spatial Planning and Regional Policy Bureau, An Overview of Spatial Policy in Asian and European Countries. Retrieved from http://www.mlit.go.jp/kokudokeikaku/international/spw/general/philippines/ on June 1, 2016. (in Japanese)

Rammed earth construction. Retrieved from http://www.rammedearthconstructions.com.au/ on June 1, 2016.

Takiyama, N., Toshiyuki, T., and Hayashi, Y., Evaluation of out-of-plane vibration characteristics for unreinforcement walls of historic masonry constructions, *Journal of structural and constuction engineering*, Architectural Institute of Japan, 77(673): 475-482., 2012. (in Japanese).

Integrated Solutions for Infrastructure Development
Edited by Lau, H. H., Tang, F. E., Ng, C. K., and Singh, A.
Copyright © 2016 ISEC Press

SEISMIC DESIGN VERIFICATION OF STEEL PRE-ENGINEERED BUILDINGS

SATISH KUMAR S. RAJARAM and NIBEDITA SAHOO

Dept of Civil Engineerinng, Indian Institute of Technology Madras, Chennai, India

Steel built-up I sections, composed of plates with high width-to-thickness ratios (slender sections), are commonly used in pre-engineered buildings under the premise that the design is governed by wind. However, in the event of a severe earthquake, the sections are susceptible to local buckling and may exhibit a non-ductile behavior. Therefore it is imperative to check the performance of such structures under the maximum credible earthquake (MCE). As a first step towards this objective, it is necessary to evaluate the post-buckling strength and ductility of such sections. In this study, a Finite element model is developed to analyze the inelastic post-buckling response of semi-compact and slender plates. The information can be used to predict the moment-rotation curves for I-sections with slender webs. A parametric study was carried out on a total of 54 pinned-base PEB frames of varying spans and heights. The elastic seismic demand under severe earthquake was estimated and compared with the design lateral capacity of PEB frames. From the results, it is concluded that even for higher seismic zones, low ductility sections (1.5 to 2) are adequate to survive MCE. Alternatively, if the design is verified for a response reduction factor of 2, then non-ductile sections can also be used.

Keywords: Single-storey industrial buildings, Slender sections, Ductility, Pushover analysis, Response reduction factor.

1 INTRODUCTION

Single-storey Pre-engineered Building (PEB) structures are preferred because of faster construction, superior quality and economy compared to the conventional steel structures. These buildings have relatively light roofs and are characterized by low axial load ratios. Hence they are designed with built-up I-sections having high plate slenderness (width-to-thickness ratios) which makes them stiff enough for controlling deformations. Being light-weight structures, the lateral load due to design basis earthquake is often less as compared to the design wind force. The ratio of wind to earthquake load is also influenced by factors such as the span to height ration of the structure and its geographical location and has been found to vary from 1 to 17 as shown later in this study. Hence, designers often assume that there is no need to adhere to ductile detailing as stipulated in section 12 of the code for structural steel design IS800:2007. However, in the event of a severe earthquake, the sections are susceptible to local buckling and may exhibit a non-ductile behavior. Therefore it is imperative to check the performance of such structures under the maximum credible earthquake (MCE).

Structures are designed for seismic force which is reduced by a factor 'R' known as response reduction factor. The R value accounts for the level of structural ductility as well as the energy dissipation capacity of the structure. Ductility of the cross-section is ensured in design by section classification criteria based on the flange width to thickness (b/t_f) ratio and web depth to thickness (d/t_w) ratio. Since Rafters and columns in PEBs are subjected to predominant bending and very little axial loads, they are designed as plate girders with slender webs. Standard codes permit the use of slender sections in design, provided the extra depth of web and width of flange, in excess of the semi-compact limits, is considered as ineffective in calculating the section properties. Hence effect of plate slenderness ratios on the seismic performance of PEBs needs to be assessed to justify the R values used in design.

2 MOMENT-ROTATION CURVES FOR SLENDER WEB SECTIONS

Finite Element Analysis was carried out to obtain the moment-rotation curve for beams with semi-compact and slender webs well beyond the point of local bucking. A cantilever beam was modeled in the finite element software ABAQUS (2011). The four-noded, S4R5 shell element was used in this model since the objective was to capture the local buckling behavior of the beam cross-section. Cyclic loads tend to reduce or remove the yield plateau and work-harden the steel. Hence, a bi-linear elastic-plastic-hardening (EPH) stress-strain relationship was used with Young's modulus E of 200000 MPa and a strain hardening stiffness Et of E/100 as prescribed by Smitha and Satish Kumar (2013). Initial imperfections were applied by first carrying out a buckling analysis and then including the local buckling shape with of amplitude of 1/250th of the depth of the web. To avoid the lateral-torsional buckling of the member, the movement at the intersection of web with top and bottom flange was restricted in the lateral direction.

The validation of the FEM model was obtained by analyzing two wide-flange cantilever beams of span 3700 mm tested by D'Aniello *et al.* (2012) the properties of which are as given below:

(a) HEA 160: Flange 160x9; Web 152x6; yield stress = 337 MPa; Ultimate stress = 465 MPA.

(b) HEB 240: Flange 240x17.5; Web 240x10; yield stress fy=319MPa; Ultimate stress=431 MPA.

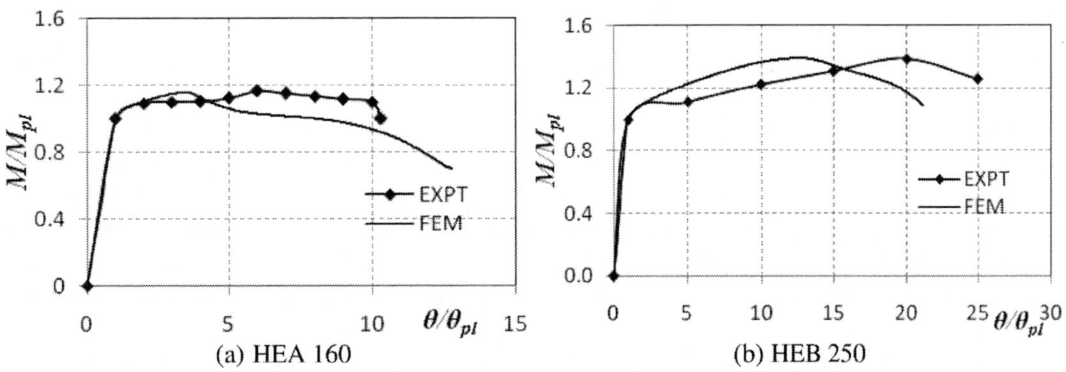

(a) HEA 160 (b) HEB 250

Figure 1. Validation of FEM results with tests by D'Aniello *et al.* (2012).

The analysis results are compared with test results in Fig. 1 for the two beams where the downward drooping of the curves is due to flange local buckling. The FEM Model was then used to generate the moment-rotation curve for various required sections and was used in the pushover analysis as explained in the next section.

3 SEISMIC PERFORMANCE OF PEB FRAMES

The Strength of PEB frames is governed by several factors such as the magnitude of the gravity, wind and seismic loads acting as well as the values of span, height and girder depths. In order to understand the effect of these on the seimic performance, 54 frames with varying parameters were designed. Pinned-base PEB frames with spans of 15, 20, 25 and 30 m length and heights of 6, 10 and 15 m were considered giving nine different L/h values. Dead load (DL), live load (LL) and collateral load (CoL) are taken as 0.1 kN/m^2, 0.6 kN/m^2 and 0.5 kN/m^2, respectively giving a total gravity load of 1.2 kN/m^2. The lateral forces such as wind and seismic loads are calculated depending on the span, height and site location of PEB frames as per the required codal provisions. Elastic analysis was used to calculate the member forces. Serviceability criteria were checked by limiting the lateral roof displacement to h/150 and vertical deflection to span/240 (IS 800:2007). Design of members was done by excel spreadsheet and complies with code provisions. In general, for spans larger than 30 m the gravity load dominates the design while for tall frames lateral drift becomes the design criteria.

Based on the Indian codes for wind (IS 875-part 3: 1985) and earthquake forces (IS 1893: 2002), it was found that depending on the frame size and location, the ratio of wind load to earthquake loads (WL/EL) varied from 1 to 17.3 thereby testifying to the fact that these frames are often governed by wind rather than earthquake loads. Another reason for this is that the earthquake loads considered is the Design Basis Earthquake (DBE) which is further reduced by a reduction factor R which is prescribed as 4 for ordinary moment frames. The reduction factor is for general building frames which have higher degree of redundancies and are assumed to have sections with moderate ductility. Thus, it is important to verify the applicability of these values to PEB frames with low redundancies and slender (non-ductile) sections.

Considering the fact that the Maximum Credible Earthquake (MCE) can be twice as large as the Design Basis Earthquake (IS 1893:2002) it can be realized that frames with ratios of Wind Load to Earthquake Load (WL/EL) greater than eight will respond elastically under MCE. Only six frames satisfied this condition. Of the remaining 48 frames, 7 were governed by drift under wind, 13 were governed by strength under wind load and 28 were governed by deflection under gravity load. For all 54 frames, the required reduction factor R_d was calculated as the ratio of design lateral capacity to the seismic strength demand under MCE and plotted in Fig. 2 against the ratio of WL/EL. It can be observed that for 17 of these frames, the value of R_d was greater than unity which means that these frames will experience significant inelastic deformations under a severe earthquake. It can also be noted that the frame designs were governed by all the three criteria, namely drift under wind, strength under wind and serviceability under gravity loads. The ductility demands on these frames under MCE were obtained by pushover analysis as below.

Pushover analysis was carried out on the 17 frames using SAP 2000 (2011) software programme. Second-order effects were considered by using the P-Δ feature available in SAP2000. For both rafter and columns,'M3 hinges' were defined based on the results obtained from the finite element model of beam. Fig. 3 shows the actual and idealized moment-rotation curve as defined in SAP2000. Point 'B' refers to the yield moment of the effective section based on the effective width of the plate elements and its corresponding rotation was obtained from the finite element analysis. Point 'C' defines the maximum moment capacity M_{max} taken as $1.1M_{eff}$ and its corresponding θ_{max} is taken as $1.4\theta_{eff}$. The moment capacity of section at point 'D' is considered to be the yield moment M_{eff} and its corresponding rotation θ_u is obtained as $\mu_s\ \theta_{eff}$, where the section ductility μ_s is given as a function of the depth-to-thickness ratio of the web by the following equation (valid only for plastic and compact flanges) as obtained from FEM.

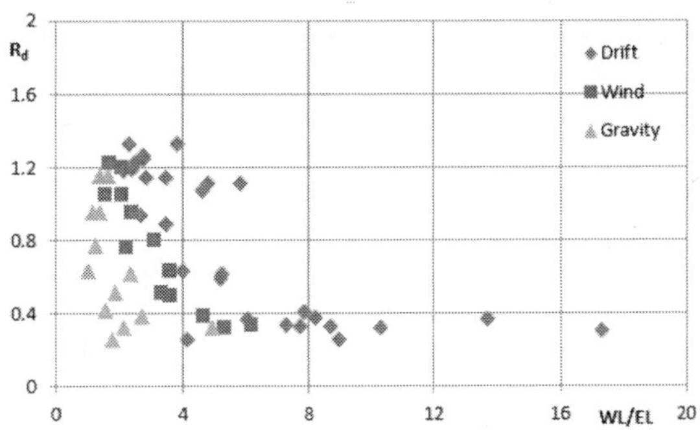

Figure 2. Variation of R_d with *WL/EL* for different governing design criteria.

$$\mu_s = 0.0004(d/t_w)^2 - 0.117(d/t_w) + 10.2 \quad (1)$$

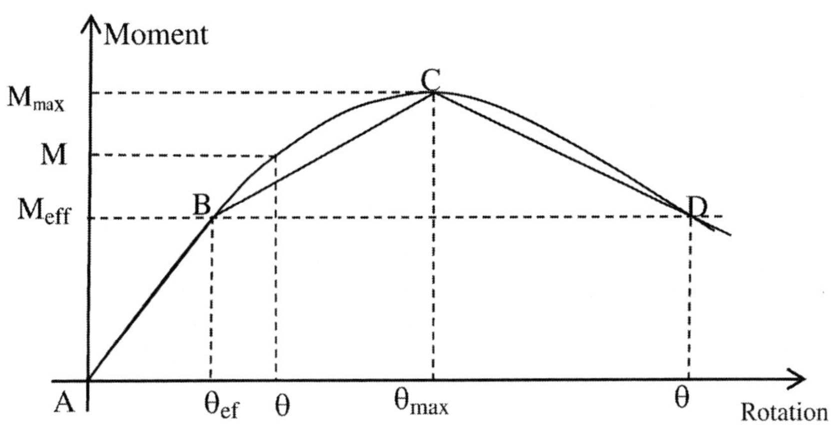

Figure 3. Actual and Idealized Moment-Rotation curves for members.

The elastic response of PEB Portal frame under the application of increasing monotonic lateral force (H) was studied to calculate the design lateral capacity of frames. To have more realistic and conservative results, the frames were subjected to constant seismic weight (2V). Fig. 4(a) shows a typical frame. Increase in lateral force leads to increase of reactions and bending moments in members. Once any section of the frame reaches its yield strength, the structure deviates from its elastic path and goes into the strain hardening region. The curve generated from pushover analysis is converted to an energy-equivalent elastic-perfectly-plastic curve (EEPP) by the bi-linear idealization method by equating the area under the actual response of frame is with the area under the bi-linear curve and yield capacity (H_y) and corresponding yield displacement (Δ_y) (see Fig. 4(b)). The ultimate displacement (Δ_u) of the frame is considered as the displacement at which the strength drops back to H_y. Structural ductility μ is defined as the ratio of ultimate displacement to the yield displacement of structure.

Permissible capacity reduction factor for a frame is denoted as R_c because it depends on the strength and ductility of the sections provided. A structure can sustain an elastic seismic demand equivalent to R_c times its yield strength by virtue of its ductility. The R_c values obtained for the frames are plotted against the section ductility in Fig. 5 and it can seen that Rc increases with ductility μ and its average value for sections with d/tw ratios from 100 to 150 is about 1.9 and corresponds to an average ductility of 1.6. Also shown are the R_N factors prescribed by Newmark and Hall as $\sqrt{(2\mu-1)}$ which are much lower than those obtained in this study.

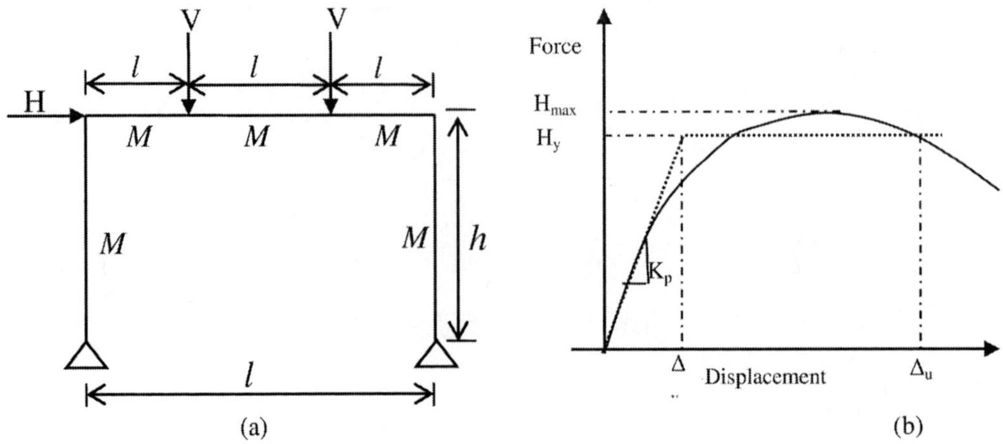

Figure 4. Typical Portal frame analyzed and its bilinear force-displacement curve.

The demand reduction factor, R_d is the ratio of elastic strength demand on the structure under a severe earthquake to the design lateral strength of the structure. R_d value depends on the seismic zone, gravity load considered as well as the importance factor of structure. In seismic design, R_c of a structure should always be greater than R_d value for collapse prevention. The values of R_c are plotted against the R_d values obtained from pushover analysis in Fig. 6. It can be observed that R_c is always higher than R_d thus testifying that the frames are safe against MCE. However, for frames with importance factor I=1.5, several frames are proving to be unsafe as shown in the same figure.

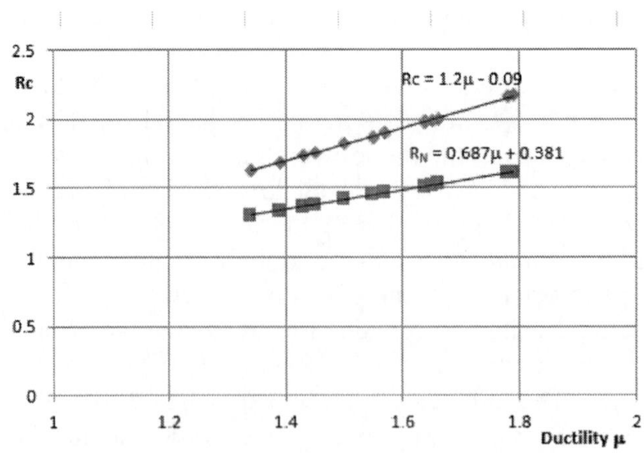

Figure 5. Variation of Rc with section ductility μ.

4 CONCLUSIONS

Based on the study, the following conclusions can be drawn:

- When the ratio of WL/EL exceeds 4 for frames with I=1, the frames will remain within their elastic limit under MCE. Similarly, frames with I=1.5 will also remain within their elastic limit for WL/EL ratios exceeding 6. So no limitations on width-to-thickness ratios need to be imposed on these frames to resist severe seismic loads.
- For WL/EL ratios less than 4 and frames with I=1, the frames will undergo inelastic deformations but will not collapse due to the limited ductility. However, for frames with I=1.5, the flanges can be plastic or compact and webs can be semi-compact so as to develop enough ductility to prevent collapse under MCE. Alternatively, the frames can be designed with R factor of 2 to ensure adequate strength to resist MCE.

The above provisions may be incorporated in the code to enable safe and economical design of PEB frames.

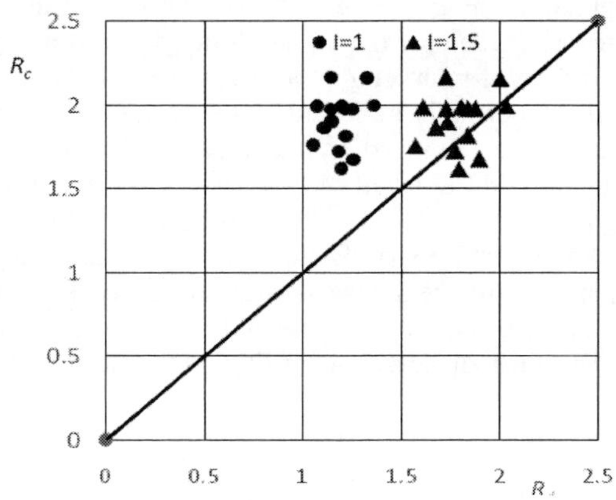

Figure 6. R_c versus R_d for *I=1* and *I=1.5*.

References

ABAQUS User's Manual, Version 6.11.(2011), Hibbitt, Karlsson and Sorensen, Dassault Systems Simulia Corp. Pawtucket, R.I.

D'Aniello, M., Landolfo, R., Piluso, V., and Rizzano, G. (2012): "Ultimate Behavior of Steel Beam Under Non-uniform Bending", *Journal of Constructional Steel Research*, 78, pp. 144-158.

IS 800:2007, Code of practice for General Construction in Steel, *Bureau of Indian Standards*, New Delhi, India.

IS 875-part 3: 1985, Code of practice for wind loading on structures, *Bureau of Indian Standards,* New Delhi, India.

IS 1893:2002 Criteria for Earthquake Resistant Design of Structures- General Provisions and Buildings, *Bureau of Indian Standards,* New Delhi, India.

SAP 2000 (2011) NL Integrated software for Analysis and Design, computers and structures, Inc., Berkeley, California, USA.

Smitha, M. S. and Kumar S. (2013): "Steel-concrete Composite Flange Plate Connections-Finite Element Modeling and Parametric Studies", *Journal of Constructional Steel Research*, 82, pp. 164-176.

Integrated Solutions for Infrastructure Development
Edited by Lau, H. H., Tang, F. E., Ng, C. K., and Singh, A.
Copyright © 2016 ISEC Press

BASE ISOLATION IN SEISMIC RETROFITTING OF R/C BUILDINGS

ALBERTO DUSI[1], MARCO MEZZI[2] and TAN TENG OR[3]

[1]*Numeria Consulting Engineers, Cremona, Italy*
[2]*Dept Civil and Environmental Engineering, University of Perugia, Perugia, Italy*
[3]*Base Isolation Technology (ASIA), Klang, Malaysia*

The paper, based on the authors' direct involvement in managing actual retrofitting design, reports on seismic vulnerability assessment, design and implementation issues related to the seismic retrofit of reinforced concrete buildings through base isolation. The retrofitting interventions design of damaged buildings involves three aspects: the damage survey, the assessment of the vulnerability of building in its original structural configuration, the design of intervention needed to reduce the building vulnerability up to a conventional level, normally indicated by the seismic standards. In the first part of the paper, the experience achieved in the retrofit of reinforced concrete buildings damaged by the Italian 2009 L'Aquila earthquake is presented by referring to a typical intervention designed by the authors. Topics related to conventional vs base isolation retrofitting strategies, structure's performance, safeguard of human life, construction efficiency and repairing cost are analyzed with reference to actual case study. Lessons learned from the Italian experience have been critically applied to the design of retrofit intervention of a building, designed according to the current Nepal set of codes and under completion at the time of the earthquake, damaged by the Gorkha 2015 earthquake. In the second part of the paper activities carried out for the definition of a specific site seismic input, for the dynamic characterization of the building and for the design of the base isolation retrofit are presented.

Keywords: Earthquake damage, Seismic isolation, Retrofitting intervention, Dynamic characterization, HVSR technique.

1 INTRODUCTION

Seismic (or base) isolation is a design technique that reduces the force demand on structures by isolating them from the damaging effect of the ground motion (Martelli *et al.* 2011). It functions primarily by lengthening the period of the structure (Skinner *et al.* 1993). This approach contrasts with conventional design schemes that rely on inelastic action of various structural elements to dissipate earthquake energy. It provides a level of performance well beyond the normal code requirements with potential for substantial life-cycle cost reduction. Base isolation offers important advantages over conventional protection methods because it reduces the earthquake forces transmitted into a structure, thus it protects not only the structure itself but also the contents and secondary structural features.

Isolation is achieved with specially designed bearings placed between the building and its foundations that provide flexibility and energy absorption capability while supporting the weight of the structure. These bearings can be replaced if such need arises. The design of the retrofitting and seismic improvement of the damaged buildings involves the vulnerability assessment of the buildings in their undamaged conditions and the identification of the most effective techniques for the seismic enhancement (Mokha *et al.* 1996). Ranking of the different seismic enhancement strategies should be based on their structural effectiveness, impact in the application, lifetime, related indirect works as well as their direct and future costs. However, also ambient and cultural conditionings and the explicit or latent reluctance to adopt innovative solutions have a significant influence on the choice of the retrofitting strategy to be adopted. Very often even minimum constraints put by the actual situation of the existing building play a decisive role in opting for a retrofit design strategy adverse to a non-traditional solution. This is also the case of base isolation (BI) that is applied in a limited number of cases despite its well-known effectiveness.

Two of the cases in which the BI technique has been proven to be more effective and convenient than conventional strengthening, and therefore adopted for the seismic improvement, are illustrated in the paper.

2 APPLICATION IN ITALY

A huge number of r/c buildings were strongly damaged by the 2009 L'Aquila earthquake (EERI 2009). Many of these buildings, designed and built more than twenty years ago, were characterized by insufficient resistance against lateral forces, as well as by insufficient ductility and resilience due to inadequate structural configuration and structural detailing. In the following are reported the conventionally based and the base isolated design retrofit of a typical Italian residential building.

2.1 Building Description and Vulnerability

The building, in its as built condition (Figure 1), has an L-shaped plan resulting from the presence of two blocks, independent from the structural point of view and separated by an insufficient seismic gap of 4 centimeters. The building has a basement with cellars and garages, five floors above ground and a loft. The structure is made of r/c beams and columns with floors made of hollow bricks and concrete.

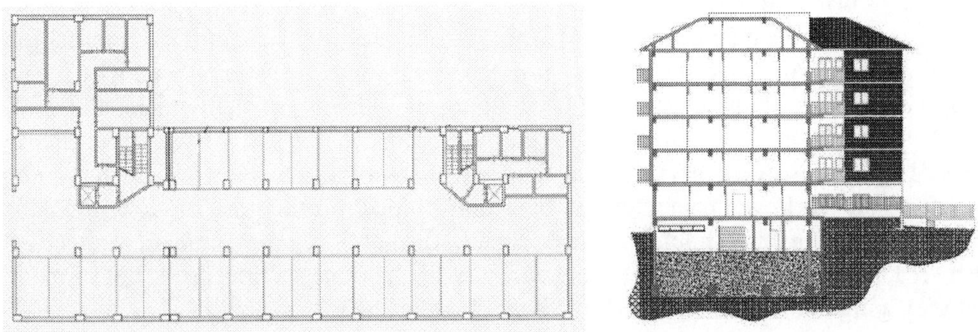

Figure 1. Basement floor plan view and building longitudinal section.

The mechanical characteristics of the materials were estimated on the basis of testing campaigns including core sampling and SonReb tests for concrete and tension tests on samples of steel bars. Tests indicated an average cubic strength of concrete R_{cm}=22.0MPa and a characteristic yield strength of steel f_{yk}=305MPa. According to the Italian code provisions on the knowledge of the existing constructions concerning geometry, constructive details and mechanical properties of materials, it results an intermediate knowledge level (adequate knowledge); this implies the application of a confidence factor FC=1.2 reducing the mechanical parameters of materials to be used in numerical calculations. The performed geological investigations pointed out that the subsoil was characterized by a reference shear wave velocity greater than 360 m/s and lower than 800 m/s and so classifiable as "Category B" according to European (EN1998-1 2004) and Italian (NTC 2008) standards. The seismic hazard is then expressed in terms of demand peak ground acceleration for different limit states: $PGA_{D,LD}$=0,125g for the Limited Damage (LD) limit state, $PGA_{D,LS}$=0,301g for the Life Safety (LS) limit state, $PGA_{D,CP}$=0,361g for the Collapse Prevention (CP) limit state.

The seismic capacity of building in its original undamaged state was assessed by nonlinear static analyses. The minimum value of bedrock acceleration for the attainment of the LS limit state is $a_{g,C,SLV}$=0.078g. The vulnerability of the construction is expressed by the ratio $a_{g,C,SLV}/a_{g,SLV}$=0,299 between the capacity and demand accelerations ("risk index" in the Italian guidelines). Since this value is less than the minimum of 0.60 prescribed by the guidelines for the repair of damaged buildings, works are required to improve the seismic capacity up to a Capacity/Demand (C/D) ratio higher than 0.60 but less than 0.80. The collapse scenario associated with the attainment of the limit state in pushover analyses shows that it is connected to the shear failure of columns together with flexural damage of beams. Actually, the building suffered damage in the 2009 seismic event, especially at the lower levels, to both nonstructural (claddings and internal partitions) and structural elements (cracking of r/c elements). In the post-earthquake survey of damaged buildings, the building was classified as class E "unfit for use – to be evacuated".

2.3 Improvement of Seismic Capacity through Traditional and Innovative Approaches

The adopted enhancement solution foresees the insertion of a base isolation system, but for the purpose of comparison, also a traditional enhancement strategy based on the reinforcement and stiffening of the structural components has been considered.

Within the framework of a traditional seismic improvement the following interventions should be made: (a) insertion of r/c walls to increase the seismic resistance and reduce the lateral and torsional deformability; (b) construction of a r/c slab to strengthen the floors, damaged by the earthquake, and ensure a diaphragm behavior; (c) steel jacketing of unconfined internal joints; (d) bonding fiber composites tapes to strengthen the external joints; (e) strengthening of local critical elements (i.e. landing beams of stairs); (f) widening of foundation beams under the walls. The traditional seismic improvement allows the structure to reach a capacitive acceleration $a_{g,C,SLV}$=0.211g, that is an acceptable C/D ratio equal 0.649>0.60. However, it has to be noted that the aforesaid works are characterized by a high impact on the construction since almost all the nonstructural elements should be demolished; furthermore, the standard protection levels cannot be reached, due to the inherent low capacity of the construction itself, not to mention the fact that they a high cost with respect to the benefit achieved.

The base isolation option was considered as the preferred solution since the very beginning of the design for the seismic improvement of the building. This was mainly due to the possibility to do most of the works at the basement level only, almost completely avoiding strengthening works at the elevation in spite of the generalized low seismic-resistant capacity of the primary structural system. The following interventions have been conceived: cutting of the top portions of the columns at basement level, insertion of isolating devices, strengthening of the columns below the cut and of the joints above the cut, demolition and reconstruction of the first flight of the stairs. Figure 2 shows some details of the base isolation works. The isolating system, consisting of one type of High Damping Rubber Bearings (HDRB) 650mm diameter and sliders, allowed to reach a base isolated period of 2.62 s and participating mass percentage almost equal to 99%.

Figure 2. Pictures of works for the base isolated retrofit.

As far as the building elevation (existing superstructure) is concerned, the resistant sections of beams and columns have been proven to be sufficient to sustain the stresses induced by the earthquake; this turned out to be the only safety check since for base-isolated constructions the detailing design rules of seismic-resistant structures are not prescribed. For the substructure the strengthening, consisting of jacketing with r/c and steel profiles of the columns below the isolators, has been designed taking into account the P-delta effect caused by the isolators' displacements. No strengthening of foundations was required. The base isolation approach allows the structure to obtain a C/D ratio equal to 0.797>0.60.

2.4 Cost Comparison

The comparison of the retrofitting costs of the two solutions is reported in Table 1 from which it can be seen that the base isolation approach allows to achieve an immediate saving of 34%, saving that in the building lifetime will be even greater. As a matter of a fact, the base-isolated building will not suffer, under the maximum expected earthquake any structural damage, while, at the same time offering full protection to the contents and secondary features (such as cladding and windows) thus improving the safety of occupants and passers-by.

On the contrary the traditionally retrofitted building will start undergoing serious consequences for an earthquake having an intensity lower than 70% of the maximum expected one and a 25% probability to be exceeded in the building life. Moreover, for the maximum expected quake, i.e. the one a probability of 10% in the lifetime, the building will suffer extensive damage with an expected reparation cost comparable with the reconstruction cost.

Table 1. Cost comparison for conventional and base isolated retrofit.

Work category	Traditional		Base-isolated		Diff.
Type A - Repair	€ 2.002.536	40.7 %	€ 199.000	32.9 %	-47.2 %
Type B - Seismic enhancement	€ 1.597.770	32.5 %	€ 646.000	36.4 %	-26.7 %
Retrofit due to Type B works	€ 620.068	12.6 %	€ 356.000	6.7 %	-65.1 %
Hygienic sanitary conformity	€ 232.783	4.7 %	€ 232.783	7.2 %	=
Energy saving conformity	€ 354.456	7.2 %	€ 354.456	11.5 %	=
Lifts	€108.863	2.2 %	€ 178.723	5.6 %	+67.1 %
Total	€ 4.916.478	100.00 %	€ 3.029.118	100.0 %	-34.3%

3 APPLICATION IN NEPAL

The April 25, 2015 M 7.8 Nepal earthquake, also known as the Gorkha earthquake had his epicenter approximately 80 km to the northwest of the Nepalese capital of Kathmandu. A number of buildings also located in the so called Kathmandu valley and designed according to the current Nepal National Building set of codes were damaged.

A damage survey has been carried out and a number of r/c buildings, representative of currently used Nepalese earthquake resistant structural schemes, have been selected by the authors for the vulnerability assessment and retrofit intervention designs. In the following the focus is on a building complex, hereinafter referred to as the "Downtown complex", made of two towers of 15 floors each resting on a common underground basement (Figure 3). The "Downtown complex" was in an advanced construction stage at the time of the earthquake and suffered damage to both structural and nonstructural elements.

Figure 3. The "Downtown building complex".

On the selected building, a set of instruments have been installed by the Seismological Research Centre of the Italian National Institute of Oceanography and Experimental Geophysics for the acquisition of accelerometric data, while estimates of site amplification of seismic ground motion have been obtained from application of HVSR technique (OGS 2016). In Figure 4 the average H/V curves at different floors are reported.

Seismic site study results showed that no amplifications are to be expected at typical isolation periods, therefore base isolation has been considered as retrofitting strategy. The seismic behavior of the buildings has also been analyzed by means of Finite Element Model that has been suitably adjusted to match the frequencies identified through the site tests. Indeed the first two frequencies identified on site are equal to 0.96Hz and 3.7Hz and the corresponding frequencies computed on the FE model are 0.97Hz and 3.6Hz. Analyses showed that the insertion of HDRBs at the top of the basement columns gives an isolated period of 3,5s with consequences in term of performance increase and cost reduction similar to those obtained in the Italian application.

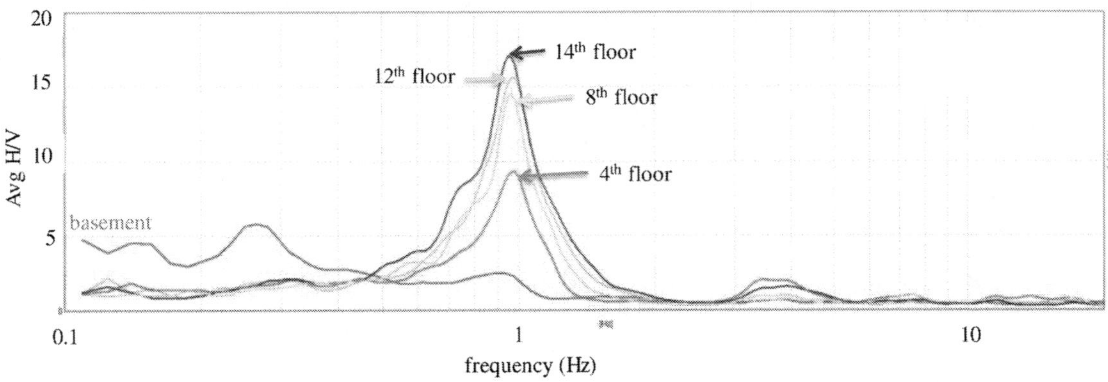

Figure 4. Average H/V curves at different floors.

4 CONCLUSIONS

Base isolation application in retrofit of r/c structures has been proved to be effective and economically convenient. The cost comparison carried out shows that base-isolation choice allows to obtain an immediate saving of 34%; savings that in the building's lifetime are actually even greater. Experiences achieved in the design and retrofit works can be extended to seismic prone area worldwide.

References

EERI, *The M_w 6.3 Abruzzo, Italy, Earthquake of April 6, 2009*, Special Earthquake Report, 2009.

EN1998-1, *Eurocode 8: Design of structures for earthquake resistance - Part 1: General rules, seismic actions and rules for buildings*. CEN, Brussels, 2004.

Martelli, A., Forni, M., Dusi, A. Seismic Isolation: Available Devices, Recent Applications and Conditions for their Correct Use, in B.H.V. Topping, Y. Tsompanakis, (Editors), "Proc. 13th Int. Conf. on Civil, Structural and Environmental Engineering Computing", Civil-Comp Press, Stirlingshire, UK, 2011.

Mokha, A., Amin, N., Constantinou, M., and Zayas, V. (1996). Seismic Isolation Retrofit of Large Historic Building. *J. Struct. Eng.*, 10.1061/(ASCE)0733-9445(1996)122:3(298), 298-308.

NTC2008, *Norme Tecniche per le Costruzioni D.M. 14 gennaio 2008*, Consiglio Superiore dei Lavori Pubblici, GU Repubblica Italiana, 2008

OGS, *Private communications*, Trieste, 2016.

Skinner, R. I., Robinson, W. H., McVerry, G. H., *An Introduction to Seismic Isolation,* John Wiley and Sons Ltd, 1993.

Integrated Solutions for Infrastructure Development
Edited by Lau, H. H., Tang, F. E., Ng, C. K., and Singh, A.
Copyright © 2016 ISEC Press
ISBN: 978-0-9960437-3-1

EXTREME ENVIRONMENTAL LOAD MODELS FOR WAVE AND CURRENT RESPONSE SURFACE METHOD

ZAFARULLAH NIZAMANI[1], YAP ENG CHING[1], MOHAMED MUBARAK BIN ABDUL WAHAB[2], and ABDULLATEEF OLANREWAJU[3]

[1]*Dept of Environmental Engineering, Universiti Tunku Abdul Rahman, Kampar, Malaysia*
[2]*Civil and Environmental Engineering Dept, Universiti Teknologi PETRONAS (UTP) Bandar Seri Iskandar, Perak, Malaysia*
[3]*Dept of Construction Management, Universiti Tunku Abdul Rahman, Kampar, Malaysia*

The three most uncertain environmental design loads acting on offshore structures are wave height, current and wind velocity. If the reliability of structure is to be determined, then we need to have limit state function for load which requires that we should transform the environmental loads into load model. Load model, which predicts the load, will ultimately affect the design resistance and thus its final impact on cost could be large. Since, there are many offshore structures located in different offshore regions in Malaysia, the calibration of the environmental load model, should be evaluated to determine which model fits best. To obtain the environmental load model, response surface technique is generally applied. Load models suggested by DNV and ISO code are analysed to determine the best model fit for local conditions of Malaysia. Base shear, wave height and current velocity are used in a linear fit to determine polynomial response surface. The results showed that due to the geography of Malaysia different regions might have to use specific load models instead of a general load model for all regions.

Keywords: Offshore structures, Linear fit, Weibull distribution, Curve and surface fit.

1 INTRODUCTION

Offshore structures are designed to resist three kinds of loads, Environmental loads i.e., wave, currents and wind, Dead loads, i.e., weight of the structure, and Live loads, i.e., weight of consumable supplies and fluids in pipes and tanks. The extreme event considered for the design and assessment of old structures is recommended to have the probability of failure of 1E-2. The ultimate strength analysis requires that the resistance should be higher than the loads. Thus, higher anticipated loads will require higher resistance. Since, there are many offshore structures located in different offshore regions in Malaysia, the calibration of the environmental load model, should be evaluated to determine which model fits best for local conditions. This is because the environmental load model, used for calibration from Gulf of Mexico (GOM) and North Sea (NS) may or may not be reasonable for local geography of Malaysia (Sigurdsson *et al.* 1994, Atkins 2001, Wahab *et al.* 2015, Nichols 2014).

2 ENVIRONMENTAL LOADS

Environmental loads are severe and unpredictable due to uncertain whether condition and they can act with unexpected severity on offshore structures. The safety of offshore structures is achieved by taking into consideration hazards produced by rare wave, current and wind events. ISO 19902 code defines this rare and extreme event of 100 years. Present day design methods depend on all energy comes from a single direction which is a long-crested waves. However, sea waves in real are multidirectional as the energy comes from many directions simultaneously. Wave loads are unsteady and exert the largest force and it is producing main contribution to the hydrodynamic load. Currents can play important role in total forces acting on offshore structures. The current generated in the ocean is steady in nature and its effect tends to vary with the water depth. When extreme waves along with super-imposed current occur in same direction velocities from both can combine and produce large wave pressure. Normally wind load will not produce much effect, especially in the design conditions and it generally accounts for only 10% of the overall design load as used by DNV code.

2.1 Environmental Load Model

Many authors have studied response surface methodology to determine the reliability of offshore structures such as Gaspar *et al.* (2015), Gaspar *et al.* (2014), Gaspar *et al.* (2013) and Gaspar *et al.* (2012). The environmental load model is determined using response surface modeling using curve and surface fitting tool of MATLAB. The custom polynomial Equation for wave and current is defined to get the parametric values for the given coefficients. The first model is suggested by SHELL for the development of load factors for ISO 19902 code are shown in Eq. (1).

$$W = aH_{max}^2 + bH_{max} + cV_c^2 + dV_c + e \tag{1}$$

W = Load effects, H_{max} = variable annual maximum wave height, V_c = variable current speed, the coefficients of *a, b, c, d* and *e* are found from curve and surface by using least square method.

Response surface fit proposed by Heidman (1980) as shown in Equation (2) is used to get the load effects from Jacket base shear and wave heights Gerhard (2005). To establish the relationship between wave and current load and response of Jacket curve fitting is done to find the coefficients of the response surface Equation.

$$W = a_1\,(H_{max} + a_2V_c)^{a_3} \tag{2}$$

W = load effects, H_{max} = maximum wave height and V_c = current velocity. Coefficients a_1, a_2, and a_3 are found from the curve and surface. The third model was suggested during development of API LRFD code by Moses and Stahl (2000), as shown in Eq. (3),

$$W = X_{hydro}(a.H^b) \tag{3}$$

W= Environmental load model, X_{hydro} = Uncertainty model for base shear, H = maximum wave height, while coefficients *a* and *b* representing structure dependent parameter, fitted from analysis results of base shear and wave height.

2.2 Linear Polynomial Response Surface Fit

Table 1 contains the platform specific load parameters for one region of Malaysia, which shows the basic values of load variables used for conversion into random input for load values in FEM software. Using these values along with Weibull distribution, 100 random values is generated for wave, current and time period using MATLAB. These 100 random load values are used to get 100 stresses from FEM output file for each component. The corresponding 100 component output stresses are produced by FEM software. These are converted to load model using curve and surface fit. To get the environmental load effect for the reliability analysis, 100 stress values for wave, time period and current are determined using Weibull distribution using Table 1.

Table 1. Offshore jacket platform data.

Parameter	10 year	100 year	Weibull Distribution Parameter Scale	Weibull Distribution Parameters Shape
H_{max}	6.8 (m)	7.7 (m)	5.86	5.58
Tp	10.6 (s)	11.0 (s)	18.71	10.14
Current	0.78 (m/s)	0.94 (m/s)	0.62	3.71

Random variables obtained through Weibull distribution, are used to determine the base shear. The load is applied at each 45° i.e., from eight different directions as these jackets were four and six legged. From initial analysis, it was determined that 0° was producing maximum load effects. Therefore, this direction is selected for further analysis. The process is repeated until one hundred values of base shear are obtained. Base shear obtained is used for the response surface fit of the jacket platform. The relationship of the dependent and independent variables can be represented by curve fitting through the regression analysis, so that, we can understand how much uncertainty is present in the curve fit with random error. Curve fit also shows the accuracy of the analysis by displaying the value of R-square. Random variables such as wave height are assigned into x-axis, while base shear is set as the input of y-axis. Then, a model equation is inserted by choosing the custom equation option. Whenever the equation is applied, fitting tool will analyze the data from the set as the axis component. Coefficient of the model equation can be determined by using least-mean square method. Since the model equation applied is a two- parameter polynomial fit where the wave height and base shear are substituted into this equation, only coefficient will remain unknown and therefore it could be determined as the analysis is run. These coefficients are displayed together with the R-square. To evaluate effects of current velocity on load model surface fit tool is used which involves three dimension variables which are wave height, current velocity and base shear. Response surface of jacket platform can be analyzed based on this quadratic polynomial response surface fit. The accuracy is measured based on how close surface area created by data point linked to the surface created by equation. In the end sensitivity analysis is made to check the results obtained by using environmental load model and base shear from FEM software.

3 RESULT AND DISCUSSION

The coefficients are derived using the least square method. Data is analyzed at 95% confidence level. Here the design wave height and current speed used are based on 100 years as required by ISO/API codes. Figures 1-2 represent the surface fitting of Polynomial fit for model equations 1-2 and Figure 3 presents curve fitting of load model Eq. (3). The curve fitting for load model Eq. 1-2 is not shown here, but results are presented and compared against base shear. Figure 1 shows a plotted surface fitting of base shear vs. wave height and current velocity with the model of Eq. (1) and the coefficients are obtained where a= 0.1133, b= -0.4702, c= 0.1886, d= 4.199 and e= -0.1356. Now when current is not part of random variable a curve fit is made and the coefficients determined are a= 0.1304, b= -0.003236, c= 0.4719, d= -0.008059 and e= -0.8972. When model Eq. (2) is used with curve fitting i.e., no current is used then the coefficients determined are a_1=

0.4188, a_2= -0.003236 and a_3= 0.4719. When current is used as a variable and surface fit is used as shown in Figure 2, the coefficients determined are a_1= 0.02824, a_2= 6.099 and a_3= 2.141. Figure 3 illustrates a graph of base shear vs. wave height by the application of model Eq. (3) using curve fitting and has coefficients of a= 0.04354 and b= 2.498 and R^2 is 0.9923. Table 2 provides the parameters for wave and current for the platforms obtained from curve fitting. Now it is clear that Eq. (1) is providing the best fit model and this is due to the effects of current velocity are significant in this region of offshore Malaysia.

3.1 Environmental Load Model Analysis

The Environmental load applied to the structure is directly proportional to the base shear exerted on the platform. The closer the value of the environmental load model to base shear, more accurate will be the model for reliability analysis. Using Eq. (1), the base shear of the platform, when analyzed using the curve fit, is very close to its environmental load model since the R^2 displayed is 0.9969 as shown in Table 3. By comparing both average base shear and environmental load model of this platform, the average base shear is 3.06 MN and the maximum environmental load model is 3.09 MN which mean the difference between them are only 0.03. Hence, the relationship of base shear and environmental load model represented by the graph shows the high value of R^2. However, when it is analyzed by using surface fitting, the value of R^2 gives 0.9999 as shown in Figure 4 and Table 3 when current velocity is considered into the environmental random variables which shows that current velocity plays a significant role and it should be incorporated in developing the load model. The R^2 values for curve fit for Eq. (2), is 0.926 as recorded in Table 2. Figure 5 and Table 3 illustrates a fit plotted based on the base shear vs. environmental load model where the environmental load models are generated from the substitution of coefficient obtained from the surface fitting of Figure 2 into the Eq. (2). This is the lowest value among the three equations used and this shows that the second equation applied may not appropriate for this platform. Figure 5 also displays the function of the fit and its R^2 values which is 0.974 as recorded in Table 3. In Figure 6, a graph of base shear vs. environmental load model is plotted based on the environmental load model generated from the substitution of coefficients into Eq. (3). Those coefficients are obtained from the curve fitting of Figure 3. From Figure 6, the function of the fit can be found and its R^2 is recorded in Table 3. The value of the R^2 is 0.9927.

Table 2. Base shear vs. wave height and current velocity, R^2.

Curve fitting model			Surface fitting model	
1st	2nd	3rd	1st	2nd
0.9960	0.9961	0.9923	0.9999	0.9971

Table 3. Base shear vs. environmental load model, R^2.

Base shear vs. Environmental load model				
Curve fitting			Surface fitting	
1st equation	2nd equation	3rd equation	1st equation	2nd equation
0.9969	0.9260	0.9927	0.9999	0.9974

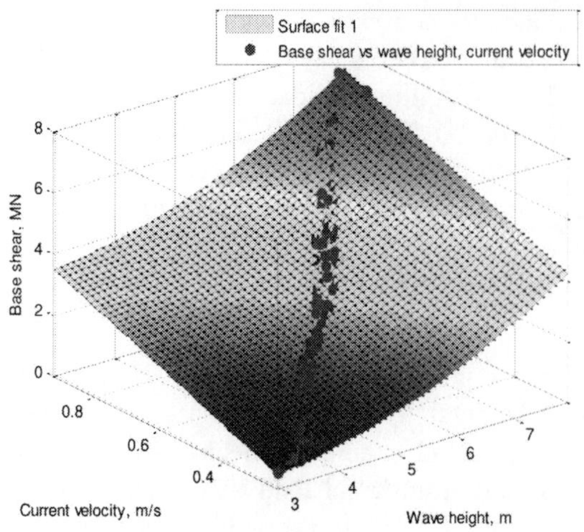

Figure 1. Base shear vs. wave height and current velocity Model 1_surface fit.

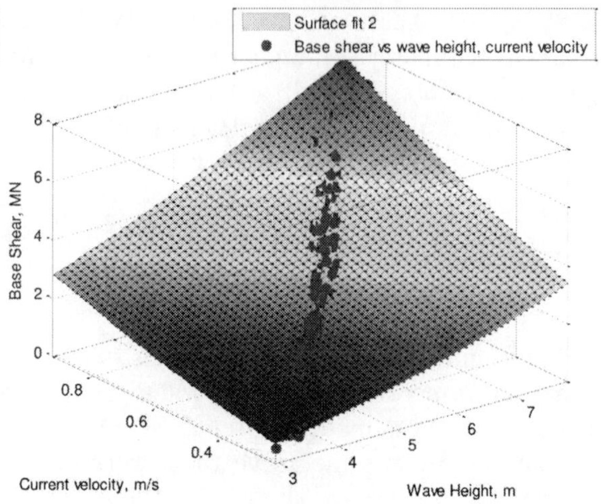

Figure 2. Base shear vs. wave height and current velocity Model 2_surface fit.

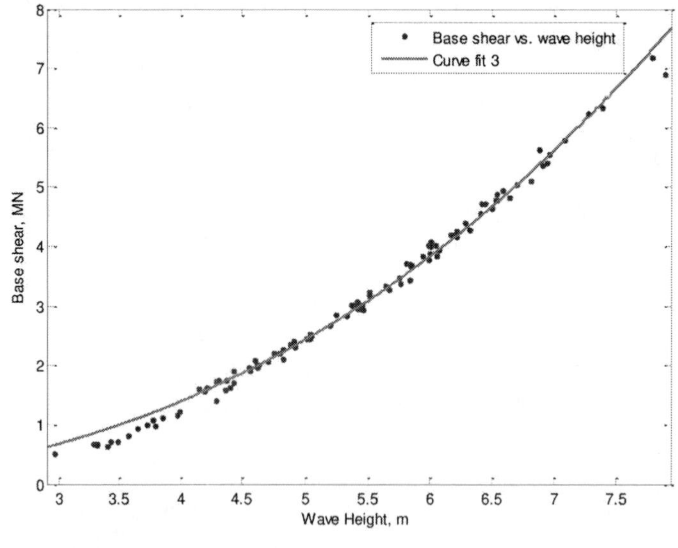

Figure 3. Base shear vs. wave height 3_curve fit.

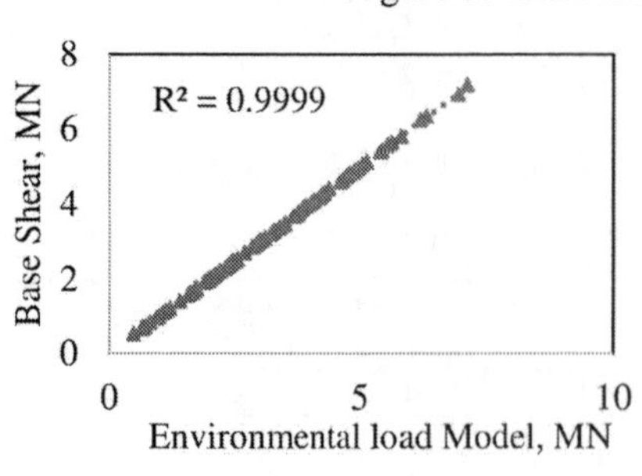

Figure 4. Base shear vs. environmental load Model 1 (Surface Fitting).

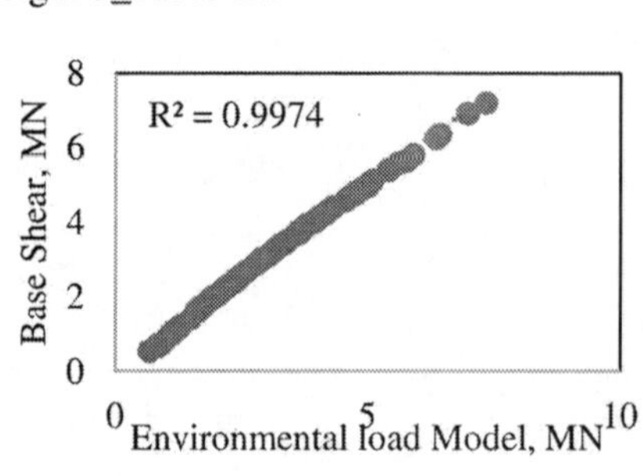

Figure 5. Base shear vs. environmental load Model 2 (Surface Fitting).

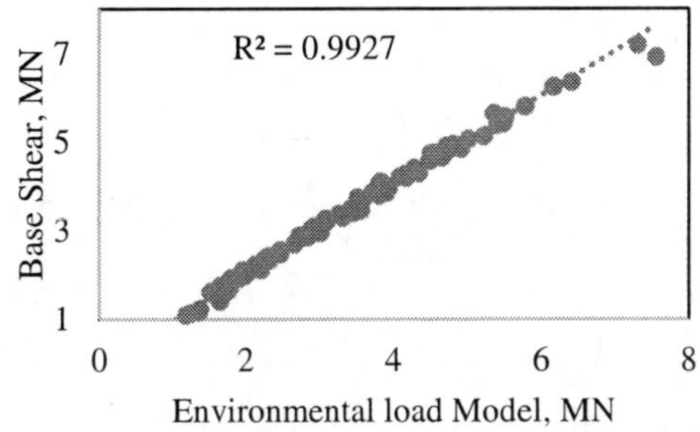

Figure 6. Base shear vs. environmental load Model 3 (Curve Fitting).

4 CONCLUSION

Wave height and current speed data from two platform specific locations are used. Regression analysis is used here to find the coefficients for response surface fit. Three polynomial models are used for converting the wave and current forces into the environmental load model. Best fit is assumed to predict as nearly as possible to the actual base shear from the analysis of jacket. The results show that first equation predicts very well to actual base shear and should be applied for the platforms lying in this region.

Acknowledgments

The research was supported by UTAR-RF, Project no. IPSR/RMC/UTARRF/2014-C2/Z01 and authors would like to thank support fund provided by the Universiti Tunku Abdul Rahman.

References

Atkins, W.S., Consultants Ltd., Comparative evaluation of minimum structures and jackets. Health and Safety Executive, offshore technology report, UK, 2001.

Gaspar, B., Bucher, C., and Guedes Soares, C., Reliability analysis of plate elements under uniaxial compression using an adaptive response surface approach, *Ships and Offshore Structures*, 2015.

Gaspar, B., Naess, A., Leira B. J., and Guedes Soares, C., System Reliability Analysis by Monte Carlo Based Method and Finite Element Structural Models, *Journal of Offshore Mechanics and Arctic Engineering*, 2014.

Gaspar, B. and Guedes Soares, C., Hull girder reliability using a Monte Carlo based simulation method. *Probabilistic Engineering Mechanics*, 2013.

Gaspar, B., Naess, A., Leira B. J., and Guedes Soares, C., System reliability analysis of a stiffened panel under combined uniaxial compression and lateral pressure loads, *Structural Safety*, 2012.

Gerhard, E., "Assessment of existing offshore structures for life extension," Doctor of Philosophy, Department of Mechanical and Structural Engineering and Material Science, University of Stavanger, 2005.

Heideman, J., "Parametric Response Model for Wave/current Joint Probability," Report Submitted to *API TAC 88-20 for API LRFD*, 1980.

Nichols, N. W., Khan, R., Rahman, A. A., Akram, M. K. M., and Chen, K., Load resistance factor design (LRFD) calibration of load factors for extreme storm loading in Malaysian waters. Journal of Marine Engineering & Technology, 2014.

Moses, F. and Stahl, B., "Calibration issues in development of ISO standards for fixed steel offshore structures," *Journal of Offshore Mechanics and Arctic Engineering*, 2000.

Sigurdsson, G., Skallerud, B., Skjong, R., and Amdahl, J., Probabilistic collapse analysis of Jackets. OMAE conference, 1994.

Wahab, M. M. A., John, K. V., Liew, M. S., Nizamani, Z., and Kim, D. K., Structural reliability analysis using quadratic polynomial response surface methodology, OMAE, 2016.

Integrated Solutions for Infrastructure Development
Edited by Lau, H. H., Tang, F. E., Ng, C. K., and Singh, A.
Copyright © 2016 ISEC Press

DYNAMIC ANALYSIS OF OFFSHORE PILES EMBEDDED IN ELASTOPLASTIC SOFT CLAY

ANIS MOHAMAD ALI[1], MOHAMAD J. K. ESSA[1],
and ABDULAMEER QASIM HASSAN[2]

[1] *Civil Engineering Dept., Basrah University / Basrah, Iraq*
[2] *Environment and Pollution Engineering Dept., South Technical University, Basrah, Iraq*

This work deals with the dynamic behavior of offshore piles embedded in soft clay, and an attempt is made to estimate the critical embedded pile length. ABAQUS finite element program is used to simulate the problem. The soil was modeled as an elastic state and elastoplastic state and represented by cam-clay model. Three dimensional elements were used to represent the interaction between pile and soil, laboratory tests are used to obtain the real properties of soil and to describe interface. The results obtained are used to develop the elastic equation used by Matlock and Reese to calculate the critical embedded pile length for pile embedded in elastoplastic soil. Also, show that the critical embedded pile length is increased by about (20 % to 40 %) due to changing soil model from elastic to elastoplastic. The pile embedded in an elastoplastic soil is dependent on soil strength, interface properties and pile rigidity. The pile head displacement is increased about 90 % while the bending moment is deceased by 10 % at pile head.

Keywords: Dynamic analysis, Pile critical embedded length, Elastoplastic soft clay.

1 INTRODUCTION

Offshore piles are subject to a certain amount of lateral loads due to wave, current, wind and impact loads in addition to overturning moments and vertical loads. In an onshore structure the lateral loads are about (10% to 15%) of vertical loads, but in the offshore and coastal structures the lateral loads are very important and may exceed 30% of the axial loads (Brebbia 1979). Piles are usually selected and designed as a cost effective for supporting the raised structures for offshore platform. Soil pile interaction system governs the response of the structure. According to the defining soil state and interface, the soil-structure interaction system is basically controls estimating the contact pressure distribution at the interaction. When the contact pressures are calculated, the displacements, moments and shear forces in the piles and the corresponding stresses and deformations in the soil medium are possible to estimate (Muthukkumaran and Arun 2015). Matlock (1970) developed curves for soft clays, and Reese *et al.* (1974) developed curves for sands and for stiff clay (Reese 2007). Those curves were obtained from actual field measurements, which represent non-linear stiffness of soil springs which later adopted by (API 2005).

In this paper the elastic and elastoplastic solutions for soil models are used to estimate the critical embedded pile length and find the variation in response of the structure and piles when the piles embedded in elastic and elastoplastic soil.

2 FINITE ELEMENT ANALYSES

ABAQUS was used to carry out the analysis. It is used to solve engineering problems based upon the finite element method applicable to linear and nonlinear solutions (Hibbitt *et al.* 2010).

2.1 Elements Selection

Two types of elements were used to represent the structure and soil element; the first element is beam element (B32), 3-node quadratic beam in space, this type of element have three node each node have six degree of freedom, three displacement and three rotations in x, y and z direction, this element was used to represent the structure element and pile element. The second element is (C3D20RP), 20-node brick with pore pressure, quadratic displacement, linear pore pressure and reduced integration which was used to represent the soil element; each node has three displacements in x, y and z. (Hibbitt *et al.* 2010).

2.2 Interface Element

12-node displacement and pore pressure three dimensional cohesive elements were used to model the interface. The local 1- and 2-directions are normal to the thickness direction and, by default, are defined by the standard ABAQUS convention for local directions on surface. Transverse shear behavior is defined in the local 1–3 and 2–3 planes for these elements. Figure 1 shows the model of interface element between soil and pile (Hibbitt *et al.* 2010) and (Naylor *et al.* 1981).

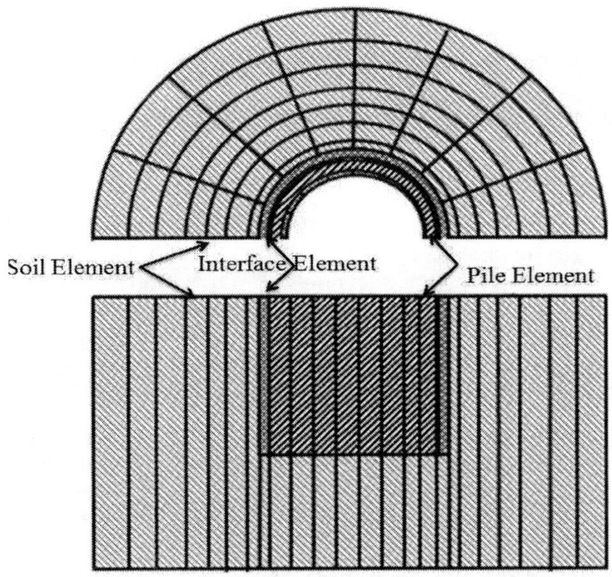

Figure 1. Interface element between soil and pile.

3 CONSTITUTIVE SOIL MODEL

Modified cam clay model is an elastoplastic model based on few and simple postulates that predicts the stress-strain behavior of soils. The model was developed initially from triaxial tests data and later extended for three-dimensional stress space. In critical state mechanics, the state of a soil sample is characterized by three parameters, effective mean stress p', deviatoric (shear stress) q, and specific volume v. Under general stress conditions, the mean stress, p', and the deviatoric stress, q, can be calculated in terms of principal stressesσ'_1, σ'_2 and σ'_3 as

$$p' = \frac{1}{3}(\sigma'_1 + \sigma'_2 + \sigma'_3) \quad (1)$$

$$q = \frac{1}{\sqrt{2}}\sqrt{(\sigma'_1 - \sigma'_2)^2 + (\sigma'_2 - \sigma'_3)^2 + (\sigma'_3 - \sigma'_1)^2} \quad (2)$$

The specific volume is defined as $v = 1 + e$ where e is the void ratio. The yield functions of modified cam clay model are determined from the following equation (Wood 1990).

$$\frac{q^2}{p'^2} + M^2 \ln\left(1 - \frac{p'_0}{p'}\right) = 0 \quad (3)$$

4 SOIL TESTS

4.1 Triaxial Test

In a conventional triaxial compression test, a cylindrical core sample is loaded axially to failure, at constant confining pressure. Conceptually, the peak value of the axial stress is taken as the confined compressive strength of the sample. In addition to axial stress, axial and radial strains may be monitored during this test, to determine basic elastic constants (Young's Modulus, E, and Poisson's ratio, v) and shear strength of soil (cohesive and internal soil friction). Test results of three samples as shown in figure 2 and figure 3 (BS5930).

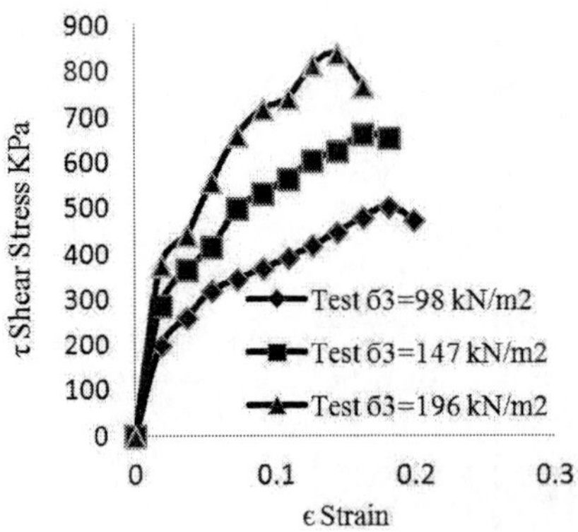

Figure 2. Stress strain curve from triaxial test of soil.

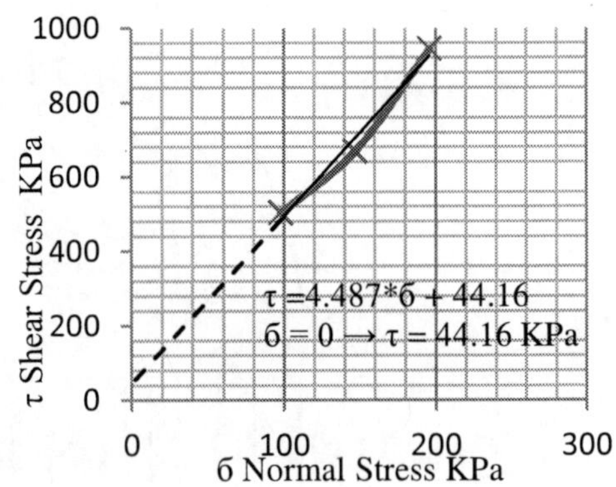

Figure 3. Soil strength parameter from triaxial test of soil.

4.2 Consolidation Tests

This test is performed to determine the magnitude and rate of volume decrease that a laterally confined soil specimen undergoes when subjected to different vertical pressures. From the measured data, the consolidation curve can be plotted. This data is useful in determining the compression index, the recompression index and the reconsolidation pressure of the soil. Test results are shown in figure 4 (BS5930).

4.3 Shear Box Test (Interface Test)

The interface shear tests were conducted at the Geotechnical Engineering Laboratory. A square base direct shear box (100 mm x 100 mm) split horizontally at mid-height was used in contact with steel plate. Normal stress is applied by placing dead weights on a hanger, horizontal displacements are monitored. Most importantly, the purpose of the test was to investigate the mechanical properties of interface between clay soil and steel plate which represent the pile wall, Figure 5 (BS5930).

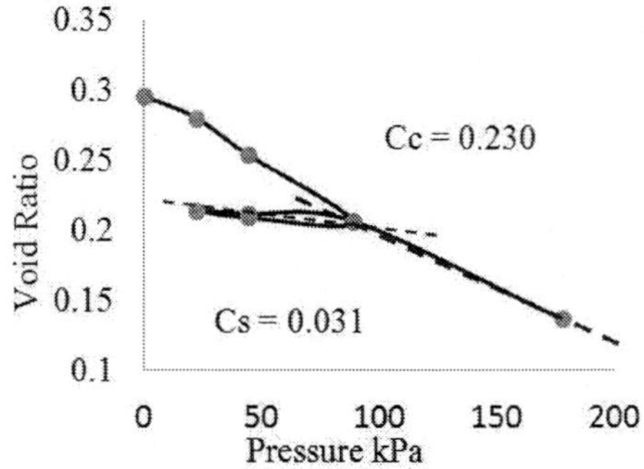

Figure 4. Consolidation test for soil sample.

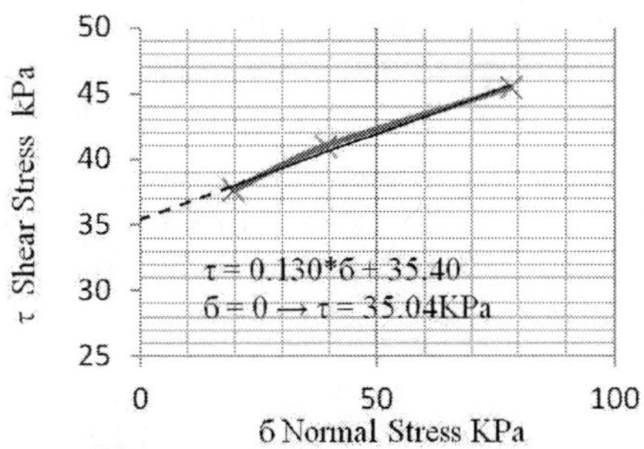

Figure 5. Shear and normal stress for interface from direct shear test.

4.4 Cam Clay Parameter

Figure 6 shows the initial or past yield surface. The three mean stresses computed above are marked on the diagram. The maximum q value occurs where the yield surface passes through the CSL (critical state line), as it properly should. Figures 7 show the deviatoric relation with volumetric strain.

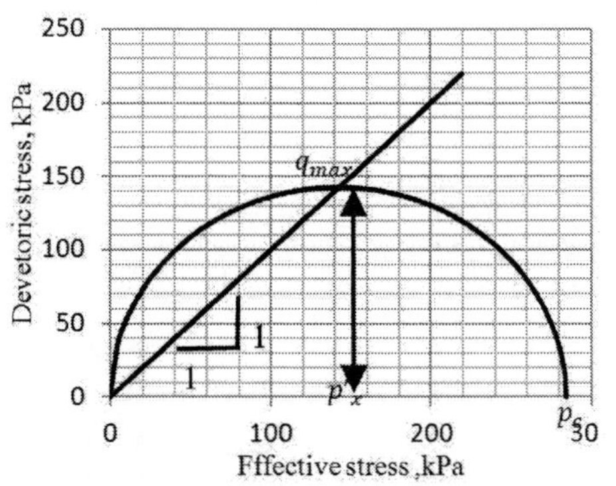

Figure 6. Initial yield surface for the past stresses.

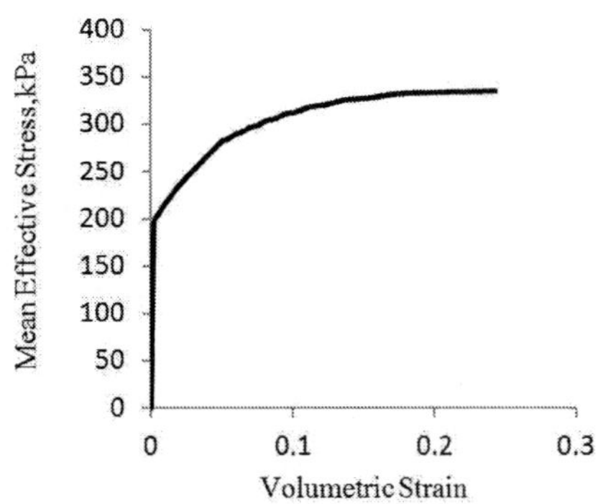

Figure 7. Mean effective stress-volumetric strain relationship.

5 DYNAMIC OF STRUCTURES

The dynamic response of the offshore structure under regular wave is given below

$$[M]\{\ddot{U}\} + [C]\{\dot{U}\} + [K]\{U\} = \{F\} \tag{4}$$

where [M] is the total mass matrix of the system, [C] is the damping matrix,[K] is the structural stiffness matrix, {U} is the nodal displacements, {U} is the nodal velocities, {U} is the nodal accelerations, and {F} is the hydrodynamic force (Brebbia 1979).

6 RESULTS AND DISCUSSION

In this paper, Um- Qaser container terminal Figure 8 is taken as a case study to analyze impact load that is given by (Al-Jasim 2000). The results show that the deck displacements at point A as shown in Figure 9 for pile embedded in elastoplastic clay is large than the pile embedded elastic clay by 10 %, but the pile head displacement is increased from 14.3 mm for pile embedded in elastic soil model to 29.1 mm in elastoplastic condition due to a reduction in the domain stiffness of soil. The critical embedded pile length is the length from the seabed to the first zero deflection, this length is increased by 40 % from pile embedded in elastic soil to the pile embedded in elastoplastic soil, Figure 10. For piles embedded in the elastic soil the critical length about 5 pile diameter and this ratio increased to reach 8 pile diameter for pile embedded in elastoplastic soil, while the bending moment is decreased about 10 %, for same reason, Figure 11.

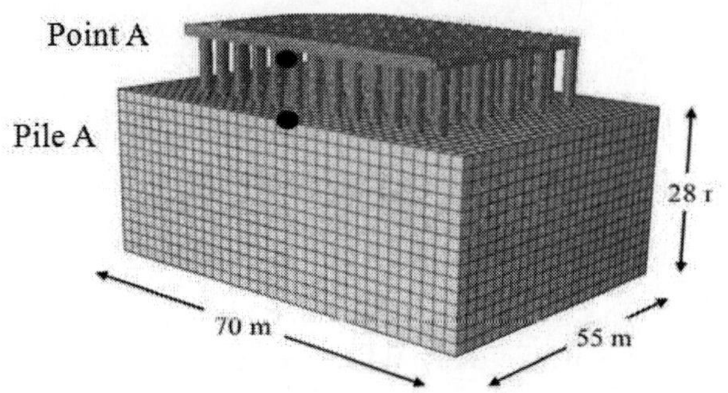

Figure 8. 3D Modeling of Um-Qaser container terminal with ABAQUS.

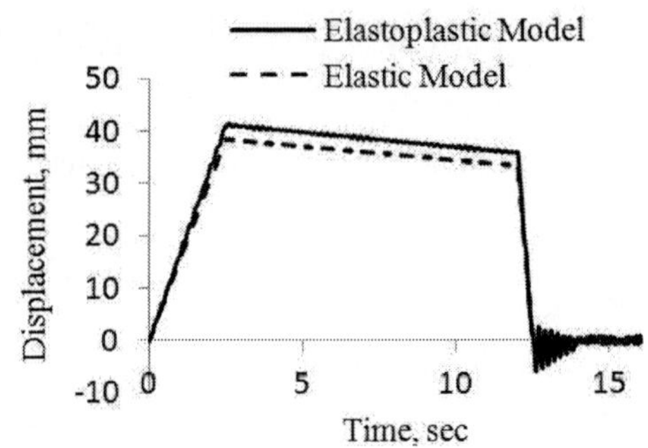

Figure 9. The variation of deck displacement at point A for two soil models.

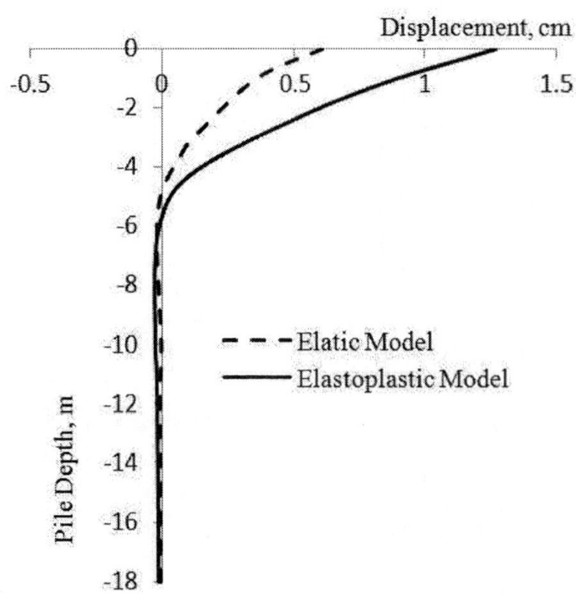

Figure 10. Pile displacement.

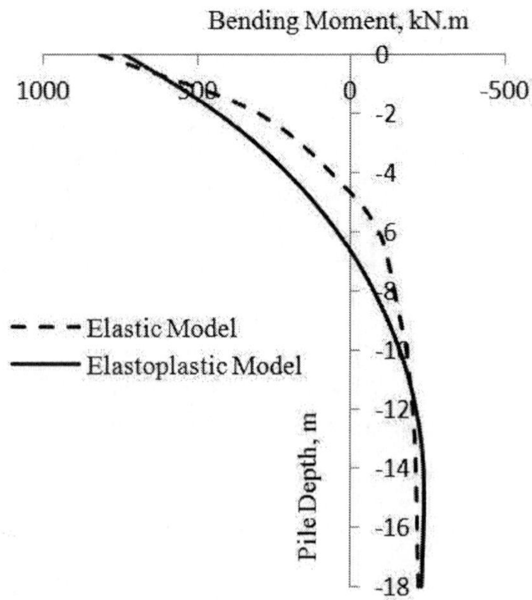

Figure 11. Pile bending moment.

7 CRITICAL EMBEDDED PILE LENGTH

The proposed equation to estimate the critical embedded pile length for soft clay is based on modifying the Reese solution. The proposed equation is based on the followings:

1. The critical embedded pile length is expressed by the undrained shear strength between the clay and steel pile, soil cohesion, and the properties obtain from the elastoplastic model of soil represented by cam clay model.
2. The elastic modulus of pile and second moment area of pile cross section remain unchanged in proposed solution as in Reese equation. According to the above assumptions, the proposed critical embedded pile length becomes (Wood 1990);

$$Lcr_{plastic} = 0.21 * \sqrt[3]{\frac{Ep\ Ip}{Cu}} \tag{5}$$

Where:
$Lcr_{elastoplastic}$ = Proposed critical embeded pile length for pile.
0.21 = Factor is found from obtained results.
Cu = The undrained shear strength between the clay and steel pile, soil cohesion.

It is found that a very good agreement is obtained for the proposed critical length which may be used to introduce a new solution for bending and deflection of pile foundation.

8 CONCLUSIONS

The main conclusions are:

1. The dynamic structure response is increased by changing soil state from elastic to elastoplastic.
2. The deck displacement is small and pile head displacements are sensitive to the soil state.
3. The critical embedded pile length is increased for elastoplastic soil model.
4. The elastoplastic soil model is decreasing the damping of structure.

References

Al-Jasim, *Dynamic Analysis of Offshore Template Structures with Soil-Structure Interaction*, Ph. D., Eng. thesis, Engineering College University of Basrah, 2000.

API Recommended Practice for Planning, Designing, and Constructing Fixed Offshore Platforms, API RP2A, 21th.Ed, 2005.

Brebbia, C. A. and Walker, S., *Dynamic Analysis of Offshore Structure*, Butterworths Co., 1979.

BS5930, Code of Practice for Site Investigations, British Standards Institution, London.

Hibbitt, D., Karlsson, B., and Sorensen, P., ABAQUS User's Manual, Version 10.1, 2010.

Matlock, H., *Correlation for design of laterally loaded piles in soft clays*, 2nd Offshore Technology Conference, Houston, Texas, 577-594, 1970.

Muthukkumaran, K. and Arun, K., *Effect of seabed slope on pile behaviour of fixed offshore platform under lateral forces,* Journal of Ocean Engineering and Marine Energy, 1(3), 207–221, August 2015.

Naylor, D. J., Pande, G. N., Simpson, B., and Tabb, R., *Finite Elements in Geotechnical Engineering,* Pineridge Press Swansea, 1981.

Reese, L. C. and Van Impe, W. F., *Single Pile and Pile Groups under Lateral Loading*, London, UK, Taylor & Francis Group Plc, 2007.

Reese, L. C., Cox, W. R., and Koop, F. D., *Analysis of laterally loaded piles in sand*, 6th Offshore Technology Conference, Houston, 473-483. 1974

Wood, D. M., *Soil Behaviour and Critical State Soil Mechanics*, Cambridge University Press, 1990.

Integrated Solutions for Infrastructure Development
Edited by Lau, H. H., Tang, F. E., Ng, C. K., and Singh, A.
Copyright © 2016 ISEC Press

THE DEVELOPMENT OF ANATOMY OF CONSTRUCTION DISPUTE

FELIX HIDAYAT[1] and BIEMO W. SOEMARDI[2]

[1]*Faculty of Engineering, Parahyangan Catholic University, Indonesia*
[2]*Faculty of Civil and Environmental Engineering, Bandung Institute of Technology, Indonesia*

The advancement of the construction industry in Indonesia is often not accompanied with adequate infrastructure system, especially in terms of regulations and legal system. The enactment of Law no 80–1999 has contributed to the expansion of construction market by allowing the private sector, including those from overseas, to play bigger roles in the industry. Although strongly acknowledged by stakeholders in the industry, the progress made by the government and the practitioner in the industry to response to the needs for more appropriate actions to cope with this ever increasingly complex system has been very slow. The dynamic of the construction industry in the region has changed the formerly rigid government-dominated construction sector to become an intricate one. The impact of such dynamic in the industry has been apparent. In particular, construction disputes have been experiencing changing in the dynamic of the way the disputes were handled; from the traditional litigation approach toward modern alternative dispute resolution mechanism. While litigation remains the dominant construction dispute resolution mechanism, non-litigation approaches such as arbitration and dispute review board are beginning to gain recognition in industry. This paper presents a study on the development of anatomy of construction dispute in Indonesia, which will serve as a framework for analyzing the dynamic of construction disputes in Indonesia. Such a framework will help to identify factors affecting construction dispute mechanism, from the initiation of disputes to the resolution. The anatomy is developed based on analysis of construction dispute cases that have been recorded in judicial courts as well as from Indonesian Arbitration Board.

Keywords: Indonesia, Construction industry, Dispute resolution framework.

1 INTRODUCTION

In a construction project there are two parties that is equally have agreed to put off each other in a form of agreement, providing services known-named construction contract. In the implementation of the construction contract, conflict often arise. For example, costs arising as a result of change of work, etc. The conflict resulted one party submits a request (claims) against the other party. If the claim isn't facilitated, then the dispute will arise (Yasin 2002). Construction industries have risk factors with the level of uncertainty higher than in other industries (Flanagan and Norman 1993). Construction dispute most commonly caused by three factors (Pang 2011): (a) task factors; (b) contract factors; and (c) human factors. The greater value and duration of the a project, will increasingly higher probability of the dispute occurrence (Pang 2011, Gebken 2006, Love 2005). Dispute incurs a loss to warring parties, which is: (a) *Cost and Time* (Allen, 2010); (b) *Productivity* (New South Wales Department of Commerce, 2008; Econtech, 2008); and (c) *Popularity and Relation* (Gebken, 2006; Love, 2005).

2 THE HANDLING OF CONSTRUCTION DISPUTE

In facing construction dispute, there are two efforts that can be made (Kumaraswamy 1997): *dispute resolution and dispute prevention.* Dispute resolution in the construction project which is quite popular in Indonesia is through arbitration/BANI (Badan Arbitrase Nasional Indonesia). This can be showed by the increasing number of cases of 56 cases (1987-1996) became the 215 cases (1997-2006), in which 40% of that number are cases of construction, as shown by Figure 1. In effort to prevent dispute, there are several way that can be done, i.e., (1) through the efforts of an agreement between the parties prior to the execution of the project (e.g., partnering, alliancing, early contractor involvement, etc. (Love 2005)), and (2) development of dispute knowledge (e.g., identification of dispute characteristic (Shin 2000) and arrangement of construction dispute (Pang 2011)).

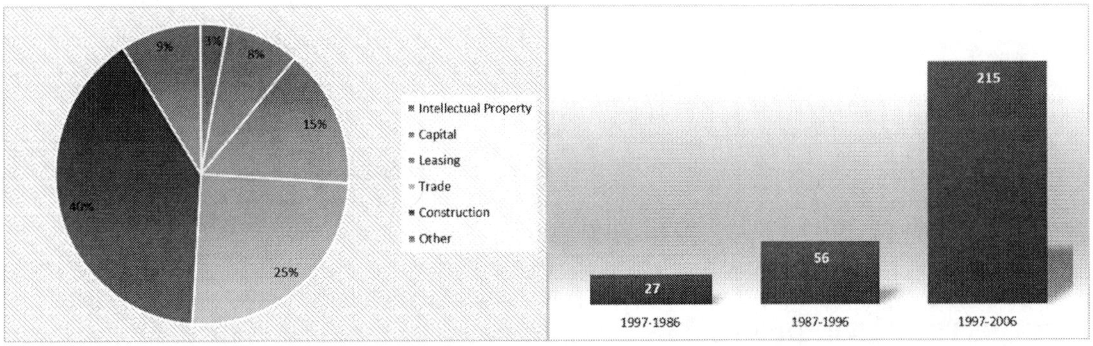

Figure 1. The number and type of cases were registered in the BANI (BANI Arbitration Center).

3 RESEARCH RATIONALITY

Disputes that occur during the execution of the construction, is an issue that must be resolved. When the potential dispute emerged as a new issue, and supported by anatomical exploration about the dispute, will support the understanding of a dispute and produce solutions in the form of identification of potential disputes, a better strategy, negotiation, dispute resolution in a better situation, and avoid the occurrence of a dispute. The development of dispute knowledge can be done through the development of anatomy, in the form of in-depth study of the structure and the mechanism construction of the dispute cases. Research on the anatomy of a construction dispute ever conducted by Pang (2011), focuses on the causes of construction disputes, while the study of anatomical concepts construction dispute thoroughly, including the process of the main three aspects, i.e., (a) causes of construction disputes anatomy, (b) construction dispute process anatomy, and (c) construction dispute resolution anatomy; has not yet been done. It can be seen from Figure 2. Research with object study in Indonesia is rarely done. It showed that dispute knowledge has not been well-developing in Indonesia. It is interesting to do further studies through preliminary studies about the description of construction disputes that occur in Indonesia. The purpose of this research is to get a preliminary picture of the anatomy of construction disputes in Indonesia.

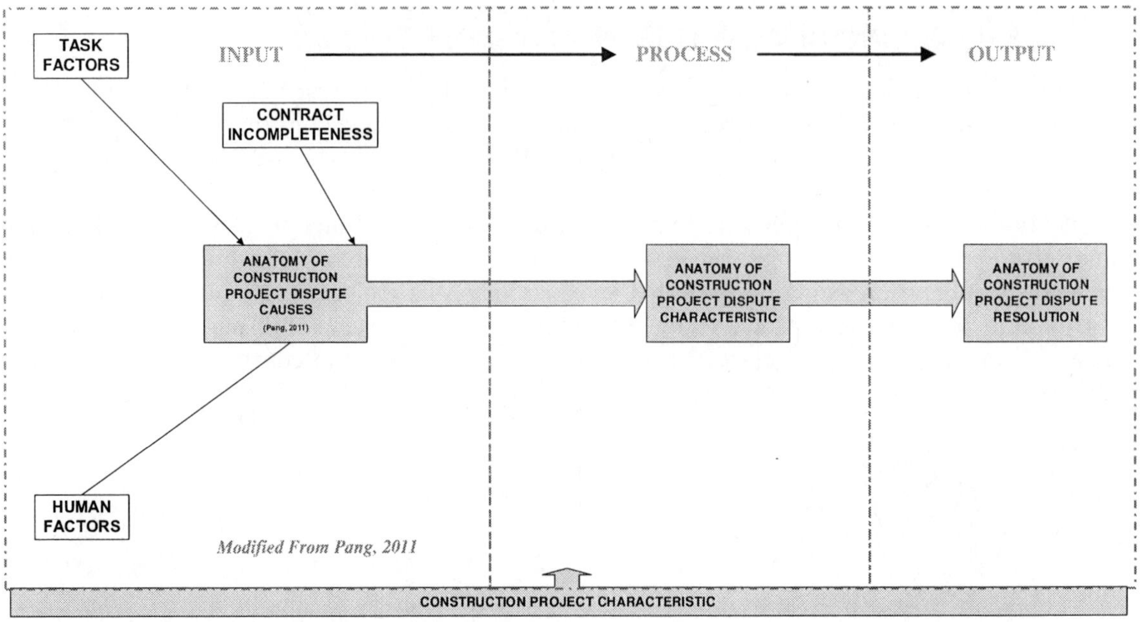

Figure 2. A conceptual description of the anatomy of construction disputes (modified from Pang, 2011).

4 A PRELIMINARY STUDY : THE DEVELOPMENT OF ANATOMY OF CONSTRUCTION DISPUTE IN INDONESIA

A preliminary study of case studies carried out, in order to get a simple overview of the disputes on a construction project in Indonesia. Case studies carried out by doing a search of fourteen cases (nine cases is the decision of the arbitration and five cases is the decision of the Supreme Court). There are fourteen cases were reviewed, including seven cases of building projects, one case of housing/residential building projects, three cases of heavy engineering building construction projects, and three cases of industrial building projects. These fourteen cases reflects different type of cases and their dispute settlement, that occurred in Indonesia. Research will be conducted through a descriptive qualitative study of some cases disputes in construction projects in Indonesia, through the review of documents, and then be prepared to describe the fact of the dispute regarding the phenomenon that occur in industrial construction in Indonesia.

4.1 Findings

The parties involved include disputes of BUMN/local government private parties, local private parties, and foreign private parties. The projects varied between years 1992 until 2010. Implementation time varied starting from 48 (forty eight) days until 1,451 (one thousand four hundred and fifty one) days. Dispute resolution times varied from 35 (thirty five) days until 2,889 (two thousand eight hundred and eighty nine) days. The value of the projects varied from USD. 85.000,- until USD. 16.270.000,-. Claims ranged from USD. 15.000,- until USD. 14.820.000,-. Values obtained ranged from USD. 6.000,- until USD. 13.250.000,-. Dispute resolution is implemented at the level of the negotiations until Supreme Court. The type of contract used Fixed Lump Sump Price and the Unit Price. Payment used Monthly Payment and Contractor's Full Pre Financed.

4.1.1 Cause of construction dispute

The cause of the occurrence of a dispute in cases that are reviewed:

a. Task Factors : (i) External Factors, consist of (a) natural causes, i.e., weather factor; and (b) interventions, i.e., monetary crisis and the increase of fuel price; (ii) Internal Factors), consist of (a) changes, i.e., acceleration demand, contract termination, work termination, change desain, and change volume; (b) delays, i.e., delays of work; (c) incomplete informations, i.e., delays approval design and delays in work progress report; (d) late/non payment, i.e., delays of payment, payment problem from main contractor to sub contractor, and non payment; (e) defective works, i.e., the use of labor that does not fit the job qualifications, the work is not according to specifications, and low achievement. Internal Factors are causative factors most frequently occurring disputes, where a common pattern is the occurrence of a change of work resulted in jobs being late and the difference in calculation of the volume at the end of the job.

b. Human Factors : Behavioral Factors, consisting of evation of obligation, i.e., the difference calculation, the difference in the volume of realization of payments with the achievements of the work, and the lack of agreement in adjustment the price.

The result showed the pattern about the cause of construction dispute, that is when there is a change of work, then it will result in delay of work. At the time of the preparation of addendum contracts, there is a difference in perception (e.g., difference calculation volume of work), that will cause problem in payment. These finding is equal to occurring in the Nigeria (Olanrewaju 2014). Unfair behavior of the parties to a construction contract agreement and psychological defense mechanisms have been also identified as likely causes of conflicts in the construction industry (Mitkus 2014).

4.1.2 Comparison of the value of the project, the value of claims, and the value obtained with the parties

Based on the preliminary studies, it can be seen that the highest ratio of comparison between the value obtained with the value of the project, is found in POV Project at Jakarta, because the process of settlement of the dispute protracted until 1,290 (one thousand two hundred ninety) days. The interesting result shown in the second rank of the comparison between the value of the claim with the value granted i.e.,LPG Cilacap Refinery Project, which is valued at USD. 16.270.000,-. The project has a highest value among projects and the parties are local government parties and foreign private parties. The value of claims filed is USD. 14.973.000, or is equal to 90,52% of the contract value and the value granted is USD. 13.250.000,-, or is equal to 78,40% of the contract value, or the value of the% 86,1 registration demands. The study found the correlation between the value of the contract, the fee which is claimed, and the cost which is granted, with the parties to the dispute, where the value of the claim and the value of the granted, which is filed by a foreign contractor, approaches the value of the contract This shows the weakness of the claim and dispute management system of Indonesia. Foreigners bid low at the time of the quote, but at the time of implementation, they sought to make a claim, and bring it to the level of dispute. The offender who committed the strategy referred to as claim artist.

4.1.2 Comparison of implementation time and dispute resolution time

The highest ratio between dispute settlement with implementation time is presented on PP Project at Jatinangor, which amounted to 680.69% (dispute resolution time 6,8x compared to implementation time). The smallest ratio between dispute settlement with implementation time, is presented on Refinery of LPG Project at Cilacap, which amounted to 8.20% (dispute resolution time 0,08x compared to implementation time). The longest dispute resolution time caused by continuing the process started from arbitration, District Court, and ultimately to the Supreme Court in cassation level. Table 1 shows a diagram of the ratio a comparison between dispute resolution and implementation time.

Table 1. Ratio between dispute resolution time/value.

PROJECT NAME	A	B	C	D	E	F	G	H	I	J	K
A. Ratio : Time Settlement of Dispute / Time of Execution Work	681%	632%	154%	73%	64%	52%	49%	23%	18%	10%	8%
B. Ratio : The Claim / Project Value	66%	n/a	n/a	152%	3%	n/a	114%	27%	n/a	1602%	91%
B. Ratio : Value Granted / Project Value	0%	n/a	n/a	21%	3%	n/a	21%	13%	n/a	317%	78%
D. Dispute Resolution											
Negotiation											√
Mediation											
BANI Negotation			√		√		√	√	√		
BANI Mediation				√		√	√	√			
Arbitration	√	√		√	√		√			√	√
District Court	√	√									
Supreme Court	√	√									
E. Ownership											
Client	Local Private	Local Private	Local Private	Local Private	Local Private	Local Private	Local Private	Local Governor	Local Private	Local Governor	Local Governor
Contractor	Local Governor	Local Governor	Foreign Private	Local Private	Local Private	Foreign Private	Local Private	Local Governor	Local Private	Local Private	Foreign Private

note :

A = PP Project at Jatinangor
B = Condominium Project at Jakarta
C = Hook Up and Work Modification PGF Project at Java Sea
D = SS II Project Renovation at Jakarta
E = Revitalizing J Market Project at Jakarta
F = WCAB Reclamation Project at Jakarta
G = Plumbing NDD LPG Project at Riau
H = International Airport Project at Lombok
I = BCS Project at Batam
J = POV Project at Jakarta
K = Refinery Project at Cilacap

The shortest dispute resolution time is caused by the attempt of internal negotiation between the parties to the dispute at the early phase of dispute resolution, although in the end, the dispute resolved through arbitration lines. It can be seen that there is a tendency where dispute resolution in construction projects pursued through litigation, the path takes longer compared to the resolution of disputes in construction projects is through the path of arbitration. With the negotiations at the beginning of the dispute resolution process, it will shorten the time of dispute resolution. Besides that, it can be seen that on the highest ratio, a dispute occurred between local private parties with the local government parties, while for the lowest ratio, a dispute occurred between the local government parties with foreign private parties. This shows that foreign private parties, has a better dispute resolution management, where dispute resolution is done early by negotiation, although in the end the dispute settled through the arbitration. The other thing that was found was the existence of an agreement resolving disputes in some cases, where the dispute resolution process occurs through negotiation and mediation at the arbitration level (hybrid arbitration), where the parties to the dispute involving foreign private party. This method of dispute resolution by mediation is not too popular in Indonesia. These findings are equal to the ones occurring in England and Scotland (Trushell 2016, Gregory-Stevens 2016). This also shows that foreign private party has a better awareness about dispute resolution. When the resolution is done at a lower level, it will bring advantages to all parties to the dispute. The main results of this are that, in addition to litigation being more expensive in money and time than ADR methods (Gill 2016). Diagram of the ratio a comparison between dispute resolution and implementation time/value can be seen in Table 1.

5 CONCLUSION

Based on a review on 14 (fourteen) case of dispute on the construction project then found the existence of a particular pattern relating to the anatomy of a construction dispute : (1) The result showed the pattern about the cause of construction dispute, that is when there is a change of work, then it will result in delay of work. At the time of the preparation of addendum contracts, there is a difference in perception (e.g., difference calculation volume of work), that will cause problem in payment. (2) In addition, it found the correlation between the value of the contract, the fee is claimed, and the cost is granted, with the parties to the dispute, where the value of the claim and the value of the granted, filed by a foreign contractor approaches the value of the contract; and (3) Besides, it is found also the relatedness of time settlement of the dispute and time the construction execution, with a method of settlement of the dispute, where dispute resolution through the litigation, takes longer compared to the resolution through the arbitration.

Reference

Allen, M. (2010). “Construction Dispute on The Rise”. Forum Conctruction and Contract News. EC Harris Global Construction Disputes Report.

Econtech (2008). “Economic Analysis of the Building and Construction Industry Productivity”. Econtech, Canberra.

Flanagan and Norman (1993). “Risk Management and Construction”. Blackwell. Oxford.

Gebken, R. (2006). “Quantification of Transactional Dispute Resolution Costs for the U.S. Construction Industry”. Dissertation. University of Texas.

Gill, et al, (2016). “Comparison of the effects of litigation and ADR in South-East Queensland”. International Journal of Construction Management. Volume 15, 2015 - Issue 3.

Gregory-Stevens, J, et al, (2016). "Mediation in construction disputes in England", International Journal of Law in the Built Environment, Vol. 8 Iss: 2, pp.123 – 136.

Kumaraswamy, M. (1997). “Conflicts, Claims and Disputes in Construction Engineering”. Construction and Architectural Management, vol. 4, no. 2, pp 95-111.

Love, P., Tse, R. and Edwards, D. (2005). “Time-cost Relationships in Australian Building Construction Projects”, Journal of Construction Engineering and Management, vol. 131, no. 2, pp. 187-194.

Mitkus, S and Mitkus, T. (2014). “Causes of Conflicts in a Construction Industry: A Communicational Approach”. Procedia - Social and Behavioral Sciences. Volume 110, 24 January 2014, Pages 777-786. The 2-dn International Scientific conference „Contemporary Issues in Business, Management and Education 2013“

Olanrewaju A. and Anavhe, P., (2014) "Perceived Claim Sources in the Nigerian Construction Industry", Built Environment Project and Asset Management, Vol. 4 Iss: 3, pp.281 - 295

Pang, H.Y. (2011). “Anatomy of Construction Dispute”. Theses. City University of Hongkong.

Shin, K.C. (2000). “Identification of Critical Dispute Characteristic during Construction Project Operations”. Dissertation. University of Georgia.

Sugiyono. (2012). “Metode Penelitian Kombinasi”. Alfabeta. Bandung.

Yasin, N. (2002). “Mengenal Klaim Konstruksi dan Penyelesaian Sengketa Konstruksi”. PT. Gramedia Pustaka Utama . Jakarta.

Trushell, I et. al, (2016) "Construction mediation in Scotland: An investigation into attitudes and experiences of mediation practitioners", International Journal of Law in the Built Environment, Vol. 8 Iss: 2, pp.101 – 122.

Integrated Solutions for Infrastructure Development
Edited by Lau, H. H., Tang, F. E., Ng, C. K., and Singh, A.
Copyright © 2016 ISEC Press

PRACTICE EXPERIENCE AND MULTIDISCIPLINARY COLLABORATION IN PROJECT MANAGEMENT: A CASE STUDY

SILVIA MAZZETTO

Dept of Architecture and Urban Planning, Qatar University, Doha, Qatar

This paper formulates an insight into the importance of practice experience in project management success and the apparent lack of emphasis placed on this within the context of university education. Against the background literature, a review will be conducted into how well university education prepares the future project managers in the area of the practical project learning approaches. The paper provides an example of approaches currently being implemented in the Department of Architecture and Urban Planning at College of Engineering in Qatar University and models, strategies of effective practices for embedding enterprise and employability in higher education, in respect to the specific project experiences which are mandatory for the management of the running projects in Doha, Qatar. The paper describes the primary components of the theoretical research including the course program in Construction and Project Management, its integration with other courses, the cooperation required to support the interdisciplinary emphasis, and the establishment of an innovative academic/industry partnership to provide a state-of-the-art physical and technological infrastructure to support the program goals. The research concludes that the efficiency of interaction between theory and practice in the construction industry needs to be further examined, and some limitations and debates remain awaiting further investigation.

Keywords: University education, Professional training, Leadership, Construction industry, Road safety, Theory, Practice.

1 INTRODUCTION

1.1 Project Management Context

An investigation of the literatures reveals that projects often comprise of people and organizations from different nationalities and disciplines such as architects, engineers, clients, contractors, suppliers, and banks which generates complexity in the managerial procedures, due to different multidisciplinary approaches inside multicultural teams, foreign managers, and international partners. In the literature it is easy to see that different disciplines can influence a variety of project management issues such us: teams, leadership, communication, performance, risk assessment, business negotiations, international project management and planning. The success and failure of any project depends upon many factors, the project manager is considered to be the key contributor to the success of any project, as well as a guide for the team members to achieve the project completion and the client satisfaction (Cost, Time and Quality) (Nguyen *et al.* 2004). For Chileshe (2009), it is really crucial understanding the importance of the project manager to the success of the construction projects due to the fact that the managerial roles, responsibilities and duties have direct impact to the success of the construction projects. In the construction field the project manager is considered to be one of the most important people who can lead and drive the projects in the right direction and conclude construction projects successfully. In addition the project manager can play the most important role regarding the improvement of the organization's performance, the organization's profitability and the client's satisfaction (Chileshe *et al.* 2009). According to Lientz and Rea (1995) the project manager has three clearly defined managerial responsibilities: resource manager, control manager and coordinator; additionally, as stated by Kerzner (1995) the project manager is responsible for ensuring the accomplishment of group and organizational goals and objectives. To affect this, he must have a deep knowledge of standards and cost control policies and procedures; finally he must understand organization behavior in order to effective and most have strong interpersonal skills. This is especially important during the controlling function. To succeed, project managers are required to deliver products and services that meet the needs of the organization and that deliver value to their clients. They must master the hard skills, such as understanding processes, tools and techniques, as well as people skills (Simsarian and Maria 2004). The latter allow us to highlight the importance of human skills in project management success and the apparent lack of emphasis placed on this within the context of university education.

1.2 Practice Experience in Project Management Education

While a great deal is written about traditional project management we know very little about the "actuality" of project based working and management, and still need to be investigated considering many controversial positions in the literature about what is needed to improve project management in practice and what should be done to enhance the value of practitioner's lived experience of projects, enriching and extending the subject of project management beyond its current theoretical foundations. The growing critiques of project management theory and the need for new research in relation to the developing practice represent a synthesis of ideas for how the current conceptual theoretical base needs to evolve in relation to the developing world of practice experience in the field of project management (Rae 2007). The major problem relates to balance: too much of an emphasis on knowledge and not enough on competence; too much emphasis on information transfer learning methods and not enough on individual small group learning methods such as project teams, peer exchange, individual counseling and workshops. There is a lack of evidence on how learning strategies influence the development of students' entrepreneurial competences and how these competences transfer into new project formation (Garavan and O'Cinneide 1994).

2 CHALLENGES IN QATAR FOR PROJECT MANAGERS

2.1 Projects and Multidisciplinary Collaboration

In Qatar there are many ongoing projects performed in collaboration with foreign companies and local partners such as Doha Metro, Lusail City, Hamad International Airport, Education City, new stadiums and infrastructures for the FIFA World Cup and a correct project management practice is fundamental to the success of all these initiatives. Project managers face numerous challenges, which include multi-disciplinary factors such as managing mixed teams pertaining to different disciplines and additionally many cultural differences within the same team. A recent report regarding Qatar construction market has announced Qatar's plan to invest many billion dollars in transport infrastructure, in addition to the new roads, metro systems and airport expansion to support the influx of football fans expected for the mega event of the FIFA World Cup 2022. The Qatar Labor Ministry estimated a massive number of additional workers required to complete the World Cup and related infrastructure projects. Actually there is a predicted increasing scarcity of both technical and labor staff which probably will continue to be one of the biggest challenges for the construction sector in Qatar. World Cup 2022 will be a whole new experience for most of the involved construction companies as well as Qatar nation itself, which needs to meet the challenges of completing all the announced projects within the remaining period of time. The only possible solution to achieve the national target is in the quality of the people working on the projects and particularly in the quality of the project managers guiding the projects' completion. In Qatar nowadays there are many projects which are more complicated than any other project completed in the past years, requiring fast tracking procedures and multidisciplinary-collaboration between construction companies, architects, engineers, project managers and owners. The multidisciplinary team, guided by the project manager, is generally involved in many tasks such as regulation, approvals, design, tender documents, manufacture, installation and maintenance of buildings and other structures. In the 21st century the skills requested for a good project manager are mainly the knowledge and experience to execute projects profitably while managing risk and enhancing the opportunities for future business. In the construction field, people are the most important resource for construction organization's and the completion of the projects depends on knowledge and skills of key people, planning and executing the work, having talented project manager in place to guide and direct operations (Sharma 2013). The construction industry particularly in Qatar, has 100% relied upon immigrant or foreign labor to varying degrees through the course of history. There are however many complications involved with foreign employment, namely language and cultural barriers and the obvious problems in communication can lead to confusion that can hamper productivity and coordination. In order to maximize long term and perfect performance is important to provide the project managers with the requested training and practice experience necessary to hasten the assimilation into the workforce.

2.2 Education and Professional Training in Project Management

Research indicates the level of education and training does influence productivity. A high level of professional training yields correspondingly higher productivity. Education and training of project managers is one of the keys to improving construction performance and leads to high levels of commitment, enthusiasm, self-motivation, productivity and innovation. Practice experience also enables project managers to make easily decisions for which they are accountable and responsible. Leadership and collaboration are two of the most important characteristics for the project managers and must be raised from among the team members to aid in effectively coordinating activities by providing the communication link between management and work crew. Professionalism starts at the first steps of a student's education in university. Different disciplines, from their initial stages of training, can model and affect the student's attitude and behavior in collaborative work, and the way they act and perform as the new Professional Managers (PMs) of the future.

2.3 Bridging the Gap

Academic institutions and construction companies jointly can play a vital role in this direction and in many countries efforts are being done to bridge the gap between the theory and practice, between academia and industry. Academic institutions, in close coordination with industry representatives are making workshop and professional training courses which can easily enhance the working capacity of the management involved in various activities of the construction industry. In the same way, in Qatar also, many new initiatives are taken up by the industry and academic and the government is supporting them by implementing and introducing new laws and regulation in the professional experience training. For a more targeted approach to training and continuing practice education, more intensive workshop or additional internship relations need to be developed, to achieve a more practical approach in project manager education and fulfill the needs of the industry. If this step will be implemented with commitment in Qatar, the shape of the construction industry will perform better within few time, which will automatically ensure the completion of all projects, in targeted time-frame with full quality assurance.

3 PROJECT MANAGEMENT AND PRACTICE EXPERINCE IN QATAR UNIVERSITY

3.1 The Research Project Proposal

In Qatar University a multidisciplinary research project has been proposed and developed to involve students and instructors specializing in diverse disciplines and aspects of Architecture and Urban Planning (AUP) and Industrial and Systems Engineering (ISE) in order to improve the students practice experience in project management learning approaches. The aim is to get them to focus on a common assignment as part of their courses (a class project in this case), using the tools of project management to investigate and develop the students' capabilities and attitudes when working in a multidisciplinary project. The extent that project managers and their teams interact effectively within the project environment is not exclusively an issue of multi-collaboration; it is influenced by different national cultures, and past experiences, and can affect a variety of managerial roles and activities, for instance, leadership schemas (the criteria of a good or a bad leader) and behavior scripts (greetings) vary between those of different cultural backgrounds. The project topic was linked to the active competition of the Qatar Road Safety Studies Center (QRSSC), which is a prominent research center in Qatar University, in the form of prize money for the winning team. The aim of the project was to increase road safety awareness in Qatar (NRSS 2013-2022). As the youth are the basis of the Qatari society, the competition target the youth of Qatar University to engage them in road safety awareness and its importance.

3.2 The Project Deliverables

The students were working towards producing the project deliverables requested, including reviewed pedestrian safety and road regulations. In addition, their work aims to propose a re-design of some dangerous areas, improving the quality of the road layout in Qatar in order to meet the required road safety measures. With the proposed students' guidelines for re-design, Qatar's road users will benefit from improved road safety, reduced traffic congestion and more environmentally friendly driving, all enabled by better developed roads. Moreover, the research project strives to develop awareness-raising messages regarding road safety to help prevent all road fatalities. The aim of the multidisciplinary project was to promote the development of Qatar road safety educational materials and target road safety publicity campaigns. The competition of Qatar Road Safety Studies Center (QRSSC) requested the students to produce an awareness video which will develop, diffuse, collate, and disseminate information on good practice in road safety education, training, and publicity throughout Qatar, not only in Doha but also in other areas- Al Khor, Lusail and Dukhan.

3.3 The Research Methodology

A structured research method and templates were selected to assess differences in PM strategies of the multidisciplinary research project team, in using PM tools (Gray and Larson 2002). A convenience sample was used to develop the project and collect data from the students, and a report template, a minute of meeting template, a poster and video template. The research items covered specific areas of project planning and were then analyzed to compare Architecture and Engineering students in their approach to the multidisciplinary project and in the proper application of managerial tools (Meredith and Mantel 2006) for planning scope, time, cost, risk, quality, integration, innovation and technology, and communication. Instructors provided course delivery to both batches of students through integrated lectures during the course and through readings, educational audio and video materials, and parts of course assignments. From the initial phase of site analysis, students developed their report, selecting one specific area in Doha, the project title, the group name, and arranging some meetings to define the multidisciplinary collaboration between batches such as project objectives, project methodology and confirmation of each member's role, demonstrating leadership. Students met in their group approximately twice in a week. They developed a proper agenda each time and discussed it during the meeting. Weekly meeting minutes prepared by the groups described the events of the meeting, starting with a list of attendees, a statement of the issues considered by the participants, and related responses or decisions for the issues. The meeting minutes were collected in a report to monitor the progress of the work; the report was checked against a rubric on written reports.

4 RESULTS

The deliverables showed the students' ability to understand the basic principles of project and practice management, the initial evaluation of project costs and budgets, the proper use of activity scheduling, ability to allocate resources to the project, and to demonstrate understanding and appropriate use of methods for controlling projects, including cost and time. All the students submitted their project report respecting the deadlines. Many groups of students interacted collaboratively for the whole duration of the project, not only demonstrating their ability to work in interdisciplinary teams and with other multidisciplinary groups to successfully complete their projects.

Particularly for the QRSSC competition, the submitted videos addressed objectives such as raising awareness about the danger of speeding on the road, the importance of wearing seat belts, the importance of using baby car seats and other measures relevant and useful to road safety in Qatar. Students provided conclusions and recommendations regarding guidelines and standards for road design that arose from their projects:

- Proposing homogenous road sections in the redesign of the road to avoid conflicts with oncoming traffic.
- Proposing street furniture and signs that will avoid unpredictable behavior will separate vehicle types and will reduce speed at potential conflict points.
- Limiting the number of design elements in each category and making them uniform for the roads safety purpose.
- Creating a sustainable safe road network having flow function generated by uninterrupted road and slip roads having an access function which need to facilitate appropriate entrances.

5 CONCLUSION

In this paper we have set out to show how the practice experience and multidisciplinary collaboration, between students, in project management can enhance research practice in a field dominated by many theoretical disciplines approaches. We have also described the methodology used and some of the issues which have arisen in developing this approach, especially because the research is still facing an initial phase of development and investigation. The multidisciplinary research included a wide range of activities, detailed review of the awareness state regarding the road safety in Qatar, site surveys, data analysis and students' project proposals for many areas in Doha, in order to contribute to the national road safety strategy. In conclusion we would like to emphasize that the managerial tools, deliverables and outcomes applied for the case study are similar to those which might be encountered in normal professional practice. The concept of re-designing for the purpose of road safety as a research activity may be difficult to recognize given this similarity, which establishes the relationship, analogies and differences between theory and practice in project management. However, in giving this partial view of a complex, long-term investigation, there are some limitations in describing the full process of evaluation and the outcomes for every single group of students, due to the fact the research project is still on process and most of the deliverables are still in their preliminary stages. All the outcomes need to be further investigated and compared with other similar outcomes from other multidisciplinary projects which are under further development in Qatar University.

References

Chileshe N. and Ali M. A. M., *The influence of the project manager on the success of the construction project*, Conference: ICCEM-ICCPM, 2009.

Garavan T. N. and O'Cinneide B., *Entrepreneurship Education and Training Programs: A Review and Evaluation* Journal of European Industrial Training, Vol. 18 (8), 3–12, 1994.

Gray, C. F. and Larson, E. W., *Project Management: The Managerial Process.* McGraw Hill, 2002.

Kerzner, H., 1995, *Project Management a systems Approach to Planning, Scheduling and controlling*, New York, Van Nostrand Reinhold, 5th Ed, 2002.

Lientz, P. B. and Rea, P. K., *Project Management for the 21st Century*. UK, Academic Press, INC, 1995.

Meredith, J. R. and Mantel, S. J., *PM: A Managerial Approach*, Fourth edition. John Wiley and Sons, New Jersey, 2006.

Nguyen, L. D., Ogunlana, S. O., and Lan, D. T. X., *A Study on Project Success Factors in Large Construction Projects in Vietnam*. Engineering, Construction and Architectural Management, 11 (6), 404–413, 2004.

NRSS National Road Safety Strategy in Qatar (2013-2022).

Rae, D., *Connecting enterprise and graduate employability: Challenges to the higher education culture and curriculum*? Education & Training, Vol. 49 Iss: 8/9, 605 – 619, 2007.

Sharma T. C., *Training and education of construction professionals*, in Construction Sites, 77, 7–8, 2013.

Simsarian, S. W. and Maria, T., *Project Managers Doubling as client account Executives,* Academy of Management Executive, 18 (1), 60–71, 2004.

Integrated Solutions for Infrastructure Development
Edited by Lau, H. H., Tang, F. E., Ng, C. K., and Singh, A.
Copyright © 2016 ISEC Press

WILL TRADESMAN COLLECTIVE WAGE AGREEMENT BE EXEMPTED FROM THE HONG KONG COMPETITION ORDINANCE?

SAI ON CHEUNG

Dept of Architecture and Civil Engineering, City University of Hong Kong, Hong Kong

Tradesman Collective Wage Agreement (TCWA) has been used for construction tradesmen for many years. This would ensure the daily wage for qualified tradesmen will not fall below the agreed level. On 14th December 2015, the Hong Kong first ever Competition Ordinance (HKCO) came into effect. Under the First Conduct Rule (Section 6 of the HKCO), an undertaking must not (a) make or give effect to an agreement; (b) engage in a concerted practice; or (c) as a member of an association of undertakings, make or give effect to a decision of the association, if the object or effect of the agreement, concerted practice or decision is to prevent, restrict or distort competition in Hong Kong. Moreover, Section 15 of HKCO allows exemption for agreements that fulfil certain criteria such as economic efficiency and public policy. This study finds that TCWA will likely violate the First Conduct Rule due to its price fixing effect. Moreover, TCWA can apply for block exemption under Section 15 of the HKCO on the ground of engendering economic benefits. Reference is also made to the first application for exemption under the HKCO to illustrate the key arguments.

Keywords: First conduct rule, Wage agreement, Exemption, Competition law.

1 INTRODUCTION

Competition is considered to be a process of economic rivalry among market players. The effect of competition is efficiency whereby production cost can be lowered. As a result, it is believed that customers will be benefitted from competition (Chen and Lin 2002). Thus, when competition is distorted, resources will not be allocated as efficiently as they would otherwise (HKCC 2015). Consequently, the inefficiency will lead to a reduction in social welfare. More recently, it has been advocated that competition will drive innovation through creativity that is now considered as the most pivotal element of a successful venture (Audretsch *et al.* 2001, Amabile 2000, 2012). Against these concepts, competition legislations are common in many parts of the world and are used as a means to ensure sufficient competition.

The Hong Kong Competition Ordinance was enacted in 2012 and came into full force on 14 December 2015 (CAP 619). It has to be emphasized that the spirit of a competition law is not against big players nor to protect small and medium sized enterprises (Whilsh and Bailey 2012). Instead, competition law seeks to prohibit conducts that lessen competition. Thus agreements that have such effect will be prohibited. This study posits to examine if the collective tradesman wage agreement will contravene the HKCO. If yes, would exemption allowed under the HKCO be available. The relevant HKCO sections will firstly be introduced. That include the conduct rule and the provisos for exemption. The TCWA will then be analyzed in the light of these sections. A response strategy is proposed for the construction industry in response to the HKC.

2 ANTI-COMPETITIVE CONDUCT UNDER THE HONG KONG COMPETITION ORDINANCE (HKCC)

After the enactment in 2012, the Competition Commission developed six guidelines in 2015. These are guideline for 1) First Conduct Rules; 2) Second Conduct Rule; 3) Merger Rule; 4) Complaints; 5) Investigations; and 6) Applications for a decision under Section 9 and 24 (Exclusions and Exemptions) and Section 15 Block Exemption Orders. With these guidelines, the HKCO came into full force on 14 December 2015. The relevant sections that will put TCWA into scrutiny is Section 6 and 8.

Section 6:

(1) An undertaking must not (a) make or give effect to an agreement; (b) engage in a concerted practice; or (c) as a member of an association of undertakings, make or give effect to a decision of the association, if the object or effect of the agreement, concerted practice or decision is to prevent, restrict or distort competition in Hong Kong.

(2) Unless the context otherwise requires, a provision of this Ordinance which is expressed to apply to, or in relation to, an agreement is to be read as applying equally to, or in relation to, a concerted practice and a decision by an association of undertakings (but with any necessary modifications).

(3) The prohibition imposed by subsection (1) is referred to in the Ordinance as the 'first conduct rule'.

Section 8:

The first conduct rule applies to an agreement, concerted practice or decision that has the object or effect of preventing, restricting or distorting competition in Hong Kong even if:

(a) The agreement or decision is made or given effect to outside Hong Kong;

(b) The concerted practice is engaged in outside Hong Kong;

(c) Any party to the agreement or concerted practice is outside Hong Kong; or

(d) Any undertaking or association of undertakings giving effect to a decision is outside Hong Kong.

Under TCWA, the associations of the key trades would review annually the daily wages for their members. The associations would come up with suggested wages for the respective trades. The contractors or subcontracts employing these tradesmen according to the agreed rates.

First, are trade associations undertakings as stated in section 6(1). The term undertaking is defined in section 2(1) of HKCO as any entity which is engaged in an economic activity. With reference to the item 2.2 of the guideline on the First Conduct Rule (HKCC 2015), examples of undertaking include individual companies, groups of companies, partnerships, individuals operating as sole traders or subcontractors, co-operatives, societies, business chambers, trade associations ad non –profit organization. Furthermore, Section 2(1)(a) of the HKCO provides that an anti-competitive conduct is that is defined as any conduct that consists of fixing, maintaining, increasing or controlling the price for the supply of goods or services.

In addition, reference is made to the the fee scales offered by many professional bodies such as the architects, engineers and surveyors before the enforcement of the HKCO whereby. Notwithstanding that in reality, fee scale has been followed for quite a while, it is not until the late 2015 that nearly all professional bodies in Hong Kong took out fee scale from their web sites. This is part of the compliance measures advised by the legal profession.

In sum, the tradesman associations are undertakings and TCWA can be interpreted as fixing price as the daily wages of tradesmen are price to those contracting employing the tradesmen. Moreover, in the event that TCWA is considered as just recommending a price, item 6.81 of the Guideline for FCR states that "Recommended or maximum resale price arrangements, when they are combined with measures that make them work in reality as fixed or minimum prices, will be assessed in the same manner as Resale Price Maintenance (RPM)." In other words, the competition effect will be assessed. As such, prima facie, TCWA will contravene the first conduct rule of the HKCO. As such, the next part of this study examines the applicability of exemption for TCWA.

3 EXCLUSIONS AND EXEMPTIONS UNDER THE HKCO

With due regard that certain anti-competitive conduct may also bring benefits to consumers and in view of the detriment if these conducts are prohibited, the HKCO allows for a limited regime of exclusions and exemptions. The Ordinance provides for the following exclusions and exemptions from the first conduct rule as summarized in Table 1. Out of the grounds for exemption listed in Table 1, only grounds for economic ground may be used by TCWA. For this reason, the discussion here follows will be directed to the relevant Sections.

Table 1. Summary of exemptions under the HKCO [Source: The Guideline on Applications for a Decision under Sections 9 and 24 (Exclusions and Exemptions) and under Section 15 Block Exemption Orders (HKCC 2015)].

<table>
<tr><th colspan="2">Relevant exclusion or exemption</th><th>Exclusion or exemption from the First Conduct Rule</th></tr>
<tr><td rowspan="6">General Exclusions under Schedule 1of the HKCO</td><td>s1 of Schedule 1- Agreements enhancing overall economic efficiency</td><td>√</td></tr>
<tr><td>s2(1) of Schedule 1 - Compliance with legal requirements</td><td>√</td></tr>
<tr><td>s3 of Schedule 2 - Services of general economic interest</td><td>√</td></tr>
<tr><td>s4 of Schedule 2 -Mergers</td><td>√</td></tr>
<tr><td>s5 of Schedule 2 - Agreements of lesser significance</td><td>√</td></tr>
<tr><td>s6 of Schedule 2 - Conduct of lesser significance</td><td></td></tr>
<tr><td>Block Exemption Orders (s1 of Schedule 1)</td><td></td><td>√</td></tr>
<tr><td>Public Policy Exemption (S31)</td><td></td><td>√</td></tr>
<tr><td>International Obligations Exemption (S32)</td><td></td><td>√</td></tr>
<tr><td>Statutory Obligations Exemption (s2(1) of Schedule 1)</td><td></td><td>√</td></tr>
</table>

3.1 General Exclusion

Under Schedule 1 of the General Exclusions from Conduct Rules, agreements enhancing overall economic efficiency can enjoy general exclusion from the conduct rules. Thus, the first conduct rule does not apply to any agreement that:

(a) Contributes to:

(i) Improving production or distribution; or

(ii) Protecting technical or economic progress,While allowing consumers a fair share of the resulting benefits;

(b) Does not impose on the undertakings concerned restrictions that are not indispensable to the attainment of the objectives stated in paragraph (a); and

(c) Does not afford the undertakings concerned the possibility of eliminating competition in respect of a substantial part of the goods or services in question.

Item 2.5 of the Guideline further adds that the efficiency exclusion applies when four separate conditions are met.

- *First condition: The agreement contributes to improving production or distribution or promoting technical or economic progress.*
- *Second condition: Consumers receive a fair share of the efficiencies.*
- *Third condition: The agreement does not impose on the undertakings concerned restrictions that are not indispensable to the attainment of the relevant efficiencies.*
- *Fourth condition: The agreement does not afford the undertakings concerned the possibility of eliminating competition in respect of a substantial part of the goods or services in question.*

4 WILL TCWA QUALIFY FOR GENERAL EXCLUSION

This section discusses whether TCWA qualifies for general exclusion. As such the four conditions listed in the above section are discussed seriatim.

Efficiency in this context refers to any improvement that may be obtained in production, distribution or promoting technical or economic progress. It has been argued that a healthy construction industry is vital in Hong Kong due to the voluminous of construction activities currently undertaken. This will continue up to 2020 at least. TCWA is of great economic value as it gives an invaluable reference on the cost of production. With such benchmarks, the daily wages of tradesmen can be kept under reasonable control. These will enable better cost estimation and also control. These benchmarks are then used for claim evaluation as well. These can be viewed as industry norms and can be used as reference/standard set by an undertaking independent of the contracting parties. This approach has been well recognized to serve as gap filling for contract adjustment (Macneil 1977). Furthermore, the use of TCWA has been proved successful in the last 20 years. TCWA can have a stabilizing effect on construction cost and ensure construction tradesmen would have a fair remuneration. The whole set up is definitely not for price fixing. Lastly, the daily wages are subject to market fluctuation. The 1st condition should be fulfilled.

The fulfilment of the 2nd condition is a natural flow as a result of accomplishment of the 1st condition. If construction cost is kept under control and dispute over wages are minimized, the consumers, buyers of constructed facilities, would be benefitted. Moreover, it is also noted that the selling price of building flats is more market driven than cost driven. The overall impact may just be neutral which is still considered as meeting the 3rd condition.

The 4th condition addresses the concern over the possibility of competition being eliminated as a result of the agreement. As mentioned, TCWA is only setting a reference or benchmark. With limits, the users of these tradesmen can still negotiate with due regard to the market condition. The vice versa is also equally true. It can therefore be concluded that TCWA will not eliminate competition.

This part of the study concludes that TCWA for construction tradesmen satisfy the four conditions and are qualify as agreements that enhance overall economic efficiency.

5 CONSIDERATION FOR BLOCK EXEMPTION

Block exemption is dealt with under Section 15:

- 15(1) If the Commission is satisfied that a particular category of agreement is an excluded agreement, it may issue a block exemption order in respect of that category of agreement.
- 15(5) In this section "excluded agreement" means an agreement that is excluded from the application of the first conduct rule or as a result of section 1 (Agreements enhancing overall economic efficiency) of Schedule 1.

As concluded in Section 4 of this paper, TCWA is a ype of excluded agreement under Section 15 of HKCO, it is suggested that the trade associations should apply for block exemption for TCWA accordingly. Reference can be made to an application of block exemption under the HKCO. The Hong Kong Liner Shipping Association (HKLSA) on behalf of the carriers in Hong Kong submitted an application for block exemption for two types of liner shipping agreements. The application was made on 17th December 2015, three days after the HKCO came into full force.

The two types of liner shipping agreements are 1) voluntary discussion agreements and 2) vessel sharing agreements. Voluntary discussion agreements are agreements under which the carriers exchange and review market data, trade flows, supply and demand forecast and business trends (ONC 2016). Parties may agree to recommend voluntary guidelines for rates, charges, service contract or tariff terms. These activities are likely be viewed as price-fixing and thus fall foul of first conduct rule. Vessel sharing agreements concern the technical and operational aspects of the provisos of liner shipping services. As such, coordination or joint operation of vessel services may be resulted. These actions may prevent, restrict or distort competition that first conduct rule seeks to control.

The main argument presented by the HKLSA is on the adverse effect on the shipping industry should a lot of transshipment cargo may choose to be transshipped in other nearby ports, particularly in the pearl delta region. In addition, these agreements do not restrict competition as there are other competitive constraints such as the overwhelming market power of the buyers. The low barrier of entry and wide and fragmented market. There is no decision by the HKCC yet but the arguments for TCWA would be quite similar. ONC (2015) commented that similar exemption has been granted in Singapore and European Commission. Since the HKCO has been modelled on the European and Singaporean counterparts, similar exemption grant is expected for the liner agreements and so does TCWA should an application for block exemption is made.

6 CONCLUSIONS

TCWA has been used in the Hong Kong construction industry and proved to serve well as a stabilizing platform for cost estimation and control. Akin to the minimum hourly rates for workers, TCWA serves ensure the contributors of the construction industry are rewarded with competitive wages so that the industry will continue to prosper. This study finds that TWCA will likely contravene the first conduct rule of the HKCO. However, the characterizing features of TWCA would satisfy the four conditions that qualify them as excluded agreements. S such it is suggested that the trade associations should apply for block exemption for TWCA like the HKLSA for the liners in Hong Kong.

Acknowledgments

The study reported in this paper was fully supported by a HKSAR Research Grant Council GRF (Project number 11201914).

References

Amabile, T. M., "Componential Theory of Creativity", Havard Business School Publishing, Working Paper 12-096, 2012.

Amabile, T. M., "A Model of Creativity in Organisations", Research in Organisational Behavior, 22:123-167, 2000.

Audretsch, D. B., Baumol, W. J., and Burke, A. E., Competition policy in dynamic markets. *International Journal of Industrial Organization, 19*(5), 613-634, 2001.

Chen, E. K., and Lin, P., Competition policy under laissez-faireism: Market Power and its Treatment in Hong Kong. *Review of Industrial Organization, 21*(2), 145-166, 2002.

HKSAR, CAP 619, Hong Kong Competition Ordinance, CAP 610, 2012.

Hong Kong Competition Commission, *Guideline on The First Conduct Rule*, 2015.

Hong Kong Competition Commission, Guidleine on Applications for a Decision under Sections 9 and 24 (Exclusions and Exemptions) and under Section 15 Block Exemption Orders, 2015.

Macneil I.R., A Primer of Contract Planning, Southern California Law Review, 48:003-704, 1977.

ONC Lawyers, Will Liner Shipping Agreements be exempted from the Competition Ordinance?, Retrieved from http://www.onc.hk/pages/show_pub.asp?cmd=print&pid=9264 on May 27, 2016.

Whilsh, R. and Bailey, D., *Competition law*: Oxford University Press, 2012.

Integrated Solutions for Infrastructure Development
Edited by Lau, H. H., Tang, F. E., Ng, C. K., and Singh, A.
Copyright © 2016 ISEC Press

A CONCENTRATION STUDY OF THE PRIVATE NEW RESIDENTIAL UNITS MARKET

SAI ON CHEUNG[1] and YUHAN NIU[2]

[1] *Dept of Architecture and Civil Engineering, City University of Hong Kong, Hong Kong*
[2]*Dept of Real Estate and Construction, University of Hong Kong, Hong Kong*

High price of private residential units is a social issue for many economies. The property price of many developed cities has risen beyond reach of the general public. In 1996, the Hong Kong Consumer Council also commented that the residential market is quite concentrated. The enactment of the Hong Kong Competition Ordinance echoes the need to look into the market competition level so as to better protect consumer welfare. Market concentration is a major indicator to measure competition. The primary aim of this study is to evaluate the concentration level of the private new residential market in Hong Kong. The concentration level is mainly evaluated by concentration measures including the Concentration Ratio and Hirfindahl-Hirschman index. The result shows that the top five developers are holding a dominant position in the market. It is envisaged that the big developers could retain their dominance due to their substantial land bank.

Keywords: New Residential Units, Price, Land Supply System, Concentration, Competition.

1 INTRODUCTION

High residential property price has always been a focal issue of social concern in major cities. In Hong Kong, the housing of the under-privileged is mainly assisted by the Government through extensive rental public housing program. House buyers rely on the private residential market. Among others, Hong Kong has recorded unprecedented rises of property price after the 2008 financial turmoil. Complaints from society have never in this connection. By 2009, the cost for purchasing residential properties ranked the 4th among all the international metropolitans (Poon 2011). The residential property prices in Hong Kong have accented 117.2% from December 2008 to December 2012, while the climbing trend of property price continues after 2012 up to 2015 at a rate far beyond inflation (Rating and Valuation Department of Hong Kong 2015).

The new private residential market is considered as the prime determinant of the price of residential units. Used residential units are usually valued at a discount of the new ones. The overly-high price can be attributed to insufficient competition in the private new residential units market in Hong Kong. Market competition has been advocated to be the main enabler to promote and preserve efficiency. In traditional static economic analysis, welfare is assumed to be enhanced with the increment of competition intensity, with perfect competition synchronously indicating welfare maximization (Audretsch *et al.* 2001). Optimum competition would bring greater economic efficiency including more effective use of resources and greater productivity gains for the economy, since market players are spurred to be more efficient, innovative and responsive to consumer needs (Hong Kong Competition Commission 2015, Whish and Bailey 2012). In a market with de-monopolization, liberalization and privatization, market players need to win consumers by offering lower prices and better products (Hussein *et al.* 2012).

In 2012, the Hong Kong government enacted the first-ever Hong Kong Competition Ordinance (HKCO). The objective of the study is to look into the application of the conduct rules of the HKCO in evaluating the existence or otherwise of dominant market power. The new residential unit market is used as an illustration. The study is reported in the following six parts: 2) the economic and legal aspects of competition, 3) Concentration measures, 4) Data analysis and findings, and 5) Discussion and Limitations of the study, and 6) Concluding remarks.

2 THE ECONOMIC AND LEGAL ASPECTS OF COMPETITION

Hong Kong has been identified as one of the most free economies in the world (Chen and Lin 2002, HKCC 2014). The Hong Kong government has adopted the laissez faire philosophy that pins on the belief that the market force is the best way to allocate resources (Chen and Lin 2002). However, the general impression is not entirely true and it tends to confuse laissez faire with perfect competition. Anti-competition conducts, such as price fixing and bid riggings that may confine consumer choices or jet up prices, are not absent from Hong Kong (HKCC 2014).

In many developed economies, there are legislative interventions to protect trades and competition. The rationale underpins these legislations is that the abuse of dominant market power could significantly influence market competition. Market power refers to conditions where the providers of a good or a service can consistently raise prices above the price level established by a competitive market (Alvarado 1998). Accumulation of market power is indicated by the concentration of resources in the hands of a few merchandise producers/service providers or an insufficient number of producers/providers (Alvarado 1998). Therefore, the producers or providers with market power would have the ability to individually affect either the total quantity or the prevailing price in the market. Consequently, the market competition might be distorted if they abuse such ability.

In Hong Kong, the first-ever competition law, the Hong Kong Competition Ordinance (HKCO), was enacted in June 2012 and came into force on 14th December 2015. The HKCO has two conduct rules to regulate anti-competition conducts and a merger rule to prohibit merges that substantially lessen competition in Hong Kong. The Second Conduct Rule is relevant to this study:

> "An undertaking that has a substantial degree of market power in a market must not abuse that power by engaging in conduct that has as its object or effect the prevention, restriction or distortion of competition in Hong Kong"

The Guidelines on the Second Conduct Rule (GSDR) highlights market share as the initial screening device in assessing the substantial market power since higher market share of undertakings usually indicate higher degree of market power (Hong Kong Competition Commission 2015). Market share distribution largely correlates to market structure, showing the extent to which market power is distributed among incumbents (Adelman 1951, Ye *et al.* 2009). In this regard, concentration is a principal indicator of market power. The significance of concentration arises from the proposition that highly concentrated market is dominated by a few large firms. This situation could be worse if barrier to entry is high (McCloughan and Abounoori 2003). Concentrated market is a fertile ground for the abuse of market power that most competition policies object to guard against (McCloughan 2004).

3 CONCENTRATION MEASURES

The GSDR suggests two methods to measure market share concentration; i) Concentration Ratios (CR) and ii) Herfindahl-Hirschman Index (HHI). Concentration Ratio (CR) is the most widely known and used measure of industrial concentration. CR is represented by the percentage of the market share occupied by the largest firms in a certain industry (Parker 1991). Data revealing employment, net output, turnover or work done can all be used in measuring CR (McCloughan 2004). As a matter of fact, concentration Ratio is often used for its convenience in collection of data. It has also been criticized for the narrow focus on a few large firms without taking the rest of firms into consideration (Ye *et al.* 2009).

HHI was proposed by Hirschman (1980) who used this index to measure concentration of control or power an industry (Hirschman 1980). Hirfindahl-Hirschman index (HHI) is the sum of the squared market shares of the top 50 firms (or all the firms if they are less than fifty) in a particular sector (Nawrocki and Carter 2010). When Hirschman (1980) developed this index for measuring concentration, he stressed that an important condition of perfect competition implies the presence of both a relatively equal distribution and a large number of producing firms in this industry. Unlike concentration ratio, all participants are considered in HHI. Besides, HHI weights each firm by its own relative market share, giving higher weights to larger firms.

CR focuses on the market share of the leading market participants while HHI weights the share of all participants to show inequality in market share distribution. While CR leaves the rest of market participants out of consideration, using HHI could offset this demerit. Therefore, CR and HHI can be used in parallel to enhance the evaluation.

4 DATA ANALYSIS AND FINDINGS

For residential units, the Hong Kong Consumer Council of Hong Kong uses housing units annually produced by each developer to represent their annual market share. Gross Floor Area (GFA) of buildings has also been adopted to demonstrate the market share condition in this study. GFA is defined as the area contained within the external walls of the building measured at each floor level by the Census and Statistic Department of Hong Kong. GFA Production is a way to represent the output of residential units. In addition, the annual completion in GFA is generally available in the annual report of listed developers and therefore used in this study.

To evaluate the concentration of private new residential unit market, developers listed on the Hong Kong Stock Exchange are ranked by their market capitals. By looking into developers' annual report from year 2009 to 2012, their annual GFA completions of residential development are compared against the total GFA completed annually in Hong Kong. Data are obtained from the Key Statistics on Business Performance and Operating Characteristics of the Building, Construction and Real Estate Sectors by Census and Statistics Department of Hong Kong.

4.1 Calculation of Concentration Ratio (CR)

The range of concentration ratio is from 0 to 1. Zero indicates that the market is perfectly competitive while one indicates little competition (Ye *et al.* 2009). The calculation of CR, by its definition, is the percentage of the market share occupied by the largest firms in a certain industry. It can be conducted by the following formula, where S_i denotes the market share of firm i and n denotes the total number of firms whose market shares are counted in;

$$CR_n=\sum_{i=1}^{n} S_i \quad (1)$$

Four-firm concentration ratio (CR4, the percentage of market share held by the largest four undertakings) and five-firm concentration ratio (CR5, the percentage of market share held by the largest five undertakings) are mostly used in previous studies (McCloughan 2004). Therefore, CR4 and CR5 are both calculated in this study.

4.2 Calculation of Hirfindahl-Hirschman Index (HHI)

Since the market shares are measured from 0 to 1, the range for HHI is between zero (perfect competition) and one (little competition) (Chiang *et al.* 2001, Nawrocki and Carter 2010). Total number of participants in a certain market shall be considered in HHI. HHI outcomes are usually compared against the threshold index suggested in Merger Guidelines of U.S. Department of Justice and the Federal Trade Commission (Kerber *et al.* 2009). To follow the index format convention, percent sign (%) is left out for both Si and HHI expressions. For example, 19.71% is expressed as 19.71. The outcome of HHI will be between 0 and 10000. In the following formula, *Si* denotes the market share of firm *i* and *n* denotes the total number of firms whose market shares are counted in;

$$\text{HHI}=\sum_{i=1}^{n} S_i^{\,2} \tag{2}$$

One drawback of HHI is the difficulty in collecting data of every single firm (Adelman 1951). In this regard, Hirschman (1980) provided some alternatives. When data of the majority of large firms are available, the rest few can be lumped as "other firms". An approximate determination could be made by setting a maximum and a minimum limit to the remaining firms (Hirschman 1980). The limit setting of HHI are purely based on math calculations. The maximum HHI could be calculated by assuming all "other firms" are holding the same market share as the firm holding the least available market share. As for the minimum HHI, it can be calculated by assuming all the "other firms" achieves perfect competition.

4.3 Findings

The five major developers with the highest market shares are shown as A-E in Table 1, with their annual market share from 2009 to 2012 and an average market share throughout these four years. CR4 and CR5 are compared against thresholds that have been suggested in other competition guidelines. Table 2 serves as reference of thresholds of concentration ratio in overseas legislations. By referring to these thresholds, the largest 4 to 5 developers in Hong Kong residential property market are collectively holding a dominant market position.

Table 1. Market share measures.

Developer	2009		2010		2011		2012		Average	
	Annual GFA Completion (ft2)	Market Share (%)	Annual GFA Completion (ft2)	Market Share (%)	Annual GFA Completion (ft2)	Market Share (%)	Annual GFA Completion (ft2)	Market Share (%)	Annual GFA Completion (ft2)	Market Share (%)
A	1815000	14.54%	4097000	22.40%	2858000	16.20%	3645000	19.71%	2483000	18.56%
B	2958774	23.70%	3610135	19.74%	2432927	13.79%	3624918	19.60%	2525351	18.87%
C	2847667	22.81%	3543963	19.38%	2090857	11.85%	3171252	17.15%	2330748	17.42%
D	1972891	15.80%	1287464	7.04%	1149517	6.52%	1341786	7.26%	1150332	8.60%
E	691395	5.54%	1419621	7.76%	1808478	10.25%	1559149	8.43%	1095729	8.19%
Total	12486136		18287883		17642049		18492398		13381693	
CR4	76.84%		69.28%		52.09%		64.89%		63.44%	
CR5	82.38%		76.32%		58.61%		72.15%		71.63%	
HHI	1573-1670		1377-1544		741-1010		1191-1285		1145-1389	

With reference to the figures in Table 1, except for year 2011, the largest 3 participants already hold a dominant position. Adding the market share of the 4th and 5th largest developers, there is only less than a quarter market share left for the rest of all property developers. Such situation signals a highly concentrated market where a "competition problem" is highly possible to occur (McCloughan 2004).

The thresholds for HHI could be found in US Horizontal Merger Guidelines where merger cases would be assessed as non-concentrated, moderately concentrated or highly concentrated (see Table 3). Taking the US guideline as a reference, HHI steadily stayed as moderately concentrated when applying the 1992 version. And it even fell into "non-concentrated" category from the year of 2010 if the 2010 version is adopted thence.

Table 2. CR thresholds as screening tools.

Region	Legislation	Suggested Threshold
Singapore	Competition Act 2004	>60%
Malaysia	Laws Of Malaysia Act 712 Competition Act 2010	>60%
UK	Competition Act 1998	>50%
Australia	Competition and Consumer Act 2010	(No threshold suggested)
Europe	Article 102 of the Treaty on the Functioning of the European Union	<40% not likely dominant
US	The Sherman Antitrust Act (1890)	>50% and sustainable over time

Table 3. HHI thresholds as screening tools.

Version		2010	1992
US Horizontal Merger Guideline	Non-concentrated	Under 1500	Under 1000
	Moderately concentrated	1500 - 2500	1000 - 1800
	Highly concentrated	Over 2500	Over 1800

5 DISCUSSIONS

Unlike high CRs that indicates a significantly concentrated market, HHI demonstrated the competition condition of property market from another perspective. The squaring mechanism of HHI places more focus to the individual market shares, showing the extent of inequality. Although the total market share of five developers is large in absolute value, their shares are quite equally matched, especially for the top three. The outcome of HHI assessment and CR analysis are in fact complementary. When the CR analyses demonstrate that the top-3 to top-5 firms held a dominant position in the residential property market, the HHI analyses show that there is no single dominant firm. The individual competitiveness is almost equal among members within this group.

The findings in this study are subject to two caveats. Firstly, on investigating the private residential property developers, the data of private developers that are not listed in the Hong Kong Stock Exchange Market are not available. These data might have potential impacts to the HHI analysis if the market share of any one of these private developers ranked higher than the 5th biggest listed developers. Secondly, this study only focuses on the two major indicators of market power suggested by guidelines. It is acknowledged that there are other indicators such as buyers' countervailing power and economic regulations by the government, which may also have impact on the existence of dominant market power. Future study could be more comprehensive if these factors are evaluated and analyzed altogether.

6 CONCLUSIONS

The residential property price in Hong Kong has kept rising in the last decade and has been ranked within the top three metropolitan cities in the world. The rise after the 2008 global financial turmoil has been phenomenal and far beyond inflation. In this regard, there may be a case of insufficient competition that has lead to the rocketed prices. This study explores the concentration of the new private residential property market using the measures suggested by the Guidelines of the Second Conduct Rule of the Hong Kong Competition Ordinance. The result suggests that the new residential unit market is quite concentrated. For the period from 2009 to 2012, the big five developers in Hong Kong collectively held a dominant position of the potential supply. However, their respective market shares are almost equal and balanced. It is further found that the substantial land banks of big developers also facilitate the big developers to continuously hold their collective dominant position in the near future.

Acknowledgments

The study reported in this paper was fully supported by a HKSAR Research Grant Council GRF (Project number 11201914)

References

Adelman, M. A., The measurement of industrial concentration. *The Review of Economics and Statistics*, 269-296, 1951.

Alvarado, F. L., Market power: A dynamic definition. *Bulk Power Systems Dynamics and Control V,* 1998.

Audretsch, D. B., Baumol, W. J., and Burke, A. E., Competition policy in dynamic markets. *International Journal of Industrial Organization, 19*(5), 613-634, 2001.

Chen, E. K., and Lin, P., Competition policy under laissez-faireism: Market Power and its Treatment in Hong Kong. *Review of Industrial Organization, 21*(2), 145-166, 2002.

Chiang, Y.-H., Tang, B.-S., and Leung, W.-Y., Market structure of the construction industry in Hong Kong. *Construction Management & Economics, 19*(7), 675-687, 2001.

Hirschman, A. O., *National power and the structure of foreign trade* (Vol. 1): Univ of California Press, 1980.

Hong Kong Competition Commission, *Guideline on The Second Conduct Rule,* Consumer Council, 2015.

Hussein, S. M., Manap, N. A., and Nor, M. Z. M., Market Definition and Market Power as Tools for The Assessment of Competition. *International Journal Of Business And Society, 13*(2), 163-182, 2012.

Kerber, W., Kretschmer, J.-P., and von Wangenheim, G., *Market share thresholds and Herfindahl-Hirschman-Index (HHI) as screening instruments in competition law: A theoretical analysis*, discussion paper, 2009.

McCloughan, P., Construction sector concentration: evidence from Britain. *Construction Management and Economics, 22*(9), 979-990, 2004.

McCloughan, P., and Abounoori, E., How to estimate market concentration given grouped data. *Applied Economics, 35*(8), 973-983, 2003.

Nawrocki, D., and Carter, W., Industry competitiveness using Herfindahl and entropy concentration indices with firm market capitalization data. *Applied Economics, 42*(22), 2855-2863, 2010.

Parker, S. C., Significantly concentrated markets: Theory and evidence for the UK. *International Journal of Industrial Organization, 9*(4), 585-590, 1991.

Poon, A., *Land and the ruling class in Hong Kong*: Enrich Professional Pub, Hong Kong 2011.

Rating and valuation Department of Hong Kong, Price Indices for Hong Kong Property Market, 2015. Retrieved from http://www.rvd.gov.hk/doc/en/statistics/graph2.pdf on July 16, 2015

Whish, R., and Bailey, D., *Competition law*: Oxford University Press, 2012.

Ye, K., Lu, W., and Jiang, W., Concentration in the international construction market. *Construction Management and Economics, 27*(12), 1197-1207, 2009.

Integrated Solutions for Infrastructure Development
Edited by Lau, H. H., Tang, F. E., Ng, C. K., and Singh, A.
Copyright © 2016 ISEC Press

BUILDING THE EDUCATION REVOLUTION (BER) PROGRAM: GOVERNING THE IMPLEMENTATION OF PROJECT POLICIES

DURO KOLAR, DEREK H.T. WALKER and PETER S.P. WONG

School of Property, Construction and Project Management, RMIT University, Melbourne, Australia

The project world is plagued with the interest and influence of diverse stakeholders shaping major project policy strategies. From deception, strategic manipulation to illusions of control, stakeholders using power and any other means to influence project strategies to achieve their desired ends: the tribalism and Machiavellianism of the project world. Governing tribalism or Machiavellianism especially at the front-end of projects is absolutely essential for project policy success. But how does a project manager shape a major project policy into success? The purpose of this research is to investigate tribalism and Machiavellianism and implement the necessary governance mechanisms to shape the implementation of project policies to optimize its success. This requires a deep understanding of human relational behavior. A case study approach was adopted that investigated Australia's largest infrastructure program in history, the AUD $16.2 billion Building the Education Revolution (BER) program, which was plagued with power and politics and Machiavellianism of the project world. Factors such as collective institutional leadership with formal and informal governance mechanisms in rationalizing decisions were identified as essential aspects from the case study. It was also revealed that if the factors are implemented at the front-end of project policies and managed appropriately it could significantly limit the potentness of the tribal or Machiavelli plague and optimize the chances of project policy success.

Keywords: Governance, Strategic decision making, government, Infrastructure, megaprojects, Machiavellianism, Institutionalization theory.

1 INTRODUCTION

The implementation of public policy has a history of failure (Larson 1980). Time, and time again, scholars are baffled with the unexpected moves within policy processes and unanticipated outcomes (Van Buuren *et al.* 2009). An understanding of factors that influence organizational strategic decision making on the implementation of project or program policies is therefore essential to optimize its success (Flyvbjerg 2012). However, megaprojects continue to fail to achieve most of their objectives and outcomes which tends to cause tremendous loss in productivity, profitability, and impacts stakeholder morale (Flyvbjerg 2014). Unsurprisingly, these problems can be seen as Machiavellianism[1] (Heath 2009). However, there is little information as to why these phenomena occur.

[1] Machiavellianism stands for power, deceit, coercion and using any means to achieve desired ends (Bass and Bass 2008).

This paper begins with a review of the salient literature on factors that influence organizational strategic decision making on project policy implementation including an explanation of coping mechanisms to optimize its success. It then reports the findings of the case study and implications for future research and practitioners.

2 IMPLEMENTATION OF PROJECT POLICIES

The traditional view of strategy focuses on the long-term goals and objectives of a firm and plans for achieving those objectives with the allocation of the necessary resources (Mintzberg 1978). Such a view still dominates contemporary organizational theory, a strategy hierarchy in which organizational objectives filter to divisional units and guide functional tactics (Kay *et al.* 2006). This turns an organization into a rather mechanical, long-driven and engineered system of formal and informal patterns of (inter)actions (Baligh 2006). The implementation of today's project policies tends to follow *in situ*, organizational strategies cascading through strategic business units or divisions, which often fail to achieve their objectives and outcomes (Young and Grant 2015). Decision making tends to be seen in the same light as strategy, a rational and easy process (Mintzberg and Westley 2010). However, this rational process turns out to be uncommon in organizations, where decision making is seen as strategic, complex, dynamic and characterized by plurality (Tsoukas 2010). Such decision making from an organizational context is considered to significantly influence project policy outcomes (Flyvbjerg 2014), which includes external and internal environmental factors (Nutt and Wilson 2010). As with theories on organizational strategy and decision making, the literature on governance and project governance also tend to follow *in situ*, project governance mechanisms implemented to align with corporate governance mechanisms (Biesenthal and Wilden 2014). Such linear and bureaucratic governance is associated with a negative cycle of relational coordination (Gittell 2012), which often leads to high cognitive biases, such as delusions, illusions, deceptions, optimism, strategic misrepresentation and overconfidence in decision making which is commonly seen on major project policies (see Flyvbjerg 2009, Flyvbjerg *et al.* 2009). On the whole, what the literature implies is that organizational strategic decision making on the implementation of project policies needs a project-specific strategy, where such a strategy requires an understanding of *strategically governing the relational actor space* where collective understandings of organizational actors emerge that create an institutionalized project reality.

3 RESEARCH METHOD

This research adopted a case study approach from an interpretivist paradigm with an ontological stance to uncover the meaning of socially constructed reality as understood by individuals or groups. Evidence was gained through an interview process with open-ended questions. Senior executives and project managers were chosen that had detailed knowledge on the complexity and dynamics of project delivery as it unfolded throughout implementation. Data was analyzed from a grounded theory approach (Corbin and Strauss 2008). A modified Delphi technique was then adopted to obtain reliable consensus of opinions from a group of handpicked experts by a series of online questionnaires with feedback (Maijala *et al.* 2015). Termination of the Delphi from further rounds, or stopping criterion, was determined when another round would not significantly add to the results (MacCarthy and Atthirawong 2003), based on predefined criteria. Consensus was achieved over one round and thus no further rounds were necessary. Table 1 provides a structured account of the seven step analytical approach adopted for this research.

Table 1. Data analysis approach (partially adapted from Walker and Jacobsson 2014, p. 656).

No.	Steps	Data Analysis Approach
1	Familiarizing with the data	Reading, re-reading, identification of initial ideas, transcription of interviews, writing memos and drawing diagrams – capture participants' view.
2	Generating initial codes and categories (open coding)	Codes and categories abductively generated (Kelle 2007) based on grounded theory approach using NVivo.
3	Naming and making connections between categories (axial coding)	Based on codes, enriching and making connections between developed categories.
4	Refining categories and themes	On-going critical analysis to refine the specifics of each category and theme – make sense and understand the 'true' meaning of the data for theory development i.e., inductive top-down theorizing (Shepherd and Sutcliffe 2011).
5	Producing the report	Refining analysis, selection of compelling extract examples, final analysis of selected extracts, relating analysis back to the research question and literature. Extract propositions.
6	Modified Delphi technique	Define consensus on a 5-point Likert scale (Von der Gracht 2012). Obtain reliable consensus of opinions from a group of handpicked experts by a series of online questionnaires with feedback to generate consensus (Maijala *et al.* 2015).
7	Validation of findings	Adopt the 'natural science model' (Eisenhardt and Graebner 2007) based on four rigour dimensions: construct validity, internal validity, external validity (or generalizability), and reliability (Yin 2014); and 'talk the walk,' priority ordering of validity types, and necessity is the mother of rigour (Gibbert and Ruigrok 2010).

4 EMPIRICAL FINDINGS

In line with the data analysis principles adopted for this research, the following seven propositions emerged from the data analysis for the implementation of project and program policies:

- Proposition 1. Creating, maintaining and disrupting institutional project relations requires collective institutional leadership.
- Proposition 2. Creating, maintaining and disrupting institutional project relations requires a relatively stabilized and strategic governance framework of formal or regulative (e.g., contracts, agreements, guidelines, formal meetings etc.) governance mechanisms, and informal or normative (e.g., norms, values, beliefs, morals, casual meetings etc.) governance mechanisms.
- Proposition 3. The decision to implement a strategic governance framework is based on a combination of trust, power and control mechanisms. In a project environment with a relatively weak institutional governance framework, power and regulative (or formal) control tend to be the dominate governance mechanisms to coordinate expectations and (inter)actions between individuals and organizations.
- Proposition 4. The decision to implement a strategic governance framework is based on a combination of trust, power and control mechanisms. In a project environment with a relatively strong and stabilized institutional governance framework, trust and informal (or normative) control tend to be the dominate and coevolving governance mechanisms to coordinate expectations and (inter)actions between individuals and organizations.

- Proposition 5. When a relatively strong and stabilized institutional governance framework is legitimately (e.g., change in human resources, change in project scope etc.) or illegitimately (e.g., self-serving interests, hidden agendas, opportunistic behavior etc.) disrupted, collective institutional leadership and informal (or normative) control with collaborative co-creation tend to be the dominate mechanisms to coordinate expectations and (inter)actions between individuals and organizations.
- Proposition 6. A project environment with a strategic governance framework which is dominated by power i.e., powerful decision makers and regulative (or formal) control tends to increase the risk of cognitive biases (e.g., overconfidence, illusory control, strategic misrepresentation or manipulation, outcome etc.) in rationalizing decisions.
- Proposition 7. A project environment with a relatively strong and stabilized governance framework tends to alleviate the risk of cognitive biases (e.g., overconfidence, illusory control, strategic misrepresentation or manipulation, outcome etc.) in rationalizing decisions.

Figure 1 provides a simplified illustration of the seven propositions for the implementation of project and program policies.

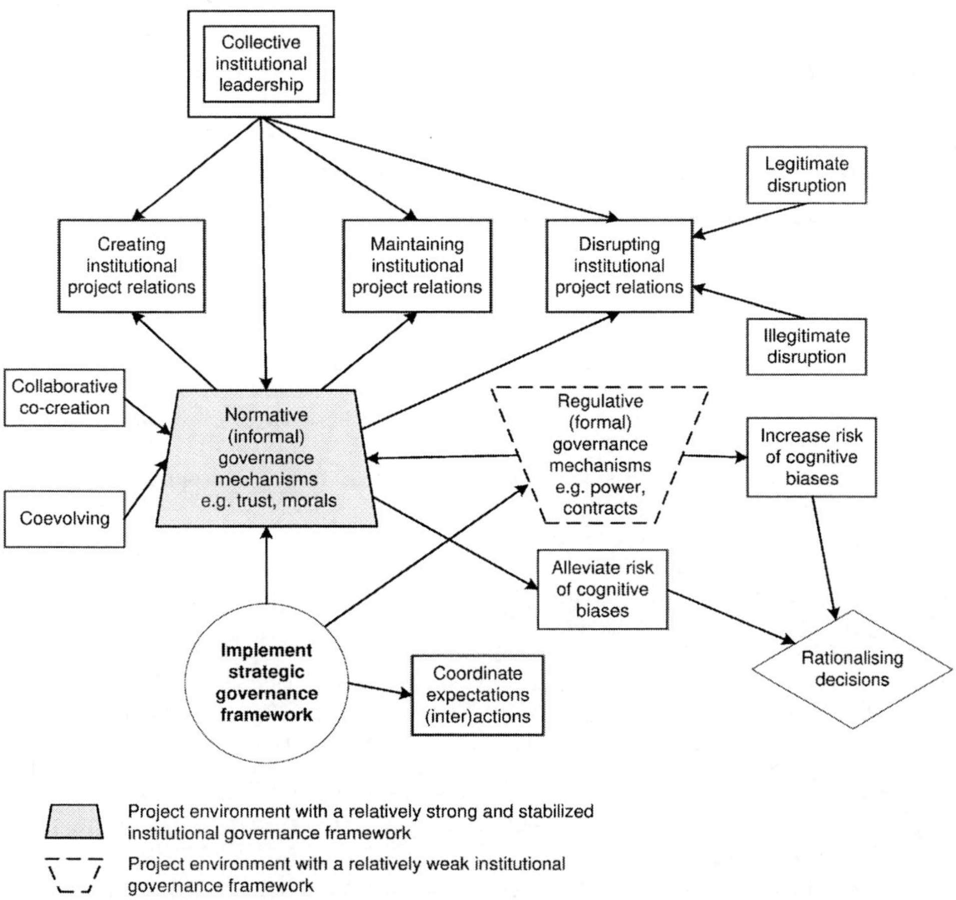

Figure 1. A simplified illustration of the propositions.

5 DISCUSSION AND CONCLUSIONS

The purpose of this research was to examine a coping mechanism for the successful implementation of project policies, especially in today's tribal and Machiavelli project world. The findings explain not only how the BER program and its subsidiary project panned out (or came to reality) but also the necessary transformations needed to optimize success for other government project and program policies. From the case study, it can be concluded that project mal-governance can significantly be tamed or even eradicated if project managers work with the project or program sponsor to implement an effective governance mechanism at the front-end of project or program policies. This is achieved through *strategically governing the relational actor space* with the processes of collective institutional leadership, normative and regulative governance mechanisms in rationalizing decisions with the aim to achieve an institutionalized project reality.

6 IMPLICATIONS FOR FUTURE RESEARCH AND PRACTITIONERS

This research reports upon a single case study within the realm of tribalism and Machiavellianism for the implementation of project and program policies. The research adds to the body of knowledge surrounding project shaping and governance in rationalizing project decisions, especially to the critical front-end phase of project policies. With its implementation, it can assist practitioners to understand how to strategically implement a governance framework to coordinate expectations and (inter)actions between individuals and organizations and achieve project policy success.

Acknowledgments

We thank the individuals associated with the BER program and its subsidiary project, whom we interviewed and who gave valuable feedback on the findings of this research.

References

Baligh, H. H. *Organization Structures: Theory and Design, Analysis and Prescription*, Springer, New York, 2006.

Bass, B. M., and Bass, R. *The Bass handbook of leadership: Theory, research, and managerial applications* (4 ed.), Free Press, New York, 2008.

Biesenthal, C., and Wilden, R., Multi-level project governance: Trends and opportunities, *International Journal of Project Management, 32*(8), 1291-1308, 2014.

Corbin, J., and Strauss, A. *Basics of Qualitative Research: Techniques and Procedures for Developing Grounded Theory* (3 ed.), SAGE Publications Inc., California, 2008.

Eisenhardt, K. M., and Graebner, M. E., Theory Building from Cases: Opportunities and Challenges, *The Academy of Management Journal, 50*(1), 25-32, 2007.

Flyvbjerg, B. Optimism and Misreprensentation in Early Project Development. In T. M. Williams, K. Samset, & K. J. Sunnevag (Eds.), *Making Essential Choices with Scant Information: Front-End Decision Making in Major Projects* (pp. 147-168), Palgrave Macmillan, Basingstoke, 2009.

Flyvbjerg, B. Over Budget, Over Time, Over and Over Again: Managing Major Projects. In P. W. G. Morris, J. K. Pinto, & J. Soderlund (Eds.), *The Oxford Handbook Of Project Management* (pp. 321-344), Oxford University Press, Oxford, 2012.

Flyvbjerg, B., What You Should Know About Megaprojects and Why: An Overview, *Project Management Journal, 45*(2), 6-19, 2014.

Flyvbjerg, B., Garbuio, M., and Lovallo, D., Delusion and Deception in large Infrastructure Projects: Two Models for Explaining and Preventing Executive Disaster, *California Management Review, 51*(2), 170-193, 2009.

Gibbert, M., and Ruigrok, W., The ''What'' and ''How'' of Case Study Rigor: Three Strategies Based on Published Work, *Organizational Research Methods, 13*(4), 710-737, 2010.

Gittell, J. H. New Directions for Relational Coordination Theory. In K. S. Cameron & G. M. Spreitzer (Eds.), *The Oxford Handbook of Positive Organizational Scholarship* (pp. 400-411), Oxford University Press, New York, 2012.

Heath, J., The Uses and Abuses of Agency Theory, *Business Ethics Quarterly, 19*(4), 497-528, 2009.

Kay, J., McKiernan, P., and Faulkner, D. O. The History of Strategy and Some Thoughts about the Future. In A. Campbell & D. O. Faulkner (Eds.), *The Oxford Handbook of Strategy: A Strategy Overview and Competitive Strategy* (pp. 27-52), Oxford University Press, Oxford, 2006.

Kelle, U. The development of categories: Different approaches in grounded theory. In A. Bryant & K. Charmaz (Eds.), *The SAGE Handbook of Grounded Theory* (pp. 191-214), SAGE Publications Ltd, London, 2007.

Larson, J. S. *Why Government programs fail: Improving policy implementations*, Praeger, New York, 1980.

MacCarthy, B. L., and Atthirawong, W., Factors affecting location decisions in international operations – a Delphi study, *International Journal of Operations & Production Management, 23*(7), 794-818, 2003.

Maijala, V., Tossavainen, K., and Turunen, H., Identifying nurse practitioners' required case management competencies in health promotion practice in municipal public primary health care. A two-stage modified Delphi study, *Journal of Clinical Nursing, 24*(17-18), 2554-2561, 2015.

Mintzberg, H., Patterns in Strategy Formation, *Management Science, 24*(9), 934-948, 1978.

Mintzberg, H., and Westley, F. Decision Making: It's Not What You Think. In P. C. Nutt & D. C. Wilson (Eds.), *Handbook of Decision Making* (1 ed., pp. 73-82), John Wiley & Sons, Ltd, Chichester, 2010.

Nutt, P. C., and Wilson, D. C. *Handbook of Decision Making*, John Wiley & Sons, Ltd, Chichester, 2010.

Shepherd, D. A., and Sutcliffe, K. M., Inductive top-down theorizing: A source of new theories of organization, *Academy of Management Review, 36*(2), 361-380, 2011.

Tsoukas, H. Strategic decision making and knowledge: A Heideggerian Approach. In P. C. Nutt & D. C. Wilson (Eds.), *Handbook of Decision Making* (pp. 379-402), John Wiley & Sons, Ltd, Chichester, 2010.

Van Buuren, A., Gerrits, L., and Marks, P. Public Policy-Making and the Management of Coevolution. In G. Teisman, A. van Buuren, & L. Gerrits (Eds.), *Managing Complex Governance Systems: Dyanmics, Self-Organization and Coevolution in Public Investments* (1 ed., pp. 154-171), Routledge, New York, 2009.

Von der Gracht, H. A., Consensus measurement in Delphi studies: Review and implications for future quality assurance, *Technological Forecasting and Social Change, 79*(8), 1525-1536, 2012.

Walker, D. H. T., and Jacobsson, M., A rationale for alliancing within a public-private partnership, *Engineering, Construction and Architectural Management, 21*(6), 648-673, 2014.

Yin, R. *Case study research: design and methods*, SAGE Publications Inc., California, 2014.

Young, R., and Grant, J., Is strategy implemented by projects? Disturbing evidence in the State of NSW, *International Journal of Project Management, 33*(1), 15-28, 2015.

Integrated Solutions for Infrastructure Development
Edited by Lau, H. H., Tang, F. E., Ng, C. K., and Singh, A.
Copyright © 2016 ISEC Press

FACTORS AFFECTING THE FULL IMPLEMENTATION OF BUILDING INFORMATION MODELING AS A CONSTRUCTION MANAGEMENT TOOL

SWAPAN SAHA and SCOTT JESSUP

School of Computing, Engineering and Mathematics, University of Western Sydney, Penrith, NSW 2751, Australia

There has been growing concern about the efficiency of the Australian construction industry. It has been identified that inefficiencies due to industry fragmentation has significant impacts on the way construction projects are delivered. It is largely believed that Building Information Modeling (BIM) has the potential to significantly improve the efficiency of a project throughout its entire lifecycle. The main focus of BIM research has, up until now, been principally focused on its ever increasing use in the design phase of a project. The construction phase of a project has been largely overlooked, particularly the implementation of BIM during this phase of a project.. A mixed method approach has been utilised in this study comprising of quantitative and qualitative techniques. The quantitative method was conducted through the distribution of a statistically analysed questionnaire survey to a sample population of construction professionals within Australia. Concurrently, the qualitative method of analysis is conducted through case studies of selected construction projects with some involvement of BIM. Through the research, it was proven that Australian construction projects are far from the full utilisation of BIM as a construction management tool. The factors affecting BIM's utilisation were also discovered in this research and validated through statistical analysis of the population data found by the questionnaire survey.

Keywords: BIM, Productivity Construction industry, Australia.

1 INTRODUCTION AND BACKGROUND

The Australian Bureau of Statistics (2012) has identified the Australian construction industry as being the fourth largest contributor to Gross Domestic Product (GDP) and is vital to Australia's economic development (Yan and Chunlu 2010). In June 2012, the industry employed 950,000 people and supported 209,783 businesses. The industry generated a total of $305.5b comprising construction services, building construction and heavy and civil engineering construction. However, the industry has been suffering from a decline in productivity due to significant inefficiencies (buildingSMART Australasia 2012). As shown in figure 1, the construction industry in Australia has been growing, however, the growth is at a slower rate than the cumulative productivity in Australia (The Allen Consulting Group 2010).

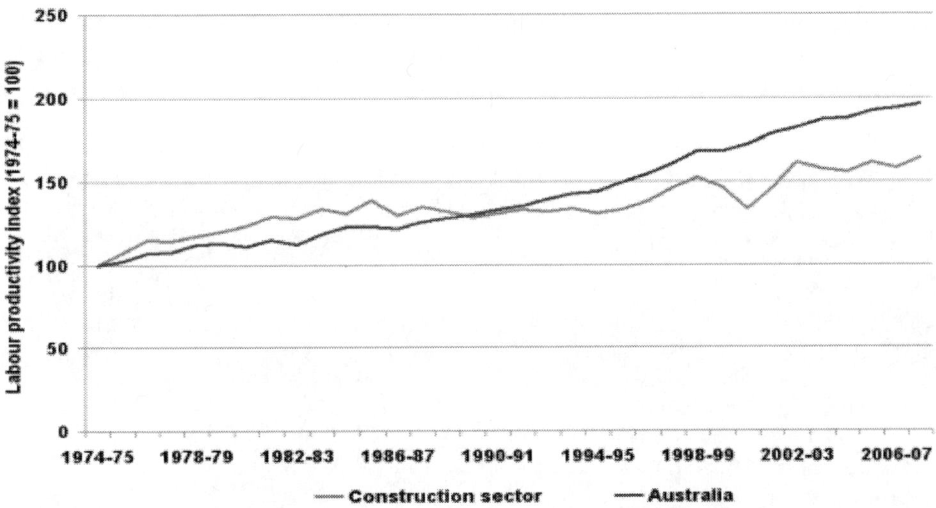

Figure 1. Construction sector labor productivity (The Allen Consulting Group 2010).

The slowing of the industry's growth in comparison to the Australian benchmark raises concerns. Given the size of the construction industry, more effective processes would lead to greater efficiencies.

The large percentage of small businesses has led to specialization of the construction industry and has therefore become more fragmented (Wu 2009). This results in linear, uncoordinated and highly variable project processes (Kagioglou *et al.* 2000). Zraika (2003) agrees and adds that the fragmentation results in construction project delays and reduced quality. Fragmentation arises from two areas within the construction process, the separation between design and construction and the construction structure itself (Mohd Nawi *et al.* 2014). As stated, a long standing problem has been the lack of coordination of the different specialist activities. Team integration or Integrated Project Delivery (IPD) is a single tri party contract between a minimum of the architect, building owner and general contractor that moves towards addressing this issue (Mohd Nawi *et al.* 2014). Other consultants or trades can join the contract but the three main parties are unchanged. IPD creates a system that integrates people, operational process, business practices and organizational systems into a collaborative framework (Furst 2010). Building Information Modelling (BIM) offers a collaborative platform to employ IPD. BIM improves productivity, efficiency and quality and ensures the whole process will become less fragmented, more streamlined and positively address communication throughout the project lifecycle (Chew and Riley 2013, The Allen Consulting Group 2010). BIM technology generates greater construction efficiency and effectiveness through increased collaboration of project participants, thus reducing the industry's fragmentation (Migilinskas *et al.* 2013).

BIM is a process that encompasses all aspects, disciplines and systems of a facility within a single, virtual model. BIM is not a simple three dimensional model, but a process to improve the performance through the whole life cycle of buildings (Lu *et al.* 2013). The model is a database providing digital information about the design, construction, project management, operations, logistics, materials and energy consumption of a building (The Allen Consulting Group 2010). BIM allows for a managed environment so that people, tools and tasks can effectively share this information throughout the building lifecycle (Migilinskas *et al.* 2013). The virtualized world of BIM allows control, information and familiarity before a shovel is put in the ground. The BIM process generates stores, manages exchanges and shares building information through a computer generated model. The model is used to simulate the project lifecycle, including planning, design, construction, operations and demolition as seen in Figure 2.

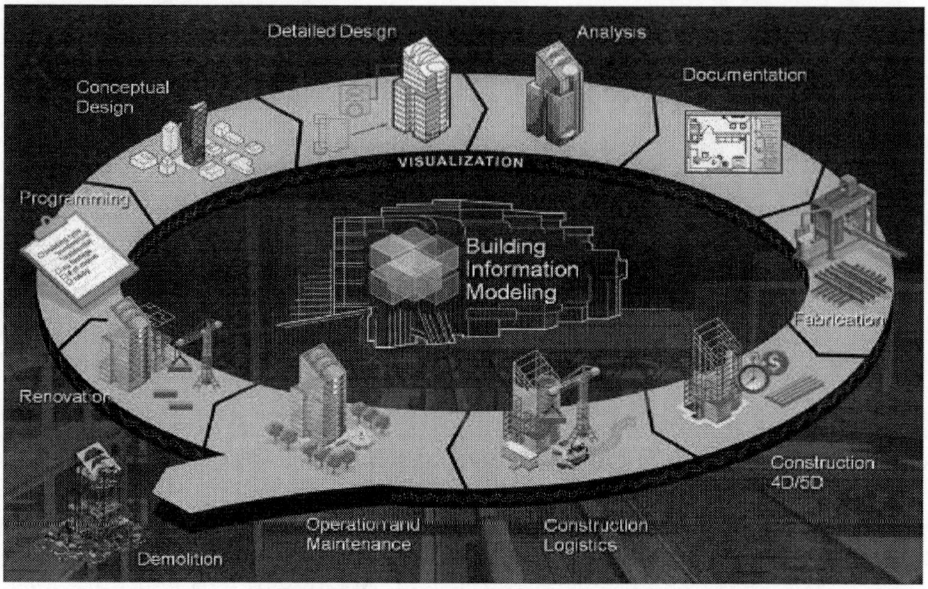

Figure 2. BIM project lifecycle (Miner 2012).

There have been numerous studies that have identified BIM as a tool that increases productivity, efficiency and quality values (Chew and Riley 2013). Therefore, why has it been identified that Australia has such a slow uptake? To help understand the lack of BIM utilization within the Australian construction industry, the research seeks to identify the barriers common to BIM implementation. Through a review of literature, it was identified that interoperability, cost, industry culture and client demand are the major contributors to the lack of implementation (Azhar 2011, Migilinskas *et al.* 2013).

2 RESEARCH METHODOLOGY

The case studies were conducted on five medium and large sized construction companies within Australia, forming a multiple-case design of research, therefore increasing validity through triangulation (Yin 2013). Additionally, the data allowed the comparison between medium and large companies. The selected construction companies, chosen by their characteristics, were contacted by email and follow up telephone call. The email allowed the researcher to describe the purpose of the research, the required areas of interest and asked for the willingness to participate in a case study interview. The collection of data used to gain background information on the specific companies and their current use of BIM was derived from electronic information available on the internet and printed information from construction journals. Observations were conducted during the interview process. This provided an insight into the current use of BIM and identified some common barriers of implementation. Observations were also made when the researcher was able to attend the construction site. Researcher observations proved invaluable due to the first-hand experience that was not filtered by what others may have self-reported. The interview was conducted in a semi-structured format. The purpose of the interview was to identify the factors restricting BIM being utilized as a construction management tool. To assist finding these factors, it was required to identify the level of BIM capability and current BIM utilization for the particular project being assessed. The mixed methodology is used to compare the two research paradigms to determine the BIM capability, utilization of BIM in construction and the factors affecting the implementation of BIM within the Australian construction industry context.

3 FINDINGS

Figure 3 presented below summarize the extent in which BIM was used as a construction management tool for both medium and large sized companies. It is clear that large companies engage BIM for construction management tasks more so than medium companies. It can also be said that large companies meet the criteria of utilizing BIM for all construction management categories. In comparison, medium companies only utilize BIM for visualization and 3D coordination tasks. This information can prove useful for medium sized construction companies to address the shortfall of BIM utilization for scheduling and planning and costing in a bid to fully utilize BIM.

The results of each factor that affects the implementation of BIM in Australian construction projects are shown overleaf. The Figure 4 clearly shows that education was the biggest factor limiting the implementation of BIM as a construction management tool. A mean score of 4.06 is described as being a major impact on implementation. Client demand was found to have the least impact with a mean of 2.7 translating into a minimal to moderate impact. However, it should be noted that all five factors present some sort of impact to the implementation of BIM for construction management purposes.

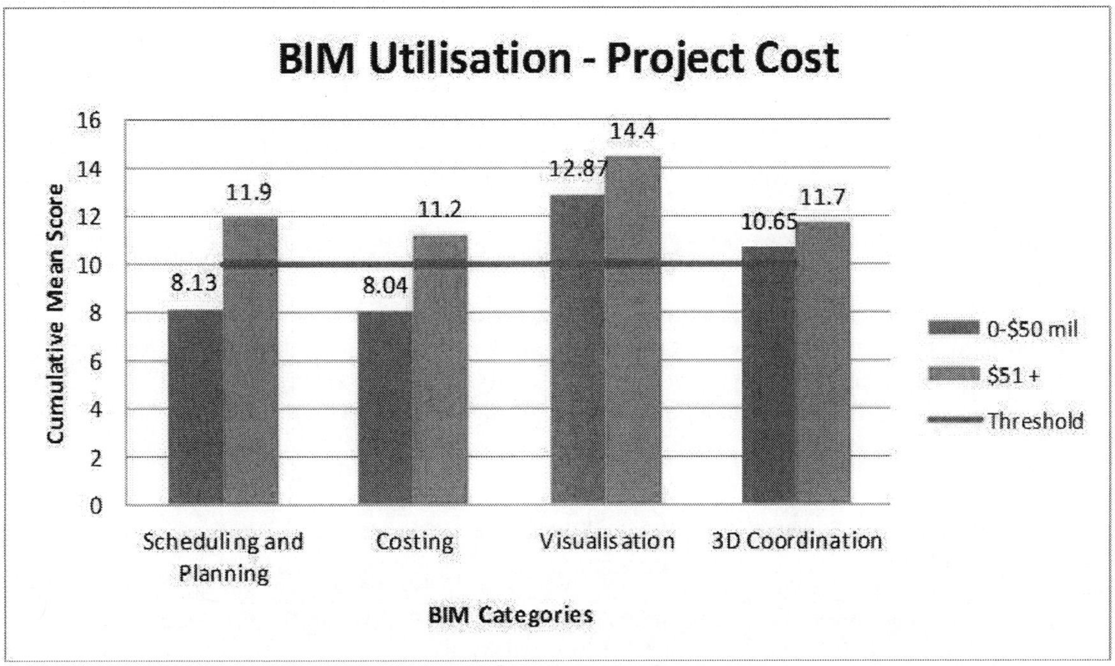

Figure 3. Project Cost 0-$50mil vs $51mil + - BIM utilisation in construction.

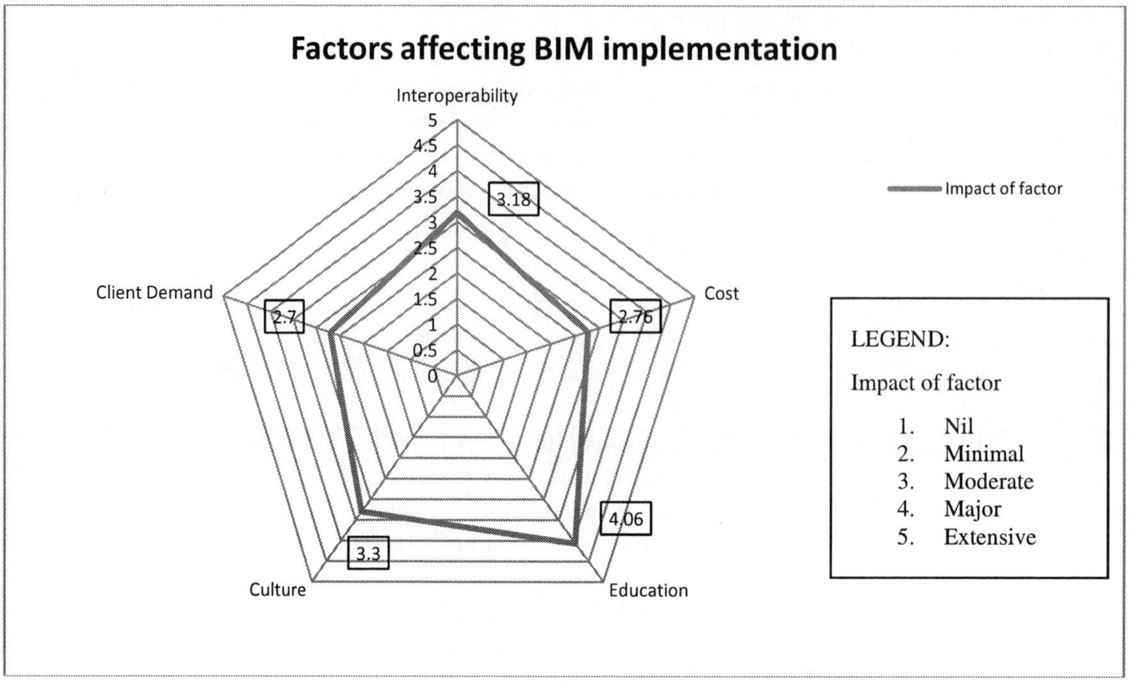

Figure 4. Factors affecting BIM implementation.

4 CONCLUSION

The results demonstrate that the Australian construction industry is capable of utilizing BIM for construction management. The results proved that generally BIM is utilized in visualization and 3D coordination only for construction management purposes. Scheduling and planning together with costing were not adopted by BIM as tools for construction management. However, sub-groups within this data showed a more detailed perspective. Medium sized construction companies were found to only utilize visualization as a BIM construction management tool. Conversely, large companies fully utilized all categories of BIM as construction management tools. Further analysis of sub-groups found that construction projects exceeding $51 million utilized all categories for BIM as a construction management tool. This was in direct contrast with projects under $50 million, as their primary utilization of BIM as a construction management tool was for visualization. The results showed that education presented the most significant factor to the implementation of BIM as a construction management tool for Australian construction companies. However, further analysis of sub groups found that medium sized companies were mostly affected by cost, education and industry culture.

References

Australian Bureau of Statistics 2012, *Private Sector Construction Industry, Australia*, 26 March 2015, Retrieved from <http://www.abs.gov.au/ausstats/abs@.nsf/Latestproducts/8772.0Main%20Features22011-12?opendocument&tabname=Summary&prodno=8772.0&issue=2011-12&num=&view=>.

Azhar, S 2011, 'Building Information Modeling (BIM): Trends, Benefits, Risks and Challenges for the AEC Industry', Leadership and Management in Engineering, vol. 11, no. 3, pp. 241-52.

buildingSMART Australasia 2012, *National Building Information Modelling Initiative. Vol.1, Strategy: A strategy for the focussed adoption of building information modelling and related digital technologies and processes for the Australian built environment sector.*

Chew, A & Riley, M 2013, 'What is going on with BIM? On the way to 6D', *The International Construction Law Review*, pp. 253-65.

Furst, PG 2010, 'Constructing integrated project delivery', *Industrial Management*, vol. 52, no. 4, pp. 19-25

Kagioglou, M, Cooper, R, Aouad, G & Sexton, M 2000, 'Rethinking construction: the Generic Design and Construction Process Protocol', Engineering Construction & Architectural Management, vol. 7, no. 2, pp. 141-53.

Lu, W, Peng, Y, Shen, Q & Li, H 2013, 'Generic Model for Measuring Benefits of BIM as a Learning Tool in Construction Tasks', *Journal of Construction Engineering & Management*, vol. 139, no. 2, pp. 195-203,

Migilinskas, D, Popov, V, Juocevicius, V & Ustinovichius, L 2013, 'The Benefits, Obstacles and Problems of Practical Bim Implementation', *Procedia Engineering*, vol. 57, pp. 767-74,

Miner, M 2012, Building Information Modeling (BIM), viewed 10 October 2014, <http://www.neuralenergy.info/search?q=building+information+modeling>.

Mohd Nawi, MN, Baluch, N & Bahauddin, AY 2014, 'Impact of Fragmentation Issue in Construction Industry: An Overview', *MATEC Web of Conferences*, vol. 15.

The Allen Consulting Group 2010, *Productivity In The Buildings Network: Assessing The Impacts Of Building Information Models, report to the Built Environment Innovation and Industry Council*, Sydney.

Wu, M 2009, 'Modeling of Fragmentation in the Construction Industry ', in *International Conference on Management and Service Science*, Wuhan and Bejing, China.

Yan, L & Chunlu, L 2010, 'Malmquist indices of total factor productivity changes in the Australian construction industry ', *Construction Management and Economics*, vol. 28, no. 9, pp. 933-45,

Yin, R 2013, *Case Study Research: Design and Methods*, 5th edn, Sage Publications, California.

Zraika, M 2003, 'Rework & Quality', Bachelor of Building thesis, University of Western Sydney

Integrated Solutions for Infrastructure Development
Edited by Lau, H. H., Tang, F. E., Ng, C. K., and Singh, A.
Copyright © 2016 ISEC Press

A DECISION SUPPORT SYSTEM FOR METHODS OF MEASUREMENT IN CONSTRUCTION PROJECTS

ABDELRAHMAN MAGDY and SAMER EZELDIN

Dept of Construction Engineering, The American University in Cairo, Egypt

Cost estimation of any construction project depends on issuing the Bill of Quantities which defines and quantifies the work items of any project to be priced. When preparing these bills, a set of guiding rules can be used to ease the preparation which is known as the Method of Measurement. Some institutions issued standard Method of Measurements such as the CESMM, NRM and POMI. Each Method of Measurement has its own characteristics and uses. The use of these standard Methods of Measurement is studied in this paper for the civil and building projects in the Middle East by preparing a questionnaire to be distributed and returned on quantity surveyors in the Middle East. The Results are analyzed and the findings are used to construct a decision support system that automates the process of selection of the suitable standard method of measurement for civil or building projects based on some project parameters. The model was validated by comparing its results with those in real life projects. A conclusion was reached the standard Methods of Measurement are used with POMI having most uses followed by CESMM unlike NRM that has least use so the model can give suitable choice for the Method of Measurement in civil projects unlike building projects since NRM is issued recently and still not used in Middle Eastern projects.

Keywords: BOQ, Standard Methods of Measurement (MOM), CESMM, NRM, POMI.

1 INTRODUCTION

Cost management is an important factor in the success of any construction project. In every project there is a cost estimation at the beginning of the project and cost control throughout the project. Accurate cost estimation facilitates the cost control process so it is important to prepare an accurate Bill of Quantities (BOQ) for every project. It is worth mentioning that the BOQ discrepancies are usually interpreted against the entity who developed the document. Standard Methods of Measurement (MOMs) facilitate the process of preparing the BOQ by providing guidance rules for the description and quantification of the project work items. Each standard MOM has different characteristics and uses, so it is important to investigate the suitability of these methods for the construction projects according to each project parameters. This is to facilitate the cost control of the project and to help avoid potential conflicts that can arise between the project parties.

This paper aims at investigating three different MOMs, namely; Civil Engineering Standard Method of Measurement (CESMM) for civil engineering projects, New Rules of Measurement (NRM) for building works, and Principles of Measurement International (POMI) for most project works. These three MOMs are selected since they are international methods that are made to accommodate the needs of the construction projects worldwide.

These MOMs are compared from the perspective of their purpose, characteristics, advantages and disadvantages. An expert system is developed to act as a decision model that helps to choose the most suitable MOM for a project.

2 RESEARCH METHODOLOGY

In order to be able to choose the most suitable MOM for a construction project, the literature is first explored to obtain findings about each of the three methods investigated which are the CESMM, NRM and POMI. The findings are then used to collect data through a questionnaire distributed on a sample of quantity surveyors in the Middle East. The data is analyzed to help construct a decision model that can automate the process of the MOM selection depending on some project parameters. This model is then verified and validated to check its results and finally reach a conclusion.

2.1 Literature Review

The literature is explored to investigate each of the three MOMs as follows:

2.1.1 Civil engineering standard method of measurement (CESMM)

This MOM was issued by the ICE (Institute of Civil Engineers) in 1976 with its last edition as the fourth edition issued in 2012. It is issued for use in heavy civil projects and it is complex so it is used to prepare detailed BOQs in terms of the items' breakdown and description. It divides the project works into 26 work divisions for the works of civil projects in addition to a provision for simple building works (CESMM3 1991). The advantages of CESMM is that it helps in preparing a detailed BOQ for the civil projects whose works are usually hard to anticipate before the execution of the project. Also these detailed BOQs can help in the evaluation of variations. CESMM also has some additional cost information such as the method related charges and it also continues to update its versions to accommodate for the new construction technologies. However, CESMM has some drawbacks since the method related charges can cause problems in case of variations and also companies need to train their employees on how to use CESMM (Eadie *et al.* 2013).

2.1.2 New rules of measurement (NRM)

This MOM was issued by the RICS (Royal Institute of Chartered Surveyors) with NRM2 as the volume used in BOQ preparation issued in 2012. It is issued for building works and it is used to produce very detailed BOQs since it is broken down into 41 work divisions (NRM2 2012). The advantages of NRM are that it gives accurate project pricing as well as being flexible in variations. However, NRM can be a waste of time in small projects. Also, NRM can increase the tender price of a project if there is double counting when pricing an item. In addition, contractors may reject pricing the detailed items and include their prices on major items. Also, the use of NRM needs a lot of training (Towey 2013).

2.1.3 Principles of measurement international (POMI)

It is a simple MOM issued by RICS Business Services Limited (a wholly owned subsidiary of the RICS) in 1979 and it is suitable for most project types. It produces abstract BOQs since it is divided into 15 work items (POMI 1979). The advantages of the POMI are represented in the idea that it is simple so most quantity surveyors are familiar with using it especially in the Middle East. However, POMI can be considered to ignore some project details which may give false idea about the project scope as well as inaccurate pricing. Also POMI is considered outdated (BCIS Report 2012).

3 DATA COLLECTION

The data collection process is conducted by preparing a questionnaire that was distributed to 140 quantity surveying professionals in the Middle East. The population was categorized into three (3) main categories; clients, consultants and contractors. The target of the questionnaire was to: a) check the recognition of each of the three methods of measurement by Middle Eastern professionals, b) gather information about the characteristics of detailed and brief BOQs, c) understand how each project parameter can affect the choice of the suitable standard MOM for each project, d) rank the project parameters according to their percentage impact on the decision making process, e) investigate which MOM would be preferred for a certain type of project if it is to be constructed in the Middle East.

4 DATA ANALYSIS

The replies received were 45, 15 from clients, 15 from consultants and 15 from contractors forming a percentage response of 32.14%. The findings of the questionnaire came addressing its 5 objectives as follows: a) among the three MOMs investigated, the majority of respondents showed experience with POMI followed by CESMM and then NRM with the least percentage recognition. b) The types of BOQs produced by the three MOMs were divided into two types; Detailed and Brief BOQs. The characteristics of each BOQ type were investigated and it was obvious that detailed BOQs can enhance project pricing since they clarify the exact items that need to be priced. However, detailed BOQs are useless and a waste of time in case of small projects since the work items are few and don't need that detailed breakdown for the project works. Also, detailed BOQs give clear description of the project works, so they are easily understood by contractors decreasing the probability of pricing errors. In addition, they can enhance the quantification of variations since the items affected directly by the variations can easily be identified. However, they can increase the tender price as it increases the possibility of double counting since contractors may double price an item. For Brief BOQs, their characteristics were investigated and it was obvious that since they are abstract, they help in quick issuance of tender documents when time is of essence. Also, brief BOQs neglect necessary details leading to pricing mistakes; however, an opposing opinion disagrees with that point and 70% of the disagreeing responses came from the side of the contractors who claim they are experienced with dealing with the brief BOQs. This may suggest that contractors prefer brief BOQs to help compensate for their losses in a project by using BOQ misinterpretations to claim extra money. In addition, brief BOQs are believed to obstacle the quantification of variations since they use the inclusive approach making it hard to define exactly what items are affected by variations. In addition brief BOQs are believed to cause pricing inaccuracies. c) The project parameters assumed were the nationality of the client and the contractor, project type, contract value and contract type. If the client was international; meaning he belongs to a foreign country other than the country where the project is being executed, then the client may be willing to choose a detailed BOQ to avoid project risks unlike local clients who may be willing to save time as they are more likely to be experienced with the conditions within their country so they may go for brief BOQs and the same trend is applicable for contractors. For projects with high contract value, it is preferred to go for detailed BOQs since the project would be important and it is preferable to avoid project risks that may arise from using brief BOQs. This is unlike the case of projects with small values where brief BOQs might satisfy their purpose. For the contract type, in re-measured or unit price contracts the BOQ has high priority among the contract documents and so going for detailed BOQs might be suitable unlike lump sum contracts where the BOQs are only used in the case of variations and so brief BOQs might be suitable. For the percentage preference according to the project type, it is addressed in the following point e) in the data analysis section. d) For the effectiveness of the project parameters on the decision making process, table 1 shows the ranking of the respondents to the parameters affecting the choice of the MOM for any project to be used in the model.

Table 1. Effectiveness of each project parameter on the choice of the MOM.

Project Parameter	Average Percentage Weight
Project Type	23%
Client's Nationality	26%
Contractor's Nationality	12%
Contract Value	21%
Contract Type	18%

For the choice of the most suitable MOM according to the project types, the projects were classified into two categories; building projects and civil projects. Each category of the two had a list of projects with different types such as residential buildings and office buildings for the building category and as the tunnels and bridges for the category of civil projects. The responses came preferring POMI for building projects over NRM and preferring CESMM over POMI for civil projects.

5 MODEL DEVELOPMENT

The approach followed in constructing the model is the weighted average approach through getting the sum of multiplying the percentage weight by the percentage preference of each BOQ type for the corresponding project parameter. A sample of how the model works is shown in table 2 below:

Table 2. Mechanism of how the model works.

Project Parameter	Percentage Preference		Weight
	Brief	Detailed	
Office Building	68%	32%	23%
Local client	49%	51%	26%
Local contractor	64%	36%	12%
Low contract value	71%	29%	21%
Lump sum contract	75%	25%	18%
Result	65%	35%	

In this example shown in table 2, the preferred BOQ type for this combination of parameters is the Brief BOQ type, so an equation is used to define the basis of selecting the suitable MOM is used and it is explained in the following figure 1 below:

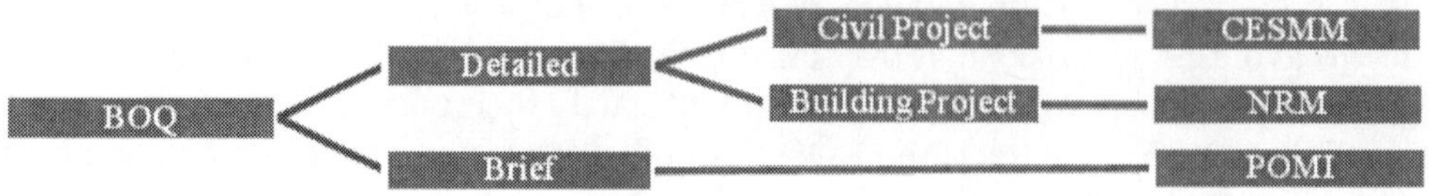

Figure 1. Mechanism of choice for suitable MOM based on the BOQ type chosen by the model.

A user interface for the model was prepared using Visual Basic for Applications as shown in the following figure 2 below:

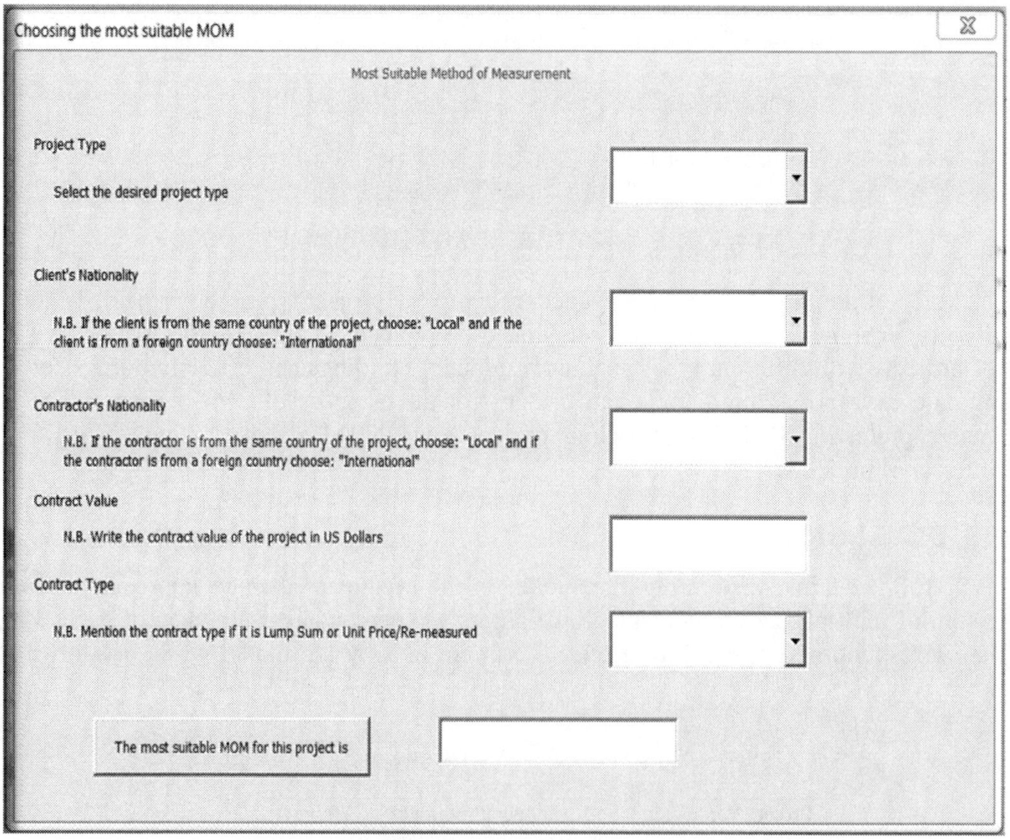

Figure 2. User Interface for the model.

As shown in figure 2, each project parameter to be chosen has some guiding tips that explain on what basis the choice should be made.

6 MODEL VERIFICATION

In order to check that the model implements the assumptions correctly, two verification steps were conducted. The first simple check is to input the extremes of the project parameters that will result in a previously known answer for the standard MOM to be chosen and the model gave the expected results. The second step in the verification was to let a sample of quantity surveyors to use the model and give their feedback. Their feedback was that the model is simple to use and the guiding tips help avoid confusion.

7 MODEL VALIDATION

The model validation is performed to check the convergence of the model's results to the real life projects cases. The model's results are compared with the MOMs chosen actually for the project. Three (3) civil projects were studied and the model gave an output of CESMM as the most suitable MOM which was the same choice in real life. However, (2) two building projects were used to test the model and the model produced NRM as the most suitable choice. This was unlike that chosen in real life since POMI was actually chosen. A justification was that the NRM has been recently issued and still not used in Middle Eastern building projects.

8 CONCLUSIONS

It can be stated that the standard Methods of Measurement investigated are recognized in the Middle East with POMI having the most recognition followed by CESMM while NRM is still not well recognized. Also, the sector of the clients and consultants are more biased towards preparing a detailed BOQ to avoid project risks that can arise from brief BOQs unlike the sector of contractors who consider brief BOQs as a chance to compensate for their losses in a project by using its discrepancies to claim extra money. Moreover, choosing the most suitable standard Method of Measurement depends on the project parameters. Finally, the proposed model produces better results for the civil projects when compared to building projects.

References

CESMM3: Civil Engineering Standard Method of Measurement, Institute of Civil Engineers, London, 1991.

Eadie, R., Millar, P., and Harte, N., A comparison between CESMM3 and MCHW as methods of measurement for Civil Engineering Work, *International Journal of Procurement Management*, Elsevier, 6(5), 523-543, August, 2013.

NRM2: Detailed Measurement for Building Works, Royal Institute of Chartered Surveyors, Reading, 2012.

POMI: Principles of Measurement International, RICS Business Services Limited, London, 1979.

RICS 2011 Principles of Measurement International Survey Report, BCIS Building Cost Information Service, RICS, London, 2011.

Towey, D., *Cost Management of Construction Projects*, John Wiley & Sons, Somerset, 2013.

Integrated Solutions for Infrastructure Development
Edited by Lau, H. H., Tang, F. E., Ng, C. K., and Singh, A.
Copyright © 2016 ISEC Press

MITIGATING CONSTRUCTION PROJECT SCHEDULE OVERRUNS

MULENGA MUKUKA, CLINTON AIGBAVBOA, and WELLINGTON THWALA

Dept of Construction Management & Quantity Surveying, University of Johannesburg, Johannesburg, South Africa

There are many factors that contribute to the causes of schedule overruns in construction projects. This ranges from factors inherent in the technology and its management, to those resulting from the physical, social, and financial environment. Schedule overruns can give rise to disruption of work and loss of productivity, late completion of project, increased time related costs and third party claims and abandonment or termination of contract. Schedule overruns are costly and often result in disputes and claims. Hence, the need to identify mitigation measures of construction projects overruns that will bring about the timely delivery of construction projects. This paper assesses the measures to mitigate against construction projects schedule overruns in the Gauteng Province construction industry in South African. The data used in this paper were derived from both primary and secondary sources. The primary data were collected through a questionnaire distributed to construction professionals in the study area. Data received from the questionnaires were analyzed using descriptive statistics procedures. Findings revealed that proper project planning and scheduling, effective strategic planning, site management and supervision, amongst others, were the major mitigation measures of construction projects schedule overruns in the Gauteng Province of South Africa. This study contributes to the body of knowledge on the subject of the measures of mitigating against construction project schedule overruns in the Gauteng Province construction industry.

Keywords: Schedule overruns, Mitigation, Construction industry, Gauteng, South Africa

1 INTRODUCTION

Aiyetan, Smallwood and Shakantu (2011) show that the principle of Right-First-Time holds great value. Right-first-Time requires accuracy and precision. Accuracy means reflecting the realities (specifications), whereas precision implies meeting the specific dates. The processes of construction demand accuracy and very high precision. However, the problem of projects schedule overruns in the construction industry is a global phenomenon (Sambasivan and Soon 2007). Alkhathami (2004) defines schedule overruns as extra time required to finish a given construction project beyond its original planned duration, whether compensated for or not. Mohamad (2010) says schedule overruns are an act or event that extends the time to complete or perform an act under the contract. Also, Assaf and Al-Hejji (2006) defined schedule overrun as the time overrun either beyond completion date specified in a contract, or beyond the date that the parties agreed upon for delivery of a project. It is basically a project slipping over its planned schedule and is considered as common problem in construction projects worldwide.

Abedi, Fathi, and Mohammad (2011) state that there are many factors that contributed to causes of schedule overruns in construction projects. These range from factors inherent in the technology and its management, to those resulting from the physical, social, and financial environment. Schedule overruns can give rise to disruption of work and loss of productivity, late completion of project, increased time related costs and third party claims and abandonment or termination of contract. Schedule overruns are costly and often result in disputes and claims. Hence, the need to identify mitigation measures of construction projects overruns that will help with timely delivery of construction projects.

2 CONSTRUCTION SCHEDULE OVERRUNS – MITIGATION MEASURES

Wei (2010) illustrates that when construction schedule overrun occurs, there is no question that the owner suffers financially. But the extent to which the owner can recover loss of income from the contractor, and more importantly minimize the risk that such overruns will occur, depends largely on how the construction contract is drawn up. Tabish and Jha (2011) identify five successful criterions that can be used to deliver construction projects within schedule. The study further revealed that there are independent measures that can be taken in each category to reduce schedule overruns. These criteria are as follows: schedule performance criterion, cost performance criterion, quality performance criterion, safety performance criterion and no-dispute performance criterion (Tabish and Jha 2011).

Aiyetan *et al.* (2011) identified twelve factors that would influence project delivery time, these includes: construction planning and control techniques, management style, economic policy, the quality of management during construction, site access conditions, site ground conditions, motivation of workers, constructability of designs, socio and political conditions, client understanding of the design, procurement and construction processes, the quality of management during design, and physical environmental conditions. Saleh, Abdelnaser, and Abdul (2009) reveal that construction schedule overruns can be mitigated by applying the flowing factors on construction projects: making risk management, proper planning, proper payment from client, prepare insurance claims, good scheduling program, client representative for project, selecting expert understand their assignment, clear contract and BOQ and compute the amount of financial damages. Wei (2010) identify fifteen ways of mitigating schedule overruns and ranks them as follows: site management and supervision, effective strategic planning, clear information and communication channels, collaborative working in construction, proper project planning and scheduling, frequent coordination between the parties involved, complete and proper design at the right time, use appropriate construction methods, accurate initial cost estimates, proper material procurement, proper emphasis on past experience, frequent progress meeting, compressing construction durations, use proper and modern construction equipment and use up-to-date technology utilization. These measures of mitigating construction schedule overruns were in agreement with the study by Majid (2006).

Abedi, Fathi, and Mohammad (2011) identified thirty measures that can be taken to minimize construction schedule overruns in Malaysia: accurate initial cost estimates, adopting a new approach to contract award procedure by giving less weight to prices and more weight to the capabilities and past performance of contractors, perform a preconstruction planning of project tasks and resource needs, selection of a competent consultants and reliable contractors to carry out the work, allocation of sufficient time and money at the design phase, availability of adequate resources, commitment to projects, competent project managers, comprehensive contract documentation, ensure adequate and available source of finance until project completion, frequent progress meetings and enforcing liquidated damage clauses. Furthermore, Ade-ojo and Babalola (2013) identify six measures that can reduce schedule overruns on construction projects, these include: accurate design, prompt payment, financial capability of client, financial capability of contractor, favorable site condition, favorable weather and availability of subcontractor and supplier. Jonathan (2013) states that effective project planning, adequate funding to finance project to completion and ensuring that project parties play their role on project execution are the most effective ways of reducing schedule overruns on construction projects in Nigeria. However, Aibinu and Jagboro (2002) state that acceleration of site activities and contingency allowance would reduce schedule overruns on construction projects in Nigeria.

3 RESEARCH METHODOLOGY

The data used in this paper were derived from both primary and secondary sources. The primary data was obtained through the survey method, while the secondary data was derived from the review of literature and archival records. The primary data was obtained through the use of a structured questionnaire survey. This was distributed to a total of 200 construction professionals that included; Architects, quantity surveyors, civil engineers, construction mangers and project managers who are currently involved in construction works in Gauteng, South Africa. This yardstick was considered vital for the survey in order to have a true reflection of the effects of construction project schedule overruns. All professionals in Gauteng province had an equal chance to be drawn and participate in the survey. Out of the 200 questionnaires sent out, 146 were received back representing a 73% response rate. This was considered adequate for the analysis based on the assertion by Moser and Kalton (1971) that the result of a survey could be considered as biased and of little value if the return rate was lower than 30–40%. The data presentation and analysis made use of frequency distributions and percentages of all the respondents.

3.1 Mean Item Score (MIS)

A five point Likert scale was used to determine the causes of construction project cost overruns in Gauteng province with regards to the identified factors from the reviewed literature. The adopted scale was as follows: 1 = Extremely unlikely; 2 = Unlikely; 3 = Neutral; 4 = likely; 5 = Extremely likely.

The five-point scale was transformed to mean item score (MIS) for each of the factors of causes of cost overruns as assessed by the respondents. The indices were then used to determine the rank of each item. The ranking made it possible to cross compare the relative importance of the items as perceived by the respondents. This method was used to analyze the data collected from the questionnaires survey. The mean item score (MIS) was calculated for each item as follows:

$$MIS = \frac{1n1 + 2n2 + 3n3 + 4n4 + 5n5}{\sum N} \quad (1)$$

Where: n1 = Number of respondents for extremely unlikely; n2 = Number of respondents for unlikely; n3 = Number of respondents for neutral; n4 = Number of respondents for likely; n5 = Number of respondents for extremely likely; N = Total number of respondents.

After mathematical computations, the factors were then ranked in descending order of their mean item score (from the highest to the lowest).

An internal consistence check was conducted using the consistence and reliability statistic measure of the Cronbach's Alpha. According to Tavakol and Dennick (2011) the Cronbach's alpha measures the internal consistency of a test or scale, it describes the extent to which all the items in a test measure the same concept or construct and hence it is connected to the inter-relatedness of the items within the test. The current study adopted the Cronbach's Alpha to check internal consistence, the results of the internal consistence for measure of mitigating construction projects schedule overruns was found to be 0.929. Gliem and Gliem (2003) state that an alpha of

0.8 is probably a reasonable goal. George and Mallery (2003) further note that the value of the Cronbach's Alpha above 0.7 is acceptable. Hence the current study proceeded with analysis as the internal consistence tests revealed that the Cronbach's Alpha was within the acceptable values, hence all items in the study were measuring the same concept and all the items were inter-related.

4 FINDINGS AND DISCUSSION

Findings from the 146 usable questionnaires revealed that 59.6% of the respondents were male and 40. 4% were female. Findings relating to the respondents' age group revealed that 29.5% of the respondents were in the age group of 20-25 years old, 26.7% of the respondents were in the age group 26-30 years old, 17.1% were in the age group 31-35 years old, 11. 6% were in the age group 36-40 years old, 6.2% of the respondents were in the age group 41-45 years old, 4.1% were in the age group 46-50 years old, 3.4% of the respondents were above 50 years old and 1. 4% of the respondents were between 51-55 years old. Further, results showed that 76.7% of the respondents were Black, 14.4% were White, 4.8% were either Indian or Asian and 4.1% of the respondents were Colored. Findings relating to respondent's work professional qualification, results showed that 38.4% were quantity surveyors, 20.5% were civil engineers, 11% were project managers, 10.3% were construction managers, 8.9% were construction project managers, 5.5% were architects and 5.5% selected others, which included an artisan, a building inspector, electrical contractors, a safety consultant and a site agent.

Furthermore, results revealed that 56.8% of the respondents had experience that ranged from 1-5 years, 20.5% had experience in the range of 6-10 years, 11.6% had experience that ranged between 11-15 years, 6.2% had more than 20 years' experience and 4.8% had experience that ranged from 16-20 years in the construction industry. Further, 46.6% of the respondents had bachelor's degrees, 32.2% had diplomas, 12.3% of the respondents had masters degrees, 4.1% of the respondents had doctorate degrees and 4.8% of the respondents only had metric (grade 12) certificates. Furthermore, it was revealed that 35.6% of the respondents were employees of contractors, 34.9% of the respondents were employees of consultants and 20.5% were government employee, 8.2% were employed by clients and 0.7% of the respondents were self-employed.

4.1 Measures of mitigating construction projects schedule overruns

The respondents were asked to ranks the measures that can be undertaken to mitigate construction projects schedule overruns. The survey revealed the following factors as the top ten measure of mitigating construction projects schedule overruns in Gauteng: proper project planning and scheduling (SD=0.686, $\overline{x}$=4.57, R=1), effective strategic planning (SD=0.678, $\overline{x}$=4.53, R=2), site management and supervision (SD=0.738, $\overline{x}$=4.48, R=3), frequent coordination between the construction team (SD=0.677, $\overline{x}$=4.44, R=4), availability of clear information and communication channels (SD=0.734, $\overline{x}$=4.42, R=5), complete and proper design at the right time (SD=0.732, $\overline{x}$=4.39, R=6), frequent progress meeting (SD=0.794, $\overline{x}$=4.39, R=7), adherence to construction specifications (SD=0.727, $\overline{x}$=4.37, R=8), building according to the construction drawings (SD=0.800, 4.37, R=9) and collaboration working by the entire construction team (SD=0.799, $\overline{x}$=4.36, R=10). These findings were similar to the findings of Jonathan (20013) where effective project planning was identified as the most effective way of minimizing construction projects schedule overruns. However, the study by Wei (2010) identified site management and supervision as the most effective measure of minimizing construction projects schedule overruns.

Table 1. Mitigating measures for construction projects schedule overruns.

Mitigating measures	σX	$\bar{x}$	R
Proper project planning and scheduling	0.686	4.57	1
Effective strategic planning	0.678	4.53	2
Site management and supervision	0.738	4.48	3
Frequent coordination between the construction team	0.677	4.44	4
Availability of clear information and communication channels	0.734	4.42	5
Complete and proper design at the right time	0.732	4.39	6
Frequent progress meeting	0.794	4.39	6
Adherence to construction specifications	0.727	4.37	7
Building according to the construction drawings	0.800	4.37	7
Collaboration working by the entire construction team	0.799	4.36	8
Proper material procurement	0.720	4.29	9
Appropriate construction methods	0.847	4.24	10
Accurate initial cost estimates	0.847	4.19	11
Use proper and modern construction equipment	0.980	3.94	12
Use up-to-date technology utilization	0.951	3.90	13
Fast-tracking construction	1.028	3.83	14

5 CONCLUSION AND RECOMMENDATION

Literature revealed that there are various measures of mitigating construction projects schedule overruns. These include: site management and supervision, effective strategic planning, clear information and communication channels, collaborative working in construction, proper project planning and scheduling were identified as the major measures of mitigating construction projects schedule overruns from literature. From the questionnaire survey obtained from the respondents, it was revealed that proper project planning and scheduling, effective strategic planning, site management and supervision, frequent coordination between the construction team, availability of clear information and communication channels, complete and proper design at the right time, frequent progress meeting, adherence to construction specifications, building according to the construction drawings and collaboration working by the entire construction team were the top ten identified measures of mitigating construction projects cost overruns in Gauteng, South Africa. It is therefore recommended that all members of construction teams should practice the identified measure of mitigating construction schedule overruns.

References

Abedi, M., Fathi, M. S., and Mohammad, M. F., Major mitigation measures for delays in construction projects. The First Iranian Students Scientific Conference in Malaysia: 9 and 10 Apr 2011: UPM, Malaysia, 2011.

Ade-Ojo, C. O. and Babalola, A. A., Cost and Time performance of construction projects under the due process reform in Nigeria. Research Inventy: International Journal of Engineering and Science, 3(6): 1-6, 2013.

Aibinu, A. A. and Jagboro, G. O., The effects of construction delays on project delivery in Nigerian construction industry: International Journal of Project Management, 20 (2002): 593–599, 2002.

Aiyetan, A., Smallwood, J., and Shakantu, W., A systems thinking approach to eliminate delays on building construction projects in South Africa. Acta Structilia, 2011: 18(2), 2011.

Alkhathami, M. M., Examination of the correlation of critical success and delay Factors in construction projects in the kingdom of Saudi Arabia. Doctor of Philosophy Thesis: University of Pittsburgh, 2004.

Assaf, S. A. and Al-Hejji, S., Causes of delay in large construction projects. International Journal of Project Management, 24: 349–357, 2006.

George, D., and Mallery, P., SPSS for Windows step by step. A simple guide and reference. 11.0 update (4th ed.), Boston: Allyn and Bacon, 2003.

Gliem, J. A. and Gliem, R. R., Calculating, Interpreting, and Reporting Cronbach's Alpha Reliability Coefficient for Likert-Type Scales. Midwest Research to Practice Conference in Adult, Continuing, and Community Education, 2003.

Jonathan, I., Effects of delay on building project delivery. Department of Building Technology: Auchi Polytechnic, Auchi, 2013.

Majid, L. A., Causes and Effects of Delays in Aceh Construction Industry. Master of Science in Construction Management. Universiti Teknologi: Malaysia, 2006.

Mohamad, M. R. B., The factors and effect of delay in government Construction project, Case study in kuantan. Bachelor's degree thesis: University Malaysia Pahang, 2010.

Moser, C. A. and Kalton, G., Survey methods in social investigation. Heinemann Educational: UK, 1971.

Saleh, A. H. T., Abdelnaser, O., and Abdul, H. K. P., Causes of delay in construction industry in Libya. The International Conference on Economics and Administration, Faculty of Administration and Business, University of Bucharest, Romania: ICEA – FAA Bucharest, 14-15th November 2009, 2009.

Sambasivan, M. and Soon, Y. W., Causes and effects of delays in Malaysian construction industry. International Journal of Project Management, 25: 517–526, 2007.

Tabish, S. Z. S. and Jha, K. N., Important factors for success of public construction projects. 2011 2nd International Conference on Construction and Project Management: vol.15 (2011) IACSIT Press, Singapore, 2011.

Tavakol, M. and Dennick, R., Making sense of Cronbach's alpha. International Journal of Medical Education, 2011, 2:53-55: ISSN: 2042-6372, 2011.

Wei, S. K., Causes, effects and methods of minimizing delays in construction projects. Bachelor's degree thesis: Universiti Teknologi Malaysia, 2010.

Integrated Solutions for Infrastructure Development
Edited by Lau, H. H., Tang, F. E., Ng, C. K., and Singh, A.
Copyright © 2016 ISEC Press

AN ASSESSMENT OF CONSTRUCTION PROJECTS COST OVERRUNS IN THE PUBLIC SECTOR

JERMINAH KHABISI, CLINTON AIGBAVBOA, and WELLINGTON THWALA

Dept of Construction Management & Quantity Surveying, University of Johannesburg, Johannesburg, South Africa

The purpose of this study was to identify factors that influence project cost overrun within public sector construction projects with specific reference to Gauteng Province of South Africa. The primary data was collected by means of structured questionnaires which were distributed to practicing construction professionals who are engaged in various dimension in public sector construction projects. The secondary data was derived through reviewed literature. Out of 120 questionnaires sent out, 119 were received, which represented 99% response rate. Data received from the questionnaires was analyzed using descriptive statistical procedures. Findings from questionnaire survey revealed that the most dominant factors of cost overruns in public sector project include: variation orders, change in scope of the project, cash flow and financial difficulties faced by contractors, delays in decision making by professional consultants representing the clients, inadequate planning, frequent design changes, lack of coordination between parties, amongst others. The study thus revealed that it is necessary to identify factors that influence construction cost overruns at the start of the project in order to minimize cost overruns and to improve the cost performance on public sector construction projects. This is because of the high level of accountability necessary required for the management of tax payers money.

Keywords: Schedule overruns, Mitigation, Construction industry, Gauteng, South Africa.

1 INTRODUCTION

Cost overrun occurs when the contract sum at project completion is greater than what was originally budgeted at the inception stage. However, there are fluctuations in costs during the life of the project. Avots (1983) shows the significance of producing a thorough estimate of the project during the planning stage to avoid any additional work which may lead to cost overruns. It is important to identify cost-effective measures to avoid cost overruns (Rahman *et al.* 2012). Various projects encounter cost overruns and construction projects are no exception (Baloi and Prince 2003). Cost overruns are a main challenge in different countries (Angelo and Reina 2002). In developing countries, the challenges are worse when the cost overruns are twice as much as the initial budget (Durdyev *et al.* 2012). Cost overruns have negative impact on all parties involved in the project for example the client loses trust on the design team and the design team earns bad reputation (Park and Papadopoulou 2012). Is it significant to understand cost overruns so that they can be avoided in the future (Angelo and Reina 2002). Hence the aim of this paper was to ascertain the factors that have an impact the project cost overrun within public sector construction projects in South African.

2 CAUSES OF COST OVERRUNS

Generally, the causes that influence project cost performance and cause cost overruns exist from the initial phase to the final project phase (Baloi and Price 2003). Winch (2010) identified common causes of cost overruns as follows: main organization's strategic priorities not being aligned with the project, including established successful procedures; non-existent top management and ministerial ownership and governance; no-existent effective stakeholders' involvement; non-existent expertise and established methodology to project management and risk management; assessment of proposals determined by preliminary estimates rather than long-term value for money; a lack of understanding of, and agreement with the supply industry at top management in the organization; and a lack of effective project team integration between clients, the supplier team and the supply chain.

According to Memon *et al.* (2010) the most severe factors that affect cost overruns in Malaysia include: contractor's cash flow problems, contractors' poor site management and supervision, inexperienced contractors, lack of site workers, inappropriate planning and scheduling by contractors while project scope changes and frequent design changes are the least factors affecting construction cost. Tat-fai (1999) identified major causes of cost overruns at the post-contract stages in mega infrastructure projects in Hong Kong as the following: design changes, scope changes, financial claims, poor client's project management, and unforeseen site conditions, poor interface between contracts, inaccurate pre-tender estimates, and late instructions. Fugar and Agyakwah-Baah (2010) revealed the causes of cost overruns in Ghana as the following: delay in honoring certificates, underestimation of the costs of projects, underestimation of the complexity of projects, difficulty in accessing bank credit, poor supervision, underestimation of time for completion of projects by contractors, shortage of materials, poor professional management, fluctuation of prices or rising cost of materials and poor site management. According to Mukuka *et al.* (2014) the causes of cost overruns in South Africa include: contractors' project inexperience, poor project management, inadequate planning, contractors' inefficiency, and inadequate financial provision, a shortage of skilled site workers, poor workmanship, inaccurate estimates, project complexity and site conflicts. These findings should encourage substantial pre-tender design completion as far as possible in order to alleviate the relatively high likelihood of subsequent post-contract cost overruns.

3 RESEARCH METHODOLOGY

The data used for this study was derived from both primary and secondary sources. A structured data collection approach was used to collect primary data, while the secondary data derived from the review of related literature. The primary data was collected by means of structured questionnaire survey. This was distributed to 120 construction professionals such as architects, contractors, project managers, construction project managers, construction managers, quantity surveyors and other professionals who worked on public sector construction projects in Gauteng Province in South Africa. This was a yardstick considered significant for the study in order to obtain a clear reflection of the causes and effects of cost overruns, measures to mitigate their impact and the critical success factors for public sector construction projects in Gauteng Province, South Africa. From the 120 questionnaires distributed to respondents, 119 participated in the main survey, which equals a 99% response rate. The questionnaires consisted mostly of closed-ended questions and a few open-ended questions. The data presentation and analysis made use of frequency distributions and percentages of all the respondents.

3.1 Mean Item Score (MIS)

A five point Likert scale was used to determine the causes and effects of cost overruns in Public Sector construction projects in South Africa with specific reference to Gauteng Province with regard to the factors identified from the related reviewed literature. The adopted scale for causes of cost overruns in public sector construction project was: 1 = Extremely unlikely (EU); 2 = Unlikely (U); 3 = Neutral (N); 4 = Likely (L) and 5 = Extremely likely (EL).

The five-point scale was transformed to a mean item score (MIS) for each of the factors of causes and effects as assessed by the respondents. The indices were then used to determine the rank of each item. These rankings made it possible to cross-compare the relative importance of the items as perceived by the respondents. These rankings made it possible to cross-compare the relative importance of the items as perceived by the respondents. Data collected from the questionnaires survey was analyzed by the use of this method. The mean item score (MIS) was derived from the following formula:

$$MIS = \frac{1n1 + 2n2 + 3n3 + 4n4 + 5n5}{\Sigma N} \quad (1)$$

Where n1= number of respondents for 'extremely unlikely'; n2= number of respondents for 'unlikely'; n3=number of respondents for 'neutral'; n4= number of respondents for 'likely'; n5=number of respondents for 'extremely likely' and N= Total number of respondents.

Following mathematical calculations, the criteria are then ranked in descending order of their relative importance index (from the highest to the lowest).

4 FINDINGS AND DISCUSSIONS

Findings from 119 respondents revealed that 61% of respondents were male while 39% of the respondents were female. In relation to respondents' age group revealed that 8.4% were 25 years or younger, 15.1% were between 26 and 30 years of age, 26.9% were between 30 and 35 years, 30.3% were between 36 and 40 years, 13.4% were between 41 and 45 years, 3.4% were between 46 and 50 years and 1.7% were 56 years or older. Findings further revealed that 64.7% of the respondents were black, 17.6% were white, 2.5% were colored and 14.3% were Indian or Asian. In relation to their position in the company, findings revealed that 12.6% of the respondents were architects, 42% were quantity surveyors, 10.1% were project managers, 0.8% were construction managers, 14.3% were construction project managers, 10.9% were civil engineers and 8.4% held other positions in the company. These included architects/construction project managers working in rural areas, candidate architectural technologists, chief quantity surveyors at the National Department of Public Works, commercial managers, electrical technicians, junior quantity surveyors, lecturers, mining engineers, professional quantity surveyors, program managers, project supervisor, at executive level, quantity surveyors, senior quantity surveyors, specialist trainers New Engineering Contracts (NEC) conditions and lastly, students.

Regarding years of experience, findings further indicated that 5% had less than one year; 16.8% had between one and five years; 30.3% had between five and ten years; 35.3% had between 10 and 15 years, 8.4% had between 15 and 20 years and 3.4% had more than 20 years of working experience. The results of the level of education revealed that 0.8% had a post-matric certificate, 14.3% had a post- matric diploma, 65.5% had a bachelor's degree and 17.6% had a master's degree. Furthermore, in relation to work entity, findings revealed that 51.3% worked for public sector clients, 10.1% worked for private sector clients, 21.8% worked for consultant quantity surveying firms, 4.2% worked for consultant project management firms, 1.7% worked for consultant mechanical and electrical engineering firms, 2.5% worked for consultant civil engineering firms, 2.5% worked for main contractors and 5% selected other entities such as consultant architects, Development Bank of Southern Africa (DBSA) architects, Department of Public works (DPW), facilities management firms, multi-disciplinary consultants, private companies and public sector implementing agents for other public sector departments.

Furthermore, findings related to respondents' company size revealed that 13.4% worked for a very small company, 19.3% worked for a small company, 15.1% worked for a medium-sized company and 49.6% worked for a large company.

4.1 Causes of Cost Overruns in Public Sector Construction Projects.

Based on the ranking (R) using the calculated standard deviation (SD) and the mean item score (MIS) for the listed causes of cost overruns in public sector construction projects in South Africa as shown in Table 1, findings revealed the most dominant factor as: variation orders, followed by change in scope of the project and cash flow and financial difficulties faced by contractors, delays in decision making, inadequate planning, frequent design changes, lack of coordination between parties, policy in accepting lowest tender, inaccurate time and cost estimates, errors and omissions in design, inaccurate quantity take-off and contractors' project inexperience. The findings from the current study were similar to the findings by Kadir *et al.* (2005), where change in project scope, cash flow and financial difficulties faced by contractors and inexperienced contractors were revealed as the major causes of cost overruns in construction projects. The findings from the current study were further in agreement with to the findings by Mukuka *et al.* (2014) and Memon *et al.* (2010), where inadequate planning and inexperienced contractors were identified as the major factors of cost overruns. However, these results did not agree with those of the study by Frimpong *et al.* (2003), Fugar and Agyakwah-Baah (2010) and Tamakloe (2011) where monthly payment difficulties from agencies, poor contract management, and material procurement were revealed as the major causes of cost overruns in construction projects. Further the results of the current study were not in agreement with the study by Baloyi and Bekker (2011) where material cost and price fluctuations, shortage of skilled labor and the client's late contract award were identified as the major causes of construction cost overruns. Causes of cost overruns share common characteristics in projects worldwide, and they are also affected by country-specific conditions (Olawale and Sun 2010).

5 CONCLUSION

The reviewed literature revealed the following causes of cost overruns in construction projects: scope change, contractor's cash flow problems, contractual claims, design changes, increase in material cost, inaccurate material estimates, inaccurate financial provision, inaccurate planning, inexperienced contractors and fraudulent practices.

Findings from the survey outcomes indicated the following as the major causes of cost overruns in public sector construction projects in South Africa: variation orders, change in scope of the project and cash flow and financial difficulties faced by contractors, delays in decision making, inadequate planning, frequent design changes, lack of coordination between parties, policy in accepting lowest tender, inaccurate time and cost estimates, errors and omissions in design, inaccurate quantity take-off and contractors' project inexperience. It is therefore recommended that all team members must understand and be strongly committed and dedicated to achieve, maintain and fulfill project goals. Additionally, the plan should be prepared with as much detail as possible, including during the design process and throughout the project development.

Table 1. Causes of cost overruns in public sector construction projects.

CAUSES OF COST OVERRUNS	σX	MIS	R
Variation orders	1.070	4.26	1
Change in scope of the project	0.844	4.24	2
Cash flow and financial difficulties faced by contractors	1.000	4.24	2
Delays in decision making	0.881	4.18	3
Inadequate planning	0.950	4.05	4
Frequent design changes	1.070	4.03	5
Lack of coordination between parties	0.963	4.02	6
Policy in accepting lowest tender	1.059	4.01	7
Inaccurate time and cost estimates	0.991	3.98	8
Errors and omissions in design	1.042	3.97	9
Inaccurate quantity take-off	1.028	3.95	10
Contractors' project inexperience	1.007	3.95	10
Extension of time claims	1.039	3.92	11
Change in material specification and type	1.006	3.88	12
Poor contract management	0.943	3.87	13
Delay in progress payment by the owner	1.126	3.85	14
Poor project management	0.979	3.82	15
Fluctuations in material prices	0.899	3.82	15
Poor workmanship	1.013	3.80	16
Lack of communication between parties	0.972	3.79	17
Mistakes and discrepancies in contract document	0.973	3.75	18
Late delivery of materials	1.060	3.72	19
Incomplete design at the time of tender	1.181	3.72	19
Incompetent sub-contractors	1.047	3.71	20
Mistakes during construction	1.044	3.70	21
Slow information flow between parties	1.009	3.70	21
Unforeseen ground conditions	1.033	3.69	22
Delay in payment to supplier/sub-contractor	1.127	3.68	23
Delay in preparation and approval of drawings	1.027	3.68	23
Poor site supervision	1.017	3.68	23
Strikes	1.091	3.67	24
Poor site management	1.057	3.65	25
Financial difficulty by owner	1.267	3.60	26
Fraudulent practices (kickbacks, corruption)	1.384	3.59	27
Labor disputes	1.004	3.55	28
Inadequate funding of the project	1.179	3.48	29
Schedule delay	0.990	3.46	30
Delay in inspection and approval of completed works	1.128	3.43	31
Site conflicts	0.998	3.43	31
Lack of executive capacity by employer	1.184	3.41	32
Negative effects of weather	1.130	3.40	33
High cost of labor	1.126	3.37	34
Shortage of site workers	1.172	3.32	35
Lack of experience of technical consultants	1.117	3.34	36

σX = Standard deviation, MIS = Mean item score, R = Rank

References

Angelo, W. and Reina, P., Mega Projects Need More Study Upfront to Avoid Cost Overruns, 249(3), 2002.

Avots, I., Cost-Relevance Analysis for Overrun Control. International Journalof Project Management, 1(3): 142-148, 1983.

Baloi, D. and Prince, A., Modelling Global Risk Factors Affecting Construction. Cost Performance, International Journal of Project Management, Volume 21: 261-269, 2003.

Durdyey, S., Ismail, S. and Bakar, N., Factors Causing Cost Overruns in Construction of Residential Projects: Case Study of Turkey. International Journal of Science and Management, 1(1): 3-12, 2012.

Frimpong, Y., Oluwoye, J., and Crawford, L., Causes of Delay and Cost Overruns in Construction of Groundwater Projects in Developing Countries: Ghana as a Case Study). International Journal of Project Management, Volume 21: 321-326, 2003.

Fugar, F. and Agyyakwah-Baah, A., Delays in Building Construction Projects in Ghana. Australasian Journal of Construction Economics and Building, 10(1/2): 103-116, 2010.

Kadir, A., Lee, W. P., Jaafar, M. S., Sapuan, S. M., and Ali, A. A. A., Factors Affecting Construction Labour Productivity for Malaysian Residential Projects. 23(1): 42-54, 2006.

Memon, A., Rahman, I., Abdullah, M., and Azis, A. Factors Affecting Cost in MARA Projects: Perspective of Project Management Consultants. International Journal of Sustainable Construction Engineering and Technology, 1(2): 41-54, 2010.

Mukuka, M., Aigbavboa, C., and Thwala, W., 2014. ppml.url.tw. Retrieved from http://www.ppml.url.tw/EPPM/conferences/2014/download/Construction%20Experts%E2%80%99%20Perception%20on%20the%20Causes%20and%20Effects%20of%20Cost%20Overruns%20in%20Johannesburg,%20Gauteng%20Province,%20South%20Africa.pdf on 20th November 2015.

Olawale, Y. and Sun, M. Cost and Time Control of Construction Projects: Inhibiting factors and Mitigating Measures in Practice. Management and Economics, 28(5): 509-526, 2010.

Park, Y. and Papadopoulou, T., Causes of Cost Overruns in Transport Infrastructure Projects in Asia: Their Significance and Relationship with Project Size. Built Environment Project and Asset Management, 2(2): 195-216, 2012.

Rahman, I. A., Aziz, A. A. A., Memon, A. H., Nagapan, S, and Latif, Q. B., Time and Cost Performance in Construction Projects in Southern and Central Regions of Peninsular Malaysia. International Journal of Advances in Applied Sciences, 1(1): 46-51, 2012.

Tamakloe, Y., Assessment of Cost and Time Impacts of Public Sector Construction Projects in Ghana, MSc Thesis Kwane Nkrumah University of Science and Technology, 2011.

Tat-fai, M. 1999, theses.lib.polyu.edu.hk., Retrieved from http://theses.lib.polyu.edu.hk/handle/200/1358 on 18 July 2014..

Winch, G., Managing Construction projects: An Information Processing approach. UK: Blackwell Publishing Ltd, 2010.

Integrated Solutions for Infrastructure Development
Edited by Lau, H. H., Tang, F. E., Ng, C. K., and Singh, A.
Copyright © 2016 ISEC Press

ECONOMIC AND FINANCIAL DEVELOPMENT OF SOLAR PHOTOVOLTAIC TECHNOLOGY

HERAWATI ZETHA RAHMAN, PERDANA MIRAJ,
and JADE SJAFRECIA PETROCEANY

Civil Engineering Dept, Universitas Pancasila, Jakarta, Indonesia

Electricity is playing important role to give access of the people doing activities. However, with unequal distribution of electricity between western and eastern region part of Indonesia, many people in eastern region such as Tomia island in Southeastern Sulawesi still suffering about limited access to electricity for daily life. This research aims to produce economic and financial feasibility about solar photovoltaic which expected to reduce non – renewable energy usage, to improve global climate and also to increase people welfare. It will use a simulation from economic and financial approach to produce targeted output. The result shows the best scenario for Solar PV installment is by using hybrid system which integrates current fuel-based generator and solar photovoltaic. It will be placed in four rural locations: Kulati; Dete; Lamanggau; and Kahianga-Wawotimu that will be integrated in one area. The number of beneficiaries is about 987 houses. The result also shows a maximum available internal rate of return with positive benefit cost ratio and lower cost of electricity.

Keywords: Benefit cost ratio, Electricity, Feasibility, Internal rate of return, Rural areas.

1 INTRODUCTION

In Indonesia, electricity consider as one of major contributor for nation's GDP whereby its average growth reach 7.5% per year. It is considered stable and forecasted to increase in greater level in the future. However, most of electricity resources are provided from fossil energy such as coal and petroleum (Narayan 2007). Fossil energy contributes for about 66.1% and the rest 4.87% is supported from renewable energy (exclude hydro energy) (Ministry of Energy and Mineral Resources 2014a). Currently, electricity in Indonesia mostly generated and distributed from PT. PLN (Persero) – a State Owned Enterprise (SOE). From 44,124 MW of total power plant energy in the country, 32.10% is produced by PLN while the rest are produced from Independent Power Producer (IPP) with 10.29% and 1.73% from Private Power Utility (PPU).

There were also fact about 19.85% of Indonesian people still unable to properly access the electricity (Ministry of Energy and Mineral Resources 2014b). In this matter, government attempts to reduce the gaps of electricity access by initiating the development of 35,000 MW of power plant to increase the electrification ratio. However, both electricity development and accessibilty were poorly experienced in Eastern Region of Indonesia (Smith 2013). Thus align with a number of power plant development will be concentrated in Sumatera and Java Islands (84.1%) as western region of Indonesia.

It is a different story for people at Tomia Island in Southeast Sulawesi who mainly depends on diesel generator for their daily rural life. As shown in Figure 1, PT PLN only intend to develop power plant in Wangi – wangi (No. 1 in yellow box) and Bau – bau (No. 2 in yellow box) without considering Tomia Island. A low viability and high potential to endanger ecosystem of coral reef in Tomia Island National Park area are argued as two main reasons to discharge this area from the electricity network distribution plan. As the result, the people of Tomia still have to use non – renewable energy as their main power despite threatening the environment (Williams *et al.* 2012).

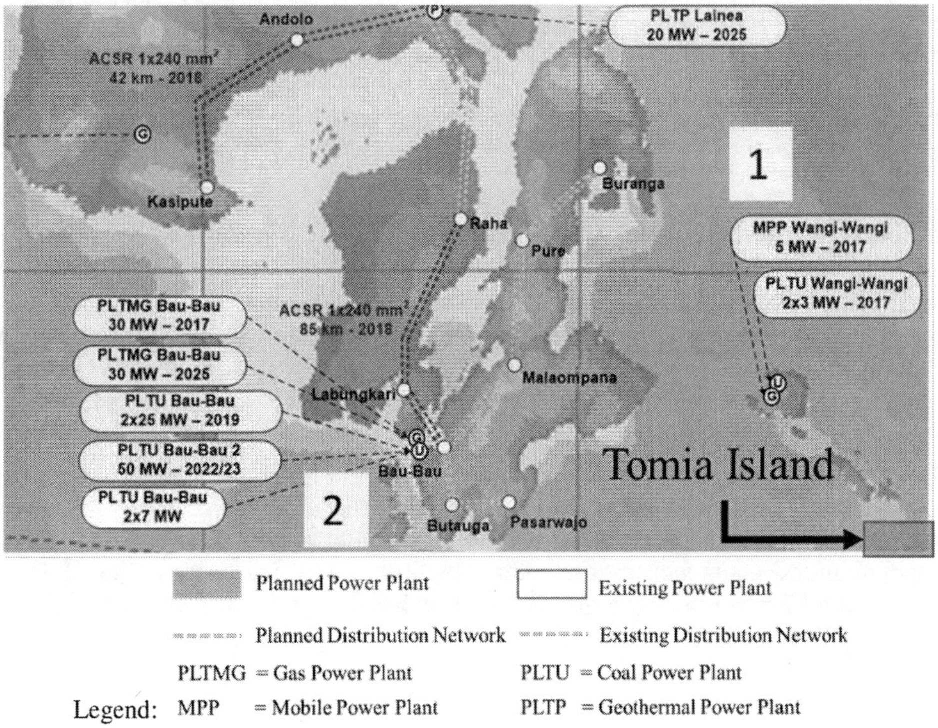

Figure 1. PT PLN plan in developing power plant and network distribution of electricity for 2015 – 2024 in southeast Sulawesi (in Indonesia).

On contrary, this area is experiencing massive solar heat for its daily life. With this potential, the development may no longer suitable using conventional electricity network system. It requires a breakthrough approach that combined with local potential to provide electricity access for the people through Solar Photovoltaic (PV). This research aims to produce economic and financial feasibility for Solar PV development in Tomia Island. It is expected to reduce non – renewable energy usage from the people in eastern region of Indonesia particularly Tomia Island by providing access to electricity to increase people welfare with cleaner energy.

2 RESEARCH PARAMETERS

In calculating the feasibility of Solar Photovoltaic (PV) development, four scenarios are proposed to accommodate the needs of Solar PV for five villages in Tomia Island. The villages consist of Kulati, Dete, Kahianga, Wawotimu and Lamanggau with number of beneficiaries estimated around 987 houses.

Table 1. Scenario of solar PV configuration.

	Configuration	Initial Cost (US$)	O & M Cost (US$)	Available Electricity
Scenario 1	Solar PV in 5 villages	6,191,250	14,520	710 kW
Scenario 2	Solar PV + diesel based generator in 5 villages	3,935,318	92,904	480 kW (PV) and 162 kW (Diesel)
Scenario 3	Solar PV in 4 villages	6,076,083	19,227	740 kW
Scenario 4	Solar PV + diesel based generator in 4 villages	3,552,407	113,094	360 kW (PV) and 128 kW (Diesel)

Scenario #1 and #3 are using only Solar PV for each villages in Tomia Island, meanwhile Scenario #2 and #4 are using hybrid system which integrate existing diesel generator and new installed Solar PV. Both scenarios in #3 and #4 only consider 4 villages where Kahianga and Wawotimu installed in one place, since the location of the two adjacent villages below 10 km thus create more efficient configuration.

IC and OM cost generated from HOMER – energy simulation software that produce optimal generator system configuration. The software automatically calculate the PV system cost by consider capital cost, replacement cost, operation and maintenance cost, fuel and profit. Further, the revenue is calculated based on electricity users, Cost of Energy (COE) and kiloWatt-hour (kWh) usage in the future.

In term of initial cost, scenario #1 has the highest cost but produce the lowest operation and maintenance cost. On the other hand, scenario #4 has the lowest initial cost but generate the highest operation and maintenance cost. This condition happened because the configuration in scenario #2 and #4 are combining Solar PV with existing diesel – based generator. It requires periodic maintenance and needs non – renewable fuel to run the system, thus O&M cost excalated five to six times higher than scenario #1 and # 3.

3 RESULT AND DISCUSSION

3.1 Financial & Economic Analysis

In calculating the analysis, transmission and distribution cost are exluded from consideration. The battery expected to have a life cycle for 20 operating years without maintenance cost. Financial simulation produces Internal Rate of Return (IRR), Net Present Value (NPV), and Benefit Cost Ratio (BCR). The project is feasible if IRR > i; where i is the interest rate, NPV shows positive value and BCR > 1.0 (Mondal and Islam 2011). Moreover, Economic Rate of Return (ERR) will consider the project cost that consist of initial cost (IC) and O&M cost; and benefit components such as direct benefit. It calculated from the potential in conducting productive business to boost individual and community/groups income. The financial feasibility analysis of the four scenarios is shown in following table.

Based on the table 1 and 2, installation of Solar PV in every village in scenario #1 and #3 (with COE of US$0.83 /kwh and US$ 0.82 /kwh) require higher cost than the hybrid system. Yet both scenarios provide greener energy and lower cost of operation and maintenance.

In general, scenario #1 shows the highest NPV and BCR, meanwhile scenario #4 rules in COE, IRR and ERR parameter. For that reason, scenario #4 is more preferable compared to other scenario. It shows a maximum available IRR and ERR with lower COE. BCR for scenario #4 also shows positive value of 1.27. Following sensitivity analysis will then be used to seek optimum result for Cost of Energy.

Table 2. Four scenarios of solar PV installation in Tomia Island.

Scenario	COE	IRR	NPV	ERR	BCR
Scenario #1	0.83	11.06%	IDR 35,348,461,462.03	20.69%	1.33
Scenario #2	0.67	11.23%	IDR 23,190,887,468.49	31.04%	1.18
Scenario #3	0.82	9.70%	IDR 23,022,995,429.58	22.70%	1.16
Scenario #4	0.68	12.91%	IDR 29,761,494,785.19	33.31%	1.27

3.2 Sensitivity Analysis

Sensitivity analysis is conducted by using Monte Carlo Simulation and Microsoft Excel Tools. The two main parameters are IRR and B/C Ratio. Cost of Energy (COE) is selected as the sensitive variable because it has direct impact to the electricity purchasement by adopting PV. Although marginal project requires B/C ratio equal to 1, this simulation is also expected to produce IRR above 10% and B/C ratio above 1 to maintain the sustainability during the 20 years of operation. Intervention of COE value is ranging from US$ 0.51 to US$ 0.68.

Table 3. COE sensitivity on scenario #4.

	COE	IRR	NPV	B/C Ratio
Scenario #4	0,68	12,91%	IDR29.761.494.785,19	1,27
Scenario #4	0,62	11,22%	IDR20.861.572.348,65	1,17
Scenario #4	0,60	10,67%	IDR18.101.906.476,85	1,14
Scenario #4	0,56	9,54%	IDR12.582.574.733,26	1,08
Scenario #4	0,52	8,34%	IDR 7.063.242.989,67	1,02
Scenario #4	0,51	8,03%	IDR 5.683.410.053,77	1,00

Initial simulation was conducted with COE equals to 0.51. At this value, B/C Ratio is equal to 1; however IRR value was about 8.03%, or below expected target. Therefore, this COE considered unfeasible. COE at 0.6 for scenario #4 was the minimum target for COE option since it produced IRR at 10.67% with B/C Ratio of 1.14. COE with value of 0.68 produced higher IRR and B/C ratio, but it will be affected the electricity price. Considering this circumstances, COE with 0.6 is selected as the best alternative. Using COE at 0.60 as baseline, the average of electricity price for Tomia Island was estimated in range of IDR 105,006 to IDR 201,537 per month.

3.3 Ability to Pay (ATP) & Willingness to Pay (WTP) Justification for Scenario #4

Income is one of the main factors that will affect the purchasing power to public utility services such as electricity and water. From field survey were held to 100 random households, the residents of Lamanggau village have been provided free electricity that runs about 12 hours/day from private entity. The other villages use communal generator that runs for about 4 – 6 hours with two types of billing: TV users and non-TV users.

Since current billing has two types distinguished from TV users and non – TV users so the price also varies from below IDR 100,000/month and IDR 100,000-250,000/month. The number of people who pay below IDR 100,000/month was higher than those who pay for IDR 100,000-250,000/month with 58.47% and 39.83% respectively. In term of willingness to pay, 85.14% of them expected to pay electricity below IDR 100,000/month. However, according to their ability to pay through electricity expenditure, 54.73% are managed to pay the bill at IDR 100,000-250,000/month and 27.03% are able to accommodate the price for more than IDR 250,000/month.

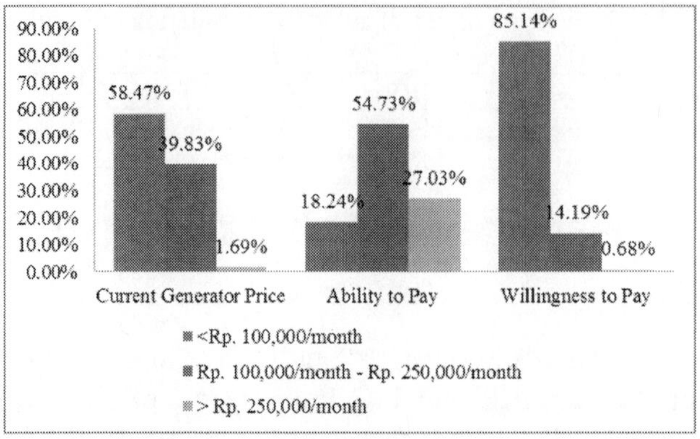

Figure 2. Comparison of current price of generator with ATP and WTP in Tomia Island.

Based on the analysis, the people tend to select a low price as their willingness to pay despite their ability to pay for much higher price. One of the reason is because dissatisfaction on low service quality, therefore they will be likely to have a low WTP, albeit higher ATP. The gap between ability to pay and willingness to pay can be reduced by improving electrical service from 4 hours to 24 hours although it might increase the tariff. It will then balance their expectation of using the electricity in daily life. Therefore, COE value about 0.60 is recommend to be implemented in rural area of Tomia Island.

4 CONCLUSION

Electricity is playing important role to give access of the people doing activities. However, with unequal distribution of electricity between western and eastern region, many people in eastern region such as Tomia Island in Southeast Sulawesi still suffering from limited access to electricity in their rural daily life.

Based on the analysis of financial and economic feasibility of four scenarios proposal, scenario #4 is selected as the best scenario. The scenario has configuration of solar PV installation with hybrid system where integrates current fuel-based generator and new solar PV generation. This scenario shows a maximum available IRR and ERR with lower COE compared to the other scenario. BCR for scenario #4 also shows positive value with 1.27.

Furthermore, sensitivity analysis showed COE with 0.6 selected as the best alternatives to be implemented which produces electricity price ranging from IDR 105,006 to IDR 201,537 per month. This value is slightly higher from the current price of electricity. However, considering the availability of 24/7 electricity, there are various prospective and productive activities that can be done by the people. Therefore, the installation of solar PV expected as initial action to improve the quality of life for the people in Eastern Indonesia particularly Tomia Island.

Acknowledgments

This paper is part of Solar Photovoltaic Electricity for Tomia Island: A Green Prosperity Model Project, funded by Millennium Challenge Account Indonesia. Authors also would like to thank PT Lemtek Konsultan Indonesia for generous support to produce significant output of the research.

References

Ministry of Energy and Mineral Resources. Energy Outlook, 2014a. Retrieved from http://prokum.esdm.go.id/Publikasi/Outlook%20Energi%202014.pdf (In Indonesian).

Ministry of Energy and Mineral Resources. Ratio Electrification, 2014b. Retrieved from http://kip.esdm.go.id/pusdatin/index.php/data-informasi/data energi/ketenagalistrikan/rasio-elektrifikasi on March 2016 (In Indonesian).

Mondal, M. A. H. and Islam, A. S., Potential and viability of grid-connected solar PV system in Bangladesh. *Renewable energy*, *36*(6), 1869-1874, 2011.

Narayan, P. K. and Singh, B., The electricity consumption and GDP nexus for the Fiji Islands. *Energy Economics*, *29*(6), 1141-1150, 2007.

Smith, P., A comparison of the merits of nuclear and geothermal energy in Indonesia. In *QiR (Quality in Research), 2013 International Conference on* pp. 160-165, IEEE, 2013.

Williams, J. H., DeBenedictis, A., Ghanadan, R., Mahone, A., Moore, J., Morrow, W. R., and Torn, M. S.,The technology path to deep greenhouse gas emissions cuts by 2050: the pivotal role of electricity. *science*, *335*(6064), 53-59, 2012.

Integrated Solutions for Infrastructure Development
Edited by Lau, H. H., Tang, F. E., Ng, C. K., and Singh, A.
Copyright © 2016 ISEC Press

EFFECTS OF MENTORING ON SOCIALIZATION OF GRADUATE EMPLOYEES IN THE BUILDING ENVIRONMENT THROUGH KNOWLEDGE TRANSFER

MORENA WIILLIAM NKOMO[1] and WELLINGTON DIDIBHUKU THWALA[2]

[1]*Construction and Quantity Surveying, University of Johannesburg, Johannesburg, South Africa*
[2]*University of Johannesburg, Johannesburg, South Africa*

Introducing graduates into the organization, thereby establishing the organizational culture and values is often under estimated for improving organizational effectiveness and efficiency, especially in a different work-force context such as South Africa. Mentoring is experiencing resurgence because business leaders not only recognize the benefits of transferring knowledge among graduates. The study describes better practices that organizations can use to address the threat of lost knowledge caused by changing labor force demographics. The study was mainly a literature review, qualitative with a special focus on the effects of mentoring on socialization of graduates within organizations. The data used in the report was mainly qualitative, based on the content analysis, and historical data. The findings indicate mentoring is a central part of socialization process and is crucial for socializing graduates into the organization. The study indicated knowledge transfer supports employees in improving their skill sets which increases their marketability and the potential for them to pursue career opportunities. The early success of the inventiveness provides useful lessons for the construction industry and executives who recognize that knowledge retention, socialization and mentoring of graduates are critical for supporting future organizational performance. This, mentoring relationships may assist organizations in simultaneously promoting effective knowledge transfer and commitment. That assists in the retention of key knowledge workers. Future research in this area is recommended to better understand how mentoring relationships may benefit organizations.

Keywords: Skills development, Construction industry, Human resource.

1 INTRODUCTION

Joining a company as a new employee, certainly, results in a degree of "culture shock" that can be very confusing if not mitigated through mentoring and orientation strategies that are intended to draw the individual out in order to add value to them as an employee and, subsequently, to the organization. In high-turnover businesses this even more serious because putting the process ahead of the individual will so often lead to discouraging to the point where they resign or are dismissed from the position which costs the company a great deal of money in terms of having to recruit and train a completely new individual employee. The increased number of job changes shortens the life cycle of an employee, putting companies at greater risk of organizational knowledge loss.

Organizational socialization refers to the mechanism through which new employees acquire the necessary knowledge, skills, and behaviors to become effective organizational members and insiders (Bauer and Erdogan 2011). Research has demonstrated that these socialization techniques lead to positive outcomes for new employees such as higher job satisfaction, better job performance, greater organizational commitment, and reduction in occupational stress and intent to quit (Kammeyer-Mueller and Wanberg 2003) .

Organizations today face a dilemma regarding the retention of key knowledge workers. Knowledge transfer amongst employees is crucial for organizational productivity. Turnover is a critical human resource issue in all sectors of the economy. Turnover affects productivity, product and service quality, and profitability. To minimize the impact of workforce turnover, companies need a proactive strategy for knowledge retention and transfer. Regardless of economic conditions, employee turnover happens. The financial impact of workforce mobility is well documented. Mentoring is considered to be the oldest form of knowledge transfer (Stephenson 1998).This requires organizations and researchers to focus more closely on processes such as mentoring that can support effective knowledge transfer and retention of critical knowledge workers. Conversely, if organizations do invest in knowledge transfer they risk increasing the marketability and job mobility of their employees which could potentially harm retention (Rousseau and Shperling 2004). Job mobility has increased in past decades because similarities in processes and technology mean that knowledge is less idiosyncratic to a particular organization and thus is more transferable (Rousseau and Shperling 2004). Mentorship has demonstrated importance in the socialization of new employees Major *et al.* (1995). Ostroff and Kozlowski (1993) discovered that newcomers with mentors become more knowledgeable about the organization than did newcomers without mentors. Mentors can help newcomers better manage their expectations and feel comfortable with their new environment through advice-giving and social support Ostroff and Kozlowski (1993).

2 LITERATURE REVIEW

Mentoring programs in organizations can be helpful in improving performance and transferring knowledge, and lead to higher job satisfaction and retention of employees, resulting in higher business productivity. In the workplace, relationships naturally develop between co-workers, clients, supervisors, and subordinates. Relationships at work can be both productive and unproductive, filled with animosity or admiration, and can foster friendships that go beyond the workplace lasting long into other careers and employers. One type of relationship that can be very beneficial in the workplace, even advancing an individual's career, is the mentoring relationship. The mentoring relationship between a mentor - a more experienced employee - and mentee can provide both parties benefits offering support and knowledge in performing a job, increased admiration in the office, and navigating the politics of an organization.

The relationship an employee has with his/her peers, colleagues and company has a significant impact on knowledge retention and transfer. A positive relationship facilitates knowledge sharing during all phases of an employee's life cycle; a negative relationship or lack of a relationship can impede knowledge sharing, especially at the point of exit. An unintended consequence of knowledge transfer via mentoring is the potential for increased job mobility for protégés and the resulting negative effect on organizational retention efforts, within construction companies. Recent findings from a longitudinal study suggest that mentoring fosters organizational retention in part because the emotional bond established between a mentor and a protégé may contribute to higher levels of organizational commitment (Payne and Huffman 2005). Mentoring relationships may operate, thus, through a dual pathway to impact organizational retention, socialization by assisting in the transfer of organizational knowledge while simultaneously developing the high-quality interpersonal relationships that strengthen a protégé's commitment to an organization. While there is an expanded interest in predicting newcomer turnover at the selection process, research on this topic has generally focused on organizational socialization and how it is used to familiarize new employees with new roles and to retain these new organizational employees. Socialization is an ongoing process that sometimes lasts for a year. It represents a sense-making process that helps new employees adjust; a typical socialization process includes three phases (Noe 2005).

- *Anticipatory socialization:* This stage occurs before new employees join the organization, through interacting with representatives of the company (for example, recruiters and managers).
- *Encounter:* When new employees begin a new job, they start to learn about job tasks and receive training. Managers can exert their influence by helping new employees understand their roles and duties.
- *Settling in:* New employees begin to feel comfortable with their job demands and social relationships. They will be interested in the company's evaluation of their performance and in learning about potential career opportunities within the company.

3 DISCUSSIONS OF FINDINGS

Though viewed as a key aspect of mentoring (Stephenson 1998), knowledge transfer has been primarily examined at the interfirm level (Carr *et al.* 2006), at the interdepartmental level (Berta and Baker 2004), and at the team level (Gibson *et al.* 2007). Knowledge management articles highlight knowledge transfer as a key mechanism for organizational success, yet a gap exists between practice (Buckman 1998) and formal research (Grover and Davenport 2001). Little research in the knowledge management literature, however, has explicitly tested mentoring as a means by which knowledge is transferred among individual's employees. Mentoring encompasses coaching, sharing perspectives, and transferring knowledge and wisdom to the mentee(s). The mentor is not a supervisor but a person with whom the employee can communicate freely and honestly without concern about being evaluated. Knowledge transfer is defined as an exchange of organizational knowledge between a source and a recipient (Grover and Davenport 2001) in which the exchange consists of information and guidance about resources and relationships. A primary mode of knowledge transfer is the direct sharing of knowledge between individuals.

3.1 Mentoring Saves Money, Retains Workers, Builds Leadership, and Growth Talent

Mentoring contributes to employee growth and tenure. In the long run, a well-organized and managed program can save the company thousands of rands.

- **Reduced turnover and recruiting costs.** Mentoring relationships can help retain talented people because they have a stronger commitment to the organization (Jacka and Quinn 2010). Talent remains much less likely to leave if they feel supported in their work and made aware, for example, of new opportunities that their mentor suggests.

- **Assistance in transferring knowledge from the retiring workforce to new workers.** Many mentoring relationships help younger employees learn from those who will retire soon. Pairing junior staff with more senior staff can reinvigorate the enthusiasm of senior employees as they transfer crucial knowledge to the next generation of workers. This reduces the loss of the tacit knowledge.

The purpose of knowledge transfer is to pass information from the more- experienced to the less-experienced employees so that the less-experienced employees can build the capabilities needed to assume future roles in the organization (DeLong 2004). Research at both the organizational and individual level of analysis support the notion that knowledge transfer mediates the relationship between mentoring and performance.

3.1.1 *Best practices in retention and knowledge transfer*

In order to address the threat of lost knowledge caused by changing workforce demographics. The heart of any knowledge-retention strategy is its knowledge sharing practices. There are many sorts of methods that contribute to knowledge capture, sharing and re-application: after-action reviews, communities of practice, face-to-face meetings, mentoring programs, expert referral services, training, video conferencing, interviews, written reports, etc. The question remains: which ones are most useful for addressing knowledge-retention problems?

A review of the HR literature suggests that sources on HR retention can be grouped into the following categories:

Based on the review of the literature, the list of retention practices that captures the main types of interventions discussed in the HR literature. They are as follows (Lochead and Stephens 2004):

Competitive and Fair Compensation is a fundamental starting point in most strategies to attract and retain employees. However, there is general agreement that compensation levels do not single-handedly guarantee employee retention. Common best practices include the use of industry surveys to benchmark and position wage and salary structures to be fair and competitive.

Training, Professional Development, and Career Planning are effective ways to enhance employee retention. Training constitutes a visible investment that the company makes in the worker, providing him or her with new skills, and greater competencies and confidence. Training often leads to work that is more intrinsically rewarding; training encourages workers to make longer term commitments to their workplace: it permits them to see a future with the company.

Knowledge Transfer – cross-training, coaching and mentoring, phased in retirement. While employee retention practices seek to retain *workers*, knowledge transfer practices seek to retain *skills*, through both formal and informal exercises in information sharing and the building of collective knowledge. Mentoring and coaching, phased-in retirements, and cross training and job rotation, are types of knowledge transfer that overlap with training.

4 CONCLUSION

Retaining organizational knowledge in the face of changing workforce demographics is a complex challenge; knowledge transfer is a key mechanism for organizational success. Knowledge management literature has tested mentoring as a means by which knowledge is transferred among employees. Mentoring encompasses transferring knowledge and wisdom to the mentee(s). Mentoring relationships help retain talented people because they have a stronger commitment to the organization. The heart of any knowledge-retention strategy is its knowledge sharing practices. Competitive and fair compensation is a fundamental starting point in most strategies to attract and retain employees .The most effective knowledge retention strategies will require a multifaceted approach and a long-term commitment on both the leadership of an organization and the employees. Knowledge transfer practices seek to retain *skills.* Organizations must understand the types of processes that facilitate effective knowledge transfer between individual employees, while simultaneously retaining knowledge in the organization so it can benefit the organization and employees. This study is one of the few empirical research efforts to shed light on this process. Results of this study suggest that a focus on mentoring simply as a vehicle to promote knowledge sharing may harm an organization. Attention must also be given to understanding the mentoring behavior's that foster a protégé's affective commitment so as to mitigate the potentially negative effect of knowledge transfer on retention.

Acknowledgments

I would like to thank my supervisor, co supervisor, department of construction management and quantity surveying for the opportunity.

References

Bauer, T. N. and Erdogan., B., Organizational socialization: s The effective onboarding of new employees. In S. Zedeck (Ed.), APA handbook of industrial and organizational psychology, Vol 3: Maintaining, expanding, and contracting the organization, APA Handbooks in Psychology (pp. 51–64). Washington, DC, US: American Psychological Association ,2011.

Berta, W. B. and Baker, G. R., Factors that impact the transfer and retention of best practices for reducing error in hospitals. Health Care Management Review, 29(2) 90-97, 2000.

Buckman, R. H., Knowledge sharing at Buckman Labs. Journal of Business Strategy, 19(1), 11-, 1998

Carr, J. C., Pearson, A. W., West, M. J., and Boyar, S. L., Prior occupational experience, anticipatory socialization, and employee retention. Journal of Management, 32, 343–359, (2006).

DeLong, D. W., Lost knowledge: Confronting the threat of an aging workforce. Oxford: Oxford University Press, 2004.

Gibson, C. B., Waller, M. J., Carpenter, M., and Conte, J., Antecedents, consequences, and moderators of time perspective heterogeneity for knowledge management in MNO teams. Journal of Organizational Behavior, 28, 1005-1024, 2000.

Grover, V. and Davenport, T. H., General perspectives on knowledge management: fostering a research agenda. Journal of Management Information Systems, 18(1), 5-2, 2001

Jacka, M. and Quinn, P., (2010) "Beware of the Hidden Costs of Staff Turnover," Absolute Recruitment Group. Retrieved from (http://www.absoluteglobal.net/absolute/globalweb.nsf/($Content)/article_staffturnover.html) on 2016/04/03.

Kammeyer-Mueller J. D., and Wanberg, C. R., Predictors and outcomes of proactivity in the socialization process. Journal of Applied Psychology, 85, 373–385, 2000.

Lochead, C., and Stephens, A. Employee retention, labour turn over and knowledge transfer: case studies from the Canadian plastics Centre, Canada, 2004.

Major, D. A., Kozlowski, S. W. J., Chao, G. T., and Gardner, P. D., A longitudinal investigation of newcomer expectations, early socialization outcomes, and the moderating effects of role development factors. Journal of Applied Psychology, 80, 418–43, 1995.

Noe, R. A., Employee training and development. New York: McGraw-Hill/Irwin, 2005.

Ostroff, C. and Kozlowski, S. W. J., The role of mentoring in the information gathering processes of newcomers during early organizational socialization. Journal of Vocational Behavior, 42, 170–183, 1993.

Payne, S. C. and Huffman, A. H., A longitudinal examination of the influence of mentoring on organizational commitment and turnover. Academy of Management Journal, 48(1), 158-168, 2005.

Rousseau, D. M., and Shperling, Z. Ownership and the changing employment relationship: Why stylized notions of labor no longer generally apply--A reply to Zardkoohi and Paetzold. Academy of Management Review, 29(4), 562-569, 2004.

Stephenson, K., What knowledge tears apart, networks make whole. Internal communication focus. Retrieved from http://www.netform.com/html/icf. Pdf. April 14, 2014], 1998.

Integrated Solutions for Infrastructure Development
Edited by Lau, H. H., Tang, F. E., Ng, C. K., and Singh, A.
Copyright © 2016 ISEC Press

CROSS TABULATION ANALYSIS: DETERMINANTS OF MATHEMATIC SUCCESS AMONG CIVIL AND BUILT ENVIRONMENT STUDENTS

JUSTUS N. AGUMBA

Dept of Construction Management and Quantity Surveying, University of Johannesburg, Johannesburg, South Africa

The socio-economic and demographic factors have been indicated to predict mathematics success. However, there is paucity of research to verify if these factors correlate with mathematics success at university level in South Africa. Hence, this paper reports on the socio-economic and demographic factors that correlate with mathematics success for civil and built environment students at a comprehensive university in South Africa. The data was obtained through, questionnaire survey from 197 students who were purposive sampled. The data was analysed using Statistical Package for the Social Sciences (SPSS) version 22. The study found that the correlation between the dependant variable and the independent variables that did not violet the assumption of each cell having more than five counts were insignificant. The cross tabulation results on frequencies should be interpreted with caution. The result indicates arithmetic difference in most of the independent variables. Male students performed better than female students in mathematics. However, the differences between the groups are not statistically significant as indicated with the Chi-square test.

Keywords: Biography, Construction, Elements, Examining, Factors, Undergraduate.

1 INTRODUCTION

South Africa fared poorly on the quality of mathematics and science education. The country was ranked 143 out of the 143 countries surveyed. On the quality of its education system it was ranked 139 out of 143 countries surveyed (World Economic Forum 2015). MacGregor (2009), indicated that vice-chancellors warned the South African government to expect more student drop outs, following shocking results of pilot National Benchmarking Test (NBT) in 2009.

Higher education institutions face greater challenges in relation to mathematics. The NBT results in 2009 revealed that only 7% of students were found to be proficient, therefore ready to study first year mathematics at the university. It was also suggested that majority of students would find it difficult to pass mathematics at university level without additional support (Dennis *et al.* 2012). With the reforms in academia since 1994, after the abolition of apartheid the government has engendered equity and access to higher education to deserving students after completing high school.

The objective of this research was therefore to undertake a prospective investigation of socio-economic and demographic factors that predict mathematics success in the current diverse student population in the department of civil engineering and the department of construction management and quantity surveying.

2 LITERATURE REVIEW

Murray (2013) found that prior academic achievement, learning styles and academic resources led to mathematics success. Rajandran *et al.* (2015) revealed that gender and place of origin as insignificant determinants, the entry qualification was a weak factor, and the student's Cumulative Grade Point Average (CGPA) of entry qualification as the strongest variable that determines the CGPA of first year students' academic performance.

Mlambo (2011), implied, gender, age, learning preferences, and entry qualifications did not cause any significant variation in the academic performance of students. The researcher suggested further research on other factors that are known to influence academic performance such as: student motivation, socio-economic status and attendance. Whannell (2013) indicated that, the background information (gender, age, ethnicity, disability, secondary school, work status, and early enrolment) gathered during the enrolment process, does not contain sufficient information for an accurately separation of successful and unsuccessful students.

According to Tewari (2014) their findings indicated that matric math score is a better predictor of academic performance of the first year courses at the Faculty of Management, University of KwaZulu-Natal. In a study by Vanthournout *et al.* (2012) they found that academic success is predicted by relating and structuring, lack of regulation, and lack of motivation on students at the end of the year.

It is evident from this discussion that studies have been undertaken to identify and analyze the numerous factors that affect academic performance in various centers of learning. However, few studies have focused on separate disciplines of studies especially in the field of engineering and the built environment. This study fills this gap in South Africa context. This study purports that socio-economic and demographic factors are predictors of mathematics success for civil engineering and the built environment students. This is a major gap in the current study.

3 RESEARCH METHODOLOGY

The research philosophy was deductive. A survey method was adopted and purposive sampling was used. The study population consisted of students from the department of civil engineering technology third years and construction management and quantity surveying third and second years. These were students registered in the 2015 academic year in the Faculty of Engineering and Built Environment. The students had been assessed in mathematics in their previous years. 199 students completed the questionnaire in the middle of the first semester of 2015. Two of the questionnaires were not suitable for analysis. The questionnaire included personal questions; age, gender, students' accommodation, education sponsor, entry to the university, final high school maths result, citizenship, parents' highest education level and weekly allowance. The outcome variable was whether students passed mathematics at first attempt in the first major assessment at the end of the first semester. SPSS version 22 was used to perform the statistics.

The independent variable (IV) and the dependent variable (DV) relationship were tested using the Pearson Chi-Square. In order for the relationship to be significant the p-value should be less than 0.05 at 95% confidence. In order to interpret the relationship of the variables, they should not violet the assumption of having more than five expected counts per cell.

4 RESULTS AND DISCUSSIONS

Results in Table 1 indicate arithmetic differences between the different independent variables and the dependent variable. However, these arithmetic differences should be interpreted with caution. The Chi-square results should also be taken into consideration. However, based on the arithmetic differences 24.4% of female students did not succeed in mathematics in their first attempt compared to 15.1% of male students. This is an indication that female students underperformed in their first major mathematics assessment in the first semester of their study. Students who are 21 years and below outperformed students who were in the age bracket 22-25 years and those who were 26 years and above. The students who were accommodated in the university residence performed better than those who lived at home.

86.6% of Civil engineering students passed their mathematics at the end of the first semester compared to 78.5% of building students. This may suggest that civil engineering students are better equipped in mathematics than building students. Students who had allowances from R600.00 and above performed better than the students who had allowances less than R600.00. In relation to mathematics results in high school, those who got less than 50% during admission to the university 26.9% of students them did not pass at their first attempt. In line with the students who attained more than 50% in mathematics in high school 17.2% of the students did not succeed in their first attempt at the university.

Table 1. Cross tabulation demographic variables & mathematics success.

		Mathematics No	**Mathematics Yes**	**Total**
Gender	Male	15.1% (18)	84.9% (101)	100% (119)
	Female	24.4% (19)	75.6% (59)	100% (78)
Age group	18-21 years	13.7% (10)	86.3% (63)	100% (73)
	22 - 25 years	27.1% (19)	72.9% (51)	100% (70)
	26 years and above	16.7% (2)	83.3% (10)	100% (12)
English First Language	No	17.8% (34)	82.2% (157)	100% (191)
	Yes	50.0% (3)	50.0% (3)	100% (6)
Major course	Building construction	21.5% (28)	78.5% (102)	100% (130)
	Civil engineering	13.4% (9)	86.6% (58)	100% (67)
Entry to university	High school	16.6% (28)	83.4% (141)	100% (169)
	College	34.6% (9)	65.4% (17)	100% (26)
Ethnicity	Asian/Indian	33.3% (1)	66.7% (2)	100% (3)
	Black	18.6% (35)	81.4% (153)	100% (188)
	Colored[1]	0.0% (0)	100.0% (3)	100% (3)
	White	33.3% (1)	66.7% (2)	100% (3)
Citizen	South African	19.2% (34)	80.8% (143)	100% (177)
	Permanent South African resident	14.3% (1)	85.7% (6)	100% (7)
	Non-South African	15.4% (2)	84.6% (11)	100% (13)
Accommodation	Home	25.0% (11)	75.0% (33)	100% (44)
	On campus	17.0% (9)	83.0% (44)	100% (53)

[1] A people Mixed ethnicity in South Africa.

Table 1. (Continued) Cross tabulation demographic variables & mathematics success.

		Mathematics No	Mathematics Yes	Total
	Off campus	17.3% (17)	82.7% (81)	100% (98)
Highest education level of parent/guardian	No schooling	21.4% (3)	78.6% (11)	100% (14)
	Elementary school	0.0% (0)	100.0% (5)	100% (5)
	Secondary/high school	16.9% (10)	83.1% (49)	100% (59)
	College	20.0% (9)	80.0% (36)	100% (45)
	University	20.3% (13)	79.7% (51)	100% (64)
	Do not know	25.0% (2)	75.0% (6)	100% (8)
Education sponsor	Myself	10.5% (2)	89.5% (17)	100% (19)
	Parent/guardian	21.9% (21)	78.1% (75)	100% (96)
	Government/bursary	17.1% (37)	82.9% (68)	100% (82)
Weekly income	Less than R200	19.2% (20)	80.8% (84)	100% (104)
	Between 200 and 399	17.8% (8)	82.2% (37)	100% (45)
	Between 400 and 599	29.2% (7)	70.8% (17)	100% (24)
	600 and above	0.0% (0)	100.0% (18)	100% (18)
Final mark high school	Less than 50	26.9% (7)	73.1% (19)	100% (26)
	50 and above	17.2% (29)	82.8% (140)	100% (169)

*The number in the parenthesis indicates the actual number of respondents.

Table 2 presents the results of the relationship of the students' socio-economic and demographic characteristics of the students' and mathematics performance. Chi-square test on mathematics performance with student characteristics was performed. The researchers only made inferences to the results where the expected observation counts per cell were 5 counts and over. Hence, no violation of the statistical assumption set for the study. The results interpreted here were: gender, accommodation and major course the students are attending.

The relationship of this independent variables with the dependent variable were insignificant as there p-value was greater than 0.05. This is an indication that no relationship exists in mathematics success whether you are male or female. However, Table 1 indicates that male students outperformed female students in mathematics by a difference of 9%. Civil engineering students performed better than building students as indicated in Table 1. However, the corrected Chi i.e., continuity correction indicated an insignificant value of $p > 0.05$. These results therefore suggest that the arithmetic differences in Table 1 were minimal to conclude that differences exist. In other words, the difference is not statistically significant.

Table 2. Correlation between demographic & mathematics success.

Variables	Pearson chi-square	Asymp. Sig. (two-sided)	Observations
Gender	2.063	0.151	a. 0 cells (0.0%) have expected count less than 5. The minimum expected count is 14.65.
Age group	4.127	0.127	a. 1 cells (16.7%) have expected count less than 5. The minimum expected count is 2.40.
English first language	2.125	0.145	a. 2 cells (50.0%) have expected count less than 5. The minimum expected count is 1.13.
Ethnicity/Race	1.530	0.675	a. 6 cells (75.0%) have expected count less than 5. The minimum expected count is .56.
Citizenship	0.212	0.899	a. 2 cells (33.3%) have expected count less than 5. The minimum expected count is 1.31.

Table 2. (Continued) Correlation between demographic & mathematics success.

Variables	Pearson chi-square	Asymp. Sig. (two-sided)	Observations
Major course	1.410	0.235	a. 0 cells (0.0%) have expected count less than 5. The minimum expected count is 12.58.
Highest education level of one parent/guardian	1.677	0.892	a. 4 cells (33.3%) have expected count less than 5. The minimum expected count is .95.
Entry level to university	3.672	0.055	a. 1 cells (25.0%) have expected count less than 5. The minimum expected count is 4.93.
Education sponsor	1.608	0.448	a. 1 cells (16.7%) have expected count less than 5. The minimum expected count is 3.57.
Weekly income/allowance	5.989	0.112	a. 2 cells (25.0%) have expected count less than 5. The minimum expected count is 3.30.
High school final mark in mathematics	0.852	0.356	a. 1 cells (25.0%) have expected count less than 5. The minimum expected count is 4.80.

*Independent variable significance at 95% confidence interval

5 CONCLUSION AND RECOMMENDATION

The study found that the relationship of the independent variables and dependant variable i.e., mathematics success were insignificant based on the socioeconomic and demographic factors that did not violet the assumption of the number of expected counts per cell. These factors were; gender, major course the student is pursuing and accommodation. Gender was insignificant in relation to mathematics success. However, in relation to percentages, the findings suggest that female students were less likely to pass mathematics at their first attempt in the university compared to male students. Further, civil students' outperformed building students in mathematics success at this university. Contrary to accommodation being insignificant, it can be suggested that students' living on campus perform better than those living out of campus especially those living at home. Based on these findings, the researchers recommend that:

Despite insignificant relationship based on the Chi-square test. The percentage difference suggests that, the university should inform the female students to be studious in the quest to pass mathematics at their first attempt in their course. Support is required for female students in order to succeed in mathematics. The additional support to female students will ensure timely completion of their studies and not to waste their finances. This will assist students to operate in close proximity to their study environment and hence not waste time in travelling to the university. Furthermore, building students will need more support than civil students in order to pass mathematics at their first major examination.

This study is timely considering the history of South Africa higher education. There is demand for education among high school learners, hence the diversity of the number of students registered at the university has increased. To fully support the diversity of these students, universities must implement strategies and interventions based on sound research studies that will ensure students are given a fair chance to succeed. Hence, the need to continue to identify factors that would influence student academic performance in mathematics and other disciplines is crucial. The researcher suggests the need to use other determinants in relation to the socio-economic and demographic factors as these factors are not exhaustive. A further study is also recommended for all Engineering and Built Environment courses at this University.

References

Dennis, R. C. and Murray, D. M., Success in first-year mathematics: school-leaving examinations and first-year performance. *South Africa Journal of Science,* 108(7/8), Article #1325, 3 pp, 2012 http://dx.doi.org/10.4102/sajs.v108i7/8.1325

MacGregor, K., South Africa: Shocking results from the University tests, 2009 http://www.universityworldnews.com/article.php?story=20090816082047397 last accessed on the 20/05/2015.

Mlambo, V., An analysis of some factors affecting student academic performance in an introductory biochemistry course at the University of the West Indies, *Caribbean Teaching Scholar,* (2), 79-92, 2011.

Murray, J., The factors that influence mathematics achievement at the Berbice campus", *International Journal of Business and Social Science,* 4(10), 150-164, 2013.

Rajandran, K., Hee, C.T., Kanawarthy, S., Soon, K.L., Kamaludin, H., and Khezrimotlagh, D., Factors affecting first-year undergraduate students academic performance, *Scholars Journal of Economics, Business and Management*, 2(1A), 54-60, 2015.

Tewari, D. D., Is matric math a good predictor of student's performance in the first year of university degree? A case study of faculty of management studies, university of KwaZulu-Natal, South Africa, *International Journal Education Science*, 8(1), 233-237, 2014.

Vanthournout, G., Gijbels, D., Coertjens, L., Donche, V., and Van Petegem, P., Students' persistence and academic success in a first-year professional bachelor program: The influence of students' learning strategies and academic motivation, *Educational Research International,* 2012, Article ID 152747, 10 pages doi:10.1155/2012/152747.

Whannell, R., Predictors of attrition and achievement in a tertiary bridging program, *Australian Journal of Adult Learning,* 53(2), 280-301, 2013.

World Economic Forum, The global information technology report 2015, ICTs for inclusive growth, 2015. Edited Dutta, S., Geiger, T., & Lanvin, B, 2015. http://www3.weforum.org/docs/WEF_Global_IT_Report_2015.pdf last accessed on the 20/05/2015.

Integrated Solutions for Infrastructure Development
Edited by Lau, H. H., Tang, F. E., Ng, C. K., and Singh, A.
Copyright © 2016 ISEC Press

DEVELOPING LEADERSHIP SKILLS OF PROJECT AND CONSTRUCTION MANAGERS IN THE CONSTRUCTION INDUSTRY

MURENDENI LIPHADZI and CLINTON AIGBAVBOA

Dept of Construction Management & Quantity Surveying, University of Johannesburg, Johannesburg, South Africa

The construction industry is one of the major industries in South Africa. Like any other industry, strong leadership is vital for its growth. Moreover in order to stay competitive, South African construction firms must find a way to train their project and construction managers to become competent leaders in their specific disciplines. It is to this end, that the study presents findings on how leaders can be developed in the South African construction industry. The data used in this research were derived from both primary and secondary sources. The secondary data was collected via a detailed review of related literature. The primary data was collected through a structured questionnaire aimed at 150 projects and construction managers in the South African construction industry. From the 150 distributed questions, only 110 questionnaires were received and subsequently data based on the received questionnaire which was analyzed using descriptive statistics. Findings revealed that, education and qualification can help develop leaders in the South African construction industry. This was followed by leadership training courses, taking responsibility, new challenges, and internal motivation, professional and personal development. The study explores leadership and leadership development in the South African construction industry. The study presents a robust background about the construction industry and the importance of leadership.

Keywords: Leadership, Project management, Leadership development, Development strategies.

1 INTRODUCTION

The Construction industry is a key sector in every country as it is heavily interconnected with the economy as a whole. It affects, and is affected by; the gross domestic product of a nation. Moreover, the government policies directly affect the construction industry. The construction industry is dependent on different types of input such as human capital, financing and products from other industries. This makes leadership and management important. Without leadership a group of human beings quickly degenerates into argument and conflict because they see things in different ways and lean towards different solutions. Leadership helps to point us in the same direction and harness our efforts jointly.

Moreover most leadership definitions assume that leadership is a process whereby intentional influence is exerted by one person over others in order to guide, structure and facilitate organizational activities and relationship (Yukl 2002). Kouzes and Posner (2002) understand leadership as a process, which ordinary people use when they are bringing forth the best from themselves and others. Leaders feel, are credible, inspire a shared vision, and enable others to act, model the way and recognize their employees' contributions. According to Havenga (2002), leaders empower others to make decisions. The aim of the study is to have an understanding in developing leaders within the construction industry, and further find out important factors of leadership development. The study will present findings on leadership development in the South African construction industry.

2 WHAT IS LEADERSHIP?

If one word could describe the essence of construction management, it is responsibility. Thus, project leaders are responsible for all that happens in a project. This doesn't mean that the project leader should or could do everything associated with the project. However, it does mean that they have ultimate responsibility for the project. In other words, the buck stops with the project leader. It is also evident that the construction industry has a greater need for leadership than arguably any other industry. Many reasons support this notion and it is evident in the nature of the construction projects, and constructed products (Hillebrandt 2000). Undoubtedly, construction projects are large and technically complex and they involve a combination of specialized skills. Thus, the construction teams are not only large but are also multi-disciplinary and the members are from several different construction disciplines and this makes leadership important in this industry.

3 LEADERSHIP IN THE CONSTRUCTION INDUSTRY

Allen and Roberts (2011) states that leadership development is a continuous, systemic process designed to expand the capacities and awareness of individuals, groups, and organizations in an effort to meet shared goals and objectives. Hence, it is important to develop leadership in the construction industry. According to Toor and Ofori (2008), development of leadership in the construction industry is a trip which requires fixed attention. In Addition, Toor and Ofori (2008) stated that leaders are born or can learn to lead however, it is important for modern leaders in the construction industry to take initiative and make future construction leaders. Moreover, leadership training and leadership development is paramount in the construction industry.

Nowadays, the construction industry faces a difficult environment of socio-economic, cultural, political, and business challenges, and it is important to note that effective leadership can help minimize the challenges (Toor and Ofori 2008). Therefore, effective leadership interventions should help to accelerate the development of new leadership. The construction industry will be more successful if we can develop leaders who can have an understanding of skills, knowledge and characteristics needed of an effective project leader. Moreover, certain construction companies have now considered the importance of developing its employees for improving their competitive and dynamic abilities (Moore 2006). It is also important for the project leaders to develop originality, initiative, leadership and ethical standards. In addition to insuring good problem solving techniques, decision making method and technical competencies are keys to leadership development.

Jarad (2012) suggested that the potential sources of leadership development include observing, mentoring or coaching by seniors, reading or self-study, education courses during university, education courses since college, company training, and job experience. Bogus and Rounds (2006) suggest that employees can be better leaders from self-education, attending seminars, reading, watching, experience, and from having active organizational mentors.

In addition, Jarad (2012) is of the view that construction organizations can develop leadership and management skills by developing a culture of teaching, mentoring, self-study, and frequent job changes. Companies can also use their own professional personnel to offer formal leadership and management training. Leadership development initiatives set the stage for organizational success by empowering employees to develop their skills and competencies. It improves retention, provides a foundation for succession planning and trains the next generation of leaders to focus on how managers lead, develop and partner with their employees.

4 RESEARCH METHODOLOGY

Research methodology considers the context of the research and the desired results in order to achieve meaningful research outcomes. Moreover, the selection of an appropriate research design involves several steps, beginning with identifying the problem, purpose of the study and in depth literature review. This research adopted a quantitative approach which involved the use of a questionnaire.

Burns and Grove (1993) describe quantitative research as a formal, systematic process that describes and test relationships and examines causes among variables. In addition, Polit and Hungler (1993) were of the view that quantitative research can be used to obtain information from a sample of people by means of self-report, whereby people respond to a sequence of questions posed to them by the researcher. The data used in this paper were derived from both primary and secondary sources. The primary data was obtained through the questionnaire survey method through the use of a purposive sample, targeting the project and construction managers; while the secondary data was derived from the review of literature and archival records. Respondents had to rank the importance of table 1 on a Likert scale of 1 to 5, whereby 1 is strongly disagree and 5 being strongly agree. Moreover, 9 factors were used to develop the elements of construction leadership development.

The questionnaire survey method was decided upon for this study because of the following reasons: they require less time and energy to administer and they offer the possibility of anonymity because the respondent's names are not required on the completed questionnaires. The questionnaire in relation to leadership development consisted of the following sections; ways to develop leaders in construction and leadership development factors that are important to developing construction leaders.

The analysis for this paper was based on the collected questionnaires and from the 150 questionnaires sent out, 110 were received representing a 74% response rate. This was considered adequate for the analysis based on the assertion by Moser and Kalton (1971) that the result of a survey could be considered as biased and of little value if the return rate is lower than 30–40%. Table 1 was analysed via a percentage and table 2 used mean rankings.

5 RESULTS AND DISCUSSIONS

Respondents were asked to rank methods and ways that can help develop leaders in the South African construction industry (See Table 1). Results obtained include the following: Education/skill/qualifications (89% of the respondents), leadership training courses (73.4% ranked 2nd), taking responsibility (72.5%), new challenges (55%), inspiration from within the current leadership (54.1%) and clear guidelines or direction by current leadership (51.4%). Financial benefits and rewards were ranked last as a way that can develop leaders (42.2% ranked 7th).

Given that it is the subordinates within firms who market, create products, make decisions and sustain construction projects, the development of human resources is vital to the success of the construction organization. Literature reviewed indicated that leadership development is a continuous, systemic process designed to expand the capacities and awareness of individuals, groups, and organizations in an effort to meet shared goals and objectives (Allen and Roberts, 2011). Jarad (2012) was of the view that, the primary purpose of most leadership development interventions was to improve individual managerial skills and on-the-job performance.

In addition, Jarad (2012) was of the view that construction companies can develop leadership and management skills by developing a culture of teaching, mentoring, self-study, and being exposed to new challenges.

Table 1. Ways to develop leaders in the construction industry.

	No/%	Rank
Education / skills / qualifications	N=97 (89.0%)	1
Leadership training courses	N=80 (73.4%)	2
Taking responsibility	N=79 (72.5%)	3
Learn new skills that will enhance capability.	N=62 (56.9%)	4
New Challenges	N=60 (55.0%)	5
Inspiration from within the current leadership	N=59 (54.1%)	6
Clear guidelines or direction by current leadership	N=56 (51.4%)	7
New experiences	N=55 (50.5%)	8
Financial benefits, rewards and incentives	N=46 (42.2%)	9

Furthermore, respondents were asked to rate their level of agreement with regard to leadership development factors (See Table 2). Results obtained include the following: Internal motivation is required to develop yourself as a leader ($\overline{x}$-Mean = 4.32; R-Ranked=1), on-going personal and professional development ($\overline{x}$=4.12; R=2), it is important for someone to demonstrate leadership skills in order to be promoted to a leadership position ($\overline{x}$=4.11; R=3), succession planning and strategy ($\overline{x}$=4.10; R=4), it is important for a leader to have formal leadership training ($\overline{x}$=4.07; R=5) and personal gains have an impact on developing leaders ($\overline{x}$=3.63; R=6). Leadership development must be a priority for both an individual and an organization, moreover developing leaders requires patience, vision and commitment.

Literature review indicated that the first step in developing a leadership development strategy is to determine why the organization believes that this is a strategy that should be pursued, in other words, what it hopes to achieve with the strategy, moreover leadership development initiative cannot be successful unless it clearly targets a specific business goal (Jarad 2012). Bogus and Rounds (2006) suggest that employees can be better leaders from self-education, attending seminars, reading, watching, experience, and from having active organizational mentors.

Table 2. Leadership development factors Leadership development factors.

	σX-SD	$\overline{x}$-Mean	R
Internal motivation is required to develop yourself as a leader.	0.667	4.32	1
On-going personal and professional development.	0.683	4.12	2
It is important for someone to demonstrate leadership skills in order to be promoted to a leadership position.	0.649	4.11	3
Succession planning and strategy.	0.701	4.10	4
It is important for a leader to have formal leadership training.	0.832	4.07	5
Personal gains have an impact on developing leaders.	0.956	3.63	6

6 CONCLUSIONS

There is a dire need to develop leadership skills of construction project leaders in the South African construction industry. Regardless of the fact that majority of the respondents agreed to having attended leadership development trainings, literature revealed that the development of leaders in the construction industry is a trip which requires fixed attention. Literature further revealed that leaders can be developed by self-education, attending leadership seminars, job experience, being coached and mentored.

From the survey results obtained from the respondents, it was observed that education and qualification can help develop leaders in the construction industry. Leadership training courses followed education. Taking responsibility and starting new challenges were some of the important factors. Moreover, internal motivation, professional and personal development, succession planning and strategy were seen as vital. It is worth noting that few respondents chose financial benefits as significant to helping them improve as a leader, yet companies use this as the most common way to retain and reward leaders.

References

Allen, S. J. and Roberts, D. C., Our Response to the question: Next steps in clarifying the Language of Leadership Learning, Journal of Leadership Studies, 5. (2), 65-70, 2011.

Bogus, S. M. and Rounds. J. L., Incorporating leadership skill development in construction training programs, Proceedings of the 2nd Specialty Conference, 2006

Burns, N. and Grooves, K. G., The practice of Nursing Research Conduct, Critique and Utilization. Second Edition, Philadelphia: WB Saunders Company, 1993.

Havenga, R., The new leadership challenge: Teaming principles with practices in times of change. Management Today, 2002.

Hillebrandt, P. M., Economic Theory and the Construction Industry. 3rd Edition, Basingstoke: Macmillan, 2000.

Jarad, G. H., The construction manager leading characteristics for the success of construction projects in the Gaza Strip, Master's Thesis, The Islamic university of Gaza, 2012.

Kouzes, J. M. and Posner, B. Z., Leadership challenge. San Francisco: Jossey-Bass, 2002.

Moore, J., Where do you start with leadership development, Development and Learning in organizations 18(5):7-9, 2006.

Polit, D. and Hungler, B. P., Essentials of nursing research. Methods, appraisal and utilization. 3rd edition. Philadelphia: Lippincott, 1993.

Toor, S. R. and Ofori, G., Leadership for future construction industry: Agenda for authentic leadership', International Journal of Project Management, 26. (6): 620–630, 2008.

Integrated Solutions for Infrastructure Development
Edited by Lau, H. H., Tang, F. E., Ng, C. K., and Singh, A.
Copyright © 2016 ISEC Press

JOB DISSATISFACTION INFLUENCE ON CONSTRUCTION EMPLOYEE ABSENTEEISM AND TURNOVER

PORTIA LEASILANE and CLINTON AIGBAVBOA

Dept of Construction Management & Quantity Surveying, University of Johannesburg, Johannesburg, South Africa

This study examines the influence of job dissatisfaction on absenteeism and employee turnover. This is because it has been observed that many construction projects are facing challenges relating to the delivering of defective work, delayed project completion and cost overruns in most cases; and job dissatisfaction has been identified as one of the factors contributing to these challenges encountered in the construction industry. The study was conducted with reference to existing theoretical literature (secondary data) and the use of a structured questionnaire survey for the primary data collection. The questionnaire survey was distributed amongst construction professional in the South African construction industry. The findings from the questionnaire survey revealed that the aspects associated with job dissatisfaction which brings about absenteeism and employee low productivity include: salary issues, incentives, professional growth, working relationship as well as enforcement of fair policies. These factors were rated by the study respondents as the most influential factors of absenteeism and employee turnover in the South African construction industry. This study contributes to the body of knowledge relating to the influence of job dissatisfaction on absenteeism and employee turnover in the South African construction industry.

Keywords: Construction psychology, Employee turnover, Job satisfaction.

1 INTRODUCTION

Within each and every sector in South Africa, new companies are established every year and this may lead to an increase in competition within those sectors. Therefore, it is the manager's task to ensure that employees within a given organization have the drive to perform their tasks to reach the organizational goals in order to have a competitive advantage over other organizations. According to Deeprose (2003), as a manager, there are many ways that one can use to create an environment where employees are most likely to feel motivated to take actions that contribute to organizational goals. However, senior management in a given organization tends to ignore the issue of motivation of employees in the workplace. Moreover, some find it difficult to address the issue of employee motivation in their organization since it entails studying behavior of each and every employee and identifying causes of such behaviors.

Bernold and AbouRizk (2010) emphasized that construction is performed by different employees including laborers, engineers, specialists, etc. which have their own reasons to perform in a particular way with varying degrees of energy and enthusiasm. The ability to have a comprehensive understanding of basic things that keep an employee motivated thereby improving their level of satisfaction, will have a significant positive impact in their performance of job tasks. The study illustrates that when employees are happy and satisfied in their jobs, their level of motivation is high and they perform at peak all the time (Tyilana 2005).

Moreover, an employee will only be committed to their job if they are happy about their duties and feeling content with rendering job tasks. If employers want to inspire people to give their best and most committed effort in support of the organizational goals, they should relate those goals to something more compelling than simple satisfaction, something that people spend a lifetime pursuing: happiness (Walters 2010). In their book, Bernold and AbouRizk (2010) argue that personal desires are the fundamental motivation behind all human actions. In that regard, it can be said that individual needs, serve as a form of motivation that will determine a level of satisfaction an employee may have with regard to their job. This is because Lavender (1996) has argued that an individual's behavior is affected by their motivation. Hence, it is important to fulfill those needs as this will help to stimulate an individual's interest in their job and it will in turn serve as motivation.

In the research conducted by CIDB (2013), it was observed that many construction projects in South Africa are faced with challenges such as client's dissatisfaction with work rendered by contractors in respect of delivering defective work not completed within budget and on time. This may be due to experiences of absenteeism and employee turnover within a given project. As a result; these two factors according to Rojas (2008) tend to reduce productivity as construction industry has traditionally paid little attention to these problems.

Rojas (2008) mentioned job satisfaction among other factors as determinants of absenteeism and turnover. In that regard, Schaffer (2001) argued that in order to achieve high productivity, members of management as well as the entire workforce need to be highly motivated, positive, and totally committed. It is because of the aforementioned challenges that this paper attempts to explore the concept of motivation and job satisfaction. Therefore, the aim of this paper is to explore the determinants of job satisfaction in the construction industry as well as the influence of factors leading to dissatisfaction on absenteeism and employee turnover. Firstly, this paper starts with an overview of determinants of job satisfaction. Subsequently, a brief discussion of the influence of dissatisfaction on absenteeism and employee turnover will be provided.

2 INFLUENCE OF JOB DISSATISFACTION ON ABSENTEEISM AND EMPLOYEE TURNOVER

Labor costs for contractors can reach as much as 40% to 60% of total construction costs (Rojas 2008). In that regards, it is of paramount importance that contractors ensure that each employee has the ability to perform their tasks to the contractor's expectations, as their performance contributes to productivity. According to Rojas (2008), research has found that absenteeism and employee turnover reduce productivity, while the construction industry is traditionally paying little attention to these problems. In order to address this issue, contractors need to understand why employees are deciding not to come to work and the reason for some of them quitting their jobs.

According to Hadebe (2001), when employees are dissatisfied with their jobs, they try to find ways of reducing exposure. That is, to stay away from the jobs. In addition, Rojas (2008) has informed that researchers agree that companies can divide the factors causing absenteeism and turnover into controllable and uncontrollable. Controllable factors include: site management capabilities, teamwork, excessive rework, and job satisfaction (Rojas 2008). Similarly, Horner and Talhouni (1984) identified factors such as unsafe working conditions and lack of job security as having effect on absenteeism.

In addition, job satisfaction also emerges as an important factor in turnover (Rojas 2008). Therefore, it can be deduced that job dissatisfaction have an influence on absenteeism and employee turnover. This is because, Rojas (2008) has identified factors such as organizational factors (pay, company size, responsibilities, tasks, and worker commitment) and worker's personal factors (age and tenure) as having effects on absenteeism and employee turnover.

Furthermore, the work of Sichani (2010) considered the following as the top 10 causes of absenteeism: long commuting hours / distance to jobsite, transportation issues to site, lack of recognition/incentives; personal safety concerns, child care, illness/school, other family responsibility, already planned time off, bad weather for working, personal appointment, personal injury/illness. Whilst, Cheney (1984) identified the following (structure, participation, consideration, and feedback) as the organizational factors that affected productivity and job satisfaction. Likewise, Tnay *et al.* (2013) concluded that pay and supervisory support act as a crucial factor in affecting employee's turnover intention. In this regard, it is important that employees get involved in the decision making process, thereby making them to feel like part of the team as this is important in reducing absenteeism and employee turnover within the organization.

3 RESEARCH METHODOLOGY

The study was conducted with reference to existing theoretical literature and the use of a structured questionnaire survey. The existing literatures were obtained from published and unpublished sources. Since the population surveyed was already identified, the researcher adopted the stratified sampling. In this type of random sampling, a researcher first divides the population into subpopulations (strata) on the basis of supplementary information and subsequently, the researcher then draws a random sample from each subpopulation (Neuman 2000). This allows each respondent within each subpopulation to have equal chance to be selected. The population comprised of professional team members within both consulting and contracting firms, working in Johannesburg, South Africa. The respondents were randomly selected from each subpopulation to make up a sample of 50. However, due to time limitation and low response rate, the final sample size used for analysis purpose, based on the acceptable questionnaires completed by the respondents, was 40. This was considered adequate for the analysis based on the assertion by Moser and Kalton (1971) that the result of a survey could be considered as biased and of little value if the return rate was lower than 30–40%. The data presentation and analysis made use of frequency distributions and percentages of all the respondents.

3.1 Mean Item Score (MIS)

A five point Likert scale was used to determine the causes of construction project cost overruns in Gauteng province with regards to the identified factors from the reviewed literature. The adopted scales were as follows:

1 = Strongly disagreed; 2 = Disagreed; 3 = Neutral; 4 = Agreed and 5 = Strongly agreed

The five-point scale was transformed to mean item score (MIS) for each of the factors of causes of cost overruns as assessed by the respondents. The indices were then used to determine the rank of each item. The ranking made it possible to cross compare the relative importance of the items as perceived by the respondents. This method was used to analyze the data collected from the questionnaires survey. The mean item score (MIS) was calculated for each item as follows:

$$MIS = \frac{1n1 + 2n2 + 3n3 + 4n4 + 5n5}{\sum N} \quad (1)$$

Where: n1= Not at all influential; n2= Slightly /Somewhat influential; n3= Neutral; n4= Very influential; n5= Extremely influential; n1= Not important at all; n2= Unimportant; n3= Neutral; n4= Important; n5= Very important; N = Total number of respondents

After mathematical computations, the factors were then ranked in descending order of their mean item score (from the highest to the lowest). An internal consistence check was conducted using the consistence and reliability statistic measure of the Cronbach's Alpha.

4 FINDINGS AND DISCUSSION

4.1 Biographical Data

The findings from the analyzed questionnaire revealed that 52.5% of the respondents were female while 47.4% were male. Also, in terms of age group, it was found that that the majority (47.5%) of respondents were between the age of 21 and 25 years; 20.0% were between 26-30 years; 17.5% were between 31-35 years with 17.50%; 2.5% were between 41-45 years; and 5.0% were above 55 years. Furthermore, the findings reveal that 50.0% of the respondents have diploma, while 45.0% have bachelor's degree, and 5.0% have master's degree. Findings relating to the years of experience revealed that 7.5% have less than a year of experience, 62.5% have 1-5 years of experience, 17.5% have 6-10 years of experience, 5.0% have 11-15 years of experience and 17.5% have more than 16 years of experience. Lastly, of 40 respondents that participated, 70% of them were quantity surveyors, 5% were construction managers, 5% were construction project managers, 2.5% were structural engineers, 10% were construction project managers and 7.5% were architects.

4.2 Influence of Job Dissatisfaction on Absenteeism and Employee Turnover

Table 1 reveals that the first seven highly ranked factors having influence on absenteeism, are: too low wage level, lack of incentive; lack of development opportunities, not having enough work to do, lack of enforcement of fair policies for all; lack of monitoring of being absent; lack of motivation to come to work. On the other hand, personal safety concern, poor relation with co-workers, lack of clear target (deadlines), lack of recognition from boss; repetitive work, poor working condition, and unchallenging work, were ranked the least. Regardless of some factors being ranked the least, it should be noted that all factors recorded mean item score falling under "somewhat influential". Meaning that, all factors associated with dissatisfaction were perceived to have influence on absenteeism.

These findings are supported by the work of Horner and Talhouni (1984), where factors such as unsafe working conditions and lack of job security were identified as factors that influence employee absenteeism. Similarly, Cheney (1984) identified organizational factors such as participation, consideration, feedback; while Rojas (2008) identified pay, company size, responsibilities, tasks and worker commitment as factors that influence absenteeism in the construction industry. As such, it can be deduced that construction job dissatisfaction have negative influence on absenteeism. According to Hadebe (2001), when employees are dissatisfied with their jobs, they try to find ways of reducing their exposure to them, and this will be to stay away from those jobs.

Table 1. Job dissatisfaction-related factors that contribute to absenteeism.

Job satisfaction related factors	MIS	SD	Rank
Too low wage level	2.50	1.18	1
Lack of development opportunities	2.48	1.01	2
Lack of incentive	2.48	1.06	2
Not having enough work to do	2.48	1.09	2
Lack of enforcement of fair policies for all	2.45	0.96	3
Lack of monitoring of being absent (feeling invisible)	2.43	1.03	4
Lack of motivation to come to work	2.4	0.98	5
Personal safety concerns	2.38	1.23	6
Poor relation with co-workers	2.33	0.97	7
Lack of clear target (deadlines)	2.31	1.08	8
Lack of recognition from boss	2.25	1.03	9
Repetitive work	2.23	1.03	10
Poor working conditions	2.18	1.10	11
Unchallenging work	2.13	0.97	12

4.3 Factors that have Influence on Employee Turnover

Table 2 indicates that the first seven highly ranked as the influencers of employee turnover associated with dissatisfaction in the construction industry are: better benefits package elsewhere; lack of development opportunities; unchallenging work; poor relationship with boss; lack of recognition from boss; and lack of enforcement of fair policies for all. On the other hand, inadequate resources, lack of advanced technology, poor relationship with co-workers, safer site elsewhere, and clean job versus dirty job, were ranked the least.

Table 2. Job dissatisfaction-related factors that contribute to employee turnover.

Job satisfaction related factors	MIS	SD	Rank
Better benefits package elsewhere	3.23	1.06	1
Lack of development opportunities	2.87	1.08	2
Poor relationship with boss	2.64	1.01	3
Unchallenging work	2.64	1.18	3
Lack of recognition from boss	2.59	1.04	4
Lack of enforcement of fair policies for all	2.5	1.13	5
Inadequate resources	2.44	1.10	6
Lack of advanced technology	2.08	1.02	7
Poor relationship with co-workers	2.00	1.00	8
Safer site elsewhere	1.83	0.98	9
Clean job versus dirty job	1.82	1.06	10

In support of these findings, Tnay *et al.* (2013) concluded that pay and supervisory support act as a crucial factors affecting employee's turnover intention; while Sichani (2010) identified lack of recognition/incentives as other factors that influence employee turnover. Also, the findings were supported by the work of Rojas (2008) that, company, responsibilities, tasks and worker commitment, better benefit package elsewhere, lack of adequate tools and equipment, are factors that have a negative influence on employee turnover.

5 CONCLUSION AND RECOMMENDATION

This paper explored the determinants of job satisfaction in the construction industry as well as the influence of factors leading to dissatisfaction on absenteeism and employee turnover. The findings revealed that, aspects associated with job dissatisfaction, such as: salary, incentives, professional growth, working relationship as well as enforcement of fair policies, were the most influential factors of absenteeism and employee turnover in the South African construction industry. Hence, it is recommended that, employers should from time to time, assign employees to different types of projects, such as residential, commercial, and civil works amongst others to gain a wider experience. This will help equip employees with different experiences; thereby ensuring continuous professional and personal development. Also, it is crucial for employees to be involved in projects from the start until completion. This will allow employers to have the holistic opportunity to monitor the performance and contribution of employees involved in such project. As a result, the employer personal's responsible for mentorship will be able to give feedback on employee's performance where necessary. This can assist with the rewarding of employee's through promotion (associated with salary increase), incentive bonus or an award as a way of appreciation. Also, the employers should improve the working environment to enable employees to perform tasks with comfort. This can be achieved by modifying and improving working environment through provision of adequate lighting, comfortable temperatures and ventilation, as well as access to eating area.

References

Bernold, L. E and AbouRizk, S. M., Managing performance in Construction: The Complexity of Human Motivation. New Jersey: John Wiley and Sons, Inc., 2010.

Cheney, P. H., Effects of Individual Characteristics, Organizational Factors and Task Characteristics on computer Programmer Productivity and Job Satisfaction. PHD. University of Minnesota, 1984.

CIDB and University of Free State., 2013, The CIDB Construction Industry Indicators: Summary Results. Retrieved from http://www.cidb.org.za/Documents/KC/cidb_Publications/Ind_RpsOther/ind_reps_cii_2013.pdf on 27 March 2015.

Deeprose, D., Smart Things to Know about Motivation. UK: Capstone Publishing Limited, 2003.

Hadebe, The Relationship between Motivation and Job Satisfaction of Employees at Vista, Information Services. Degree of Masters in Administration.,University of Johannesburg, 2001.

Horner, R. M. W. and Talhouni, B. T., Effects of Accelerated Working, Delays and Disruption on Labour Productivity: Absenteeism. UK. Publisher Unknown, 1984.

Lavender, S., Management for the Construction Industry: Individual Motivation. Addison Wesley Longman. Limited. England, 1996.

Neuman, W. L., Social Research Methods: Qualitative and Quantitative Approaches. USA: A Pearson Education Company, 2000.

Rojas, E. M., Construction Productivity: Absenteeism and Turnover. United State of America: ELECTRI International, 2008.

Schaffer, R. Managing Productivity. Mumbai: Jaico Publishing House, 2001.

Sichani, M. S., Understanding Absenteeism in Construction: A Pilot Study on Industrial Construction Alberta. Masters of Science in Construction Engineering and Management. University of Alberta, 2010.

Tnay, E., Othman, A.E.A, Siong, H.C and Lim, S.L.O. The Influences of Job Satisfaction and Organisational Commitment on Turnover Intention. Proceedings of Social and Behavioral Sciences, 6 November 2013, Universiti Malaysia Sarawak, pp. 201-208, 2013.

Tyilana, E. X., Impact of Motivation on Job Satisfaction Amongst employees Of A National Broadcaster. Degree of Master Commerci. University of Johannesburg, 2005.

Walters, J., Positive Management: Increasing Employee Productivity: Positive Management, Strategy, and Productivity. New York: Business Expert Press, 2010.

Integrated Solutions for Infrastructure Development
Edited by Lau, H. H., Tang, F. E., Ng, C. K., and Singh, A.
Copyright © 2016 ISEC Press

PERFORMANCE OF REQUEST FOR INFORMATION (RFI) MANAGEMENT

MUHAMAD ABDUH and FARREL ADAM

Faculty of Civil and Environmental Engineering, Institut Teknologi Bandung, Bandung, Indonesia

Request for information (RFI) is considered as one of the important administrative tools in construction projects. It can be used as an indicator on how well the coordination between parties, especially between contractors and engineers and/or supervisors, performed in a project. Delays in a construction project could be identified by analyzing the RFI documents, but on the other hand, they could also be caused by poor administrative in handling the RFI documents and by delays in processing RFIs. This paper reviews the performance of four building projects in handling their RFI documents. The four buildings were constructed by a national joint-operation contractor, designed and supervised by an international joint-operation consultant. The RFI responses and wok-in-process were used to measure the project's performance in handling the RFI documents; from the contractor's and the engineer's point of views. The analysis results show that the engineer had been very responsive in processing RFIs, while the contractor had difficulties. In general, the contractor was lack of capable personnel in handling RFIs for all four buildings. Moreover, communication problem was the key of the project's performance due to different language and culture. The authors believed that the results of the analysis would influence to the overall project performance.

Keywords: Contractor, Documentation, Engineer, Response, Work-in-process.

1 INTRODUCTION

According to Lee and Larry (2009), a good project management practice is determined by its organization structure and information system. Communication and distribution of data and information related to all project activities in the field to all stakeholders of the project is then crucial to its success. The fragmentation problem of the construction industry makes good communication and robust information management in a project is a must (Dave and Koskela 2009).

Moreover, it is not uncommon that construction design documents, i.e., design drawings, technical specification, and quantities, are far from perfect to be built by the contractors. It is a contractor's duty to identify the imperfections existed in the construction documents and then to coordinate with the engineer, or the owner's representative in the field to find the solutions. Sometimes, the designer should be involved in solving such problems. In this kind of situation, the RFI or Request for Information procedure is widely used in construction projects. It is a formal procedure raised by contractor to confirm the interpretation of specification and drawings, or to secure a documented directive from the engineer, or the owner's representative before continuing the works (Tadt *et al.* 2012). RFI could be considered as one of the important administrative tools in construction projects. It can be used as an indicator on how well the coordination between parties, especially between contractors and engineers and/or supervisors, performed in a project. Delays in a construction project could be identified by analyzing the RFI documents, but on the other hand, they could also be caused by poor administrative in handling the RFI documents and by delays in processing RFIs. RFI processes need time to conduct and will influence the schedule performance of a project (Mohamed *et al.* 1999).

This paper reviews the performance of a project, constructing four buildings in a university complex, in handling the RFI documents. The need for the review was coming from the fact, as the research was taken, that the contractor's performance was considered already behind schedule and the contractor claimed that the reason for the delay was from the engineers' side in responding the contractor's RFIs. The authors were informally involved in this investigation to give impartial information to the owner.

2 THE FOUR-BUILDING PROJECT

The case study used in the research was a four-building project in a university campus, in Bandung city, Indonesia. There were four different buildings located on different locations at the campus complex and constructed by a national joint-operation contractor. In general, the buildings have 6 to 7 stories and similar functions for accommodating lecture rooms, laboratories, faculty rooms, and offices. The contractor decided to have four different field and engineering teams; while for administrative and management matters were centralized. The contractor had a contract of 18.75 Million USD that should be finished in 18 months. The project was considered as an international construction project and the English was used as the formal language in the project.

On the other side, the owner was supported by an engineering and supervision services consultant to design the buildings as well as to supervise the construction process. The consultant was an international joint-operation consultant (further referred as engineer); a Japanese consultant and two Indonesian consultants. For engineering works and supervision, the Japanese consultant held a role as the leader.

During the construction process, the contractor frequently complained on the stringent supervision by the engineer. The contractor also found some imperfect construction drawings and specification that led to issuance of RFI to the engineer. The communication related to RFI was considered a lengthy process and might cause a delay to the actual works. Misunderstanding and different views of the issues discussed in RFIs were common. Delays and slowing down progresses of some construction works were obviously noticed. The contractor claimed that RFI process might cause the project performance and the lengthy process of RFIs were caused by inability of the engineer to respond them. On the other hand, the engineer pointed the cause of it to the contractor.

3 METHODOLOGY

The authors were informally engaged to the project to investigate the situation and to provide recommendation to the owner to solve the problem after the project underwent 375 days or about 13 months, with only less than 50% progress. In order to achieve the objective of the investigation, 1,876 RFI documents were collected and the following analyses were used:

1. Categorize all RFIs using an expanded-classification suggested by Tilley *et al.* (1997). They are:
   (a) Design Alternative
   (b) Approval
   (c) Confirmation
   (d) Clarification
2. Measure the Response Performance Indicator (PI_2) as suggested by Tilley *et al.* (1997), where:

$$PI_2 = \frac{1}{N_c}\sum\frac{T_a - T_r}{T_a} \tag{1}$$

   Where
   N_c = Number of RFIs
   T_r = Maximum response time of RFI as written in the contract (day)
   T_a = Actual response time of each RFI (day)

3. Measure the Work-in-Process Indicator (*Average WIP*) as suggested by Chin (2009), where:

$$Average\ WIP = \frac{Total\ Number\ of\ WIP}{Project\ Duration} \tag{2}$$

4. Measure the On-Time Rate Indicator (*OTR*) as suggested by Chin (2009), where:

$$OTR = \frac{Number\ of\ On\text{–}Time\ and\ Earlier\ Delivered\ RFIs}{Total\ Number\ of\ RFIs} \tag{3}$$

5. Measure the Documentation Quality Indicator (PI_1) as suggested by Tilley *et al.* (1997), where:

$$PI_1 = \frac{N_c}{CV \times D} \tag{4}$$

   Where:
   N_c = Number of RFIs in 'Clarification' category
   CV = Contract Value (in 100.000 AUD)
   D = Initial project duration (month

4 ANALYSIS AND DISCUSSION

4.1 RFIs Documents

Based on 1,876 RFIs, as depicted in Table 1, there were only few RFIs that could be categorized as design alternative (0.37%), confirmation (0.05%), and clarification (0.43%). Most of the RFIs were approvals related RFIs (99.15%). Based on further analysis, the approvals related RFIs consisted of document for approving the drawings (60.43%), submitted materials (9.25%), and work permits (30.32%) as depicted in Table 2.

It can be said that if the project were delayed because of RFIs, then late approval of work permit due to unready site for the work and efforts related to detailing the drawings for construction would be the causes. The findings raised an issue of technical personnel's' and sub-contractors' competences in fulfilling the requirements given by the engineer. The issue of poor communication between contractor and engineer would also be the case due to different culture and language.

Table 1. RFIs category of the case study.

RFIs	Frequency	Percentage
Design Alternative	7	0.37%
Approvals	1,860	99.15%
Confirmation	1	0.05%
Clarification	8	0.43%
Total	1,876	100.00%

Table 2. Approval RFIs category.

RFIs	Frequency	Percentage
Drawings	1,124	60.43%
Materials	172	9.25%
Work Permit	564	30.32%
Total	1,860	100.00%

4.2 Response Performance

It was informed from the agreement between parties in the project, that the maximum response time for RFI was 3 days (T_r). The actual response time of all RFIs were calculated (T_a), and by using Eq. 1, the Response Performance Indicator (PI_2) was found to be 0.094. Based on a graphic provided in Tilley *et al.* (1997), the $PI_2 = 0.094$ was plotted on the graphic and found that the response performance could be categorized as a 'Good'. It means that the response from the engineer is timely, and therefore there should be no complain on engineer's performance in responding the RFIs.

4.3 Work-in-Process of RFIs

In order to calculate the Work-in-Process Indicator, the total number of work-in-process (*WIP*) should be calculated first. The RFI's dates of submission and return were identified, and the daily number of RFIs that were being processed by the engineer could be calculated. For 375 days, it was found that total number of work-in-process RFIs was 6,297 RFI-day. Using Eq. 2, the average RFIs that were being processed by the engineer was 16.8 RFIs daily. The distribution of daily work-in-process RFIs is depicted in Figure 1. The maximum WIP was 67 RFIs daily.

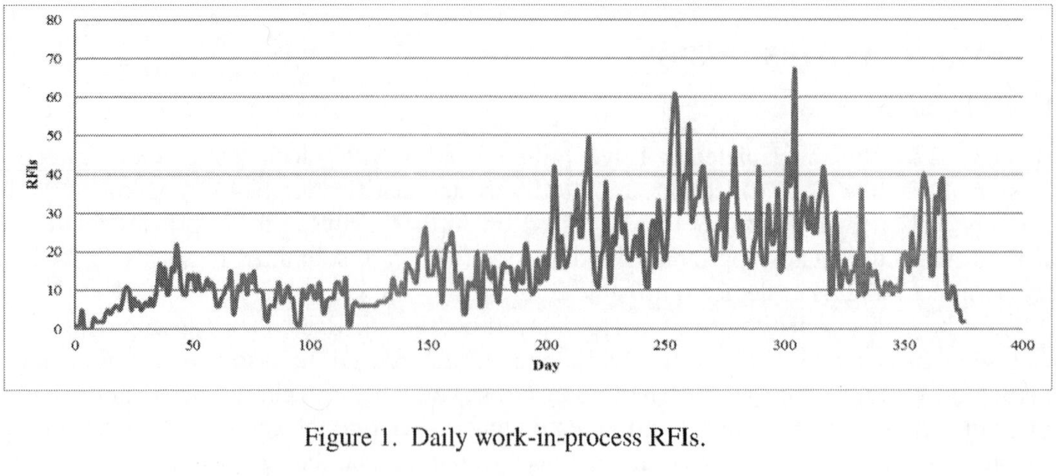

Figure 1. Daily work-in-process RFIs.

4.4 On-Time Rates Performance

Moreover, to get the information on how good the engineer's response to each RFI the contractor gave, the on-time rates of RFIs was calculated as suggested in Eq. 3. For that purpose, the Figure 2 is important to be shown. It can be seen that there were 1,506 RFIs that were responded on time or sooner by the engineer. And therefore, the OTR was 80.28% with average time needed to respond RFI was 2.45 days. This is still less than the maximum days needed to respond the RFI as written in the contract, which is 3 days. As previously stated, the engineer did a good job in responding RFIs from the contractor by processing 16.8 RFIs in average daily. From the interview with the engineer, maximum capacity of the engineer to process the RFIs was 60 RFIs per day since there were 6 technical personnel that could process 12 RFIs daily. It also means that the utility of technical personnel of the engineer was still low, only 28%.

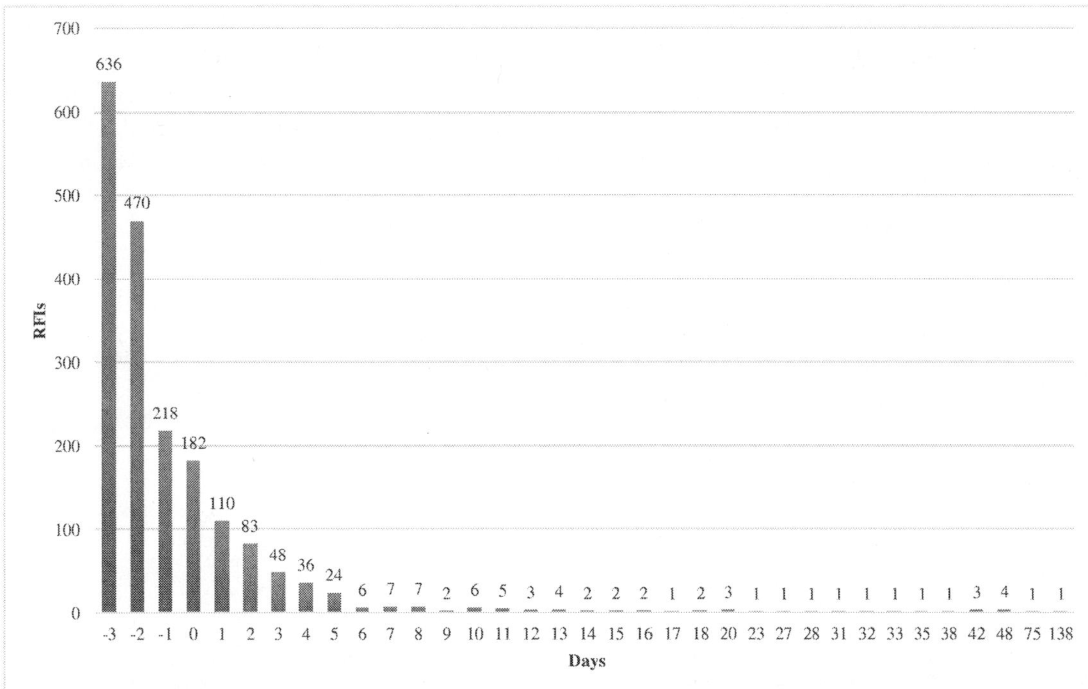

Figure 2. Response time of the RFIs by engineer.

4.5 Documentation Quality

Besides the above mentioned analyses, it is worth to know how good the project documentation that was given to the contractor before the project began. Number of RFIs issued by the contractor could be influenced by the quality of the project documents given to the contractor (Tilley *et al.* 1997). By using Eq. 4, the Documentation Quality Indicator (PI_1) can be calculated. It was found that PI_1 of this project was 0.008. Based on a graphic provided in Tilley *et al.* (1997), the PI_1 = 0.008 can be plotted to the graphic and found that the documentation quality could be considered as an 'Excellent' one. It means that the products of the engineer in designing the four buildings were categorized as excellent; imperfect documents were merely found by the contractor.

5 CONCLUSION

The contractor's claim on the cause of delays of the project due to late response of the engineer's side in processing the contractor's RFI was not verified; the response performance indicator was good, the work-in-process indicator was still far below the engineer's capacity, the on-time-rate indicator was also good, and lastly the documentation quality was considered excellent. Those analyses have shown the problem faced by the project was not on the engineer's side. On the other hand, the ability of the contractor in managing the project documentation was questionable.

As also shown in the analysis, if the project were delayed because of RFIs as the contractor claimed, then the causes would be late approval of work permits due to unprepared site for the work and low capabilities in preparing shop drawings for construction. Improvement of contractor's personnel for site supervision as well as the technical personnel for preparing drawings would be the solution for that. The contractor's capability in managing the subcontractors to comply with the demanding engineer's requirements would also be the cause of the delays. Therefore, the coordination with the subcontractors should be improved.

The authors believe that the root cause for the problem in this project would be poor communication between the engineer's and the contractor's personnel due to different culture and language. Introducing a Japanese-Indonesian interpreter as well as personnel in the engineer's side who are able to speak Indonesian would help the project since the use of English as the bridging language seemed to be not effective.

Acknowledgments

The authors would like to thank the Head of Project Implementation Unit of the ITB JICA Development Project III, the Engineering Service Consultant, and the contractor for providing data and access for the investigation of the delays that occurred in the project.

References

Chin, Chang-Sun, Work-in-Process and Construction Project Information Flows, *Proceedings for the 17th Annual Conference of the International Group for Lean Construction,* Taipei, Taiwan, 15-17 Jul 2009, pp 257-266.

Dave, B., and Koskela, L., Collaborative Knowledge Management, A Construction Case Study, *Automation in Construction*, 18 (7) 2009, 894 – 902.

Lee, J. K., and Larry, P. R., *Operation Management (9th Edition)*, Boston: Pearson Education, 2009.

Mohamed, S., Tilley, P. A., and Tucker, S. N., Quantifying the Time and Cost Associated with The Request for Information (RFI) Process in Construction, *The International Journal of Construction Information Technology*, 7(1) 1999, 35-50.

Tadt, E., Hanna, A., and Whited, G., *Best Practices from WisDOT Mega and Arra Projects – Request for Information: Benchmarks and Metrics*, University of Wisconsin – Madison: Construction and Materials Support Center, 2012.

Tilley, P. A., Wyatt, A., and Mohamed, S., *Indicators of Design and Documentation Deficiency*, 1999. Retrieved from www.fep.up.pt/disciplinas/PGI914/Ref_topico1/indicators_of_Design_Deficiency.pdf February 12, 2015.

Integrated Solutions for Infrastructure Development
Edited by Lau, H. H., Tang, F. E., Ng, C. K., and Singh, A.
Copyright © 2016 ISEC Press

USE OF UNMANNED AERIAL VEHICLES AND COMPUTER VISION IN CONSTRUCTION SAFETY INSPECTIONS

MOHAMAD ABBAS, BAHAA EDDINE MNEYMNEH, and HIAM KHOURY

Civil and Environmental Engineering Dept, American University of Beirut, Beirut, Lebanon

Construction is unarguably one of the most dangerous industries whereby many hazardous tasks and conditions exist, which may pose injuries, risks and fatalities to the workers. Hence, safety inspections are continuously carried out to maintain a safe environment. These typically involve safety officers who circulate around the construction site to detect unsafe working conditions, and ensure compliance with health and safety regulations. However, the task of efficiently supervising a large number of workers and consistently identifying all possible violations is still considered manual and tedious. Therefore, this paper takes the initial steps and presents work targeted at automating the safety inspection process using Unmanned Aerial Vehicles (UAVs). These are commonly known as drones and are small, aerial camera-equipped robots capable of rapidly visualizing spacious environments. More specifically, in this study, a UAV system is used to capture real-time videos from a construction site. The videos are then streamed to a central automated system and analyzed using digital image processing techniques to check whether construction workers are wearing personal protective equipment (PPE), in particular hard hats. The components of the proposed system were created and preliminary results highlighted the potential of using camera-equipped UAVs and computer vision to automate safety inspections in construction environments.

Keywords: Hazards, PPE, Automation, Drone, Image processing.

1 INTRODUCTION AND PERTINENT LITERATURE

According to the United States Department of Labor, 874 worker deaths were recorded in the construction sector in 2014, out of which 73 (8.4%) were caused by being struck by objects (OSHA 2016). Hence, inspections have been carried out over the years to check whether workers are wearing personal protective equipment (PPE), in particular hard hats. However, this inspection task is still considered manual and time-consuming (Ham *et al.* 2016). Therefore, the objective of this paper is to improve and automate the construction safety inspections using UAVs or drones. Drones have been recently used on construction sites in many different applications such as: progress monitoring, site monitoring and building inspection (Ham *et al.* 2016). More specifically, Gheisari *et al.* (2014) evaluated the potential applications that unmanned aerial systems (UAS) have for improving safety in construction environments. In this case, safety officers were provided with captured videos from construction sites and the number of hard hats was manually counted.

However, this paper aims at automating this process and identifying workers not wearing hard hats by analyzing captured drone videos using several digital image processing techniques (Chdid *et al.* 2011, Oueiss *et al.* 2012).

2 METHODOLOGY

To achieve the objective identified above, the paper addresses issues in two main task areas: 1) Hardware, and 2) Software and Algorithms.

2.1 System Hardware: UAV Specifications

The UAV used to capture videos from a construction site is a 2.4 GHz Hubsan drone equipped with a camera. It is characterized by a lightweight airframe with nice durability. Its six-axis flight control system with adjustable gyro sensitivity permits stable flights. The drone hovers over the site, and capture videos without affecting construction activities nor disturbing construction workers.

Figure 1. Hubsan drone (Hubsan 2015).

2.2 System Software and Algorithms

The videos captured by the UAV were analyzed using image processing algorithms. More specifically, hard hats were detected using feature detection, extraction, and matching, and the total number of workers was obtained using motion-based multiple object tracking. As such, subtracting the number of hard hats obtained from the number of workers returns the number of workers not wearing hard hats. The following subsections describe both algorithms.

2.2.1 *Feature detection, extraction and matching*

The implemented algorithm detects a hard hat by finding point correspondences between a reference image (image of a single hard hat) and the target image (image of a construction site captured by a drone). According to MathWorks (Matlab 2016), the object detection method works best for objects that display non-repeating texture patterns to allow unique feature matches. Since the helmet is a uniformly-colored object, a sticker was plastered on its top, as shown in Figure 2. This greatly increased the number of features detected by the algorithm as shown in Figure 3.

Subsequently, outliers were removed and the transformation matrix was calculated, using Random Sample Consensus (RANSAC) algorithm (Khoury *et al.* 2015). This helped identifying the hard hat having the best match with the reference image. The number of hard hats was then computed by hiding identified ones from the target image so that the next best match hard hat can be detected in the next iteration of the algorithm. This counting iterative process halts when no more hard hats can be detected in the target image.

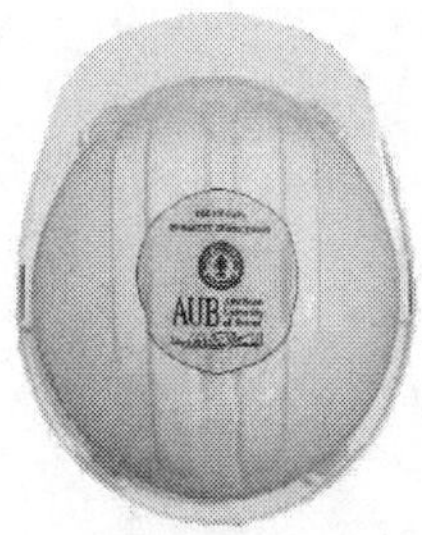

Figure 2. Sticker plastered on the hard hat.

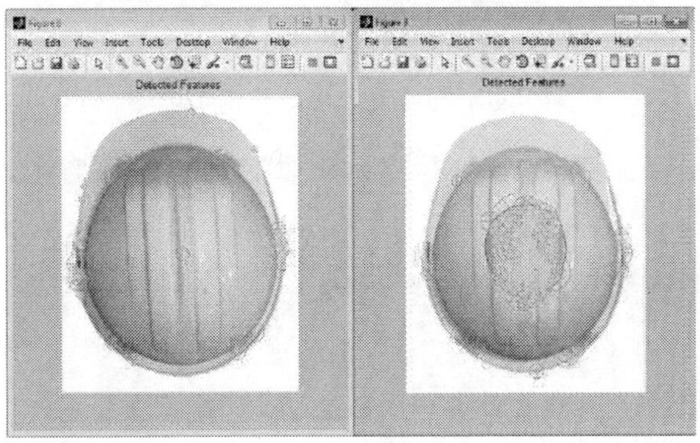

Figure 3. (Left) 49 detected features without sticker and (right) 282 with sticker.

2.2.2 *Motion-based multiple object tracking*

This algorithm performs automatic detection and motion-based tracking of moving objects in a video (Veeraraghavan *et al.* 2006). As a matter of fact, mobile workers were detected using a background subtraction algorithm based on Gaussian mixture models. Morphological operations are applied to the resulting foreground mask to eliminate noise. Finally, blob analysis detects groups of connected pixels, which are likely to correspond to moving objects.

For this algorithm, the motion-based multiple object tracking, to work properly, the drone needs to remain in a fixed location for a period of time.

3 EXPERIMENTS AND RESULTS

In order to demonstrate the feasibility of the aforementioned methods and evaluate their performance, experiments were conducted on a construction site using the drone (Figure 4).

The hard hat counting algorithm was tested on images extracted from the captured video. In the first iteration, 63 matching features were found between the reference and the target image (Figure 5).

The first detected hard hat was then hidden from the target image and in the second iteration, 44 matching features were found between the reference and the new target image (Figure 6).

After the second hard hat was detected then hidden from the target image, no more matching features were detected and the algorithm stopped.

On the other hand, the second algorithm (motion-based multiple object tracking algorithm) was tested on videos captured from the same construction site. Accurate results were obtained while the drone was in a fixed location and expectedly static workers could not be detected (Figure 7).

Figure 4. Drone capturing videos on construction site.

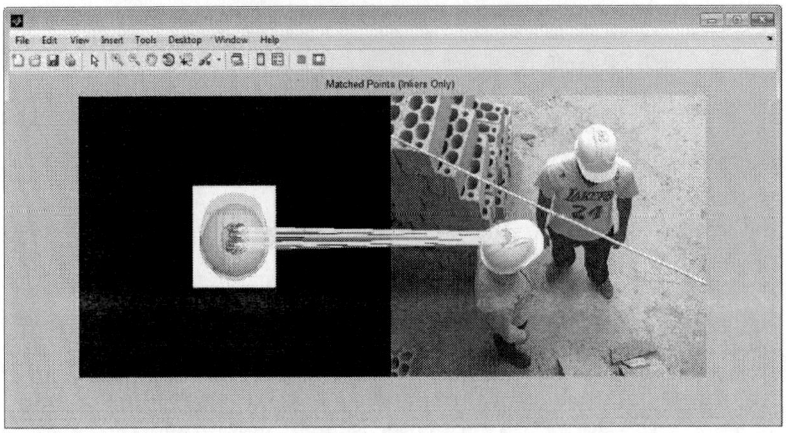

Figure 5. Matching features for the first hard hat.

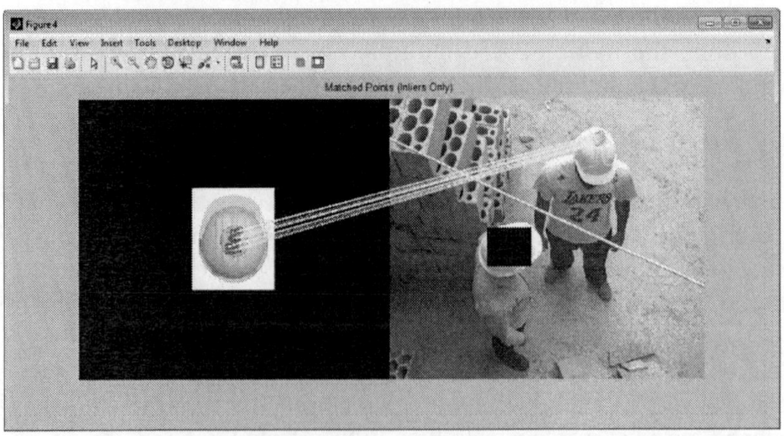

Figure 6. Matching features with the second hard hat.

Figure 7. Motion-based multiple object tracking.

4 CONCLUSION

This study focused on creating an intelligent system targeted at automating safety inspections on construction sites, by detecting workers not wearing hard hats. This was achieved by capturing videos using a drone or UAV and analyzing them using digital image processing techniques. More specifically, the number of workers not wearing hard hats was computed by subtracting the number of hard hats detected using feature detection, extraction, and matching from the total number of workers obtained using motion-based multiple object tracking.

One limitation of the proposed approach is that hard hats placed on the ground can be sometimes detected instead of those actually worn by workers. As such, future work will improve the developed algorithms to rapidly and accurately detect the mobile hard hats and workers. Further research will opt for an autonomous drone with better stability and equipped with a higher resolution camera. Additional work is needed as well to determine the exact location of workers not wearing hard hats.

References

Chdid, D., Oueis, R., Khoury, H. M., Asmar, D., and Elhajj, I., Inertial-Vision Sensor Fusion for Pedestrian Localization, *Proceedings of the 2011 IEEE International Conference on Robotics and Biomimetics (ROBIO)*, 1695 – 1701, Pucket, Thailand, 2011.

Gheisari, M., Irizarry, J., and Walker, B. N., UAS4SAFETY: The Potential of Unmanned Aerial Systems for Construction Safety Applications, *Construction Research Congress*, Castro-Lacouture, D. (ed.), 1801-1810, Atlanta, Georgia, 2014.

Ham, Y., Han, K. K., Lin, J. J., and Golparvar-Fard, M., Visual monitoring of civil infrastructure systems via camera-equipped Unmanned Aerial Vehicles (UAVs): a review of related works, *Visualization in Engineering*, 4:, DOI: 10.1186/s40327-015-0029-z, January, 2016.

Hubsan, Porducts, The Hubsan X4, Hubsan.com, Retrieved from http://www.hubsan.com/product/index413.html on December 15, 2015

Khoury, H., Chdid, D., Oueis, R., Elhajj, I., and Asmar, D., Infrastructureless approach for ubiquitous user location tracking in construction environments, *Automation in Construction,* Elsevier, 56, 47-66, August, 2015.

MATLAB, Product Help, Object Detection and Tracking. Retrieved from http://www.mathworks.com/help/releases/R2014a/vision/gs/object-detection-and-tracking.html#btt5qyu on January 28, 2016

Occupational Safety & Health Administration, United States department of labor, Retrieved from https://www.osha.gov/oshstats/commonstats.html on January 26, 2016

Oueiss, R., Chdid, D., Khoury, H. M., Asmar, D., and Elhajj, I., Infrastructureless Inspection Process Automation in Construction, Journal: Gerontechnology, 11(2), International Society for Gerontechnology, 92-98, Eindhoven, Netherlands, 2012.

Veeraraghavan, A., Srinivasan, M., Chellappa, R., Baird, E., and Lamont, R., Motion based correspondence for 3D tracking of multiple dim objects, *2006 IEEE International Conference on Acoustics Speech and Signal Processing Proceedings*, 696 – 672, Toulouse, France, 2006.

Integrated Solutions for Infrastructure Development
Edited by Lau, H. H., Tang, F. E., Ng, C. K., and Singh, A.
Copyright © 2016 ISEC Press

MODIFIED CONCEPTUALIZED HEALTH AND SAFETY COMPLIANCE MODEL FOR SMALL AND MEDIUM-SIZED CONTRACTORS IN GHANA

ZAKARI MUSTAPHA, CLINTON AIGBAVBOA, and WELLINGTON DIDI THWALA

Dept of Construction Management & Quantity Surveying, University of Johannesburg, Johannesburg, South Africa

The majority of the construction industries in Ghana are dominated by the Small and Medium Sized (SMEs) contractors. The SMEs inability to manage Occupational Health and Safety (OHS) effectiveness has been the contributing factor to the accident. A review of the literature was conducted for the development of Health and Safety (H&S) compliance model for SMEs contractors in Ghana. The study adopted various constructs from Accident Root Causes Tracing Model (ARCTM) and Domino theory. Both objective and subjective attributes were combined during the assessment. The Domino Theory form the basis for the theoretical and conceptual framework of this paper. The hypothesized H&S compliance model is presented in this paper based on an in-depth review of the previous models. The gaps identified in the literature are government support and contractor's organizational culture. Discussions also included the integrated holistic model and the variables of the model, identification of the model and justification for the selected variables. The identified gaps serve as variable constructs in H&S compliance research. The theorized conceptual model is a multidimensional structure composed of six latent constructs.

Keywords: Challenges, Justification, Measurement, Selection, Specifications, Variables.

1 INTRODUCTION

Small and Medium-Sized Enterprise (SMEs) contractors form bulk of the constrictors' in Ghana (Ofori and Toor 2012; Frempong and Essegbey 2006; Laryea 2010; Kheni *et al.* 2007; Ghana Health Service, GHS, 2007). This limitation prevents the implementation of H&S standards on construction sites. Improving Occupational Health and Safety (OHS) has always been a concern for intensive actions to be taken. The prevalence of OHS challenges in Ghana might also be attributed to limited financial resources and attention given to OHS. Kheni and Braimah (2014) asserted that the attitudes of construction companies towards H&S in Ghana have been affected by institutional structure responsible for H&S implementing standards at workplaces. Kheni and Braimah (2014) cited the following as the major problems: poor coordination of the activities of the many institutions responsible for implementing H&S standards, lack of specific H&S regulation and undesirable level of compliance with relevant H&S legislation. It was further reported that for OHS to be implemented, it is necessary for construction companies to have a positive change in their attitudes. The implementation of OHS will be achieved by re-structuring of the OHS administration system in Ghana (Kheni and Braimah 2014). Various attempts by previous research works to substantiate the relationship between compliance with H&S seem elusive. The issue of H&S compliance in the construction industry is important because of the high rate of accidents. This study attempts to develop H&S compliance model for SMEs contractors' in Ghana to assist in accident minimization. The discussions of the H&S model for SME's contractors are based on the selection of the variables for H&S compliance model. The conceptual model latent features, specification and justification of the models and structural component of the model were discussed. This is followed by H&S compliance model and measurement component of the model.

2 SELECTION OF VARIABLES FOR HEALTH AND SAFETY COMPLIANCE

This section provides details on the selection of variables for Health and Safety (H&S) compliance model. The International Labor Organization (ILO) (in National Occupational Health and Safety Policy of South Africa (2003) indicated that safe work creates no obstacles to being competitive and successful. Health and safety in the workplace is about preventing work-related injury and disease, and designing an environment that promotes well-being for everyone at work (Safe work Australia 2013; Heinrich *et al.* 1980). The recklessness and undesirable traits leading to accident can be prevented when employers ensure the safety of employees at work environment. Jamal Khan (in Mat Zin and Ismail 2012) opined that ignorant behavior and attitude of employers and employees contribute to rise of issue on behavioral safety non - compliance to requirements of Occupational Safety and Health Act (OSHA) 1994". "Most of the accident causation theories addressed the human (worker) as the main problem that makes an accident happen such as permanent characteristic of human, the combination of extreme environment and overload of human capability and conditions that make human tends to make mistake" (Abdelhamid and Everett, 2000). Safety behavior or acts can be achieved through the following processes: working with safety devices such as personal protective equipment, use of equipment that are in good condition, follow the correct work procedure at any time work is to be carried out, employees should have good knowledge level of work and they should also obey work procedures whenever they are carrying out any activity. Smallwood (2010) identified workers attitude as one of the factors leading to unsafe act of a worker. Workers safety behavior will contribute to safety practices therefore. The worker conducts safe act under the condition that he has undergone safety training and been provided with health and safety equipment to protect him from any harm. Hosseinian and Torghabeh (2012), Fang, Choudhry and Hinze, (2006) and Abdul Hamid, Yusuf and Singh (2003) are of the view that a worker must perform safe acts and this should be in relation to safety standards, by working with protective equipment and when they have enough rest before the day's work to prevent any accident occurring. The resultant of unsafe acts or unsafe condition is accident (Heinrich *et al.* 1980). It is therefore mandatory for employers to provide their employees with safe working conditions to enable them abide by health and safety regulations.

Therefore, the Health and Safety (H&S) compliance study models for the assessment of H&S compliance was made up of objective and subjective attributes. The Domino Theory by Heinrich and that of Adams had similar concept, but the elements were different (Heinrich *et al.* 1980). Weaver had similar concepts of elements or factors as Henrich's (Heinrich *et al.* 1980). Petersen's model developed in the 1971 had different concept with the Domino Theory of 1959 which had influence on many researchers during Heinrich's time. It is believed that the contributing factors, causes, and sub-causes are the main culprits in an accident scenario as inspired by the model (Abdelhamid and Everett 2000). Behavior model, human factor model, and Ferrel theory relate to human error theory (Hosseinian and Torghabeh 2012; Hughes and Ferrett, 2007; Taylor, Easter, and Hegney 2004; Abdul Hamid, Yusif, and Singh 2003). Most of these theories address the human (worker) as the main problem that makes an accident happen

(Abdelhamid and Everett 2000). Abdelhamid and Everett (2000) indicated that there is every tendency of humans to make error under various conditions and situations, but finally, the blame will fall on human most often (unsafe). Many important rules of Accident Root Causes Tracing Model (ARCTM) have been derived from the effort of Heinrich in the year 1959. ARCTM insist on specific issues such as worker training, worker attitude and management procedure problems should be recognized and modified to avoid reoccurrence of accident. Research conducted by Addelhamid and Everett (2000) in identifying root causes of construction accidents concluded that the application of ARCTM should serve as a complement to accident investigation process and should be able to give solutions to accident occurrence and preventive measures in the industry. The three constructs proposed by ARCTM to the two construct from Heinrich are supported and adopted for the current study. But both models have one construct in common. This study considers the H&S compliance bundle in a typical construction industry to contain SE with 6 variables; SAW with 20 variables; SWC with 7 variables; RWSC with 5 variables. Almost all the H&S compliance studies have these constructs conceptualized on frequent basis. However, the current study brings into focus GS with five variables and COC with 11 variables each. The two additional constructs are the identified gaps in the literature and were found to be peculiar to Ghana as a developing country.

3 DESIGN/METHODOLOGY

The assessment of H&S compliance model for the study was carried out through the combination of objective and subjective attributes. The Domino Theory form the basis for theoretical and conceptual framework of this paper. The study adopted various constructs from ARCTM and Domino theory. The selection of the variables were based on the theoretical framework built from literature review. Based on the fundamental underpinning of two models, and the incorporated theoretical perspectives, which has been adopted in other similar studies. The model to be conceptualized within the broad theoretical framework is based on the approach used by Heinrich and ARCTM. The model or conceptual framework model for the paper looks at the safe environment (SE), safe acts of worker (SAW), safe work condition (SWC) and reaction of worker to safe condition (RWSC). Government support (GS) and contractor's organizational culture (COC) as the identified gaps will in turn predict the construction industry health and safety compliance. The structural components of the model are: SE, SAW, SWC, RWSC, GS and COC. The measurement component of the hypothesized model comprises of the following health and safety compliance (HSC) factors: SE = 8 measurement variables; SAW = 17 measurement variables; SWC = 17 measurement variables; RWSC = 8 measurement variables; GS = 5 measurement variables; COC= 11 measurement variables and HSC = 7 measurement manifest variables.

4 MODEL SPECIFICATION AND JUSTIFICATION

The theoretical conceptual framework for this paper was built on the work of Heinrich *et al.* (1980) and ARCTM which was also built on the previous accident models. Heinrich *et al.* (1980) conceptualized that ancestry and social environment, fault of a person, unsafe acts and condition lead to accident. Because the cause of accident is people and management handles the prevention of accident. The reason being that majority of accidents are due to human error and the accident can only be prevented if management provides conducive environment for the employees to work. Adams shared similar view with Heinrich. Adams emphases was on management organizational structure and reflect the relationship involved with the causes and effects of all incidents and accidents which has management involvement directly (Heinrich *et al.* 1980). Five elements were stated in both Heinrich and Adams and have similar concept, but the elements were different (Heinrich *et al.* 1980). The role of management in accident prevention was also emphasized in a broader sense taking into consideration the root of unsafe acts or conditions (Heinrich *et al.* 1980). ARCTM conceptualized that unsafe condition, reaction of worker to unsafe condition and unsafe acts of worker lead to accident. The non-compliance level of H&S in the construction industry are related to the environment, unsafe acts, unsafe condition, reaction of worker to unsafe condition and unsafe acts of worker. Both Heinrich and Perterson, as well as ARCTM, emphasized on unsafe acts and unsafe condition as the main causes of accident in the construction industry. The two basic components of the model are: safe acts and safe condition. Based on the fundamental underpinning of two models, and the incorporated theoretical perspectives, which has been adopted in other similar studies, they are therefore useful for conceptualizing the present study as a variety of health and safety studies and health and safety compliance being conceptualized within the broad theoretical framework.

Therefore, the conceptual framework for this paper is primarily based on the approach used by Heinrich and ARCTM. Based on the fundamental factors and constructs associated with all the previous models in my research. The present model or conceptual framework model for the study looks at the safe work environment, safe acts, safe condition, reaction of worker to safe condition and safe acts of worker. These factors have been measured in a majority of the previous studies, but consideration has not been given to government support and contractor's organizational culture; which have been classified as the exogenous variables and their role in predicting overall health and safety compliance, which is the endogenous variable. These will in turn, predict the construction industry health and safety compliance. The reason for combining both objective and subjective indicators within the proposed model is supported by Campbell *et al.* (1976) and Falah *et al.* (1995) who stated that: by themselves, objective indicators are often misleading and will remain so until indicators that human beings attached to them, are obtained. Likewise, by themselves, subjective indicators are insufficient as guides to policy. The conceptual model theorizes that health and safety compliance is established by the relationship that exists between the exogenous variables, which include the basic elements by which the subjective and objective measurements are linked. These variables identified from the review of literature are considered the major determinants of H&S compliance. The determinants identified have been adopted to fit with the H&S compliance in the Ghanaian construction industry.

5 STRUCTURAL COMPONENT OF THE MODEL

The integrated health and safety (H&S) compliance model for the Ghanaian construction industry in the case of developing countries, is derived from safe work environment (SE), safe acts of Workers (SAW), safe work condition (SWC), reaction of worker to safe condition (RWSC), government support (GS) and contractor's organizational culture (COC) in the process of achieving safety in the construction industry. The theoretical underpinning of this priori is derived from the works of Heinrich *et al.* (1980) and ARCTM. Most of the important rules of ARCTM was derived from the efforts of Heinrich, Peterson, Bird, Ferrell and Peterson (Hosseinian and Torghabeh 2012); Jha (2011); Fang, Choudhry and Hinze (2006) as discussed in chapter 3 of my research. The conceptualized model is the notion that compliance of health and safety is related to the evaluation of many variables, such as SE, SAW, SWC, RWSC, GS and COC. It is difficult to discuss the principal variable without reference to variables of government support and contractor's organizational culture and inclusion of the other exogenous variables. The evaluation will depend on the compliance assessment of several indicator variables under each of the exogenous variables. In this study, the objective evaluation of health and safety compliance will be assessed by measuring the actual condition of the construction industry that is an exogenous variable in the model as shown in Figure1.

5.1 Measurement Component of the Theoretical Model

The measurement component of the hypothesized model comprises of the following health and safety compliance factors: SE = 8 measurement variables; SAW = 17 measurement variables; SWC = 17 measurement variables; RWSC = 8 measurement variables; GS = 5 measurement variables; COC= 11 measurement variables and HSC = 7 measurement manifest variables. The success for the consideration of health and safety compliance for the benefit of the construction industry has been theorized in the present model. The health and safety compliance (HSC) model has seven measurement manifested variables as shown in Figure 1.

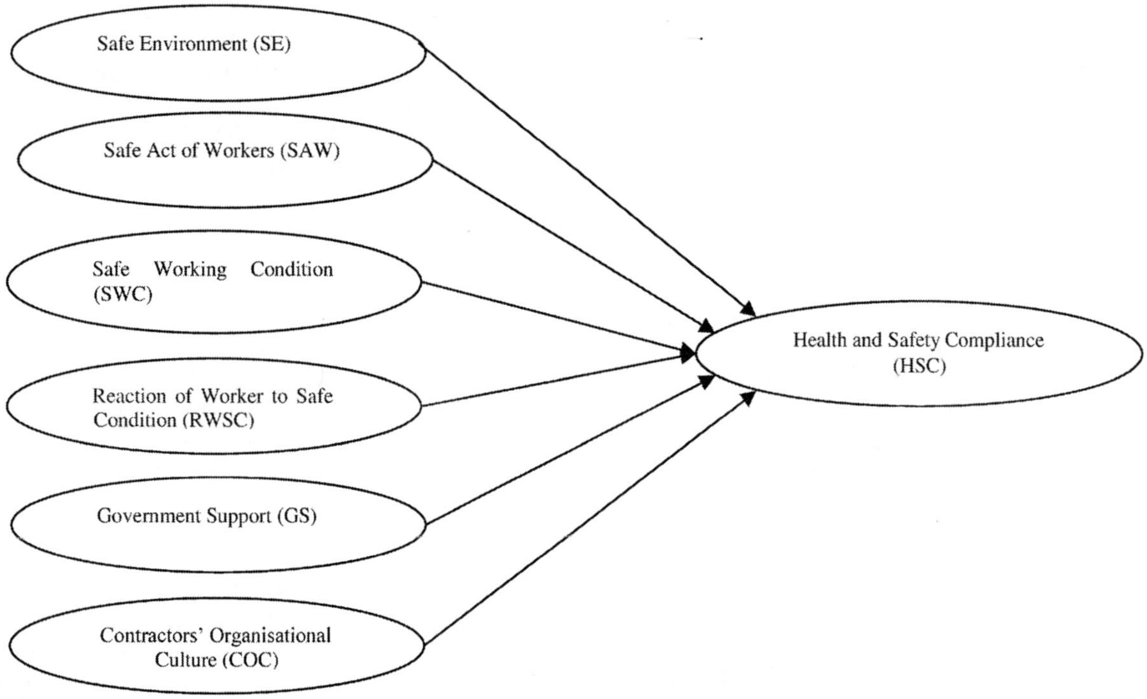

Figure 1. Conceptualized mode for health and safety (H&S) compliance.

6 DISCUSSION OF RESULTS

The main problem that leads to accidents on construction sites as posited by Abdelhamid and Everett (2000) and in most theories were the workers. Worker training, worker attitude and management procedure problems were emphasized as the means of minimizing the reoccurrence of accidents on sites. These findings concur with the findings of Abdelhamid and Everett (2000). However, the H&S compliance study models used for the conceptualization of the H&S compliance model in Ghana was based on Domino Theory by Heinrich and Adams which had similar concept. Three constructs proposed by ARCTM to two construct proposed by Heinrich *et al.* (1980) were adopted in addition to the two gaps identified.

7 CONCLUSION AND FURTHER STUDIES

The theorized conceptual HSC model for SMEs contractors in Ghana is composed of six latent constructs. Further studies should be conducted among large contractors in Ghana to validate the developed conceptual model.

References

Abdelhamid, T. S. and Everett, J. G., "Identifying root causes of construction accidents", Journal of Construction Engineering and Management, 126 (1), 52-60, 2000.

Abdul Hamid, A. R., Yusuf, W. Z. W., and Singh, B., "Hazards at construction sites", Proceedings of the 5th Asia-Pacific Structural Engineering and Construction Conference (APSEC 2003), Johor Bahru, Malaysia, 26-28 2003.

Campbell, A., Converse, P. E., and Rogers, W.J., *The quality of the America life: Perceptions, evaluations, and satisfaction,* New York: Russell Sage Foundation, 1976.

Falah, M., Al-Abed, A., and Stan, W., A model for assessing the effectiveness of public housing in Sana'a (Republic of Yemen). *Construction Management and Economics, 13, 457-465*, 1995.

Fang, D., Choudhry, R. M., and Hinze, J. W., "Proceedings of CIB W99 International Conference on Global Unity for Safety & Health in Construction", Beijing, China, 28-30, 2006.

Frempong, G. and Essegbey, G., Towards an African e- Index, SME e-ACCESS AND USAGE across 14 African countries, 25-27, 2006. Available at: http://www.researchictafrica.net. (Accessed 30 September 2014).

Ghana Health Service (GHS), The health sector in Ghana and figures. Accra. Ghana, 2007.

Heinrich, H. W., Peterson, D., and Roos, N., Industrial accident prevention. McGraw-Hill, 1980.

Hosseinian, S. S. and Torghabeh, Z. J., Major Theories of Construction Accident Causation Models: A Literature Review. Intl Journal of Advances in Engineering & Technology, 4 (2), 53-66, 2012.

Hughes, P. and Ferrett, E., Introduction to Health and Safety in Construction, Oxford, Elsevier Butterworth, Heinemann, 2005.

Jha, K. N., "Construction Project Management: theory and practice", Dorling Kindersley (India) Pvt. Ltd., Licensees of Pearson Education in South Asia, 2011.

Kheni, N. A. and Braimah, C., Institutional and Regulatory Frameworks for Health and Safety Administration: Study of the Construction Industry of Ghana. International Refereed Journal of Engineering and Science (IRJES), 3 (2), 24-34, January, 2014. Available at: www.irjes.org (Accessed 24 December 2015).

Kheni, N. A, Dainty, A. R. J., and Gibb, A. G. F., Influence of political and socio-cultural environments on health and safety management within SMEs: a Ghana case study. *In:* Boyd, D (ed) *Proceedings of the 23rd Annual ARCOM Conference,* 3-5 September 2007, Belfast, UK, Association of Researchers in Construction Management, 159-168, 2007.

Laryea, S., Challenges and opportunities facing contractors in Ghana. In: Laryea, S., Leiringer, R. and Hughes, W. (eds) Procs West Africa Built Environment Research (WABER) Conference, 27-28 July 2010, Accra. Ghana, 215-226, 2010

Mat Zin, S. and Ismail, F., Employers' Behavioural Safety Compliance Factors toward Occupational, Safety and Health Improvement in the Construction Industry, ASEAN Conference on Environment-Behaviour Studies, Savoy Homann Bidakara, Bandung Hotel, Bandung, Indonesia, 15-17, 2011.

National Occupational Health and Safety Policy, Available at: http://www.kznhealth.gov.za (Accessed 15-10-2014), 2003.

Ofori, G. and Toor, S.R., Leadership Development for Construction SMEs, Proceedings - POC 2012 Conference, Working Paper Proceedings, Engineering Project Organizations Conference, Rheden, Netherlands July 10-12, 2012.

Safe work Australia, Available at: http://www.jobaccess.gov.au (Accessed 05 October 2014), 2013.

Smallwood, J. J., "Excavation health and safety (H&S): a South African perspective", In Egbu, C. (Ed) Procs 26th Annual ARCOM Conference, 6-8 September 2010, Leeds, UK, Association of Researchers in Construction Management, 233-241, 2010.

Integrated Solutions for Infrastructure Development
Edited by Lau, H. H., Tang, F. E., Ng, C. K., and Singh, A.
Copyright © 2016 ISEC Press

DEVELOPING ALTERNATIVE CONCEPT FOR RAILWAY OPERATION AND MAINTENANCE CONTRACT

HERAWATI ZETHA RAHMAN, PERDANA MIRAJ,
and JADE SJAFRECIA PETROCEANY

Civil Engineering Dept, Universitas Pancasila, Jakarta, Indonesia

Most of Indonesia's railway investment is supported from the government budget where the operation and maintenance conducted by PT KAI – a sole state owned enterprise in railway industry. Currently, the government attempt to attract private sectors in railway infrastructure by contracting out operation and maintenance. However, limited experience in formulating suitable contract makes the progress relatively slow. The objective of this research is to examine alternative contract in operation and maintenance for Indonesia's railway transportation by taking Greater Jakarta Light Rail Transit (LRT) as the case study. It will consider financial components that consist of net present value, internal rate of return and payback period to produce targeted output. This research generates two simulations; (1) a simple contract where the government covers investment cost; procurement of rolling stock; and construction of stabling yard and depot. Meanwhile, the winner from tendering process have to conduct operation and maintenance for both the infrastructure and rolling stock; (2) a lease contract where a winner from tendering process require to pay lease fee through selected government body to run the operation and maintenance. The result shows simple contact where the operator only responsible for operating and maintaining the infrastructure and rolling stock is more favorable to attract private investment.

Keywords: Internal rate of return, Management contract, Lease contract, Simulation.

1 INTRODUCTION

Railway sector as one means of transportation has significant role to support economic life for the people particularly in developing countries such as Indonesia (Berawi *et al.* 2015). Railway enabling people to travel from one place to the other with greater speed compared to other modes of transportation such as private vehicles in urban area. It also has the ability to transport goods and individuals for greater volume in single travel time (Jones *et al.* 2014). According to statistics of Indonesia (2012), railway passenger volume has been increase for 2.16% or about 102.1 million passengers in 2012, while freight transport around 19.59% or about 11.5 million ton compared to the same period from previous year.

Government of Indonesia attempt to increase railway contribution in future years by publishing strategic program through National Railway Master Plan. It is expected in 2030, railway market increase for about 13% and 17% for passenger and freight transport respectively. In order to achieve those target, the needs for investment in railway transportation in 2015 – 2019 period estimated around 234.33 trillion rupiah or equal to US$ 16.74 billion. It will consist of the construction and management of railway infrastructure and its supported facilities, rolling stock procurement, formulating safety in railways and other related significant programs. Government of Indonesia has limited funds to support railway infrastructure network in the country. Therefore, collaboration with related stakeholders is encouraged to ensure public demands in railways development (Berawi *et al.* 2014, Patil and Laishram 2016).

Indonesian Law No 23 in 2007 has fundamentally created opportunities for other parties from either private sector or local government to participate in railway industry. Their investment will create multi – operator system and therefore gradually improve services for the users. Government as the regulator has the obligation to generate supporting regulation and compact investing mechanism.

Currently, the government attempt to attract private sectors in railway infrastructure by contracting out operation and maintenance. However, limited experience in formulating suitable contract makes the progress relatively slow. The objective of this research is to examine alternative contract in operation and maintenance for Indonesia' railway transportation by taking Greater Jakarta Light Rail Transit (LRT) as the case study.

LRT Jakarta is one of promising urban rail – based transportation that aims to reduce congestion and travel time uncertainty in Greater Jakarta. Government of Indonesia select PT. Adhi Karya – a state owned enterprises through a direct appointment supported by presidential law no 98/2015 to construct only the infrastructure. It then given out to the government prior to completion for tendering process for its operation and maintenance.

2 METHODOLOGY

In order to achieve targeted output this research was combining both desk research and in – depth interview as research methods. Desk research is a technique to collect related data according to the project that being studied. The data gained from various resources from statistics of Indonesia, Government report, manuals, handbooks, previous studies and research as well as other related public records. It will be compared, digested and combined to generate semi – structured instrument for in – depth interview.

The semi – structured interview comprises of three parts. The information about respondents, railway operation and maintenance contract in Indonesia and public – private partnership scheme in railway contract. The first part of instrument identified respondents' general information about institution, academic background, position and work experience. Second part of the instrument collected information about type of contract, benefits, and optimum contract. Last part provides information about government support and division of responsibility in operation and maintenance contract.

The interview were conducted by the semi – structured instruments to 10 respondents that ranging from academics, Indonesia – China High Speed Train Corporation, government bodies and practitioners. All respondents hold a minimum of master degree in transportation, experienced in railway project for more than 10 years and have been involved in public private partnership in Indonesia. With such qualifications, it enables them to validate the research findings and to formulate robust contractual framework for railway transportation.

3 RESULT AND DISCUSSION

There are various type of collaboration in public private partnership (PPP) scheme from build-operate-transfer (BOT), design-build-operate and maintain (DBOM), design and build (DB), contract management, as well as operation and maintenance (O-M). From those varied concept of PPP, private investors has the rights to own, maintain, or operate the infrastructure, facilities and services that mostly dominated by public entity. The privilege in managing the railway transportation depends on PPP contract considering numerous factors such as scope of task, payment, and revenue generation (Evenhuis and Vickerman 2010).

Operation and maintenance contract is conducted where government built infrastructure then operated and maintained by the private sector. This type of contract is part of public private partnership scheme that aims to improve railway competitiveness especially in developing countries such as Indonesia. In this research the contract will simplified into a simple contract and a lease contract of operation and maintenance.

In calculating the financial analysis, several assumptions and supporting data has been made as follows:

- Investment cost to develop the LRT system is about US$ 1,226,800,000.
- Rolling stock procurement and depot estimated for about US$ 146,870,000.
- Overhaul activity conducted every 10 years. In 2027 the cost is about US$ 1,628,000 for 37 unit rolling stock and US$ 2,464,000 in 2042 for 56 unit of rolling stock.
- Operation and maintenance cost estimated around US$ 24,515,154 in the first year of operation and increase 5.46% every year.
- The analysis use pessimist scenario of passenger demand with 35% from existing planned (210,000passengers/day) and estimated for about 73,500 passengers per day.
- Ticket price is about US$ 1.07 with exchange rate 1 US$ = Rp. 14,000.

A simple contract assumed that rolling stock depot and rolling stock procurement supported by the government through a sunk cost. Meanwhile operator expected to operate and maintain the infrastructure and rolling stock. In case, cost of renewal is required, it will be subjected to the operator. Simple contract produce internal rate of return for more than 20% with payback period in the fourth year of operation. Timeline of the project and cash flow diagram visualize in Figure 1.

On the other hand, operation and maintenance cost simulation using lease contract is a BOT contract variation where the government lease their asset to private sector. They oblige to manage and maintain the infrastructure according to mutual agreement between two parties. This type of contract has been implemented in Japan following JNR (Japan National Railways) reformation.

Prior 2003, the construction of railway in Japan particularly for its high speed train networks both for existing and new track was developed through Japan Railway Construction, Transport and Technology Agency (JRTT). They not only manage to construct the infrastructure but also given ownership rights to be leased for other interest parties (Mizutani and Shoji 2004, Kurosaki 2013). Operator shall pay for lease fee around 30% from total budget of investment. Meanwhile, one third portion will be covered by local government and the rest supported by central government (Transport Analysis 2014).

In this research, lease fee scheme to operator comprises of 5 scenarios including 10% up to 30%. It also assumed that the operator has rights of rolling stock, therefore the procurement for about US$ 126 million dollar subjected to the operator. In these schemes rolling stock depot for about US$ 20.87 million supported by government. More detail simulation of tariff and lease fee simulation to generate expected IRR can be seen in following table.

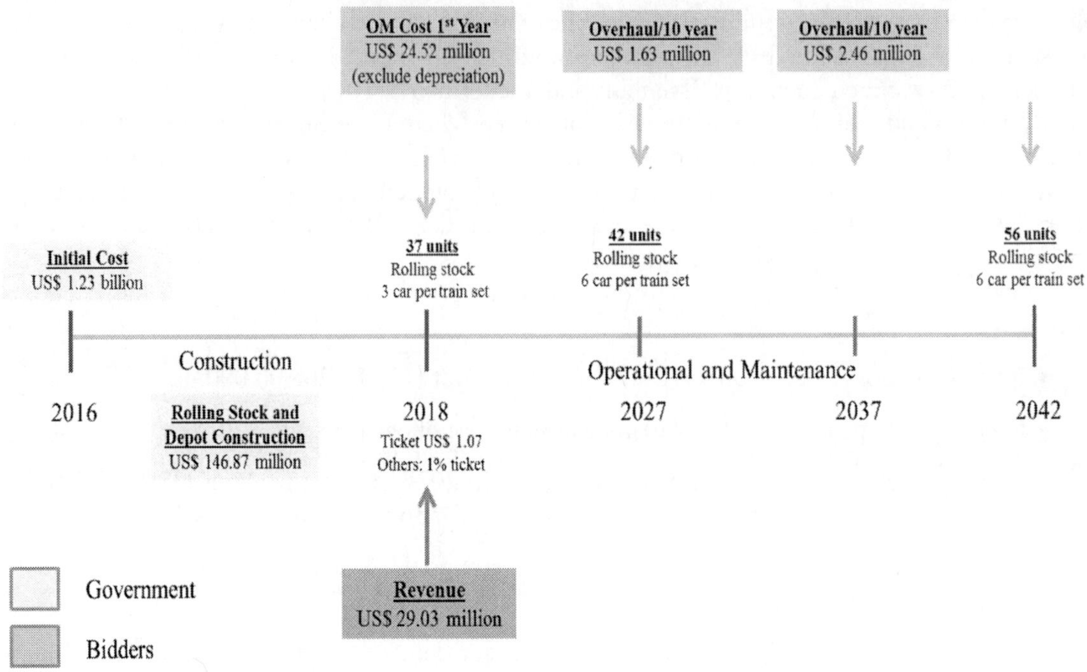

Figure 1. Cash flow diagram with project timeline.

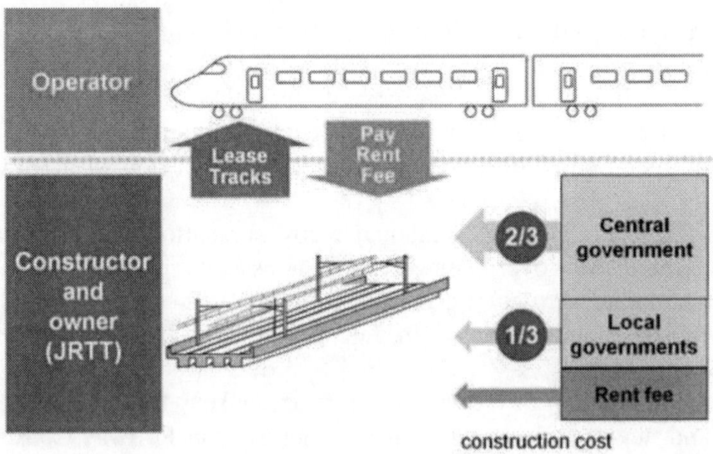

Figure 2. Financial scheme in Japanese high speed train using lease fee. (Transport Analysis 2014).

Table 1. Lease fee scheme for LRT.

Lease fee scheme	Tariff (US$)			
	1.07	1.43	1.79	2.14
	Internal Rate of Return (%)			
Simulation 10%	2.75	10.05	15.63	20.63
Simulation 15%	1.56	8.24	13.20	17.51
Simulation 20%	0.60	6.86	11.38	15.25
Simulation 25%	-0.21	5.73	9.95	13.50
Simulation 30%	-0.90	4.79	8.78	12.09

Based on table above, the tariff of about US$ 1.07 were unable to provide optimum feasibility for the project. Maximum IRR that can be achieved is 10% lease fee with 2.75% of IRR. Similar situation also applied to US$ 1.43 tariff. Although, it produce higher IRR, yet still gained maximum 10.05% of IRR or below targeted minimum attractive rate of return (MARR) for investor (12%). This scheme only feasible if the tarrif had reached US$1.79 with maximum 20% of lease fee scheme and US$ 2.14 tariff.

Considering two alternatives of contract simulation, simple contract required a huge amount of investment for the public with estimated cost around US$ 1.3 billion. However this type of contract where the operator only responsible for operating and maintaining both infrastructure and rolling stock of LRT system generate more profits and surely attract more investors compared to lease fee contract. It also produces a short term contract, thus more manageable to be evaluated. World Bank (2015) also supporting the argument that simple contract commonly used by the authority for a short period of time (2 – 5 years) and can be used either for task specific and input – based or output and performance – based.

4 CONCLUSION

Railway development plays significant role in increasing productivity and creating alternative accessibility for people. Based on the analysis of two scenarios from simple and lease fee contracts, simple contract produce more benefits in term of financial feasibility.

Operation and maintenance using lease contract is conducted through mutual agreement between two parties both government and private entity. In detail, the government leases their asset to private sector. They will oblige to manage and maintain the infrastructure as the contract agreement. The contract only feasible for investor with tariff for about US$1.79 and 20% of lease fee scheme; or US$ 2.14 tariff and varied lease fee can be used to produce optimum IRR.

In simple contract, rolling stock depot and rolling stock procurement supported by the government through a sunk cost. Meanwhile operator expected to operate and maintain the infrastructure and rolling stock. Furthermore, Simple contract produce internal rate of return for more than 20% with payback period in the fourth year of operation. This type of contract is more profitable and able attract more investors compared to lease fee contract. It also produces a short term contract, thus more manageable to be evaluated by the public.

Acknowledgments

This paper is supported by Universitas Pancasila and Directorate General of Railways –Sub Directorate Investment, Ministry of Transportation, Republic of Indonesia.

References

Berawi, M. A., Susantono, B., Miraj, P., Berawi, A. R. B., Rahman, H. Z., and Husin, A., Enhancing Value for Money of Mega Infrastructure Projects Development Using Value Engineering Method. *Procedia Technology*, *16*, 1037-1046, 2014.

Berawi, M., Berawi, A., Prajitno, I., Nahry, N., Miraj, P., Abdurachman, Y., Tobing, E., and Ivan, A., Developing Conceptual Design of High Speed Railways using Value Engineering Method: Creating Optimum Project Benefits. *International Journal Of Technology*, *6*(4), 670-679, 2015.

Evenhuis, E. and Vickerman, R., Transport pricing and public-private partnerships in theory: Issues and suggestions. *Research in Transportation Economics*, *30*(1), 6-14, 2010.

Jones, H., Moura, F., and Domingos, T.,Transport infrastructure project evaluation using cost-benefit analysis. *Procedia-Social and Behavioral Sciences*, *111*, 400-409, 2014.

Kurosaki, F., "Shinkansen investment before and after JNR Reform."International Transport Forum Discussion Paper, 2013.

Mizutani, F. and Shoji, K., "Rail operation–infrastructure separation: the case of Kobe rapid transit railway." *Transport Policy* 11.3, 251-263, 2004.

Patil, N. A. and Laishram, B., Public–private partnerships from sustainability perspective–a critical analysis of the Indian case. *International Journal of Construction Management*, *16*(2), 161-174.Statistics Indonesia, Passengers Volume, 2010 – 2014, 2016.

Transport Analysis, Railway in Sweden and Japan – a comparative Study. Brita Saxton. Stockholm, Sweden, 2014.

World Bank (2015), Management/Operation and Maintenance Contracts. Retrieved from: http://ppp.worldbank.org/public-private-partnership/agreements/management-and-operating-contracts.

Integrated Solutions for Infrastructure Development
Edited by Lau, H. H., Tang, F. E., Ng, C. K., and Singh, A.
Copyright © 2016 ISEC Press

A REVIEW OF THE CHALLENGES OF RURAL ROAD NETWORK DEVELOPMENT

JUSTUS N. AGUMBA

Dept of Construction Management and Quantity Surveying, University of Johannesburg, Johannesburg, South Africa

It is a proven fact that development and maintenance of road infrastructure are prerequisites for rapid economic growth and poverty reduction. This paper aims to provide a review of challenges faced on rural road network development, the effects of poor rural road network, and the interventions proposed to improve the state of the rural road network. This research is based on literature review. The current challenges stifling the rural roads development are: technically inexperienced constituency roads committee, increased overhead costs, large conflict of interest due to bigger groups involvement, inadequate supervision capacity, transition into the devolved government structures and challenges of road classifications into national and county roads. The negative effects on the rural road network on growth and productivity are delays, and loss of sending farm produce to the market. It is recommended that governments could intervene by; increasing fuel levy, to create an enabling environment for the private sector participation under private public partnership (PPP) arrangements but to name a few. This study advocates for practical policies on rural road network improvement that will lead to the achievement of some of the Sustainable Development Goals 2030. The author views the SDGs as catalyst to correct the ills faced by the majority of the populace currently.

Keywords: Constraints, Developing, Effect, Infrastructure, Literature, Road.

1 INTRODUCTION

It is expected that rural road construction, matching those of the United States and China, will occur in Sub-Saharan Africa well before 2050. South Sudan, alone need some 20,000 to 30,000 km of new roads (Faiz 2012). Rural-road upgrading is required in all provinces in China; hence a panacea to improve access to markets and employment (Duncan 2007). The continued need in rural roads is predicated by many factors, the foremost being the food requirements of 9-billion global population by 2050 (Faiz 2012). According to the Africa Development Bank Group, (ADBG 2014), road networks provide the necessary linkages for promoting national and international trade, economic growth, poverty reduction and wealth creation.

In Kenya the transport and telecommunication infrastructure subsector contributed the highest GDP at 9.3% in 2012 (Kenya Institute for Public Policy Research and Analysis 2013). Road transport accounts for over 90% of Kenya's total passenger and freight transportation (ADBG 2014).

In spite of the stated importance of the road sector stated and the need for rapid expansion, Gbadamosi and Olorunfemi (2016) indicated that poor rural road conditions stifled the required farming achievement of the dwellers in the rural area. In Kenya, the paved rural roads are insignificant in relation to the total length of rural roads network (Kenya Rural Roads Authority 2012-2013). In Sri Lanka 69,000 kilometres of, roads used by rural people are 'unclassified'; therefore have minimal maintenance budgets (Practical Action 2008). According to Africa Development Bank (AFDB 2010), rural Africa has only 34% of road access covered as compared to 90% in the rest of the world.

In-line with the aforementioned discussions, the purpose of this paper is to establish the challenges that governments are facing in developing rural road network, the effect of rural road network and the intervention strategies implemented or proposed. Hence, the following research questions were proposed:

- To identify the challenges faced by governments on its rural road network development;
- To determine the effects/impacts of rural road network; and
- To identify the interventions instituted by the governments to improve the development of rural road networks.

2 LITERATURE REVIEW

2.1 Challenges of Rural Road Network Development

According to Kasuku and Macharia (2003) opined that lack of coordination i.e. it is difficult to coordinate the activities of the various road agencies, to determine their financial requirements, and to address the problems of the road sector in a synchronized manner. Funding estimates were far higher than available and allocations from government resulting in inadequate maintenance and rehabilitation of roads. Development support monitor, (2012) inferred that the huge backlog of rural infrastructure in Africa which includes roads is stifled by, decades of under investment, lack of maintenance, and destruction and dilapidation due to conflict. The quality of infrastructure is generally poor and infrastructure service costs are relatively high. This lack of access to infrastructure services constrains the region's agricultural growth and limits its market access.

Odero and Njenga (2005) suggested that there is no maintenance policy for Kenyan roads despite apportioning specific road classes to particular agencies. Revenue for road maintenance remains a challenge due to the level of resources in the constituency development funds which was insufficient for rural roads and other transport infrastructure and should be increased. These challenges are further exacerbated by limited finance to implement the road polices effectively.

Briceño-Garmendia and Shkaratan (2010) indicated road investments have been characterized by low rates of budget execution, cost overruns of as much as 80% over engineering estimates, and lengthy delays that tend to double the implementation period. Furthermore, inadequacies in the system for supervising construction contracts have cut quality and shortened the life of road networks.

The Ministry of Roads (2011) indicated: non implementation of the drafted Integrated National Transport Policy, led to the lack of integration of transport modes; lack of a national spatial plan resulted in haphazard development; the available funds currently for development and maintenance of roads are inadequate, therefore a substantial part of the road network is not sufficiently attended to, resulting in huge maintenance backlog.

The Kenya Rural Roads Authority report (2012-2013) inferred that the challenges experienced were; a number of technically inexperienced constituency roads committee members with little knowledge of procurement law and regulations, increased overhead costs, less cooperation from local authorities over road works, large conflict of interest due to bigger groups involvement, inadequate supervision capacity, the transition into the devolved government structures, and challenges of road classifications into national and county roads. Some of these factors are supported by Faiz (2012).

2.2 Effects/Impact of Road Network

The Ministry of State for Planning (2008) indicated that the poor and dilapidated state of road transport infrastructure in Kenya has led to low productivity, high production and distribution costs and uncompetitive products and services. This is supported by Faiz (2012). However, according to Terefe (2012) good rural roads are imperative for poverty reduction and consumption growth in rural areas, therefore, a positive impact/effect. Adedeji *et al.* (2014), indicated that poor condition of roads have negative effects on agricultural activities which is the major source of income of residents, thereby increasing the poverty rate. Abur *et al.* (2015) suggested that rural road infrastructure has positive influence on productivity and income of farmers which translate into creation of employment opportunities and better welfare for the rural populace.

The African Development Bank Group (2014) indicated that family farming has been characterized by low productivity due to inadequate rural roads including other infrastructure components. Furthermore, police road blocks and multiple weighbridges continue to slow the traffic flow and create bribery and corruption opportunities. However, Llanto (2012) indicated that rural roads provide connectivity with growing markets adjacent to rural areas; they also lessen input costs and transaction costs of rural producers and consumers. Kiprono and Matsumoto (2014), results indicate that land allocated to hybrid maize, inorganic fertilizer intensification, maize yield and milk market participation increase more in areas with better road access improvement.

2.3 Government of Response to Rural Road Network Challenges

The Ministry of Roads (MoR 2011) report, indicated that Road Maintenance Levy Fund and the establishment of Kenya Roads Board (KRB), are initiatives to fund and manage the road sector respectively. According to the Kenya Institute for Public Policy Research and Analysis report (KIPPRA, 2013) the roads sub-sector has been receiving increased budget resource allocation for construction of new roads, bridges, rehabilitation of roads and periodic maintenance as an intervention to improve the road network condition. Adedeji *et al.* (2014) suggests that in order for the existing roads to be kept in good condition, there is need to set up maintenance units at local government level who would be responsible for rural road maintenance. Abur *et al.* (2015) indicated that construction of more rural roads is very essential to farmer's productivity. This will facilitate access to inputs and product market by the rural farmers.

Kasuku *et al.* (2003) and Odero *et al.* (2005) indicated that reducing the unit cost of transportation through, infrastructure improvement by coordination of road development and maintenance to ensure that the road network is maintained, rehabilitated/upgraded and expanded to rural areas. This is supported in the report of Faiz (2012).

The ADBG (2014) affirmed that in 1999 the GoK established through an Act of Parliament, the Kenya Roads Board (KRB), with a specific mandate to oversee the development and maintenance of the country's road network. The board is imperative for all governments. The initiation of the performance based contracting, whereby newly constructed or rehabilitated roads are placed under a long term maintenance scheme, and outsourced to a private contractor, with payments based on service delivery. It is imperative to professionalise the construction industry by instituting professional bodies/associations. A public private partnership (PPP) framework has also been adopted to lay ground for private sector participation in road development and maintenance.

3 RESEARCH METHODOLOGY

Desk top research was used for this study. This involved search of secondary literature in relevant websites to be able to obtain data and information for this study. In addition, the documents listed in the reference section were reviewed. This non-experiential data was analyzed using thematic content analysis. The articles were read a number of times to obtain a sense of the content and the emerging themes were noted. The themes were further categorized into sub-themes. The review identified challenges of rural roads development, its effects/impact, finally on methods used or proposed to improve the state of the rural road network.

4 FINDINGS AND DISCUSSIONS

4.1 Challenges Faced by Governments on Rural Roads Network Development

Rural road sector challenges continue to undermine the sustainability of investments. The key challenges established were: funding shortfalls which has resulted in maintenance backlog estimated at about 20-30%; inadequate institutional capacity for road maintenance, established fragmented agencies; poor planning of rural road network which leads to inaccessibility, lack of an efficient integrated road network locally. These challenges identified corroborates with the sentiments of Faiz (2012), Kasuku and Macharia (2003), Odero and Njenga (2005), and Briceño-Garmendia and Shkaratan (2010).

4.2 The Effects/Impact of Rural Road Network

When the rural road network is in good condition there is bound to have agglomeration of industries in these areas, thus creating rural development. However, poor state of the road network causes delays, breakages and high maintenance cost for transport machinery leading to high costs of doing business. Furthermore, agricultural produce are spoilt and cost of farming escalates. These findings are in line with the Ministry of State for Planning (2008). This is in line with Faiz report (2012). According to Terefe (2012) rural roads are imperative for poverty reduction and consumption growth in rural areas, therefore, a positive impact/effect.

4.3 Government Interventions on Rural Road Network Challenges

In recognition of the various challenges, the negative and positive effects of rural road network. It is suggested that governments consider having fuel levy to generate additional revenue required to reduce the road maintenance backlog; and to create an enabling environment for the private sector participation under private public partnership (PPP) arrangements. Furthermore, floating an infrastructure bond to mobilize additional funding for road construction and maintenance is an important intervention. These proposals are supported by the ADBG (2014) to improve the rural roads.

5 CONCLUSION

In some countries road network accounts for over 90% of the total movement of passengers and freight. It is also the means of access to majority of rural communities. The importance of road infrastructure has been highlighted in the SDG 2030 blueprint. This blueprint, aims to transform the world into a newly industrialized, middle-income countries that provides a high quality of life to all citizens by 2030. In relation to the SDG, 2030, this discourse had identified challenges that government experience in rural road developments, namely; technically inexperienced constituency roads committee members with little knowledge of procurement law and regulations, increased overhead costs, large conflict of interest due to bigger groups involvement, inadequate supervision capacity, the transition into the devolved government structures, and challenges of road classifications into national and county roads. These challenges have led to negative effect on the rural road network usage i. e. delays due to poor roads, loss of sending farming produce to the market. The challenges and the negative effects of rural road network, has led some governments to increase fuel levy to generate additional revenue required to reduce the road maintenance backlog; and to create an enabling environment for the private sector participation under private public partnership (PPP) arrangements. Furthermore, governments floating an infrastructure bond to mobilize additional funding for road construction and maintenance. It is also vital that road boards and professional bodies are formed to ensure professionalism in designing and managing rural roads development globally.

References

Abur, C. C., Ademoyewa, G. R., and Damkor, M., Impact of rural roads infrastructure on the income and productivity of households farmers in North Central Nigeria, Research Journal of Agriculture and Environmental Management, 4(10) 451-458, 2015.

Adedeji, O. A., Olafiaji, E. M., Omole, F. K., Olanibi, J. A., and Lukman, Y., An assessment of the impact of road transport on rural development: A case study of Obokun local government area of Osun State, Nigeria, British Journal of Environmental Sciences, 2(1), 34-48, 2014.

African Development Bank Group (ADBG), Country Strategy Paper 2014-2018., 2014, retrived from http://www.afdb.org/fileadmin/uploads/afdb/Documents/Project-and-Operations/2014-2018_-_Kenya_Country_Strategy_Paper.pdf on the 20/03/2015.

African Development Bank (AFDB), Infrastructure, 2010, retrieved from http://www.afdb.org/en/topics-sectors/sectors/infrastructure/ on the 21/10/2016.

Briceño-Garmendia, C.M., and M. Shkaratan, Africa Infrastructure Country Diagnostic report, Kenya's Infrastructure: A Continental Perspective, Africa's Infrastructure, A time for transformation, 2010., retrived from http://siteresources.worldbank.org/INTAFRICA/Resources/Kenya-Country_Report_03.2011.pdf on the 20/03/2015.

Development support monitor, Rural infrastructure in Africa, unlocking the African moment, paper series No. 1. 2012.

Duncan, T., Findings from studies of poverty Impacts of road projects. A case study from the 2007, Sector Assistance Program Evaluation of Asian Development Bank, assistance for roads and railways in the People's Republic of China.

Faiz, A., The Promise of Rural Roads, Review of the Role of Low-Volume Roads in Rural Connectivity, Poverty Reduction, Crisis Management, and Livability, transport research circular, Number E-C167, 2012.

Gbadamosi, K. T. and Olorunfemi, S. O., Rural road infrastructural challenges: An impediment to health care service delivery in Kabba-Bunu local government area of Kogi State, Nigeria, Academic Journal of Interdisciplinary Studies, 5(2), 35-44, 2016.

Kasuku, S. and Macharia, L., Links between Transport and Poverty: A Review of Transport Policies in Kenya and the Links with Poverty Reduction, 2003.

Kenya Institute for Public Policy Research and Analysis report, Creating an Enabling Environment for Stimulating Investment for Competitive and Sustainable Counties: Kenya Economic Report, 2013, retrived from http://www.kippra.org/downloads/Kenya%20Economic%20Report%202013.pdf on the 21/04/2015.

Kenya Rural Roads Authority, Anual report 2012-2013, retrived from http://www.kerra.go.ke/index.php?option=com_phocadownload&view=category&download=107:kerra-annual-report-2013-board-input&id=17:annual-reports-2013 on the 31/11/2014.

Kiprono, P., and Matsumoto, T., Roads and farming: The effect of infrastructure improvement on agricultural input use, farm productivity and market participation in Kenya, Proceeding of CSAE Conference: Economic Development in Africa, The University of Oxford, 2014.
Llanto, M. G., The impact of infrastructure on agricultural productivity, Philippine Institute for Development Studies, discussion paper, 2012.
Ministry of Roads (MoR), Cost Estimation Manual for Road Maintenance Works, Republic of Kenya, 2011.
Ministry of State for Planning, First Medium Term Plan, Kenya Vision 2030: A globally competitive and prosperous Kenya, 2008-2012., 2008, Retrived from http://www.sida.se/contentassets/855677b831b74ea0b226ce2db4eb93a3/kenya_medium_term_plan_2008-2012.pdf on the 15/05/2015.
Odero, K., and Njenga, P., IFRTD Pro-poor transport policies: Experiences and lessons from four East and Southern African countries, a synthesis report, 2005.
Practical Action, Village Roads in Sri Lanka, 2008, retrieved from http://practicalaction.org/transport/road_building on the 20/10/ 2016.
Terefe, A. L., Impact of road on rural poverty: Evidence from fifteen rural villages in Ethiopia, (2012), unpublished Masters of Arts in Development Studies, Graduate School of Development Studies, The Hague, Netherlands.

AUTHOR INDEX

Abbas, Mohamad 437
Abduh, Muhamad 431
Abdulkadir, Nathnael 37
Adachi, Kotaro 282
Adam, Farrel 431
Agumba, Justus N. 414, 455
Aigbavboa, Clinton 67, 390, 396, 420, 425, 443
Akguner, Cem 121
Alemu, Bisrat 37
Ali, Anis Mohamad 342
Alisjahbana, Irene 174
Alisjahbana, Sofia W. 174
Ansary, Nazeem 67
Aoki, Mizuki 312
Aregawi, Samson 37
Ario, Ichiro 282
Arthur-Aidoo, Bernard Martin 67
Asao, Naoyuki 270, 294
Aso, Toshihiko 258, 288
Aziz, Abdul Rashid Abdul 204
Azmi, M. R. 151
Badaruzzman, W. H. Wan 151
Berawi, Mohammed Ali 228
Beretta, Luca 216
Busada, Rania 73
Cassano, Manuele 216
Chen, Xinyan 180
Cheung, Sai On 360, 366
Chikahiro, Yuki 282
Ching, Yap Eng 336
Christou, Petros 61
Dusi, Alberto 330
Edmonson, Luke 234
Edwards, Kisa 91
El-Sadek, Amr Khaled 210
Endut, Intan Rohani 97, 222
Essa, Mohamad J. K. 342
Ezeldin, Samer 73, 210, 384
Faheem, Ahmed 43
Fujii, Katashi 115, 133, 270, 294
Fukuda, Hiroaki 270, 294
Goshu, Abiy 37
Han, Taek Hee 7, 13
Hara, Kohei 19, 318
Hassan, Abdulameer Qasim 342
Hassan, Lokman 276
Herzanita, Ayu 228
Hidayat, Felix 348
Hidekuma, Yuya 133
Hong, Hyemin 7, 13
Horii, Hisakazu 133
Huang, He 109
Hussin, Mohd Warid 252
Idate, Naoto 157
Ikeda, Mariko 264
Ikedo, Fumiya 127
Islam, Md Saiful 43
Ismail, Mohamed A. 252
Jessup, Scott 378
Jimenez, Verdejo Juan Ramon 318
Jiundar, Ong 276
Joo, M.S. 169
Kanellopoulos, Michael Phillip 234
Kasaeian, Shahrzad 55
Kato, Yasuyuki 186
Kawaai, Keiyu 25
Khabisi, Jerminah 396
Khoury, Hiam 437
Kim, Sungwon 7, 13
Kitsutaka, Yoshinori 127
Koesalamwardi, Ario Bintang 228
Kolar, Duro 372
Lai, Jian-Yu 192
Latief, Yusuf 228, 240
Lau, H.H. 169
Law, David W. 31
Le Nguyen, Hai 103
Leasilane, Portia 425
Lee, Han-Seung 252
Li, Yeou-Fong 192
Liphadzi, Murendeni 420
Magallanes, Alvin 49
Magdy, Abdelrahman 384
Manabe, Yukio 115
Mannan, Mohammad Abdul 43
Mantoufeh, Andrew 79
Martin, Hector 91
Maruyama, Kazuto 258
Mazzetto, Silvia 354
Mesfin, Yonatan 37
Mezzi, Marco 330
Minh, Thien Vo 103
Miniotaite, Ruta 145, 246
Miraj, Perdana 402, 449
Mirasa, Abdul Karim 43
Miyamoto, Mitsuhiro 139
Mneymneh, Bahaa Eddine 437
Moores, Andrew 49
Mukuka, Mulenga 390
Mustapha, Zakari 443
Mutepfe, Millie T 61
Mydaraboina, Himabindu 1
Nagaratnam, Brabha 43
Nagata, Ryoya 288
Naoto, Mine 204
Nassif, Ayman Y 61
Nguyen, Hai Than 103, 163
Nicolini, Giada 216
Nikoonejad, Reza 55
Nishikawa, Hayato 300, 306

AUTHOR INDEX

Niu, Yuhan ... 366
Nizamani, Zafarullah ... 204, 336
Nkomo, Morena Wiilliam ... 408
Ogami, Hiroshi ... 115
Ohdo, Katsutoshi ... 312
Ohgaki, Kazuo ... 312
Okuhiro, Haruka ... 139
Olanrewaju, Abdullateef ... 204, 336
Ono, Shuichi ... 282
Or, Tan Teng ... 330
Ota, Kohei ... 115
Patnaikuni, Indubhushan ... 1
Patnaikuni, Indubushan ... 31
Petroceany, Jade ... 228
Petroceany, Jade Sjafrecia ... 402, 449
Rahman, Herawati Zetha ... 402, 449
Rahman, Muhammad Ekhlasur ... 43
Rahman, M. Mukhlesur ... 31
Rajaram, Satish Kumar S ... 324
Ramon, Verdejo Jimenez Juan ... 19
Rarasati, Ayomi ... 240
Razak, Nur Nabihah Abd ... 97, 222
Ridzuan, Ahmad Ruslan Mohd ... 222
Rlzduan, Ahmad Ruslan Mohd ... 97
Ronagh, Hamid R. ... 55
Rotimi, James Olabode Bamidele ... 85
Saaidin, Sabihah ... 97, 222
Saha, Swapan ... 79, 378
Sahoo, Nibedita ... 324
Sakano, Masahiro ... 264
Samah, Siti Akmar Abu ... 97, 222
Samasoni, Jeff Clement ... 85
Selcuk, Murat E. ... 121
Shibasaki, Naho ... 264
Sihombing, Lukas ... 240
Singh, Amarjit ... 49
Soemardi, Biemo W. ... 348
Stewart, Ann ... 61
Sullivan, Ben ... 49
Supriadi, Leni ... 228
Sy, Lam Nguyen ... 163
Tachibana, Shusaku ... 288
Takahashi, Hiroki ... 312
Takatani, Tomiya ... 198, 300, 306
Takiyama, Noriko ... 19, 157, 180, 318
Tan, Seong Yeow ... 204
Tat, Lee Lim ... 204
Thwala, Wellington ... 390, 396
Thwala, Wellington Didi ... 443
Thwala, Wellington Didibhuku ... 408
Ting, T.C.H. ... 169
Tonaroglu, Murat ... 121
Trani, Marco L. ... 216
Ujike, Isao ... 25
Wahab, M. Mubarak Bin Abdul ... 336
Walker, Derek H. T. ... 372
Wang, Bo-Shiang ... 192
Wangsadinata, Wiratman ... 174
Wibowo, Andreas ... 240
Woldaregay, Dagmay ... 37
Wong, Peter S. P. ... 234, 372
Yamada, Yukimasa ... 157
Yamaguchi, Akari ... 19, 180, 318
Yamamoto, Haruyuki ... 109
Yatim, M. Y. M. ... 151
Yokota, Ryuichi ... 133, 294
Yonekura, Asuo ... 115